This book comes with access to more content online.

Quiz yourself, track your progress, and improve your math skills!

Register your book or ebook at **www.dummies.com/go/getaccess.**

Select your product, and then follow the prompts to validate your purchase.

You'll receive an email with your PIN and instructions.

Pre-Calculus

ALL-IN-ONE

Pre-Calculus

ALL-IN-ONE

by Mary Jane Sterling

Pre-Calculus All-in-One For Dummies®

Published by: **John Wiley & Sons, Inc.**, 111 River Street, Hoboken, NJ 07030-5774, www.wiley.com

For general information on our other products and services, please contact our Customer Care Department within the U.S. at 877-762-2974, outside the U.S. at 317-572-3993, or fax 317-572-4002. For technical support, please visit https://hub.wiley.com/community/support/dummies.

Wiley publishes in a variety of print and electronic formats and by print-on-demand. Some material included with standard print versions of this book may not be included in e-books or in print-on-demand. If this book refers to media such as a CD or DVD that is not included in the version you purchased, you may download this material at http://booksupport.wiley.com. For more information about Wiley products, visit www.wiley.com.

Library of Congress Control Number: 2023943405

ISBN 978-1-394-20124-2 (pbk); ISBN 978-1-394-20125-9 (ebk); ISBN 978-1-394-20126-6 (ebk)

SKY10077368_061224

Contents at a Glance

Table of Contents

UNIT 6: SYSTEMS, SEQUENCES, AND SERIES 443

Introduction

Here you are: ready to take on these challenging pre-calculus topics — possibly on your way to calculus! Believe it or not, it was calculus that was responsible for my switching majors and taking on the exciting world of mathematics!

Pre-calculus books and classes are wonderful ways of taking the mathematics you've studied in the past and bolstering the experience with new, exciting, and challenging material. Some of what is presented in pre-calculus is review, but studying it and adding on to the topics is what will make you even more of a success in your next endeavor.

Maybe some of the concepts you've already covered in pre-calculus have given you a hard time, or perhaps you just want more practice. Maybe you're deciding whether you even want to take pre-calculus and then calculus at all. This book fits the bill to help you with your decision for all those reasons. And it's here to encourage you on your pre-calculus adventure.

You'll find this book has many examples, valuable practice problems, and complete explanations. In instances where you feel you may need a more thorough explanation, please refer to *Pre-Calculus For Dummies* or *Pre-Calculus Workbook For Dummies* by Mary Jane Sterling (Wiley). This book, however, is a great stand-alone resource if you need extra practice or want to just brush up in certain areas.

About This Book

Pre-calculus can be a starting point, a middle point, and even a launching point. When you realize that you already know a whole bunch from Algebra I and Algebra II, you'll see that pre-calculus allows you to use that information in a new way. Before you get ready to start this new adventure, you need to know a few things about this book.

This book isn't a novel. It's not meant to be read in order from beginning to end. You can read any topic at any time, but it's structured in such a way that it follows a "typical" curriculum. Not everyone agrees on exactly what makes pre-calculus pre-calculus. So this book works hard at meeting the requirements of all those curriculums; hopefully, this is a good representation of any pre-calculus course.

Here are two different suggestions for using this book:

- Look up what you need to know when you need to know it. The index and the table of contents direct you where to look.

» Start at the beginning and read straight through. This way, you may be reminded of an old topic that you had forgotten (anything to get those math wheels turning inside your head). Besides, practice makes perfect, and the problems in this book are a great representation of the problems found in pre-calculus textbooks.

For consistency and ease of navigation, this book uses the following conventions:

» Math terms are *italicized* when they're introduced or defined in the text.

» Variables are *italicized* to set them apart from letters.

» The symbol used when writing imaginary numbers is a lowercase *i*.

Foolish Assumptions

I don't assume that you love math the way I do, but I do assume that you picked this book up for a reason special to you. Maybe you want a preview of the course before you take it, or perhaps you need a refresher on the topics in the course, or maybe your kid is taking the course and you're trying to help them to be more successful.

It has to be assumed that you're willing to put in some time and effort here. Pre-calculus topics include lots of algebraic equations, geometric theorems and rules, and trigonometry. You will see how these topics are used and intertwined, but you may need to go deeper into one or more of the topics than what is presented here.

And it's pretty clear that you are a dedicated and adventurous person, just by the fact that you're picking up this book and getting serious about what it has to offer. If you've made it this far, you'll go even farther!

Icons Used in This Book

Throughout this book you'll see icons in the margins to draw your attention to something important that you need to know.

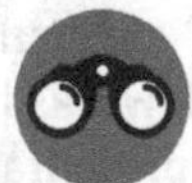
EXAMPLE

You see this icon when I present an example problem whose solution I walk you through step by step. You get a problem and a detailed answer.

TIP

Tips are great, especially if you wait tables for a living! These tips are designed to make your life easier, which are the best tips of all!

The material following this icon is wonderful mathematics; it's closely related to the topic at hand, but it's not absolutely necessary for your understanding of the material being presented. You can take it or leave it — you'll be fine just taking note and leaving it behind as you proceed through the section.

This icon is used in one way: It asks you to remember old material from a previous math course.

Warnings are big red flags that draw your attention to common mistakes that may trip you up.

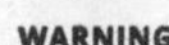

When you see this icon, it's time to tackle some practice questions. Answers and explanations appear in a separate section near the end of the chapter.

Beyond the Book

No matter how well you understand the concepts of algebra, you'll likely come across a few questions where you don't have a clue. Be sure to check out the free Cheat Sheet for a handy guide that covers tips and tricks for answering pre-calculus questions. To get this Cheat Sheet, simply go to `www.dummies.com` and type **Pre-Calculus All In One For Dummies** in the Search box.

The online quiz that comes free with this book contains over 300 questions so you can really hone your pre-calculus skills! To gain access to the online practice, all you have to do is register. Just follow these simple steps:

1. **Register your book or ebook at Dummies.com to get your PIN. Go to** `www.dummies.com/go/getaccess`.
2. **Select your product from the dropdown list on that page.**
3. **Follow the prompts to validate your product, and then check your email for a confirmation message that includes your PIN and instructions for logging in.**

If you do not receive this email within two hours, please check your spam folder before contacting us through our Technical Support website at `http://support.wiley.com` or by phone at 877-762-2974.

Now you're ready to go! You can come back to the practice material as often as you want — simply log on with the username and password you created during your initial login. No need to enter the access code a second time.

Your registration is good for one year from the day you activate your PIN.

Where to Go from Here

Pick a starting point in the book and go practice the problems there. If you'd like to review the basics first, start at Chapter 1. If you feel comfy enough with your algebra skills, you may want to skip that chapter and head over to Chapter 2. Most of the topics there are reviews of Algebra II material, but don't skip over something because you think you have it under control. You'll find in pre-calculus that the level of difficulty in some of these topics gets turned up a notch or two. Go ahead — dive in and enjoy the world of pre-calculus!

If you're ready for another area of mathematics, look for a couple more of my titles: *Trigonometry For Dummies* and *Linear Algebra For Dummies*.

1
Getting Started with Pre-Calculus

IN THIS UNIT . . .

Sharpening algebraic skills.

Reviewing number systems and their uses.

Describing basic function types.

Performing operations on real numbers and functions.

IN THIS CHAPTER

» Refreshing your memory on numbers and variables

» Accepting the importance of graphing

» Preparing for pre-calculus by understanding the vocabulary

Chapter 1
Preparing for Pre-Calculus

Pre-calculus is the bridge (drawbridge, suspension bridge, covered bridge) between Algebra II and calculus. In its scope, you review concepts you've seen before in math, but then you quickly build on them. You see some brand-new ideas, and even those build on the material you've seen before; the main difference is that the problems now get even more interesting and challenging (for example, going from linear systems to nonlinear systems). You keep on building until the end of the bridge span, which doubles as the beginning of calculus. Have no fear! What you find here will help you cross the bridge (toll free).

Because you've probably already taken Algebra I, Algebra II, and geometry, it's assumed throughout this book that you already know how to do certain things. Just to make sure, though, I address some particular items in this chapter in a little more detail before moving on to the material that is pre-calculus.

If there is any topic in this chapter that you're not familiar with, don't remember how to do, or don't feel comfortable doing, I suggest that you pick up another *For Dummies* math book and start there. If you need to do this, don't feel like a failure in math. Even pros have to look up things from time to time. Use these books like you use encyclopedias or the Internet — if you don't know the material, just look it up and get going from there.

Recapping Pre-Calculus: An Overview

Don't you just love movie previews and trailers? Some people show up early to movies just to see what's coming out in the future. Well, consider this section a trailer that you see a couple months before the *Pre-Calculus For Dummies* movie comes out! The following list presents some items you've learned before in math, and some examples of where pre-calculus will take you next.

- **Algebra I and II:** Dealing with real numbers and solving equations and inequalities.

  Pre-calculus: Expressing inequalities in a new way called *interval notation.*

  You may have seen solutions to inequalities in set notation, such as $\{x|x>4\}$. This is read in inequality notation as $x>4$. In pre-calculus, you often express this solution as an interval: $(4,\infty)$. (For more, see Chapter 2.)

- **Geometry:** Solving right triangles, whose sides are all positive.

  Pre-calculus: Solving non-right triangles, whose sides aren't always represented by positive numbers.

  You've learned that a length can never be negative. Well, in pre-calculus you sometimes use negative numbers for the lengths of the sides of triangles. This is to show where these triangles lie in the coordinate plane (they can be in any of the four quadrants).

- **Geometry/trigonometry:** Using the Pythagorean Theorem to find the lengths of a triangle's sides.

  Pre-calculus: Organizing some frequently used angles and their trig function values into one nice, neat package known as the *unit circle* (see Unit 3).

  In this book, you discover a handy shortcut to finding the sides of triangles — a shortcut that is even handier for finding the trig values for the angles in those triangles.

- **Algebra I and II:** Graphing equations on a coordinate plane.

  Pre-calculus: Graphing in a brand-new way with the polar coordinate system (see Chapter 16).

  Say goodbye to the good old days of graphing on the Cartesian coordinate plane. You have a new way to graph, and it involves goin' round in circles. I'm not trying to make you dizzy; actually, polar coordinates can make you some pretty pictures.

- **Algebra II:** Dealing with imaginary numbers.

  Pre-calculus: Adding, subtracting, multiplying, and dividing complex numbers gets boring when the complex numbers are in rectangular form $(a+bi)$. In pre-calculus, you become familiar with something new called *polar form* and use that to find solutions to equations you didn't even know existed.

Checking in on Number Basics and Processes

When entering pre-calculus, you should be comfy with sets of numbers (natural, integer, rational, and so on). By this point in your math career, you should also know how to perform operations with numbers. You can find a quick review of these concepts in this section. Also, certain properties hold true for all sets of numbers, and it's helpful to know them by name. I review them in this section, too.

Understanding the multitude of number types: Terms to know

Mathematicians love to name things simply because they can; it makes them feel special. In this spirit, mathematicians attach names to many sets of numbers to set them apart and cement their places in math students' heads for all time.

- **The set of natural or counting numbers: {1, 2, 3 . . .}.** Notice that the set of natural numbers doesn't include 0.
- **The set of whole numbers: {0, 1, 2, 3 . . .}.** The set of whole numbers consists of all the natural numbers plus the number 0.
- **The set of integers: {. . . –3, –2, –1, 0, 1, 2, 3 . . .}.** The set of integers includes all positive and negative natural numbers and 0.

TIP

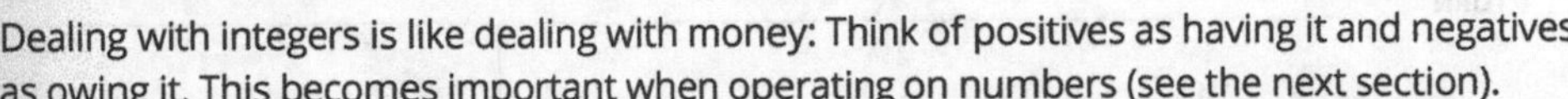

Dealing with integers is like dealing with money: Think of positives as having it and negatives as owing it. This becomes important when operating on numbers (see the next section).

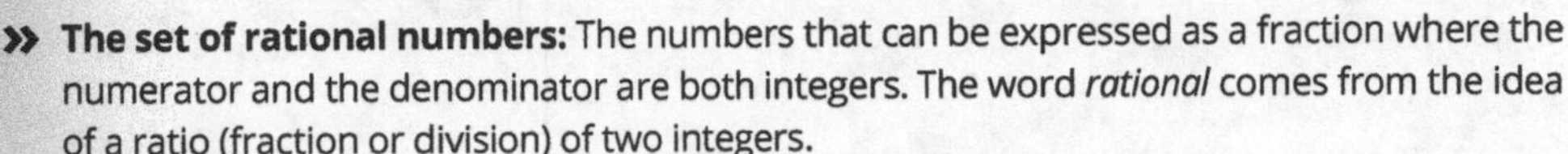

- **The set of rational numbers:** The numbers that can be expressed as a fraction where the numerator and the denominator are both integers. The word *rational* comes from the idea of a ratio (fraction or division) of two integers.

  Examples of rational numbers include (but in no way are limited to) $\frac{1}{5}$, $-\frac{7}{2}$, and 0.23. A rational number is any number in the form $\frac{p}{q}$ where p and q are integers, but q is never 0. If you look at any rational number in decimal form, you notice that the decimal either stops or repeats.
- **The set of irrational numbers:** All numbers that can't be expressed as fractions. Examples of irrational numbers include $\sqrt{2}$, $\sqrt[3]{4}$, and π.

  The decimal value of an irrational number never ends and never repeats.
- **The set of all real numbers:** All the sets of numbers previously discussed. For an example of a real number, think of a number . . . any number. Whatever it is, it's real. Any number from the previous bullets works as an example. The numbers that aren't real numbers are imaginary.

  Like telemarketers and pop-up ads on the web, real numbers are everywhere; you can't get away from them — not even in pre-calculus. Why? Because they include all numbers except the following.

  - **A fraction with a zero as the denominator:** Such numbers don't exist and are called *undefined*.
  - **The square root of a negative number:** These numbers are part of *complex numbers*; the negative root is the *imaginary* part (see Chapter 15). And this extends to any even root of a negative number.
  - **Infinity:** Infinity is a concept, not an actual number. It describes a behavior.

» **The set of imaginary numbers: square roots of negative numbers.** Imaginary numbers have an imaginary unit, like *i*, *4i*, and *–2i*. Imaginary numbers used to be considered to be made-up numbers, but mathematicians soon realized that these numbers pop up in the real world. They are still called imaginary because they're square roots of negative numbers, but they are a part of the language of mathematics. The imaginary unit is defined as $i = \sqrt{-1}$. (For more on these numbers, head to Chapter 15.)

» **The set of complex numbers: the sum or difference of a real number and an imaginary number.** Complex numbers include these examples: $3+2i$, $2-\sqrt{2}i$, and $4-\frac{2}{3}i$. However, they also cover all the previous lists, including the real numbers (3 is the same thing as $3+0i$) and imaginary numbers ($2i$ is the same thing as $0+2i$).

REMEMBER

The set of complex numbers is the most complete set of numbers in the math vocabulary because it includes real numbers (any number you can possibly think of), imaginary numbers (*i*), and any combination of the two.

YOUR TURN

In Questions 1 to 4, determine the different ways you can refer to each number. Your choices are: natural, whole, integer, rational, irrational, and real. Select all that apply.

1 -14	2 $\sqrt{15}$
3 $\frac{49}{9}$	4 $\sqrt{-16}$

Looking at the fundamental operations you can perform on numbers

From positives and negatives to fractions, decimals, and square roots, you should know how to perform all the basic operations on all real numbers. These operations include adding, subtracting, multiplying, dividing, taking powers of, and taking roots of numbers. The *order of operations* is the way in which you perform these operations.

TIP

The mnemonic device used most frequently to remember the order is PEMDAS, which stands for

1. **P**arentheses (and other grouping devices such as brackets and division lines)
2. **E**xponents (and roots, which can be written as exponents)
3. **M**ultiplication and **D**ivision (done in order from left to right)
4. **A**ddition and **S**ubtraction (done in order from left to right)

REMEMBER

One type of operation that is often overlooked or forgotten about is absolute value. *Absolute value* gives you the distance from zero on the number line. Absolute value should be included with the parentheses step because you have to consider what's inside the absolute-value bars first (because the bars are a grouping device). Don't forget that absolute value is always positive or zero. Hey, even if you're walking backward, you're still walking!

EXAMPLE

Q. Simplify the expression using the order of operations: $\frac{-3(x+4)^2+3x}{\sqrt{27}}$.

A. $\frac{-\sqrt{3}(x^2+7x+16)}{3}$. First, simplify the numerator. Do this by squaring the binomial, multiplying through by –3, and then combining like terms.

$$\frac{-3(x+4)^2+3x}{\sqrt{27}}=\frac{-3(x^2+8x+16)+3x}{\sqrt{27}}=\frac{-3x^2-24x-48+3x}{\sqrt{27}}=\frac{-3x^2-21x-48}{\sqrt{27}}$$

Next, factor the –3 from the terms in the numerator and simplify the denominator.

$$\frac{-3x^2-21x-48}{\sqrt{27}}=\frac{-3(x^2+7x+16)}{\sqrt{9\cdot 3}}=\frac{-3(x^2+7x+16)}{\sqrt{9}\sqrt{3}}=\frac{-3(x^2+7x+16)}{3\sqrt{3}}$$

Divide the numerator and denominator by 3. And, finally, rationalize the denominator.

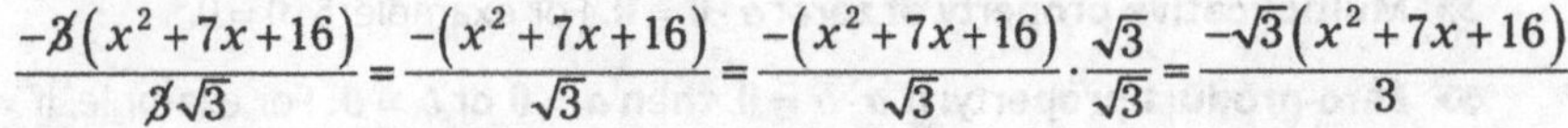

$$\frac{-\cancel{3}(x^2+7x+16)}{\cancel{3}\sqrt{3}}=\frac{-(x^2+7x+16)}{\sqrt{3}}=\frac{-(x^2+7x+16)}{\sqrt{3}}\cdot\frac{\sqrt{3}}{\sqrt{3}}=\frac{-\sqrt{3}(x^2+7x+16)}{3}$$

YOUR TURN

In Questions 5 and 6, simplify each expression using the *order of operations*.

5 $3x(x-1)^2+x$	**6** $\frac{-6\pm\sqrt{(-6)^2-4(-2)(8)}}{2(-2)}$

Knowing the properties of numbers: Truths to remember

Remembering the properties of numbers is important because you use them consistently in pre-calculus. You may not often use these properties by name in pre-calculus, but you do need to know when to use them. The following list presents properties of numbers.

- **Reflexive property: $a = a$.** For example, $10 = 10$.
- **Symmetric property: If $a = b$, then $b = a$.** For example, if $5+3=8$, then $8=5+3$.
- **Transitive property: If $a = b$ and $b = c$, then $a = c$.** For example, if $5+3=8$ and $8=4\cdot 2$, then $5+3=4\cdot 2$.
- **Commutative property of addition: $a+b=b+a$.** For example, $2+3=3+2$.
- **Commutative property of multiplication: $a\cdot b=b\cdot a$.** For example, $2\cdot 3=3\cdot 2$.
- **Associative property of addition: $(a+b)+c=a+(b+c)$.** For example, $(2+3)+4=2+(3+4)$.
- **Associative property of multiplication: $(a\cdot b)\cdot c=a\cdot(b\cdot c)$.** For example, $(2\cdot 3)\cdot 4=2\cdot(3\cdot 4)$.
- **Additive identity: $a+0=a$.** For example, $-3+0=-3$.
- **Multiplicative identity: $a\cdot 1=a$.** For example, $4\cdot 1=4$.
- **Additive inverse property: $a+(-a)=0$.** For example, $2+(-2)=0$.
- **Multiplicative inverse property: $a\cdot\frac{1}{a}=1$.** For example, $2\cdot\frac{1}{2}=1$. (Remember: $a\neq 0$.)
- **Distributive property: $a(b+c)=a\cdot b+a\cdot c$.** For example, $10(2+3)=10\cdot 2+10\cdot 3=20+30=50$.
- **Multiplicative property of zero: $a\cdot 0=0$.** For example, $5\cdot 0=0$.
- **Zero-product property: If $a\cdot b=0$, then $a=0$ or $b=0$.** For example, if $x(x+2)=0$, then $x=0$ or $x+2=0$.

REMEMBER

If you're trying to perform an operation that isn't on the previous list, then the operation probably isn't correct. After all, algebra has been around since 1600 B.C., and if a property exists, someone has probably already discovered it. For example, it may look inviting to say that $10(2+3)=10\cdot 2+3=23$, but that's incorrect. The correct process and answer is $10(2+3)=10\cdot 2+10\cdot 3=50$. Knowing what you can't do is just as important as knowing what you can do.

EXAMPLE

Q. Use the associative and commutative properties to simplify the expression: $(5x+3)+(4y-5)+(6x+11)$.

A. **$11x + 4y + 9$.** Rewrite with the variable terms grouped followed by the numbers: $5x+6x+4y+3-5+11$. Now combine the like terms: $(5x+6x)+4y+(3-5+11)=11x+4y+9$.

Q. Use the distributive property to simplify the expression: $-3(5x^2-1)-x(2x+3)-2$.

EXAMPLE

A. $-17x^2 - 3x + 1$. First, distribute the –3 and $-x$ over their respective parentheses; then combine like terms: $-3(5x^2-1)-x(2x+3)-2=-15x^2+3-2x^2-3x-2=-17x^2-3x+1$.

YOUR TURN

In Questions 7 and 8, simplify each using the distributive, associative, and commutative properties.

 $16a^2+4ab+5b^2-2a^2+3ab-9b$	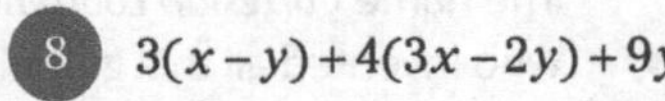 8 $3(x-y)+4(3x-2y)+9y$

Looking at Visual Statements: When Math Follows Form with Function

Graphs are great visual tools. They're used to display what's going on in math problems, in companies, and in scientific experiments. For instance, graphs can be used to show how something (like the price of real estate) changes over time. Surveys can be taken to get facts or opinions, the results of which can be displayed in a graph. Open up the newspaper on any given day and you can find a graph in there somewhere.

Hopefully, the preceding paragraph answers the question of why you need to understand how to construct graphs. Even though in real life you don't walk around with graph paper and a pencil to make the decisions you face, graphing is vital in math and in other walks of life. Regardless of the absence of graph paper, graphs are indeed everywhere.

For example, when scientists go out and collect data or measure things, they often arrange the data as x and y values. Typically, scientists are looking for some kind of general relationship between these two values to support their hypotheses. These values can then be graphed on a coordinate plane to show trends in data. For example, a good scientist may show with a graph that the more you read this book, the more you understand pre-calculus! (Another scientist may show that people with longer arms have bigger feet. Boring!)

Using basic terms and concepts

Graphing equations is a huge part of pre-calculus, and eventually calculus, so it's good to review the basics of graphing before getting into the more complicated and unfamiliar graphs that you will see later in the book.

Although some of the graphs in pre-calculus will look very familiar, some will be new — and possibly intimidating. This book will get you more familiar with these graphs so that you will be more comfortable with them. However, the information in this chapter is mostly information that you remember from Algebra II. You did pay attention then, right?

- Each point on the coordinate plane on which you construct graphs — that is, a plane made up of the horizontal (x-) axis and the vertical (y-) axis, creating four quadrants — is called a coordinate pair (x,y), also often referred to as a *Cartesian coordinate pair.*

The name *Cartesian coordinates* comes from the French mathematician and philosopher who invented all this graphing stuff, René Descartes. Descartes worked to merge algebra and Euclidean geometry (flat geometry), and his work was influential in the development of analytic geometry, calculus, and cartography.

- A *relation* is a set (which can be empty, but in this book I only consider nonempty sets) of ordered pairs that can be graphed on a coordinate plane. Each relation is kind of like a computer that expresses x as input and y as output. You know you're dealing with a relation when the set is given in those curly brackets (like these: { }) and has one or more points inside. For example, $R = \{(2,-1),(3,0),(-4,5)\}$ is a relation with three ordered pairs. Think of each point as (input, output) just like from a computer.
- The *domain* of a relation is the set of all the input values, usually listed from least to greatest. For example, the domain of set R is $\{-4,\ 2,\ 3\}$. The *range* is the set of all the output values, also often listed from least to greatest. For example, the range of R is $\{-1,\ 0,\ 5\}$. If any value in the domain or range is repeated, you don't have to list it twice. Usually, the domain is the x-variable and the range is y.

If different variables appear, such as m and n, input (domain) and output (range) usually go alphabetically, unless you're told otherwise. In this case, m would be your input/domain and n would be your output/range. But when written as a point, a relation is always (input, output).

Graphing linear equalities and inequalities

When you first figured out how to graph a line on the coordinate plane, you learned to pick domain values (x) and plug them into the equation to solve for the range values (y). Then, you went through the process multiple times, expressed each pair as a coordinate point, and connected the dots to make a line. Some mathematicians call this the ol' *plug-and-chug method.*

After a bit of that tedious work, somebody said to you, "Hold on! You can use a shortcut." That shortcut involves an equation called *slope-intercept form*, and it's expressed as $y = mx + b$. The variable m stands for the slope of the line (see the next section, "Gathering information from graphs"), and b stands for the y-intercept (or where the line crosses the y-axis). You can change equations that aren't written in slope-intercept form to that form by solving for y. For example, graphing $2x - 3y = 12$ requires you to subtract $2x$ from both sides first to get $-3y = -2x + 12$. Then you divide every term by -3 to get $y = \frac{2}{3}x - 4$.

In the first quadrant, this graph starts at -4 on the y-axis; to find the next point, you move up 2 and right 3 (using the slope). Slope is often expressed as a fraction because it's rise over run — in this case $\frac{2}{3}$.

Q. Write the linear equation $4x-5y-100=0$ in slope-intercept form.

EXAMPLE **A.** $y=\frac{4}{5}x-20$. First, subtract $4x$ and add 100 to each side.

$4x-5y-100=0 \rightarrow -5y=-4x+100$

Then, divide each term by -5: $\frac{-5y}{-5}=\frac{-4x}{-5}+\frac{100}{-5} \rightarrow y=\frac{4}{5}x-20$. The line has a slope of $\frac{4}{5}$ and a y-intercept at $(0,-20)$.

Inequalities are used for comparisons, which are a big part of pre-calculus. They show a relationship between two expressions (greater-than, less-than, greater-than-or-equal-to, and less-than-or-equal-to). Graphing inequalities starts exactly the same as graphing equalities, but at the end of the graphing process (you still put the equation in slope-intercept form and graph), you have two decisions to make:

- Is the line *dashed,* indicating $y<$ or $y>$, or is the line *solid,* indicating $y\le$ or $y\ge$?
- Do you shade under the line for $y<$ or $y\le$, or do you shade above the line for $y>$ or $y\ge$? Simple inequalities (like $x<3$) express all possible answers. For inequalities, you show all possible answers by shading the side of the line that works in the original equation. For example, when graphing $y<2x-5$, you follow these steps (refer to Figure 1-1):

1. Graph the line $y=2x-5$ by starting off at -5 on the y-axis, marking that point, and then moving up 2 and right 1 to find a second point.
2. When connecting the dots, you produce a straight dashed line through the points.
3. Shade on the bottom half of the graph to show all possible points in the solution.
4. Check your choice on the shading by checking a point. You have (5,0) in your shaded area, so inserting it into $y<2x-5$, you have $0<2(5)-5$, which is a true statement. It checks!

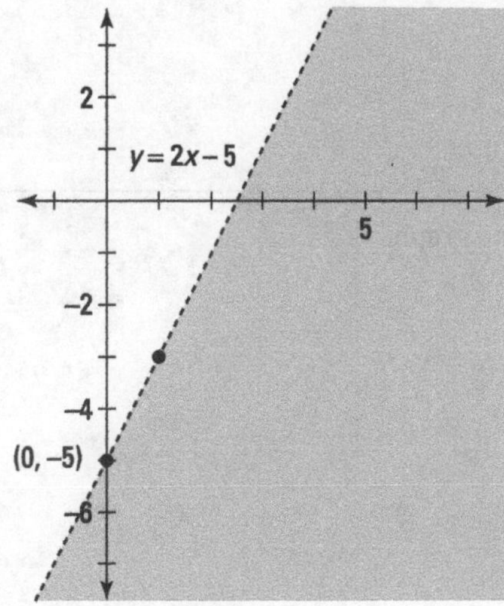

FIGURE 1-1: The graph of $y<2x-5$.

Q. Write the inequality represented by the graph.

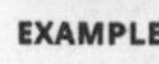

EXAMPLE

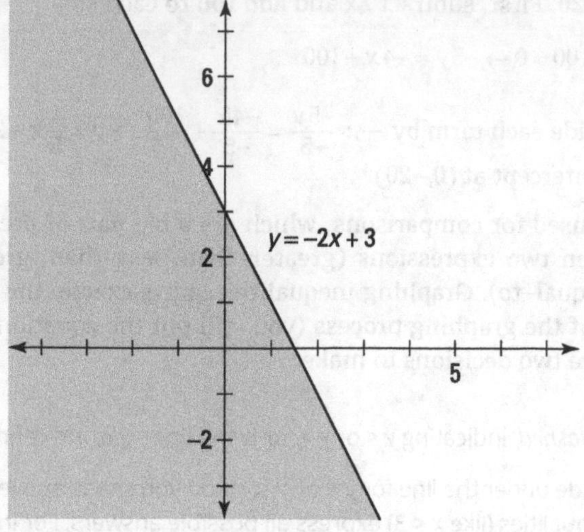

A. $y \leq -2x + 3$. The graph of the shaded area shows all the points below the graph of $y = -2x + 3$. Write the inequality $y \leq -2x + 3$, using less-than-or-equal to because the line is solid. Select a point in the shaded area such as (0,0) and check to be sure the inequality is correct. Substituting, you have $0 \leq 0 + 3$, which is a true statement.

YOUR TURN

 9 Write the equation $-11 - 3y = 7x$ in slope-intercept form.

10 Write the inequality represented by the graph.

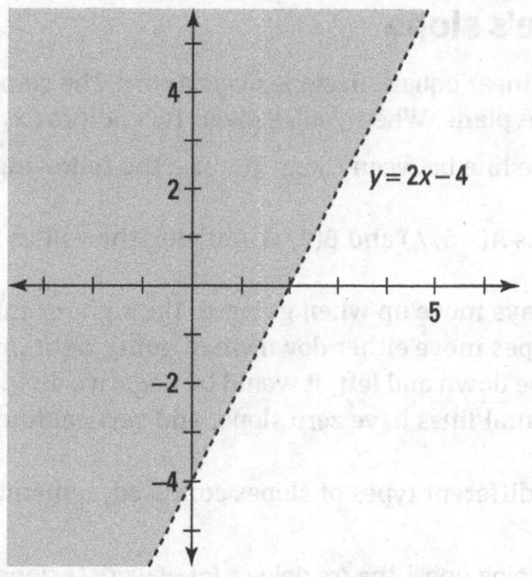

Gathering information from graphs

After getting you used to coordinate points and graphing equations of lines on the coordinate plane, typical math books and teachers begin to ask you questions about the points and lines that you've been graphing. The three main things you'll be asked to find are the distance between two points, the midpoint of the segment connecting two points, and the exact slope of a line that passes through two points.

Calculating distance

TECHNICAL STUFF

Knowing how to calculate distance by using the information from a graph comes in handy in a big way, so allow this quick review of a few things first. *Distance* is how far apart two objects, or two points, are. To find the distance, *d*, between the two points (x_1, y_1) and (x_2, y_2) on a coordinate plane, for example, use the following formula: $d = \sqrt{(x_2 - x_1)^2 + (y_2 - y_1)^2}$

You can use this equation to find the length of the segment between two points on a coordinate plane whenever the need arises. For example, to find the distance between A(−6,4) and B(2,1), first identify the parts: $x_1 = -6$ and $y_1 = 4$; $x_2 = 2$ and $y_2 = 1$. Plug these values into the distance formula: $d = \sqrt{(2-(-6))^2 + (1-4)^2}$. This problem simplifies to $\sqrt{73}$.

Finding the midpoint

TECHNICAL STUFF

Finding the midpoint of a segment pops up in pre-calculus topics like conics (see Chapter 17). To find the midpoint of the segment connecting two points, you just average their *x* values and *y* values and express the answer as an ordered pair: $M = \left(\frac{x_1 + x_2}{2}, \frac{y_1 + y_2}{2}\right)$.

You can use this formula to find the center of various graphs on a coordinate plane, but for now you're just finding the midpoint. You find the midpoint of the segment connecting the two points A(−6,4) and B(2,1) by using this formula. You have $M = \left(\frac{-6+2}{2}, \frac{4+1}{2}\right)$ or $\left(-2, \frac{5}{2}\right)$.

Figuring a line's slope

TECHNICAL STUFF

When you graph a linear equation, slope plays a role. The slope of a line tells how steep the line is on the coordinate plane. When you're given two points (x_1, y_1) and (x_2, y_2) and are asked to find the slope of the line between them, you use the following formula: $m = \frac{y_2 - y_1}{x_2 - x_1}$.

If you use the points A(−6, 4) and B(2, 1) and plug the values into the formula, the slope is $-\frac{3}{8}$.

Positive slopes always move up when going to the right or move down going to the left on the plane. Negative slopes move either down when going right or up when going left. (Note that if you moved the slope down and left, it would be negative divided by negative, which has a positive result.) Horizontal lines have zero slope, and vertical lines have undefined slope.

TIP

If you ever get the different types of slopes confused, remember the skier on the ski-slope:

- When they're going uphill, they're doing a lot of work (+ slope).
- When they're going downhill, the hill is doing the work for them (– slope).
- When they're standing still on flat ground, they're not doing any work at all (0 slope).
- When they hit a wall (the vertical line), they're done for and they can't ski anymore (undefined slope)!

EXAMPLE

Q. Find the distance between the points $(11, -2)$ and $(-1, 4)$.

A. $6\sqrt{5}$. Using the distance formula, $d = \sqrt{(-1-11)^2 + (4-(-2))^2} = \sqrt{(-12)^2 + 6^2}$ $\sqrt{144+36} = \sqrt{180} = 6\sqrt{5}$.

Q. Find the midpoint of the two points $(0, -3)$ and $(-1, 2)$.

A. $\left(-\frac{1}{2}, -\frac{1}{2}\right)$. Using the midpoint formula, $M = \left(\frac{0+(-1)}{2}, \frac{-3+2}{2}\right) = \left(-\frac{1}{2}, -\frac{1}{2}\right)$.

Q. Find the slope of the line through the two points $(4, 6)$ and $(-8, 5)$.

A. $\frac{1}{12}$. Using the slope formula, $m = \frac{5-6}{-8-4} = \frac{-1}{-12} = \frac{1}{12}$.

YOUR TURN

Given the two points $(-4, 5)$ and $(0, -3)$:

11 Find the distance between the points and the midpoint of the two points.	**12** Find the slope of the line that goes through the points.

Getting Yourself a Graphing Calculator

It's highly recommended that you purchase a graphing calculator for pre-calculus work. Since the invention of the graphing calculator, the emphasis and time spent on calculations in the classroom and when doing homework have changed because the grind-it-out computation isn't necessary. Many like doing most of the work with the calculator, but others prefer not to use one. A graphing calculator does so many things for you, and, even if you don't use it for every little item, you can always use one to check your work on the big problems.

Many different types of graphing calculators are available, and their inner workings are all different. To figure out which one to purchase, ask for advice from someone who has already taken a pre-calculus class, and then look around on the Internet for the best deal.

REMEMBER

Many of the more theoretical concepts in this book, and in pre-calculus in general, are lost when you use your graphing calculator. All you're told is, "Plug in the numbers and get the answer." Sure, you get your answer, but do you really know what the calculator did to get that answer? Nope. For this reason, this book goes back and forth between using the calculator and doing complicated problems longhand. But whether you're allowed to use the graphing calculator or not, be smart with its use. If you plan on moving on to calculus after this course, you need to know the theory and concepts behind each topic.

The material found here can't even begin to teach you how to use your unique graphing calculator, but the good *For Dummies* folks at Wiley supply you with entire books on the use of them, depending on the type you own. However, I can give you some general advice on their use. Here's a list of hints that should help you use your graphing calculator:

- **Always double-check that the mode (degrees versus radians) in your calculator is set according to the problem you're working on.** Look for a button somewhere on the calculator that says *mode.* Depending on the brand of calculator, this button allows you to change things like degrees or radians, or $f(x)$ or $r(\theta)$. For example, if you're working in degrees, you must make sure that your calculator knows that before you use it to solve a problem. The same goes for working in radians. Some calculators have more than ten different modes to choose from. Be careful!
- **Make sure you can solve for *y* before you try to construct a graph.** You can graph anything in your graphing calculator as long as you can solve for *y* to write it as a function. The calculators are set up to accept only equations that have been solved for *y.*

REMEMBER

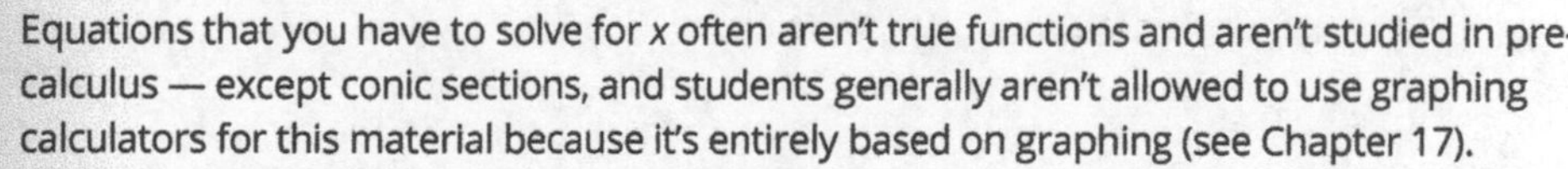

  Equations that you have to solve for *x* often aren't true functions and aren't studied in pre-calculus — except conic sections, and students generally aren't allowed to use graphing calculators for this material because it's entirely based on graphing (see Chapter 17).

- **Be aware of all the shortcut menus available to you and use as many of the calculator's functions as you can.** Typically, under your calculator's graphing menu you can find shortcuts to other mathematical concepts (like changing a decimal to a fraction, finding roots of numbers, or entering matrices and then performing operations with them). Each brand of graphing calculator is unique, so read the manual. Shortcuts give you great ways to check your answers!

» **Type in an expression exactly the way it looks, and the calculator will do the work and simplify the expression.** All graphing calculators do order of operations for you, so you don't even have to worry about the order. Just be aware that some built-in math short-cuts automatically start with grouping parentheses.

For example, most calculators start a square root off as $\sqrt{(}$ so all information you type after that is automatically inside the square root sign until you close the parentheses. For instance, $\sqrt{(4+5)}$ and $\sqrt{(4)}+5$ represent two different calculations and, therefore, two different values (3 and 7, respectively). Some smart calculators even solve the equation for you. In the near future, you probably won't even have to take a pre-calculus class; the calculator will take it for you!

Okay, after working through this chapter, you're ready to take flight into pre-calculus. Good luck to you and enjoy the ride!

Practice Questions Answers and Explanations

1. **Integer, rational, real.** The number –14 is a negative integer. It is rational because it can be written as a fraction, such as $\frac{-14}{1}$. And it is a real number.

2. **Irrational, real.** The number 15 is not a perfect square, so its decimal value continues on forever without repeating.

3. **Rational, real.** The fraction $\frac{49}{9}$ can be written as the repeating decimal $5.\overline{4}$.

4. **None of those given.** This is an imaginary number, which can be written as $4i$.

5. $\mathbf{3x^3-6x^2+4x}$**.** First, square the binomial, then distribute the $3x$. And, finally, combine like terms. $3x(x-1)^2+x=3x(x^2-2x+1)+x=3x^3-6x^2+3x+x=3x^3-6x^2+4x$

6. **1,4.** Perform the square and multiplication under the radical, and multiply the factors in the denominator. Then add the terms under the radical and take the root.
$\frac{-6\pm\sqrt{(-6)^2-4(-2)(8)}}{2(-2)}=\frac{-6\pm\sqrt{36+64}}{-4}=\frac{-6\pm\sqrt{100}}{-4}=\frac{-6\pm10}{-4}$

   Now write the two answers, one by adding 10 and the other by subtracting 10:

$$\frac{-6\pm10}{-4}=$$
$$\frac{-6+10}{-4}\text{ or }\frac{-6-10}{-4}=$$
$$\frac{4}{-4}\text{ or }\frac{-16}{-4}=$$
$$-1\text{ or }4$$

7. $\mathbf{14a^2+7ab+5b^2-9b}$**.** Rearrange the terms: $16a^2+4ab+5b^2-2a^2+3ab-9b=16a^2-2a^2+4ab+3ab+5b^2-9b$. Now combine like terms to get $16a^2-2a^2+4ab+3ab+5b^2-9b=14a^2+7ab+5b^2-9b$.

8. $\mathbf{15x-2y}$**.** First, distribute the 3 and 4. Then rearrange the terms and combine like terms. $3(x-y)+4(3x-2y)+9y=3x-3y+12x-8y+9y=3x+12x-3y-8y+9y=15x-2y$

9. $\mathbf{y=-\frac{7}{3}x-\frac{11}{3}}$**.** Add 11 to each side, and then divide each side by –3. $-11-3y=7x\rightarrow-3y=7x+11\rightarrow\frac{-3y}{-3}=\frac{7x}{-3}+\frac{11}{-3}\rightarrow y=-\frac{7}{3}x-\frac{11}{3}$

10. $\mathbf{y>2x-4}$**.** The line is dashed, so the > symbol is used. The shaded side is above the line. To check, use the point (0,0) and you have $0>0-4$, which is a true statement.

11. $\mathbf{4\sqrt{5}}$ **and** $\mathbf{(-2,1)}$**.** Using the distance formula, $d=\sqrt{(0-(-4))^2+(-3-5)^2}=\sqrt{16+64}=\sqrt{80}=\sqrt{16}\sqrt{5}=4\sqrt{5}$. Using the midpoint formula, $M=\left(\frac{-4+0}{2},\frac{5+(-3)}{2}\right)=(-2,1)$.

12. **–2.** Using the slope formula, $m=\frac{-3-5}{0-(-4)}=\frac{-8}{4}=-2$.

If you're ready to test your skills a bit more, take the following chapter quiz that incorporates all the chapter topics.

Whaddya Know? Chapter 1 Quiz

Quiz time! Complete each problem to test your knowledge on the various topics covered in this chapter. You can then find the solutions and explanations in the next section.

1. **Which of the following numbers are *integers*?** $\left\{-3, 0, 8.795, \frac{19}{5}, \sqrt{16}, \pi, \sqrt{10}\right\}$

2. **Find the midpoint of the points (3, −2) and (−5, 5).**

3. **Simplify:** $\dfrac{-5-\sqrt{(-5)^2-4(-3)(-2)}}{(-3)^2}$

4. **Simplify:** $5x(3x-4)^2+2(x+1)^2-3$

5. **Write an inequality that describes the graph shown here.**

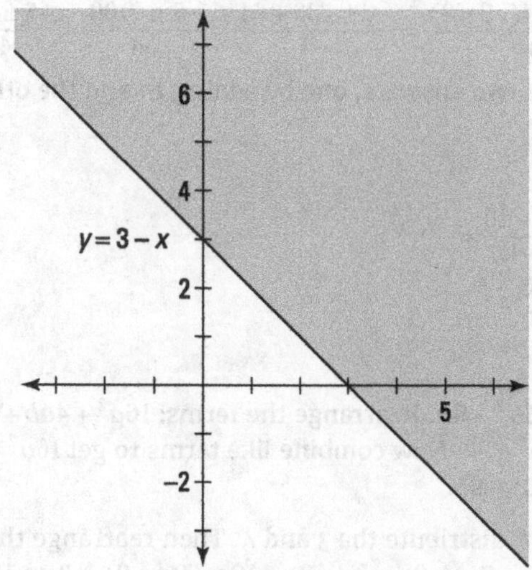

6. **Which of the following numbers are *rational numbers*?** $\left\{-3, 0, 8.795, \frac{19}{5}, \sqrt{16}, \pi, \sqrt{10}\right\}$

7. **Find the distance between the points (−8, 4) and (1, 6).**

8. **Simplify:** $\dfrac{x^2-5x-(9-5x)}{7+x-4}$

9. **Find the slope of the line through the points (−4, 0) and (6, −2).**

10. **Simplify:** $6\cdot 100\left(1+\dfrac{0.04}{4}\right)^3$

Answers to Chapter 1 Quiz

1. **-3, 0, $\sqrt{16}$.** These are integers. They can be written as a positive or negative whole number or 0. The numbers 8.795, $\frac{19}{5}$, π, $\sqrt{10}$ are not integers.

2. $\left(-1, \frac{3}{2}\right)$**.** Using the midpoint formula, $M = \left(\frac{3+(-5)}{2}, \frac{-2+5}{2}\right) = \left(-1, \frac{3}{2}\right)$.

3. $-\frac{2}{3}$**.** Working inside the radical, first square the number and multiply the three factors. Square the denominator. Simplify under the radical. Then simplify and reduce the fraction.

$$\frac{-5-\sqrt{(-5)^2-4(-3)(-2)}}{(-3)^2} = \frac{-5-\sqrt{25-24}}{9} = \frac{-5-\sqrt{1}}{9} = \frac{-6}{9} = -\frac{2}{3}$$

4. $45x^3 - 118x^2 + 84x - 1$**.** First, square both binomials and distribute the factors, multiplying them. Then combine like terms.

$$\begin{aligned} 5x(3x-4)^2 + 2(x+1)^2 - 3 &= 5x\left(9x^2 - 24x + 16\right) + 2\left(x^2 + 2x + 1\right) - 3 \\ &= 45x^3 - 120x^2 + 80x + 2x^2 + 4x + 2 - 3 \\ &= 45x^3 - 118x^2 + 84x - 1 \end{aligned}$$

5. $y \geq 3 - x$**.** The shaded area is above the line. The line is solid, so $\geq$ is used to include the points on the line. To check the accuracy of the shading, select a point in the inequality. Using the point (2,2), you have $2 \geq 3 - 2$, which is true.

6. **-3, 0, 8.795, $\frac{19}{5}$, $\sqrt{16}$.** These are all rational because each can be written as a fraction with an integer in the numerator and denominator. The numbers π and $\sqrt{10}$ are not rational; their decimal values go on forever without ever repeating.

7. $\sqrt{85}$**.** Using the distance formula, $d = \sqrt{(1-(-8))^2 + (6-4)^2} = \sqrt{9^2 + 2^2} = \sqrt{81+4} = \sqrt{85}$.

8. $x - 3$**.** Distribute the negative sign over the two terms in the parentheses. Combine like terms and simplify the denominator. Factor the numerator, and then reduce the fraction.

$$\frac{x^2-5x-(9-5x)}{7+x-4} = \frac{x^2-5x-9+5x}{7+x-4} = \frac{x^2-9}{x+3} = \frac{(x-3)\cancel{(x+3)}}{\cancel{x+3}} = x-3$$

9. $-\frac{1}{5}$**.** Using the slope formula, $m = \frac{-2-0}{6-(-4)} = \frac{-2}{10} = -\frac{1}{5}$.

10. **618.1806.** First, simplify the expression in the parentheses. Next, raise the result to the power, and then multiply by the product in front.

$$6 \cdot 100\left(1 + \frac{0.04}{4}\right)^3 = 6 \cdot 100(1.01)^3 = 6 \cdot 100(1.030301) = 618.1806$$

IN THIS CHAPTER

» **Taking a look at the number line**

» **Working with equations and inequalities**

» **Mastering radicals and exponents**

Chapter 2 Operating with Real Numbers

If you're studying pre-calculus, you've probably already taken Algebra I and II and survived (whew!). You may also be thinking, "I'm sure glad that's over; now I can move on to some new stuff." Although pre-calculus presents many new and wonderful ideas and techniques, these new ideas build on the solid-rock foundation of algebra. A bit of a refresher will help you determine just how sturdy your foundation is.

It's assumed that you have certain algebra skills down cold, but this book begins by reviewing some of the tougher ones that become the fundamentals of pre-calculus. In this chapter, you find a review of solving inequalities, absolute-value equations and inequalities, and radicals and rational exponents. There's also an introduction of a new way to express solution sets: interval notation.

Describing Numbers on the Number Line

The number line goes from negative infinity to positive infinity with 0 right in the middle. You can show it with only integers as labels or you can divide up the intervals with fractions or multiples of integers. The lines in Figure 2-1 show you some of the possibilities.

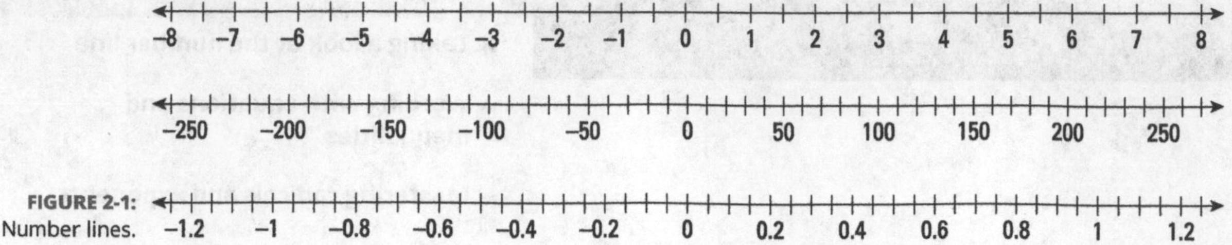

FIGURE 2-1: Number lines.

You can also describe sets of numbers on a number line in many different ways. Look at how you can say that you want all the real numbers from –3 to 7, including the –3 but not the 7:

- $-3 \le x < 7$
- $\{x \mid -3 \le x < 7\}$
- $\{x : -3 \le x < 7\}$
- $\{x \mid x \in -3 \le x < 7\}$
- $[-3,7)$

And, when you graph this set of numbers on a number line, you use a solid dot at the –3 and a hollow dot at the 7. Take a look at the example.

Q. Graph the values of x in $0 < x \le 6$ on a number line.

A. Place a hollow dot at the 0 and a solid dot at the 6. Draw a line between these two points.

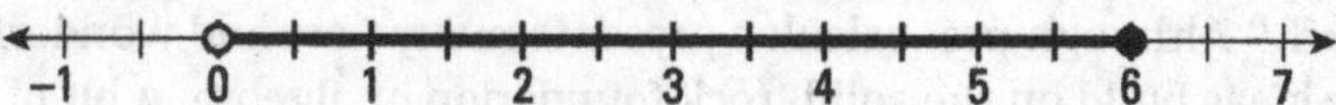

Solving Inequalities

By now you're familiar with equations and how to solve them. When you get to pre-calculus, it's generally assumed that you know how to solve equations, so most courses begin with inequalities. An *inequality* is a mathematical sentence indicating that two expressions either aren't equal or may or may not be equal. The following symbols express inequalities:

Less than: <	Less than or equal to: ≤
Greater than: >	Greater than or equal to: ≥

Recapping inequality how-tos

Inequalities are set up and solved pretty much the same way as equations. In fact, to solve an inequality, you treat it exactly like an equation — with one exception.

REMEMBER

If you multiply or divide an inequality by a negative number, you must change the inequality sign to face the opposite way.

EXAMPLE

Q. Solve $-4x+1<13$ for x.

A. **$x>-3$. Follow these steps:**

1. **Subtract 1 from each side: $-4x<12$**
2. **Divide each side by -4: $x>-3$**

When you divide both sides by −4, you change the less-than sign to the greater-than sign. You can check this solution by picking a number that's greater than −3 and plugging it into the original equation to make sure you get a true statement. If you check 0, for instance, you get $-4(0)+1<13$, which is a true statement.

WARNING

Switching the inequality sign is a step that many people tend to forget. Look at an inequality with numbers in it, like $-2<10$. This statement is true. If you multiply 3 on both sides, you get $-6<30$, which is still true. But if you multiply −3 on both sides — and don't fix the sign — you get $6<-30$. This statement is false, and you always want to keep the statements true. The only way for the inequality to work is to switch the inequality sign to read $6>-30$. The same rule applies if you divide $-2<10$ by −2 on both sides. The only way for the problem to make sense is to state that $1>-5$.

YOUR TURN

Solve $-4x+1<13$ for x. Graph your solution on a number line.

Solve $-5\leq-2x+1<11$ for x. Graph your solution on a number line.

Solving equations and inequalities involving absolute value

If you think back to Algebra I, you'll likely remember that an absolute-value equation usually has two possible solutions. Absolute value is a bit trickier to handle when you're solving inequalities. Similarly, though, inequalities have two possible solutions:

- One where the quantity inside the absolute-value bars is less than a number
- One where the quantity inside the absolute-value bars is greater than a number

In mathematical terminology (where a, b, and c are real numbers):

- $|ax \pm b| < c$ becomes two inequalities: $ax \pm b < c$ AND $ax \pm b > -c$. These can be combined into a compound inequality, $-c < ax \pm b < c$. The "AND" comes from the graph of the solution set on a number line, as shown in Figure 2-2a.
- $|ax \pm b| > c$ becomes two inequalities: $ax \pm b > c$ OR $ax \pm b < -c$. The "OR" also comes from the graph of the solution set, which you can see in Figure 2-2b.

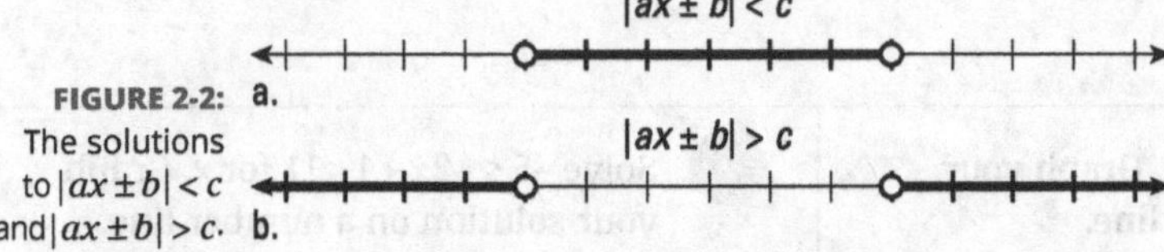

FIGURE 2-2: The solutions to $|ax \pm b| < c$ and $|ax \pm b| > c$.

REMEMBER

Here are two caveats to remember when dealing with absolute values:

- **If the absolute value is less than (<) or less than or equal to (≤) a negative number, it has no solution.** An absolute value must always be zero or positive (the only thing less than negative numbers is other negative numbers). For instance, the absolute-value inequality $|2x - 1| < -3$ doesn't have a solution, because it says that the inequality is less than a negative number.

  Getting 0 as a possible solution is perfectly fine. It's important to note, though, that having no solutions is a different thing entirely. No solutions means that no number works at all, ever.

- **If the result is greater than or equal to a negative number, the solution is all real numbers.** For example, given the inequality $|x - 1| > -5$, x is all real numbers. The left-hand side of this inequality is an absolute value, and an absolute value always represents a positive number. Because positive numbers are always greater than negative numbers, these types of inequalities always have a solution. Any real number that you put into this inequality works.

To solve and graph an inequality with an absolute value — for instance, $a|bx \pm c| < d$ — follow these steps:

1. **Isolate the absolute-value expression.**

   In this case, divide both sides by a to get $|bx \pm c| < \frac{d}{a}$.

2. **Break the inequality into two separate statements.**

   This process gives you $bx \pm c < \frac{d}{a}$ and $bx \pm c > -\frac{d}{a}$. Notice how the inequality sign for the second part changes. When you switch from positives to negatives in an inequality, you must change the inequality sign.

WARNING

Don't fall prey to the trap of changing the equation inside the absolute-value bars. For example, $|bx \pm c| < \frac{d}{a}$ doesn't change to $bx \mp c < \frac{d}{a}$ or $bx \mp c > -\frac{d}{a}$.

3. **Solve both inequalities.**

4. **Graph the solutions.**

   Create a number line and show the answers to the inequality.

EXAMPLE

Q. Solve $2|3x - 6| > 12$ for x.

A. Follow these steps:

1. **Isolate the absolute-value expression.**

   In this case, divide both sides by 2 to get $|3x - 6| > 6$.

2. **Break the inequality into two separate statements.**

   This process gives you $3x - 6 > 6$ and $3x - 6 < -6$.

3. **Solve both inequalities.**

$3x - 6 > 6$	$3x - 6 < -6$
$3x > 12$	$3x < 0$
$x > 4$	$x < 0$

4. **Graph the solutions.**

   Create a number line and show the answers to the inequality. Refer to Figure 2-3a.

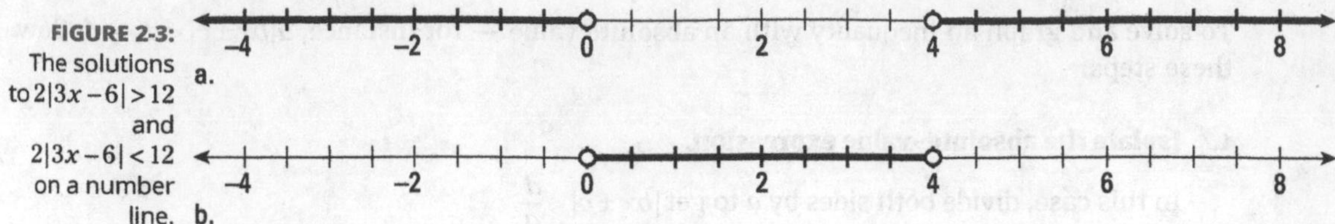

FIGURE 2-3: The solutions to $2|3x-6|>12$ and $2|3x-6|<12$ on a number line.

Q. Solve $2|3x-6|<12$ for x.

EXAMPLE

A. Follow these steps:

1. **Isolate the absolute-value expression.**

   In this case, divide both sides by 2 to get $|3x-6|<6$.

2. **Write the compound inequality.**

   This process gives you $-6<3x-6<6$.

3. **Solve the inequality.**

   $-6<3x-6<6$

   $0<3x<12$

   $\frac{0}{3}<\frac{3x}{3}<\frac{12}{3}$

   $0<x<4$

4. **Graph the solution.**

   Create a number line and show the answer to the inequality. Refer to Figure 2-3b.

YOUR TURN

3 Solve $|3x+1|\leq 16$ for x. Graph your solution on a number line.

4 Solve $5|2x-1|>35$ for x. Graph your solution on a number line.

Expressing solutions for inequalities with interval notation

Now comes the time to venture into interval notation to express where a set of solutions begins and where it ends. *Interval notation* is another way to express the solution set to an inequality, and it's important because it's how you frequently express solution sets in calculus.

TIP

The easiest way to create interval notation is to first draw a graph of your solution on a number line as a visual representation of what's going on in the interval.

If the coordinate point of the number used to define the interval isn't included in the solution (for < or >), the interval is called an *open interval*. You show it on the graph with an open circle at the point and by using parentheses in notation. If the point is included in the solution (≤ or ≥), the interval is called a *closed interval*, which you show on the graph with a filled-in circle at the point and by using square brackets in notation.

EXAMPLE

Q. Write the solution set $-2 < x \le 3$ in interval notation.

A. (–2,3]. First, graph the inequality on a number line. See Figure 2-4.

FIGURE 2-4: The graph of $-2 < x \le 3$ on a number line.

In interval notation, you write this solution as (–2,3].

The bottom line: Both of these inequalities *have* to be true at the same time.

You can also graph OR statements (also known as *disjoint sets* because the solutions don't overlap). OR statements are two different inequalities where one or the other is true.

EXAMPLE

Q. Write the solution set $x < -4$ OR $x > -2$ in interval notation.

A. $(-\infty,-4)\cup(-2,\infty)$. First, graph the inequality on a number line. See Figure 2-5.

–6 –5 –4 –3 –2 –1 0 1 2

FIGURE 2-5: The graph of the OR statement $x < -4$ OR $x > -2$.

The variable x can belong to two different intervals, but because the intervals don't overlap, you have to write them separately:

- The first interval is $x < -4$. This interval includes all numbers between negative infinity and –4. Because $-\infty$ isn't a real number, you use an open interval to represent it. So in interval notation, you write this part of the set as $(-\infty, -4)$.
- The second interval is $x > -2$. This set is all numbers between –2 and positive infinity, so you write it as $(-2, \infty)$.

You describe the whole set as $(-\infty, -4) \cup (-2, \infty)$. The symbol in between the two sets is the *union symbol* and means that the solution can belong to either interval.

REMEMBER

When you're solving an absolute-value inequality that's greater than a number, you write your solutions as OR statements.

EXAMPLE

Q. Write the solution set for $|3x-2| > 7$ in interval notation.

A. $\left(-\infty, -\frac{5}{3}\right) \cup (3, \infty)$. First, solve the absolute-value inequality. You rewrite this inequality as $3x - 2 > 7$ OR $3x - 2 < -7$.

$3x - 2 > 7$	$3x - 2 < -7$
$3x > 9$	$3x < -5$
$x > 3$	$x < -\frac{5}{3}$

In interval notation, this solution is $\left(-\infty, -\frac{5}{3}\right) \cup (3, \infty)$.

YOUR TURN

5 Solve $3x - 8 \geq -2$ for x. Write your solution using interval notation.

6 Solve $|5x + 1| < 24$ for x. Write your solution using interval notation.

Working with Radicals and Exponents

Radicals and exponents (also known as *roots* and *powers*) are two common elements of basic algebra. And, of course, they follow you wherever you go in math, just like a cloud of mosquitoes follows a novice camper. The best thing you can do to prepare for calculus is to be ultra-solid on what can and can't be done when simplifying expressions containing exponents and radicals. You'll want to have this knowledge so that when more complicated math problems come along, the correct answers also come along. This section gives you the solid background you need for those challenging moments.

Defining and relating radicals and exponents

Before you dig deeper into your work with radicals and exponents, make sure you remember the facts in the following list about what they are and how they relate to each other:

- A ***radical*** indicates a **root** of a number, and an ***exponent*** represents the **power** of a number. Radicals are represented by the root sign, $\sqrt{\ }$, and exponents are superscripts written to the upper right of the base.
- The radical equation $\sqrt[n]{a} = b$ says that $b^n = a$. The *n*th root of *a* is *b* if, when you raise *b* to the *n*th power, you get *a*.

EXAMPLE

Q. What is the value of $\sqrt[3]{8}$?

A. **2.** You find that $\sqrt[3]{8} = 2$, because $2^3 = 8$.

EXAMPLE

Q. What is the value of $\sqrt{81}$?

A. **9.** You find that $\sqrt{81} = 9$, because $9^2 = 81$. When there is no digit in the upper-left position, you assume that it's a 2: a square root.

REMEMBER

- The square root of any number represents the principal root (the fancy term for the *positive root*) of that number. For example, $\sqrt{16}$ is 4, even though $(-4)^2$ gives you 16 as well. The expression $-\sqrt{16}$ equals –4 because it's the opposite of the principal root. When you're presented with the equation $x^2 = 16$, you do state both solutions: $x = \pm 4$.
- Also, you can't take the square root of a negative number and get a real number answer; however, you can take the cube root of a negative number. For example, the cube root of –8 is –2, because $(-2)^3 = -8$.
- Other types of exponents, including negative exponents and fractional exponents, have different meanings and are discussed in the sections that follow.

YOUR TURN

7 What is the value of $\sqrt{121}$?	8 What is the value of $\sqrt[4]{625}$?

Rewriting radicals as exponents (or, creating rational exponents)

Sometimes a different (yet equivalent) way of expressing radicals makes the simplification of an expression easier to perform. For instance, when you're given a problem in radical form, you may have an easier time if you rewrite it by using *rational exponents* — exponents that are fractions. You can rewrite every radical as an exponent by using the following property — the top number in the resulting rational exponent tells you the power, and the bottom number tells you the root you're taking:

$$x^{m/n} = \sqrt[n]{x^m} = \left(\sqrt[n]{x}\right)^m = \left(x^{1/n}\right)^m$$

EXAMPLE

Q. Rewrite $8^{2/3}$ in radical form and find its value.

A. 4. You find that $8^{2/3}$ is equal to $\sqrt[3]{8^2}$. You can square the 8 and find $\sqrt[3]{64} = 4$, because $4^3 = 64$. Or, because you can also write $\sqrt[3]{8^2} = \left(\sqrt[3]{8}\right)^2$, you can find the cube root of 8 first and then square the answer: $\left(\sqrt[3]{8}\right)^2 = (2)^2 = 4$.

EXAMPLE

Q. Find the value of $\sqrt[3]{8^5}$ and rewrite it in exponential form.

A. $8^{5/3}$. First, rewrite the expression: $\sqrt[3]{8^5} = \left(\sqrt[3]{8}\right)^5$. You now have $\left(\sqrt[3]{8}\right)^5 = 2^5 = 32$. This is much preferable to raising 8 to the 5th power and then finding the cube root: $\sqrt[3]{8^5} = 8^{5/3}$.

TIP

The order of these processes really doesn't matter. When you have $x^{m/n} = \sqrt[n]{x^m}$, you choose whether to find the root first or the power — whichever makes the process easier.

EXAMPLE

Q. **Simplify the expression $\sqrt{x}\left(\sqrt[3]{x^2}-\sqrt[3]{x^4}\right)$.**

A. $\sqrt[6]{x^7}-\sqrt[6]{x^{11}}$**. First, rewrite the expression using rational exponents.**

$$\sqrt{x}\left(\sqrt[3]{x^2}-\sqrt[3]{x^4}\right)=x^{1/2}\left(x^{2/3}-x^{4/3}\right)$$

Distribute the multiplier over the terms in the parentheses. Remember, when you multiply monomials with the same base, you add the exponents, so you'll perform these additions: $\frac{1}{2}+\frac{2}{3}=\frac{7}{6}$ and $\frac{1}{2}+\frac{4}{3}=\frac{11}{6}$. You get $x^{1/2}\left(x^{2/3}-x^{4/3}\right)=x^{7/6}-x^{11/6}$

Now rewrite the result back in radical form.

$$x^{7/6}-x^{11/6}=\sqrt[6]{x^7}-\sqrt[6]{x^{11}}$$

Typically, your final answer should be in the same format as the original problem: if the original problem is in radical form, your answer should be in radical form, and if the original problem is in exponential form with rational exponents, your solution should be as well.

YOUR TURN

 9 Simplify $\sqrt[3]{x}\left(x-\sqrt{x}\right)$.

 10 Simplify $\frac{\sqrt[4]{x^3}}{\sqrt{x}}$.

Getting a radical out of a denominator: Rationalizing

Another convention of mathematics is that you don't leave radicals in the denominator of an expression when you write it in its final form; the process used to get rid of radicals in the denominator is called *rationalizing the denominator*. This convention makes collecting like terms easy, and your answers will be truly simplified.

REMEMBER

In the rationalized form, a numerator of a fraction can contain a radical, but the denominator can't. The final expression may look more complicated in its rational form, but that's what you have to do sometimes.

This section shows you how to get rid of pesky radicals that may show up in the denominator of a fraction. The focus is on two separate situations: expressions that contain one radical in the denominator and expressions that contain two terms in the denominator, at least one of which is a radical.

A square root

Rationalizing expressions with a square root in the denominator is easy. At the end of it all, you're just getting rid of a square root. Normally, the best way to do that in an equation is to square both sides. For example, if $\sqrt{x-3}=5$, then $\left(\sqrt{x-3}\right)^2=5^2$ or $x-3=25$.

Here are the steps to rationalizing when you have one square-root term in the denominator:

1. **Multiply the numerator and the denominator by the same number — the square root.**

   Starting with $\frac{b}{\sqrt{a}}$, multiply by $\frac{\sqrt{a}}{\sqrt{a}}$, which gives you $\frac{b}{\sqrt{a}}\cdot\frac{\sqrt{a}}{\sqrt{a}}$.

   Whatever you multiply the bottom of a fraction by, you must also multiply the top by; this way, it's really like you multiplied by 1 and you didn't change the fraction. Here's what it looks like:

2. **Multiply the tops, multiply the bottoms, and simplify.**

   $$\frac{b}{\sqrt{a}}\cdot\frac{\sqrt{a}}{\sqrt{a}}=\frac{b\sqrt{a}}{\sqrt{a}\sqrt{a}}=\frac{b\sqrt{a}}{\sqrt{a^2}}=\frac{b\sqrt{a}}{a}$$

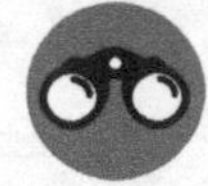

Q. Rationalize $\frac{2}{\sqrt{3}}$.

EXAMPLE **A.** $\frac{2\sqrt{3}}{3}$. Multiply the numerator and denominator by $\frac{\sqrt{3}}{\sqrt{3}}$.

$$\frac{2}{\sqrt{3}}\cdot\frac{\sqrt{3}}{\sqrt{3}}=\frac{2\sqrt{3}}{\sqrt{3}\sqrt{3}}=\frac{2\sqrt{3}}{\sqrt{9}}=\frac{2\sqrt{3}}{3}$$

A cube root

The process for rationalizing a cube root in the denominator is quite similar to that of rationalizing a square root. To get rid of a cube root in the denominator of a fraction, you must cube it. If the denominator is a cube root to the first power, for example, you multiply both the numerator and the denominator by the cube root to the 2nd power to get the cube root to the 3rd power (in the denominator). Raising a cube root to the 3rd power cancels the root — and you're done!

Q. Rationalize $\frac{6}{\sqrt[3]{4}}$.

EXAMPLE **A.** $\frac{6\sqrt[3]{16}}{4}$. Multiply the numerator and denominator by $\left(\sqrt[3]{4}\right)^2$.

You now have $\frac{6}{\sqrt[3]{4}}\cdot\frac{\left(\sqrt[3]{4}\right)^2}{\left(\sqrt[3]{4}\right)^2}=\frac{6\left(\sqrt[3]{4}\right)^2}{\left(\sqrt[3]{4}\right)^3}=\frac{6\left(\sqrt[3]{4}\right)^2}{4}$. The fraction can be simplified by writing 4 as a power of 2, finding the cube root, and then reducing: $\frac{6\left(\sqrt[3]{4}\right)^2}{4}=\frac{6\left(\sqrt[3]{2^2}\right)^2}{4}=\frac{6\left(\sqrt[3]{2^4}\right)}{4}=\frac{6\left(\sqrt[3]{2^3}\sqrt[3]{2}\right)}{4}=\frac{6\left(2\sqrt[3]{2}\right)}{4}=\frac{3\sqrt[3]{2}}{4}$.

A root when the denominator is a binomial

You can rationalize the denominator of a fraction when it contains a binomial with one or more radical terms. Getting rid of the radical in these denominators involves using the conjugate of the denominators. A *conjugate* is a binomial formed by using the opposite of the second term of the original binomial. The conjugate of $a+\sqrt{b}$ is $a-\sqrt{b}$. The conjugate of $x+2$ is $x-2$; similarly, the conjugate of $x+\sqrt{2}$ is $x-\sqrt{2}$.

TECHNICAL STUFF

Multiplying a number by its conjugate is really the FOIL method in disguise. Remember from algebra that FOIL stands for first, outside, inside, and last. So you have something like $(x+\sqrt{y})(x-\sqrt{y})=x^2-x\sqrt{y}+x\sqrt{y}-\sqrt{y}^2=x^2-y$. The middle two terms always cancel each other, and the radicals disappear.

EXAMPLE

Q. Rationalize $\frac{1}{\sqrt{5}-2}$.

A. $\sqrt{5}+2$**.** Multiply the numerator and denominator by $\sqrt{5}+2$, the conjugate of the denominator $\sqrt{5}-2$. You now have

$$\frac{1}{\sqrt{5}-2}\cdot\frac{\sqrt{5}+2}{\sqrt{5}+2}=\frac{\sqrt{5}+2}{(\sqrt{5}-2)(\sqrt{5}+2)}$$

$$=\frac{\sqrt{5}+2}{\sqrt{5}^2+2\sqrt{5}-2\sqrt{5}-2^2}$$

$$=\frac{\sqrt{5}+2}{\sqrt{25}-4}=\frac{\sqrt{5}+2}{5-4}=\frac{\sqrt{5}+2}{1}=\sqrt{5}+2$$

EXAMPLE

Q. Rationalize $\frac{\sqrt{2}-\sqrt{6}}{\sqrt{10}+\sqrt{8}}$.

A. $\sqrt{5}-2-\sqrt{15}+2\sqrt{3}$**.** Multiply by the conjugate of the denominator. The conjugate of $\sqrt{10}+\sqrt{8}$ is $\sqrt{10}-\sqrt{8}$.

$$\frac{\sqrt{2}-\sqrt{6}}{\sqrt{10}+\sqrt{8}}\cdot\frac{\sqrt{10}-\sqrt{8}}{\sqrt{10}-\sqrt{8}}$$

Multiply the numerators and denominators. FOIL the top and the bottom.

$$=\frac{\sqrt{20}-\sqrt{16}-\sqrt{60}+\sqrt{48}}{\sqrt{10^2}-\sqrt{80}+\sqrt{80}-\sqrt{8^2}}$$

Simplify. Both the numerator and denominator simplify first:

$$=\frac{2\sqrt{5}-4-2\sqrt{15}+4\sqrt{3}}{10-8}=\frac{2\sqrt{5}-4-2\sqrt{15}+4\sqrt{3}}{2}$$

This expression simplifies even further because the denominator divides into every term in the numerator, which gives you this:

$$=\sqrt{5}-2-\sqrt{15}+2\sqrt{3}$$

REMEMBER

Simplify any radical in your final answer — always. For example, to simplify a square root, find perfect square-root factors. Also, you can add and subtract only radicals that are like terms. This means the number inside the radical and the *index* (which is what tells you whether it's a square root, a cube root, a fourth root, or whatever) are the same.

A numerator

Rationalizing a numerator may seem contrary to what you want, but in calculus you'll be introduced to the *difference quotient*, where this process may be needed. Don't worry about the whys of this right now. Just notice that by rationalizing the numerator, you're able to factor the fraction. This necessity will become very apparent to you when finding derivatives in calculus.

EXAMPLE

Q. Simplify the fraction $\frac{\sqrt{x+h}-\sqrt{x}}{h}$ by rationalizing the numerator.

A. $\frac{1}{\sqrt{x+h}+\sqrt{x}}$. You multiply the numerator and denominator by the conjugate of the numerator.

$$\frac{\sqrt{x+h}-\sqrt{x}}{h}\cdot\frac{\sqrt{x+h}+\sqrt{x}}{\sqrt{x+h}+\sqrt{x}}=\frac{\left(\sqrt{x+h}-\sqrt{x}\right)\left(\sqrt{x+h}+\sqrt{x}\right)}{h\left(\sqrt{x+h}+\sqrt{x}\right)}$$

$$=\frac{\sqrt{x+h}\sqrt{x+h}+\sqrt{x}\sqrt{x+h}-\sqrt{x}\sqrt{x+h}-\sqrt{x}\sqrt{x}}{h\left(\sqrt{x+h}+\sqrt{x}\right)}$$

$$=\frac{x+h-x}{h\left(\sqrt{x+h}+\sqrt{x}\right)}=\frac{h}{h\left(\sqrt{x+h}+\sqrt{x}\right)}.$$

Now you can reduce the fraction by dividing by the common factor h:

$$\frac{\cancel{h}}{\cancel{h}\left(\sqrt{x+h}+\sqrt{x}\right)}=\frac{1}{\sqrt{x+h}+\sqrt{x}}.$$

Wonderful! You may not see how great this result is now, but you'll so appreciate it when studying calculus.

YOUR TURN

11 Rationalize the denominator: $\frac{10}{3-\sqrt{5}}$

12 Rationalize: $\frac{1}{\sqrt[3]{z}}$

13 Rationalize the denominator: $\frac{-2}{\sqrt{x}-\sqrt{5}}$

14 Rationalize the numerator: $\frac{\sqrt{x+2}-\sqrt{x}}{2}$

Practice Questions Answers and Explanations

1. $x > -3$. **First, subtract 1 from each side, and then divide each side by –4. Since you're dividing by a negative number, reverse the direction of the inequality.**

$$-4x+1<13 \rightarrow -4x<12 \rightarrow \frac{-4x}{-4} > \frac{12}{-4} \rightarrow x > -3$$

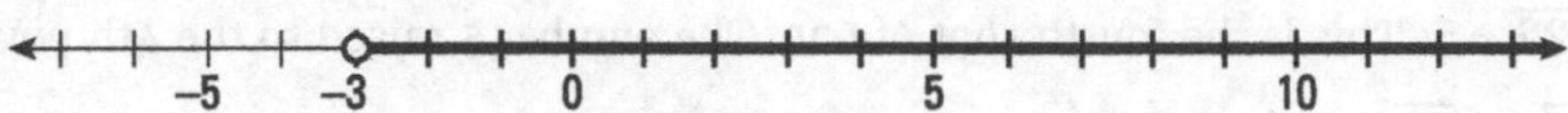

2. $-5 < x \le 3$. **Subtract 1 from each section. Then divide each section by –2. Reverse the direction of the inequalities.**

$$-5 \le -2x+1 < 11 \rightarrow -6 \le -2x < 10 \rightarrow \frac{-6}{-2} \ge \frac{-2x}{-2} > \frac{10}{-2} \rightarrow 3 \ge x > -5$$

Now rewrite the compound inequality from the smallest numbers to the largest; this means you reverse the order of the numbers and reverse the inequalities again.

$3 \ge x > -5 \rightarrow -5 < x \le 3$

3. $-\frac{17}{3} \le x \le 5$. **Write the corresponding inequality and solve for *x* by subtracting 1 from each section and then dividing each section by 3.**

$$-16 \le 3x+1 \le 16 \rightarrow -17 \le 3x \le 15 \rightarrow \frac{-17}{3} \le \frac{3x}{3} \le \frac{15}{3} \rightarrow -\frac{17}{3} \le x \le 5$$

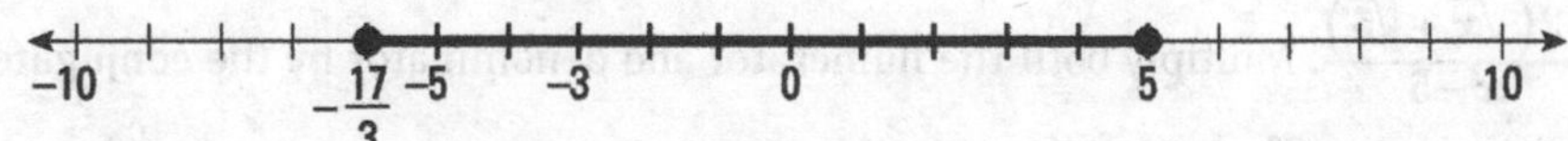

4. $x > 4$ **or** $x < -3$. **First, divide each side of the statement by 5. Then write the corresponding inequalities, and solve for *x* by adding 1, and then dividing by 2.**

$5|2x-1| > 35 \rightarrow |2x-1| > 7$

$2x-1 > 7 \rightarrow 2x > 8 \rightarrow x > 4$

Or $2x-1 < -7 \rightarrow 2x < -6 \rightarrow x < -3$ so $x > 4$ or $x < -3$

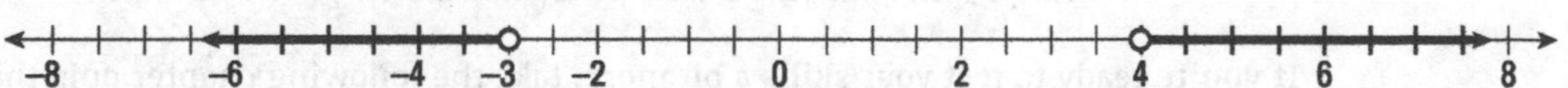

5. **$[2,\infty)$.** First, add 8 to each side, and then divide each side by 3: $3x-8\geq -2 \rightarrow 3x\geq 6 \rightarrow x\geq 2$. In interval notation, this is written $[2,\infty)$.

6. **$\left(-5,\frac{23}{5}\right)$.** Rewrite the absolute-value statement as a compound inequality.

   Then subtract 1 from each section and divide each by 5: $|5x+1|<24 \rightarrow -24<5x+1<24 \rightarrow$

   $-25<5x<23 \rightarrow \frac{-25}{5}<\frac{5x}{5}<\frac{23}{5} \rightarrow -5<x<\frac{23}{5}$. In interval notation, this is written $\left(-5,\frac{23}{5}\right)$.

7. **$\sqrt{121}=11$.** This is the square root of 121. The number 11 squared is 121.

8. **$\sqrt[4]{625}=5$.** This is the fourth root of 625. The number 5 raised to the 4th power is 625.

9. **$\sqrt[3]{x^4}-\sqrt[6]{x^5}$.** First, change each term's exponent to a whole number or fraction. Then distribute: $\sqrt[3]{x}\left(x-\sqrt{x}\right)=x^{1/3}\left(x^1-x^{1/2}\right)=x^{1/3+1}-x^{1/3+1/2}=x^{4/3}-x^{5/6}$. Changing the results back into radical notation, $x^{4/3}-x^{5/6}=\sqrt[3]{x^4}-\sqrt[6]{x^5}$.

10. **$x^{1/4}=\sqrt[4]{x}$.** First, change each term's exponent to a whole number or fraction. Then divide: $\frac{\sqrt[4]{x^3}}{\sqrt{x}}=\frac{x^{3/4}}{x^{1/2}}=x^{3/4-1/2}=x^{1/4}$. Change the result back to radical notation: $x^{1/4}=\sqrt[4]{x}$.

11. **$\frac{5}{2}(3+\sqrt{5})$.** Multiply both the numerator and denominator by the conjugate of the denominator. Then simplify.

$$\frac{10}{3-\sqrt{5}}=\frac{10}{3-\sqrt{5}}\cdot\frac{3+\sqrt{5}}{3+\sqrt{5}}=\frac{10(3+\sqrt{5})}{9-5}=\frac{\not{10}^{5}(3+\sqrt{5})}{\not{4}^{2}}=\frac{5}{2}(3+\sqrt{5})$$

12. **$\frac{\sqrt[3]{z^2}}{z}$.** Multiply both the numerator and denominator by $z^{2/3}$.

$$\frac{1}{\sqrt[3]{z}}=\frac{1}{z^{1/3}}=\frac{1}{z^{1/3}}\cdot\frac{z^{2/3}}{z^{2/3}}=\frac{z^{2/3}}{z^{1/3+2/3}}=\frac{z^{2/3}}{z^1}=\frac{\sqrt[3]{z^2}}{z}$$

13. **$\frac{-2(\sqrt{x}+\sqrt{5})}{x-5}$.** Multiply both the numerator and denominator by the conjugate of the denominator. Then simplify.

$$\frac{-2}{\sqrt{x}-\sqrt{5}}\cdot\frac{\sqrt{x}+\sqrt{5}}{\sqrt{x}+\sqrt{5}}=\frac{-2(\sqrt{x}+\sqrt{5})}{x-5}$$

14. **$\frac{1}{\sqrt{x+2}+\sqrt{x}}$.** Multiply both the numerator and denominator by the conjugate of the numerator. Then simplify.

$$\frac{\sqrt{x+2}-\sqrt{x}}{2}\cdot\frac{\sqrt{x+2}+\sqrt{x}}{\sqrt{x+2}+\sqrt{x}}=\frac{x+2-x}{2(\sqrt{x+2}+\sqrt{x})}=\frac{\not{2}}{\not{2}(\sqrt{x+2}+\sqrt{x})}=\frac{1}{\sqrt{x+2}+\sqrt{x}}$$

If you're ready to test your skills a bit more, take the following chapter quiz that incorporates all the chapter topics.

Whaddya Know? Chapter 2 Quiz

Quiz time! Complete each problem to test your knowledge on the various topics covered in this chapter. You can then find the solutions and explanations in the next section.

1. Solve for x: $5x - 11 \geq 14$. Write your answer in both inequality and interval notation.

2. Determine the value: $\sqrt[3]{64}$

3. Simplify: $\sqrt[3]{x}\left(\sqrt[4]{x^3}\right)$

4. Solve for x: $-2x + 3 < -9$. Write your answer in both inequality and interval notation.

5. Rationalize the denominator: $\dfrac{2x}{\sqrt{x} - 3}$

6. Solve for x: $-5 \leq 3x - 1 < 14$. Graph your answer on a number line.

7. Determine the value: $\sqrt{64}$

8. Rationalize the denominator: $\dfrac{\sqrt{x} + 6}{\sqrt{x} - 2}$

9. Solve for x: $4|3x + 11| \leq 20$. Write your answer in both inequality and interval notation.

10. Simplify: $\dfrac{\sqrt{x} - \sqrt[3]{x}}{\sqrt[5]{x}}$

11. Determine the value: $\sqrt[3]{1331} - \sqrt[5]{32}$

12. Solve for x: $|2 - 9x| > 7$. Graph your answer on a number line.

Answers to Chapter 2 Quiz

1. **$x \geq 5$; $[5,\infty)$.** You get $5x-11 \geq 14 \rightarrow 5x \geq 25 \rightarrow x \geq 5$. In interval notation, this is $[5,\infty)$.

2. **4.** You get $\sqrt[3]{64}=4$ because $4^3=64$.

3. **$\sqrt[12]{x^{13}}$.** You get $\sqrt[3]{x}\left(\sqrt[4]{x^3}\right)=x^{1/3}\left(x^{3/4}\right)=x^{1/3+3/4}=x^{13/12}=\sqrt[12]{x^{13}}$.

4. **$x > 6$; $(6,\infty)$.** You get $-2x+3<-9 \rightarrow -2x<-12 \rightarrow x>6$. In interval notation, this is written $(6,\infty)$.

5. **$\dfrac{2x\sqrt{x}+6x}{x-9}$.** You get $\dfrac{2x}{\sqrt{x}-3}\cdot\dfrac{\sqrt{x}+3}{\sqrt{x}+3}=\dfrac{2x\left(\sqrt{x}+3\right)}{x-9}=\dfrac{2x\sqrt{x}+6x}{x-9}$.

6. **$-\dfrac{4}{3} \leq x < 5$.**

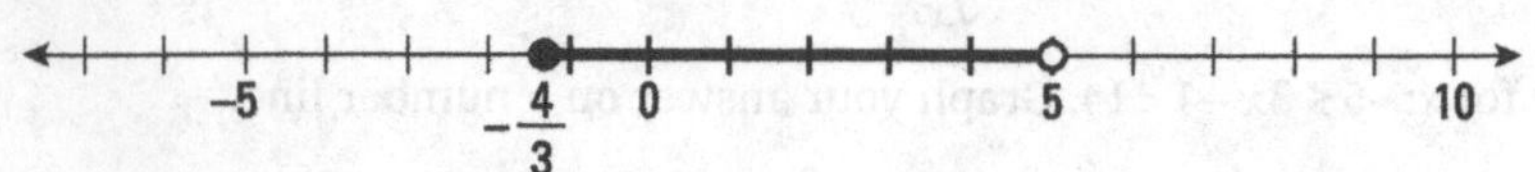

$$-5 \leq 3x-1<14 \rightarrow -4 \leq 3x<15 \rightarrow -\frac{4}{3} \leq x<5$$

7. **8.** You get $\sqrt{64}=8$ because $8^2=64$.

8. **$\dfrac{x+8\sqrt{x}+12}{x-4}$.** You get $\dfrac{\sqrt{x}+6}{\sqrt{x}-2}=\dfrac{\sqrt{x}+6}{\sqrt{x}-2}\cdot\dfrac{\sqrt{x}+2}{\sqrt{x}+2}=\dfrac{x+2\sqrt{x}+6\sqrt{x}+12}{x-4}=\dfrac{x+8\sqrt{x}+12}{x-4}$.

9. **$-\dfrac{16}{3} \leq x \leq -2$; $\left[-\dfrac{16}{3},-2\right]$.** You get $4|3x+11| \leq 20 \rightarrow |3x+11| \leq 5 \rightarrow -5 \leq 3x+11 \leq 5 \rightarrow$

$$-16 \leq 3x \leq -6 \rightarrow -\frac{16}{3} \leq x \leq -2.$$

In interval notation, this is written $\left[-\dfrac{16}{3},-2\right]$.

10. **$\sqrt[10]{x^3}-\sqrt[15]{x^2}$.** You get $\dfrac{\sqrt{x}-\sqrt[3]{x}}{\sqrt[5]{x}}=\dfrac{x^{1/2}-x^{1/3}}{x^{1/5}}=\dfrac{x^{1/2}}{x^{1/5}}-\dfrac{x^{1/3}}{x^{1/5}}=x^{1/2-1/5}-x^{1/3-1/5}$

$$=x^{3/10}-x^{2/15}=\sqrt[10]{x^3}-\sqrt[15]{x^2}.$$

11. **9.** You get $\sqrt[3]{1331}-\sqrt[5]{32}=11-2=9$, because $11^3=1331$ and $2^5=32$.

12. **$x<-\dfrac{5}{9}$ or $x>1$.**

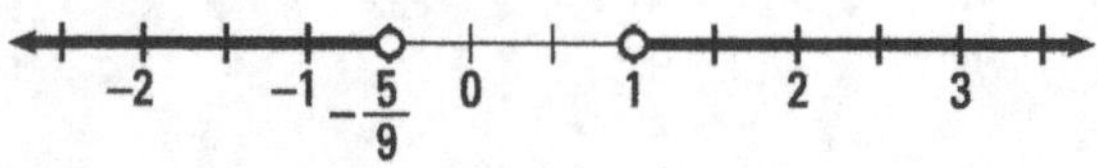

$$|2-9x|>7 \rightarrow 2-9x>7 \text{ or } 2-9x<-7$$

$$2-9x>7 \rightarrow -9x>5 \rightarrow x<-\frac{5}{9}$$

$$2-9x<-7 \rightarrow -9x<-9 \rightarrow x>1$$

IN THIS CHAPTER

» Identifying, graphing, and translating parent functions

» Defining even, odd, and one-to-one functions

» Comparing quadratic, cubic, and their root functions

» Breaking down and graphing rational functions

Chapter 3 Cementing the Building Blocks of Pre-Calculus Functions

Maps of the world identify cities as dots and use lines to represent the roads that connect them. Modern country and city maps use a grid system to help users find locations easily. If you can't find the place you're looking for, you look at an index, which gives you a letter and number. This information narrows down your search area, after which you can easily figure out how to get where you're going.

You can take this idea and use it for your own pre-calculus purposes through the process of graphing. But instead of naming cities, the dots name points on the coordinate. A point on this plane relates two numbers to each other; these are called coordinates. The whole coordinate plane really is just a big computer, because it's based on input and output, with you as the operating system. This idea of input and output is best expressed mathematically using functions. A *function* is a set of ordered pairs, where every input value x gives one, and only one, output value y.

This chapter shows you how to perform your role as the operating system, explaining the map of the world of points and lines on the coordinate plane along the way.

Identifying Special Function Types and Their Graphs

Functions can be categorized in many different ways. In the following sections, you see functions in terms of the operations being performed. Here, though, you see classifications that work for all the many types of functions. If you know that a function is even or odd or one-to-one, then you know how the function can be applied and whether it can be used as a model in a particular situation. You have a heads-up as to how the function behaves in different circumstances.

Even and odd functions

Knowing whether a function is even or odd helps you to graph it because that information tells you which half of the points you have to graph and which half are a special variation of them. These types of functions are symmetrical, so whatever appears in one half is exactly the same as the other half. If a function is even, the graph is symmetrical over the y-axis. If the function is odd, the graph is symmetrical about the origin.

- **Even function:** The mathematical definition of an *even function* is $f(-x)=f(x)$ for any value of x.
- **Odd function:** The mathematical definition of an *odd function* is $f(-x)=-f(x)$ for any value of x.

Q. Show that $f(x)=x^2$ is an even function.

EXAMPLE

A. You know that $f(3)=9$, and $f(-3)=9$. Basically, the opposite input yields the same output. Visually speaking, the graph is a mirror image across the y-axis.

Q. Show that $f(x)=x^3$ is an odd function.

EXAMPLE

A. The function $f(x)=x^3$ is an odd function because $f(3)=27$ and $f(-3)=-27$. The opposite input gives the opposite output. These graphs have 180-degree symmetry about the origin. If you turn the graph upside down, it looks the same.

One-to-one functions

A function is considered to be one-to-one if every output value is unique — it appears just once in the range. Another way of saying this is that every input value has exactly one output value (which is essentially the definition of a function), and every output value comes from exactly one input value. There are no repeats in output values.

EXAMPLE

Q. **Show that are $f(x)=2x^3$ and $g(x)=\frac{1}{x}$ are one-to-one functions.**

A. **Replacing $f(x)$ with y, the function $y=2x^3$ is one-to-one, because, if you solve for x in the function equation, you get $x=\sqrt[3]{\frac{y}{2}}$. There's just x for each y value. And the same goes for $y=\frac{1}{x}$. Solving for x, you get $x=\frac{1}{y}$.**

YOUR TURN

Determine which functions are even or odd and which are one-to-one.

1. $f(x)=\sqrt[3]{x}$

2. $f(x)=x^4+8$

3. $f(x)=x$

4. $f(x)=\frac{1}{x^2}$

Dealing with Parent Functions and Their Graphs

In mathematics, you see certain types of function graphs over and over again. There are basic functions and the variations on those original functions. The basic functions are called the *parent graphs*, and they include graphs of quadratic functions, square roots, absolute value, cubic polynomials, and cube roots. In this section, you'll find information on graphing the parent graphs so you can graduate to more in-depth graphing work in the form of transformations, found in Chapter 4.

Linear functions

A *linear* function is the simplest of all the functions involving variables. The parent linear function is $f(x)=x$ and is the line that cuts through the origin, bisecting the first and third quadrants. This parent function is odd, so it has symmetry about the origin.

Quadratic functions

Quadratic functions are equations in which the 2nd power, or square, is the highest to which the unknown quantity or variable is raised. In such an equation, the independent variable is squared in one of the terms. The equation $f(x)=x^2$ is a quadratic function and is the parent graph for all other quadratic functions. The graph of $x=y^2$ isn't a function, because any positive x value produces two different y values — look at (4, 2) and (4, −2), for example.

TIP

The shortcut to graphing the function $f(x)=x^2$ is to take advantage of the fact that it's an even function. Start at the point (0, 0) (the *origin*) and mark that point, called the *vertex.* Note that the point (0, 0) is the vertex of the parent function but not all quadratic functions. Later, when you transform graphs, the vertex moves around the coordinate plane. In calculus, this point is called a *critical point.* Without getting into the calculus definition, it just means that the point is special.

The graph of any quadratic function is called a *parabola.* All parabolas have the same basic shape (for more, see Chapter 17). To get the other points on the graph of $f(x)=x^2$, you move 1 unit horizontally from the vertex, up 1^2; over 2, up 2^2; over 3, up 3^2; and so on. Because even functions are symmetric to the *y*-axis, this graphing occurs on both sides of the vertex and keeps going, but usually just plotting a couple points on either side of the vertex gives you a good idea of what the graph looks like. Check out Figure 3-1 for an example of a quadratic function in graph form.

Square-root functions

REMEMBER

A *square-root graph* is related to a quadratic graph (see the previous section). The quadratic function is $f(x)=x^2$, whereas the square-root function is $g(x)=x^{1/2}$. The graph of a square-root function looks like the left half of a parabola that has been rotated 90 degrees clockwise. You can also write the square-root function as $g(x)=\sqrt{x}$.

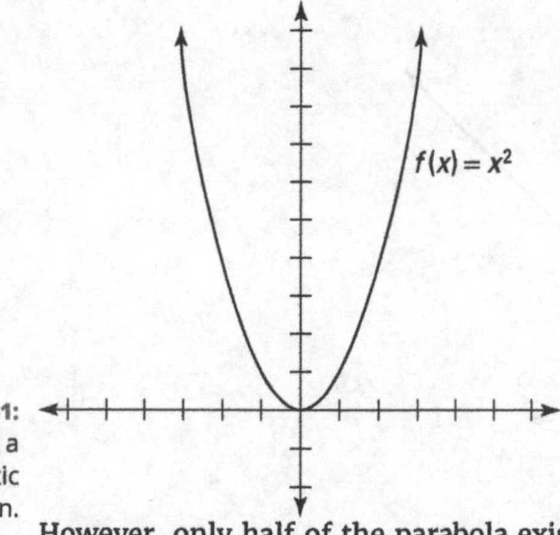

FIGURE 3-1: Graphing a quadratic function.

However, only half of the parabola exists in the square-root graph, for two reasons. First, its parent graph exists only when x is zero or positive (because you can't find the square root of negative numbers, and the domain consists of just non-negative numbers). The second reason is because the function $g(x) = \sqrt{x}$ is positive, because you're being asked to find only the principal or positive root.

This graph starts at the origin (0, 0) and then moves to the right 1 position, up $\sqrt{1}$; to the right 2, up $\sqrt{2}$; to the right 3, up $\sqrt{3}$; and so on. Check out Figure 3-2 for an example of this graph.

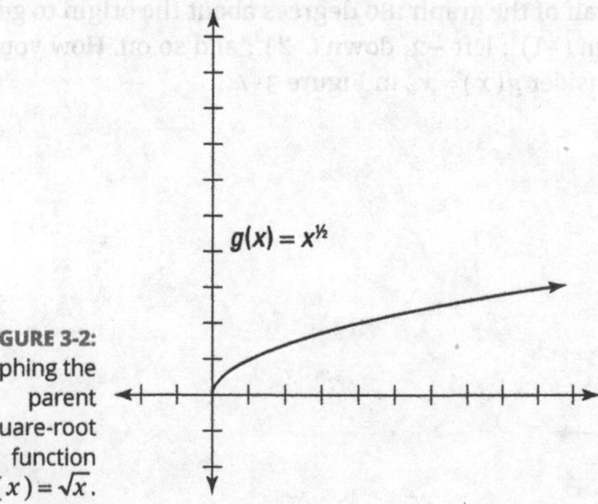

FIGURE 3-2: Graphing the parent square-root function $f(x) = \sqrt{x}$.

TIP

Notice that the values you get by plotting consecutive points don't exactly give you the nicest numbers. Instead, try picking values for which you can easily find the square root. Here's how this works: Start at the origin and go right 1, up $\sqrt{1}$; right 4, up $\sqrt{4}$; right 9, up $\sqrt{9}$; and so on.

Absolute-value functions

REMEMBER

The absolute-value parent graph of the function $y = |x|$ turns all inputs into non-negative (0 or positive) outputs. To graph absolute-value functions, you start at the origin and move in both directions along the x-axis and the y-axis from there: over 1, up 1; over 2, up 2; and on and on forever. This is another even function, and so it's symmetric to the vertical axis. Figure 3-3 shows this graph in action.

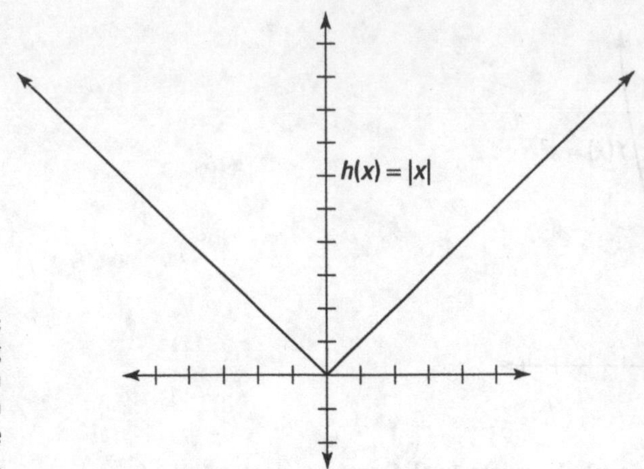

FIGURE 3-3: Staying positive with the graph of an absolute-value function.

Cubic functions

REMEMBER

In a *cubic function*, the highest degree on any variable is three; $f(x) = x^3$ is the parent function. You start graphing the cubic function parent graph at its critical point, which is also the origin (0, 0). The origin isn't, however, a critical point for every variation of the cubic function.

From the critical point, the cubic graph moves right 1, up 1^3; right 2, up 2^3; and so on. The function x^3 is an odd function, so you rotate half of the graph 180 degrees about the origin to get the other half. Or, you can move left –1, down $(-1)^3$; left –2, down $(-2)^3$; and so on. How you plot is based on your personal preference. Consider $g(x) = x^3$ in Figure 3-4.

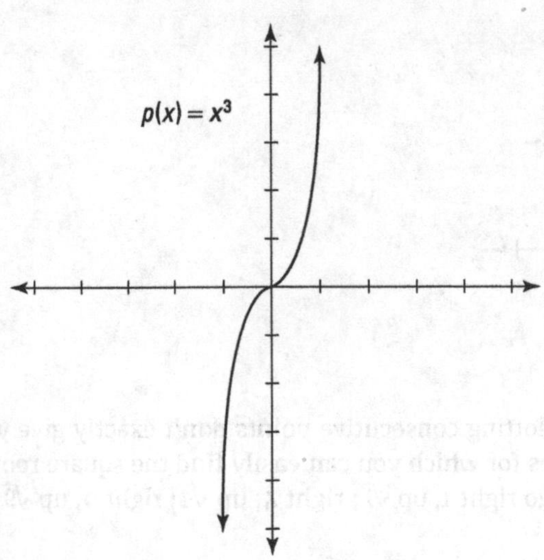

FIGURE 3-4: Putting the cubic parent function in graph form.

Cube-root functions

REMEMBER

Cube-root functions are related to cubic functions in the same way that square-root functions are related to quadratic functions. You write cubic functions as $f(x) = x^3$ and cube-root functions as $g(x) = x^{1/3}$ or $g(x) = \sqrt[3]{x}$.

Noting that a cube-root function is odd is important because it helps you graph it. The critical point of the cube-root parent graph is at the origin (0, 0), as shown in Figure 3-5.

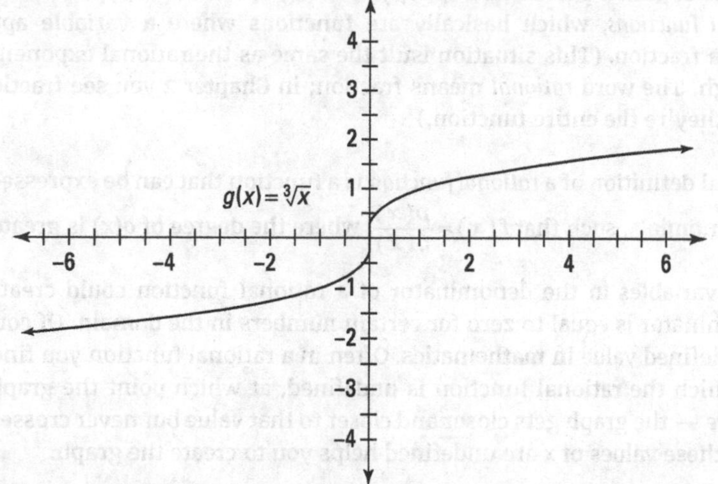

FIGURE 3-5: The graph of a cube-root function.

YOUR TURN

Identify the function type from each function equation. Your choices are: linear, quadratic, cubic, square root, cube root, and absolute value.

5 $f(x)=9-4x$	6 $g(x)=\sqrt{9-4x}$
7 $f(x)=\lvert 9-4x\rvert$	8 $f(x)=9-4x^3$

Setting the Stage for Rational Functions

In addition to the common parent functions, you will graph another type of function in pre-calculus: *rational functions*, which basically are functions where a variable appears in the denominator of a fraction. (This situation isn't the same as the rational exponents you see in Chapter 2, though. The word *rational* means fraction; in Chapter 2 you see fractions as exponents, and now they're the entire function.)

TECHNICAL STUFF

The mathematical definition of a *rational function* is a function that can be expressed as the quotient of two polynomials, such that $f(x)=\frac{p(x)}{q(x)}$, where the degree of $q(x)$ is greater than zero.

TIP

The variable or variables in the denominator of a rational function could create a situation where the denominator is equal to zero for certain numbers in the domain. Of course, division by zero is an undefined value in mathematics. Often in a rational function you find at least one value of x for which the rational function is undefined, at which point the graph will have a vertical *asymptote* — the graph gets closer and closer to that value but never crosses it. Knowing in advance that these values of x are undefined helps you to create the graph.

In the following sections, you see the considerations involved in finding the outputs of (and ultimately graphing) rational functions.

Searching for vertical asymptotes and holes

Having the variable on the bottom of a fraction can create a problem; the denominator of a fraction must never be zero. With rational functions you may find some domain value(s) of x that makes the denominator zero. The function "jumps over" this value in the graph, creating what's called a *vertical asymptote* or a *hole*. Graphing a vertical asymptote first shows you the number in the domain where your graph can't pass through. The graph approaches this point but never reaches it.

To identify vertical asymptotes and holes, do the following:

- Determine for which values the denominator of the function can be 0.
- Determine if the highest power in the numerator is greater than the highest power in the denominator. If that's the case, then move on to *oblique asymptotes*.
- Determine if there's a common factor in the numerator and denominator. If that's the case, then you have a hole.

The following examples will provide a guide for your search.

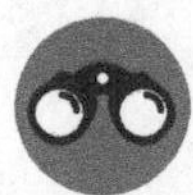

EXAMPLE

Q. Identify any vertical asymptotes in the function $f(x)=\frac{3x-1}{x^2+4x-21}$.

A. **−7 and 3.** Factoring the denominator, you have $f(x)=\frac{3x-1}{(x+7)(x-3)}$. The denominator is equal to 0 when $x=-7$ or $x=3$. Those are the equations of the two vertical asymptotes.

Q. Identify any vertical asymptotes in the function $g(x) = \frac{6x+12}{4-3x}$.

EXAMPLE **A.** $\frac{4}{3}$. The denominator is equal to 0 when $x = \frac{4}{3}$. This is the equation of the vertical asymptote.

Q. Identify any vertical asymptotes in the function $h(x) = \frac{x^2-9}{x+2}$.

EXAMPLE **A.** None. The power in the numerator is greater than that in the denominator, so there is no vertical asymptote.

Q. Identify any vertical asymptotes in the function $k(x) = \frac{x^2-9}{x^2-3x}$.

EXAMPLE **A.** $x = 0$. Factoring the numerator and denominator, you have $k(x) = \frac{(x-3)(x+3)}{x(x-3)}$. This reduces to $k(x) = \frac{x+3}{x}$. When $x = 3$, you have no vertical asymptote, because of the common factor in the numerator and denominator. Instead, you have a hole at (3, 2). You get the y value by solving for $k(3)$ in the reduced version, $k(x) = \frac{x+3}{x}$. But you do have a vertical asymptote when $x = 0$.

YOUR TURN

9 Write equations of the vertical asymptotes in the function $f(x) = \frac{4x-7}{x^2-5x-6}$.

10 Determine the equation of the vertical asymptote and coordinates of the hole in the function $g(x) = \frac{x-1}{x^2+2x-3}$.

Searching for horizontal asymptotes

To find a horizontal asymptote of a rational function, you need to look at the degree of the polynomials in the numerator and the denominator. The *degree* is the highest power of the variable in the polynomial expression. Here's how you proceed:

- If the denominator has the bigger degree, the horizontal asymptote automatically is the *x*-axis, or $y = 0$.
- If the numerator and denominator have an equal degree, you must divide the leading coefficients (the coefficients of the terms with the highest degrees) to find the horizontal asymptote.

WARNING

Be careful! Sometimes the terms with the highest degrees aren't written first in the polynomial. You can always rewrite both polynomials so that the highest degrees come first, if you prefer.

- If the numerator has the bigger degree of exactly one more than the denominator, the graph will have an oblique asymptote; see the next discussion for more information on how to proceed in this case.

EXAMPLE

Q. Find any horizontal asymptotes in the functions $f(x) = \frac{3x-1}{x^2+4x-21}$, $g(x) = \frac{6x+12}{4-3x}$, and $k(x) = \frac{x^2-9}{x^2-3x}$.

A. **−2 and 1.** The function $f(x) = \frac{3x-1}{x^2+4x-21}$ has a horizontal asymptote $y = 0$, because the degree in the numerator is 1 and the degree of the denominator is 2. The denominator wins. Rewrite the denominator of $g(x) = \frac{6x+12}{4-3x}$ as $-3x+4$ so that it appears in descending order. The function $g(x)$ has equal degrees on top and bottom. To find the horizontal asymptote, divide the leading coefficients on the highest-degree terms. You get $y = \frac{6}{-3}$ or $y = -2$. You now have your horizontal asymptote for $g(x)$. And $k(x) = \frac{x^2-9}{x^2-3x}$ also has equal degrees on the top and bottom. Make a fraction of the coefficients and you have $y = \frac{1}{1} = 1$.

YOUR TURN

11 Write the equation of the horizontal asymptote of $h(x) = \frac{9x-2}{9x^2-2}$.

12 Write the equation of the horizontal asymptote of $k(x) = \frac{3x^2-4x+7}{5-3x^2}$.

Seeking out oblique asymptotes

Oblique asymptotes are neither horizontal nor vertical. In fact, an asymptote doesn't even have to be a straight line at all; it can be a slight curve or a really complicated curve. This chapter will stick to the straight and narrow, though.

To find an oblique asymptote, you have to use long division of polynomials to find the quotient. You take the denominator of the rational function and divide it into the numerator. The quotient (neglecting the remainder) gives you the equation of the line of your oblique asymptote.

REMEMBER

Long division of polynomials is discussed in Chapter 5. You must understand long division of polynomials in order to complete the graph of a rational function with an oblique asymptote.

EXAMPLE

Q. Find the oblique asymptote in the function $h(x)=\frac{x^2-9}{x+2}$.

A. $y = x - 2$. This function has an oblique asymptote because the numerator has the higher degree in the polynomial. By using long division, dividing x^2-9 by $x+2$ you get a quotient of $x-2$ and ignore the remainder of –5. This quotient means the oblique asymptote follows the equation $y=x-2$. Because this equation is first-degree, you graph it by using the slope-intercept form. Keep this oblique asymptote in mind, because graphing is coming right up!

YOUR TURN

13 Determine the equation of the oblique asymptote of $f(x)=\frac{x^2-2x+1}{x+2}$.

Locating the *x*- and *y*-intercepts

The final piece of the puzzle is to find the intercepts (where the line or curve crosses the *x*- and *y*-axes) of the rational function, if any exist:

» To find the *y*-intercept of an equation, set $x = 0$. Then solve for *y*.

» To find the *x*-intercept or intercepts of a function, set $y = 0$. Solve for any values of *x* that satisfy the equation.

TIP

For any rational function, the shortcut is to set the numerator equal to zero and then solve. Sometimes when you do this, however, the equation you get is unsolvable, which means that the rational function doesn't have an *x*-intercept.

EXAMPLE

Q. Find the intercepts of $f(x)=\frac{3x-1}{x^2+4x-21}$.

A. **The *y*-intercept is $\left(0,\frac{1}{21}\right)$, and the *x*-intercept is $\left(\frac{1}{3},0\right)$. The *y*-intercept of $f(x)=\frac{3x-1}{x^2+4x-21}$ is found when you let all the *x*'s be 0. You get $f(x)=\frac{-1}{-21}$, meaning that the *y*-intercept is $\left(0,\frac{1}{21}\right)$. The *x*-intercept of $f(x)=\frac{3x-1}{x^2+4x-21}$ is found by solving $0=3x-1$ for *x*. You get $x=\frac{1}{3}$, so the *x*-intercept is $\left(\frac{1}{3},0\right)$.**

EXAMPLE

Q. Find the intercepts of $g(x)=\frac{6x+12}{4-3x}$.

A. **The *y*-intercept is $(0,3)$, and the *x*-intercept is $(-2,0)$. The *x*-intercept of $g(x)=\frac{6x+12}{4-3x}$ is found by solving $6x+12=0$ for *x*. You get $x=-2$. The *x*-intercept is then $(-2,0)$. To solve for the *y*-intercept, let $x=0$ and solve for *y*. You get $y=\frac{12}{4}=3$. The *y*-intercept is then $(0,3)$.**

As with vertical asymptotes, you have to be careful that the factor providing the intercept isn't also a factor of the denominator. Reducing the fraction first eliminates this, but be sure you also eliminate the number provided in the factor from the domain.

YOUR TURN

14 Find all the intercepts of $f(x)=\frac{x-6}{4-x}$.

15 Find all the intercepts of

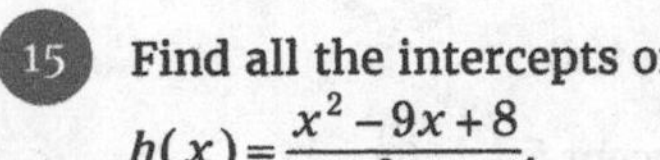

$h(x) = \dfrac{x^2 - 9x + 8}{x^2 + 2}$.

16 Find all the intercepts of

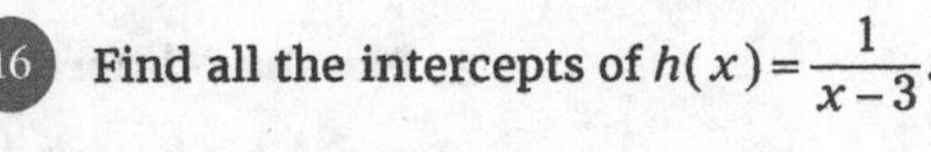

$h(x) = \dfrac{1}{x - 3}$.

Putting the Results to Work: Graphing Rational Functions

After you calculate all the asymptotes and the x- and y-intercepts for a rational function (you see that process in the preceding section), you have all the information you need to start graphing the rational function. Graphing a rational function is all about locating the intercepts and any asymptotes, and determining what the graph of the function does as it approaches a vertical asymptote from the left or right and approaches a horizontal asymptote from above or below. Three of the functions featured in the previous section will be graphed now, using the known information and finding a bit more to aid in the graphing.

EXAMPLE

Q. Graph

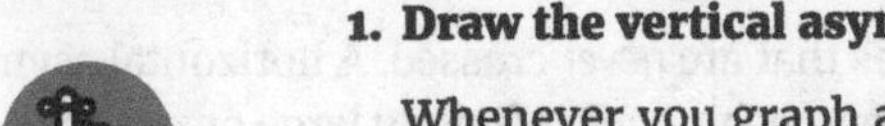

$f(x) = \dfrac{3x - 1}{x^2 + 4x - 21}$.

A. To graph the function $f(x)$, do the following:

1. **Draw the vertical asymptote(s).**

REMEMBER

Whenever you graph asymptotes, be sure to use dotted lines, not solid lines, because the asymptotes aren't part of the rational function.

For $f(x)$, you found that the vertical asymptotes are $x = -7$ and $x = 3$, so draw two dotted vertical lines, one at $x = -7$ and another at $x = 3$.

2. **Draw the horizontal asymptote(s).**

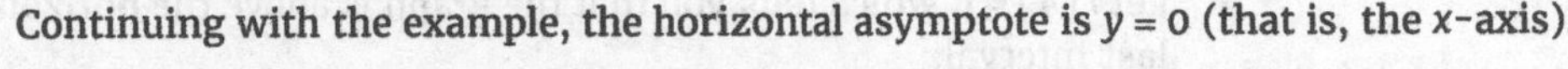

Continuing with the example, the horizontal asymptote is $y = 0$ (that is, the x-axis).

REMEMBER

In any rational function, where the denominator has a greater degree as values of x get infinitely large, the fraction gets infinitely smaller until it approaches zero (this process is called a *limit*; you can see it again in Chapter 22).

3. **Plot the *x*-intercept(s) and the *y*-intercept(s).**

   The *y*-intercept is $\left(0, \frac{1}{21}\right)$, and the *x*-intercept is $\left(\frac{1}{3}, 0\right)$.

   Figure 3-6 shows what you've found so far in the graph.

4. **Fill in the graph of the curve between the vertical asymptotes.**

   The vertical asymptotes divide the graph and the domain of $f(x)$ into three intervals: $(-\infty, -7)$, $(-7, 3)$, and $(3, \infty)$. For each of these three intervals, you must pick at least one test value and plug it into the original rational function; this test determines whether the graph on that interval is above or below the horizontal asymptote (the *x*-axis).

   a. *Test a value in the first interval.*

      In $f(x)$, the first interval is $(-\infty, -7)$, so you can choose any number you want as long as it's less than -7. You might choose $x = -8$ for this example, so now you evaluate $f(-8) = \frac{-25}{11}$.

      This negative value tells you that the function is under the horizontal asymptote on the first interval.

   b. *Test a value in the second interval.*

      If you look at the second interval $(-7, 3)$ in Figure 3-7, you'll realize that you already have two test points located in it. The *y*-intercept has a positive value, which tells you that the graph is above the horizontal asymptote for that part of the graph.

      Now here comes the curve ball: It stands to reason that a graph should never cross an asymptote; it should just get closer and closer to it. In this case, there's an *x*-intercept, which means that the graph actually crosses its own horizontal asymptote. The graph becomes negative for the rest of this interval.

      Sometimes the graphs of rational functions cross a horizontal asymptote, and sometimes they don't. In this case, where the denominator has a greater degree, and the horizontal asymptote is the *x*-axis, it depends on whether the function has roots or not. You can find out by setting the numerator equal to zero and solving the equation. If you find a solution, there is a zero and the graph will cross the *x*-axis. If not, the graph doesn't cross the *x*-axis.

      Vertical asymptotes are the only asymptotes that are *never* crossed. A horizontal asymptote actually tells you what value the graph is approaching for infinitely large or negative values of *x*.

   c. *Test a value in the third interval.*

      For the third interval, $(3, \infty)$, you might use the test value of 4 (you can use any number greater than 3) to determine the location of the graph on the interval. You get $f(4) = 1$, which tells you that the graph is above the horizontal asymptote for this last interval.

   Knowing a test value in each interval, you can plot the graph by starting at a test value point and moving from there toward both the horizontal and vertical asymptotes. Figure 3-8 shows the complete graph of $f(x)$.

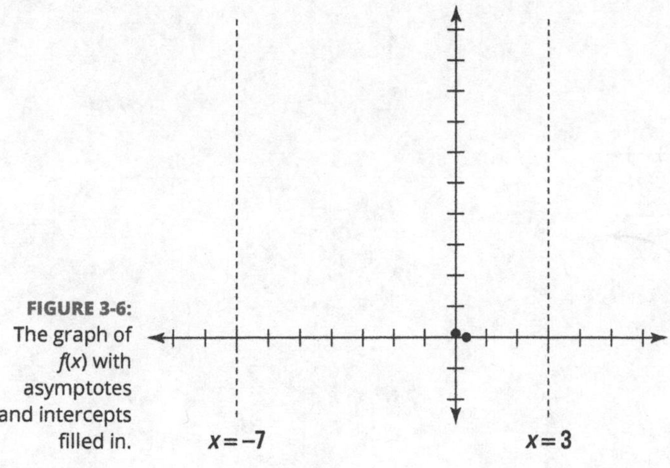

FIGURE 3-6: The graph of $f(x)$ with asymptotes and intercepts filled in.

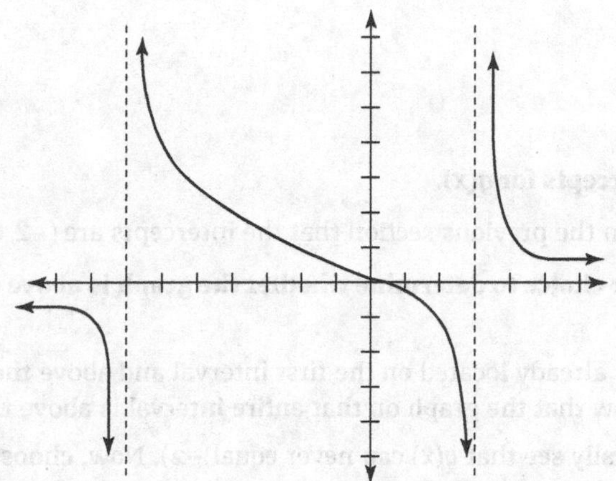

FIGURE 3-7: The final graph of $f(x)$.

Rational functions with equal degrees in the numerator and denominator behave the way that they do because of limits (see Chapter 22). What you need to remember is that the horizontal asymptote is the quotient of the leading coefficients of the top and the bottom of the function (see the earlier section, "Searching for horizontal asymptotes," for more info).

EXAMPLE

Q. Graph $g(x) = \dfrac{6x+12}{4-3x}$.

A. The function $g(x)$ has equal degrees on the variables for each part of the fraction. Follow these simple steps to graph $g(x)$, which is shown in Figure 3-8.

1. **Sketch the vertical asymptote(s) for $g(x)$.**

   From your work in the previous section, you find only one vertical asymptote at $x = \frac{4}{3}$, which means you have only two intervals to consider: $\left(-\infty, \frac{4}{3}\right)$ and $\left(\frac{4}{3}, \infty\right)$.

2. **Sketch the horizontal asymptote for $g(x)$.**

   You find in Step 2 from the previous section that the horizontal asymptote is $y = -2$. So you sketch a horizontal line at that position.

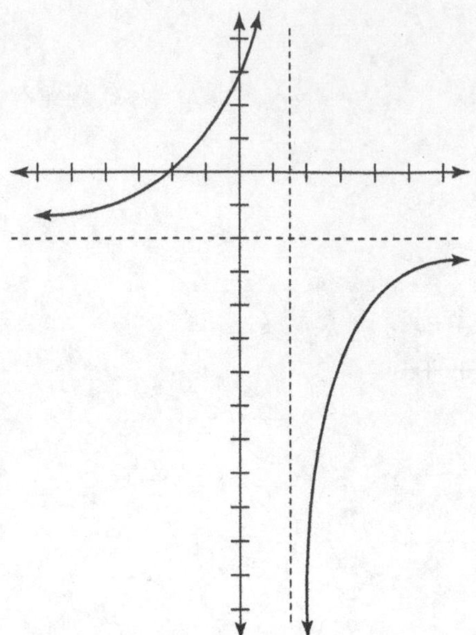

FIGURE 3-8: The graph of $g(x)$, which is a rational function with equal degrees on top and bottom.

3. **Plot the x- and y-intercepts for $g(x)$.**

   You find in Step 4 from the previous section that the intercepts are $(-2, 0)$ and $(0, 3)$.

4. **Use test values of your choice to determine whether the graph is above or below the horizontal asymptote.**

   The two intercepts are already located on the first interval and above the horizontal asymptote, so you know that the graph on that entire interval is above the horizontal asymptote (you can easily see that $g(x)$ can never equal -2). Now, choose a test value for the second interval greater than $\frac{4}{3}$. Here, you choose $x = 2$. Substituting this into the function $g(x)$ gives you -12. You know that -12 is way under -2, so you know that the graph lives under the horizontal asymptote in this second interval.

Rational functions, where the numerator has the greater degree, don't actually have horizontal asymptotes. If the degree of the numerator of a rational function is exactly one more than the degree of its denominator, it has an oblique asymptote, which you find by using long division (see Chapter 5). If the degree in the numerator is more than one greater than that in the denominator, then there is no horizontal or oblique asymptote.

Q. Graph $h(x) = \dfrac{x^2 - 9}{x + 2}$.

EXAMPLE

A. 1. **Sketch the vertical asymptote(s) of $h(x)$.**

   You find only one vertical asymptote for this rational function: $x = -2$. And because the function has only one vertical asymptote, you find only two intervals for this graph: $(-\infty, -2)$ and $(-2, \infty)$.

2. **Sketch the oblique asymptote of $h(x)$.**

   Because the numerator of this rational function has the greater degree, the function has an oblique asymptote. Using long division, you find that the oblique asymptote follows the equation $y = x - 2$. Sketch in the line using dots, not a solid line.

3. **Plot the x- and y-intercepts for $h(x)$.**

   You find that the x-intercepts are $(-3, 0)$ and $(3, 0)$; the y-intercept is $\left(0, -\frac{9}{2}\right)$.

4. **Use test values of your choice to determine whether the graph is above or below the oblique asymptote.**

   Notice that the intercepts conveniently give test points in each interval. You don't need to create your own test points, but you can if you really want to. In the first interval, the test point $(-3, 0)$, hence the graph, is located above the oblique asymptote. In the second interval, the test points $\left(0, -\frac{9}{2}\right)$ and $(3,0)$, as well as the graph, are located under the oblique asymptote.

Figure 3-9 shows the complete graph of $h(x)$.

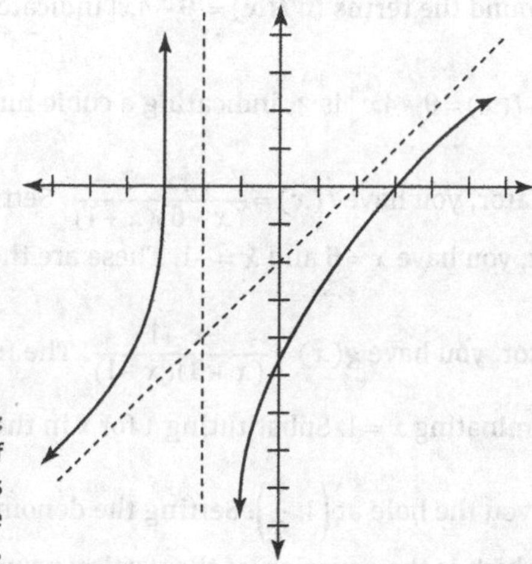

FIGURE 3-9: The graph of $h(x)$, which has an oblique asymptote.

YOUR TURN

17 Sketch the graph of $f(x) = \frac{x^2 - 2x - 15}{x^2 - 4}$.	 Sketch the graph of $h(x) = \frac{x^2 - 9}{x + 1}$.

Practice Questions Answers and Explanations

1. **Odd and one-to-one.** For every negative x in the domain of $f(x)=\sqrt[3]{x}$, $f(-x)=-f(x)$.

2. **Even.** For every negative x in the domain of $f(x)=x^4+8$, $f(-x)=f(x)$.

3. **Odd and one-to-one.** For every negative x in the domain of $f(x)=x$, $f(-x)=-f(x)$.

4. **Even.** For every negative x in the domain of $f(x)=\frac{1}{x^2}$, $f(-x)=f(x)$.

5. **Linear.** The power on the only variable in $f(x)=9-4x$ is 1.

6. **Square root.** The two terms in $g(x)=\sqrt{9-4x}$ are under the radical for a square root.

7. **Absolute value.** The two vertical lines around the terms in $f(x)=|9-4x|$ indicate the absolute value operation.

8. **Cubic.** The power on the only variable in $f(x)=9-4x^3$ is 3, indicating a cubic function: $f(x)=9-4x^3$.

9. **$x=6$ and $x=-1$.** Factoring the denominator, you have $f(x)=\frac{4x-7}{(x-6)(x+1)}$. Setting the denominator equal to 0 and solving for x, you have $x=6$ and $x=-1$. These are the equations of the vertical asymptotes.

10. **$x=-3$, $\left(1,\frac{1}{4}\right)$.** Factoring the denominator, you have $g(x)=\frac{x-1}{(x+3)(x-1)}$. The fraction reduces, $g(x)=\frac{\cancel{x-1}}{(x+3)\cancel{(x-1)}}=\frac{1}{x+3}$, eliminating $x=1$. Substituting 1 for x in the reduced fraction, you have $g(1)=\frac{1}{4}$, which gives you the hole at $\left(1,\frac{1}{4}\right)$. Setting the denominator equal to 0 and solving for x, you have $x=-3$, which is the equation of the vertical asymptote.

11. **$y=0$.** Since the power in the numerator of $h(x)=\frac{9x-2}{9x^2-2}$ is smaller than that in the denominator, the horizontal asymptote is the x-axis, $y=0$.

12. **$y=-1$.** Make a fraction of the coefficients on the two squared terms (highest powers in numerator and denominator are the same) of $k(x)=\frac{3x^2-4x+7}{5-3x^2}$. You get $y=\frac{3}{-3}=-1$.

13 **$y = x - 4$. Divide the numerator by the denominator of $f(x) = \frac{x^2 - 2x + 1}{x + 2}$.**

$$\begin{array}{r} x - 4 \\ x+2 \overline{)x^2 - 2x + 1} \\ \underline{x^2 + 2x} \\ -4x + 1 \\ \underline{-4x - 8} \\ 9 \end{array}$$

Ignore the remainder. Write the asymptote equation using the dividend: $y = x - 4$.

14 $\left(0, -\frac{3}{2}\right)$ **and $(6, 0)$.** Setting $x = 0$, $f(0) = \frac{0-6}{4-0} = \frac{-6}{4} = -\frac{3}{2}$. So the y-intercept is $\left(0, -\frac{3}{2}\right)$. Setting $y = 0$, $0 = \frac{x-6}{4-x}$, which is true when $x = 6$, so the x-intercept is $(6, 0)$.

15 **$(0, 4)$ and $(1, 0)$ and $(8, 0)$.** Setting $x = 0$, $h(0) = \frac{0 - 0 + 8}{0 + 2} = 4$. So the y-intercept is $(0, 4)$. Setting $y = 0$, $0 = \frac{x^2 - 9x + 8}{x^2 + 2} = \frac{(x-1)(x-8)}{x^2 + 2}$, which is true when $x = 1$ or $x = 8$. The two x-intercepts are at $(1, 0)$ and $(8, 0)$.

16 $\left(0, -\frac{1}{3}\right)$. Setting $x = 0$, $h(0) = \frac{1}{0-3} = -\frac{1}{3}$, so the y-intercept is $\left(0, -\frac{1}{3}\right)$. Setting $y = 0$, you have $0 = \frac{1}{x-3}$, which has no solution. There is no x-intercept.

17 **Refer to the following graph.** There are two vertical asymptotes, determined when you set the denominator equal to 0 and solve for x: $x = 2$ and $x = -2$. The horizontal asymptote is $y = 1$, determined when the two coefficients of the squared terms are made into a fraction. Letting x be 0, you have the y-intercept of $\left(0, \frac{15}{4}\right)$. And setting the numerator equal to 0, the two x-intercepts are $(5, 0)$ and $(-3, 0)$.

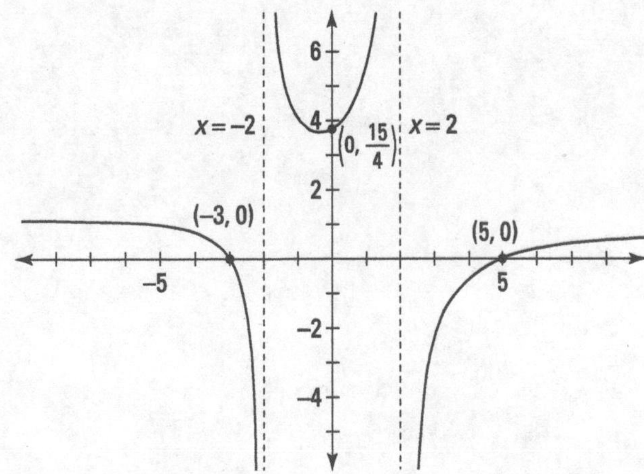

18. **Refer to the following graph.** There is one vertical asymptote at $x = -1$, which is determined by setting the denominator equal to 0. You find the oblique asymptote, $y = x - 1$, by dividing the numerator by the denominator. When x is 0, you find the y-intercept at $(0, -9)$. The two x-intercepts are found when $h(x) = 0$. They are $(3, 0)$ and $(-3, 0)$.

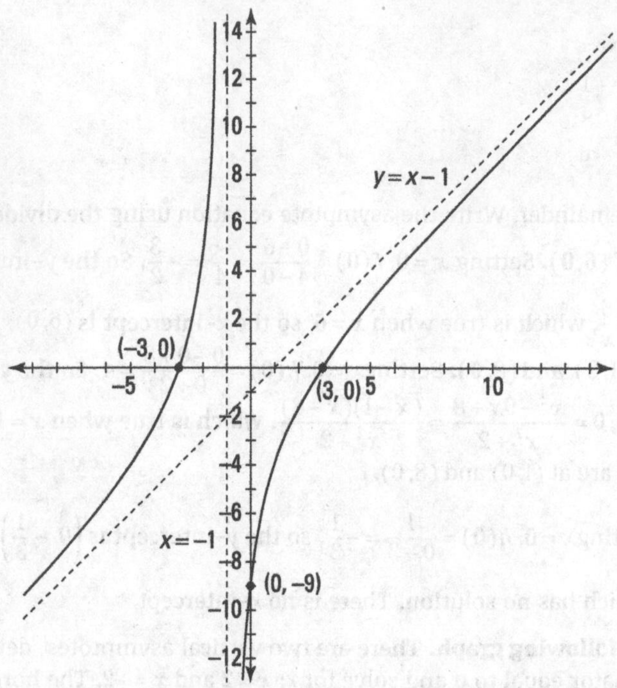

If you're ready to test your skills a bit more, take the following chapter quiz that incorporates all the chapter topics.

Whaddya Know? Chapter 3 Quiz

Quiz time! Complete each problem to test your knowledge on the various topics covered in this chapter. You can then find the solutions and explanations in the next section.

1. **Find the *y*-intercept of the function $f(x)=\dfrac{5-x}{x+1}$.**
2. **Is the function $f(x)=\dfrac{x^2}{x^2+3}$ one-to-one?**
3. **Name the oblique asymptote of the function $f(x)=\dfrac{x^2+6x-5}{x+2}$.**
4. **Name the vertical asymptote(s) of the function $f(x)=\dfrac{x-3}{x^2-4x-21}$.**
5. **Graph the function $f(x)=\dfrac{x+1}{x^2-x-6}$.**
6. **Find the *x*-intercept(s) of the function $f(x)=\dfrac{x^2-3x}{x^2+4x-1}$.**
7. **Identify the function $f(x)=\dfrac{x^2}{x^2+3}$ as even, odd, or neither.**
8. **Name the horizontal asymptote of the function $f(x)=\dfrac{x^2-4}{x^3+4}$.**

Answers to Chapter 3 Quiz

1. **(0,5).** Letting x be equal to 0, you have $f(0)=\frac{5-0}{0+1}=5$, so the y-intercept is $(0, 5)$.

2. **No.** Even and odd integers with the same absolute value result in the same function value in $f(x)=\frac{x^2}{x^2+3}$, so this is not one-to-one.

3. $\boldsymbol{y=x+4}$**.** Divide the numerator by the denominator to get a dividend of $x+4$ and a remainder of -13. Ignore the remainder and write the asymptote as $y=x+4$.

4. $\boldsymbol{x=7,\ x=-3}$**.** Factoring the denominator, you have $f(x)=\frac{x-3}{(x-7)(x+3)}$. Setting the denominator equal to 0, the two solutions are $x=7$, $x=-3$, which are the equations of the asymptotes.

5. **Refer to the following figure.** The vertical asymptotes are $x=3$, $x=-2$. The horizontal asymptote is the x-axis. The x-intercept is $(-1, 0)$, and the y-intercept is $\left(0, -\frac{1}{6}\right)$.

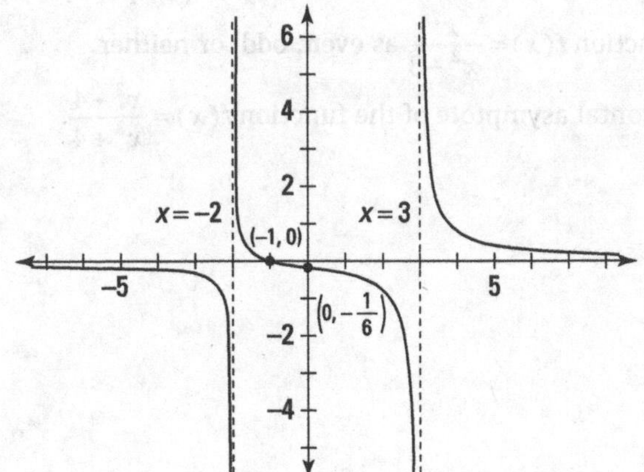

6. **(0,0) and (3,0).** Solving $0=\frac{x^2-3x}{x^2+4x-1}$, you have $0=x^2-3x=x(x-3)$. The two solutions are 0 and 3, so the intercepts are $(0, 0)$ and $(3, 0)$.

7. **Even.** For every odd number in the domain of $f(x)=\frac{x^2}{x^2+3}$, it is true that $f(-x)=f(x)$.

8. $\boldsymbol{y=0}$**.** The greatest power of the variable is in the denominator, so the horizontal asymptote is the x-axis.

IN THIS CHAPTER

» Transforming one function into many others

» Performing different operations on functions

» Finding and verifying inverses of functions

Chapter 4

Operating on Functions

A basic mathematical function serves as the model or parent for many, many other functions. By just adding or subtracting numbers, or multiplying or dividing by numbers, a parent function can be slid around, flipped, stretched, and squashed. The basic function is usually still recognizable and most of its properties are in place. It makes life easier not to have to add completely different functions to the mix; you just have to do some minor adjustments to the original.

The basic operations on rational numbers also apply to functions. But there's a whole new operation that applies only to combining functions — *composition*. It's somewhat like composing a musical number; you insert certain patterns of sounds in different parts of the arrangement to make something totally wonderful.

Inverse functions undo the operations of the original function. Just like inverse operations undo something another operation has done, inverse functions operate to undo other functions.

Transforming the Parent Graphs

In certain situations, you need to use a parent function to get the graph of a more complicated version of the same function. For instance, you can graph each of the following by *transforming* its parent graph:

$f(x)=-2(x+1)^2-3$ comes from $f_1(x)=x^2$.

$g(x)=\frac{1}{4}|x-2|$ has the parent function $g_1(x)=|x|$.

$h(x)=(x-1)^4+2$ began as the function $h_1(x)=x^4$.

As long as you have the graph of the parent function, you can transform it by using the rules described in this section. When using a parent function for this purpose, you can choose from the following different types of transformations (discussed in more detail in the following sections):

- Some transformations cause the parent graph to **stretch** or **shrink** vertically or horizontally — to become steeper or flatter.
- Translations cause the parent graph to **shift** left, right, up, or down (or to shift both horizontally and vertically).
- **Reflections** flip the parent graph over a horizontal or vertical line. They do just what their name suggests: mirror the parent graphs (unless other transformations are involved, of course).

REMEMBER

The methods used to transform functions work for all types of common functions; what works for a quadratic function will also work for an absolute-value function and even a trigonometric function. A function is always a function, so the rules for transforming functions always apply, no matter what type of function you're dealing with.

And if you can't remember these shortcut methods later on, you can always take the long route: picking random values for *x* and plugging them into the function to see what *y* values you get.

Stretching and flattening

REMEMBER

A number (or *coefficient*) multiplying in front of a function causes a change in height or steepness. The coefficient always affects the height of each and every point in the graph of the function. It's common to call such a transformation a *stretch* or *steepening* if the coefficient is greater than 1 and a *shrinking* or *flattening* if the coefficient is between 0 and 1.

EXAMPLE

Q. Determine the change in the graph $f(x) = x^3$ when you multiply by 4 and create $f(x) = 4x^3$.

A. **Over 1, up 4; over 2, up 32; over 3, up 108; and so on.** The graph of $f(x) = 4x^3$ takes the graph of $f(x) = x^3$ and stretches it by a factor of 4. That means that each time you plot a point vertically on the graph, the original value gets multiplied by 4 (making the graph four times as tall at each point). So, from the vertex, you move over 1, up $4 \cdot 1^3 = 4$; over 2, up $4 \cdot 2^3 = 32$; over 3, up $4 \cdot 3^3 = 108$; and so on. See Figure 4-1.

EXAMPLE

Q. Determine the change in the graph $f(x) = x^3$ when you multiply by $\frac{1}{4}$ and create $f(x) = \frac{1}{4}x^3$.

A. **When the multiplier is between 0 and 1 (a proper fraction), the curve flattens out.** Figure 4-1 shows how multiplying by $\frac{1}{4}$ compares to multiplying by 4.

A number multiplying a variable inside a function's operation, rather than multiplying after the operation has been performed, also affects the horizontal movement of the graph.

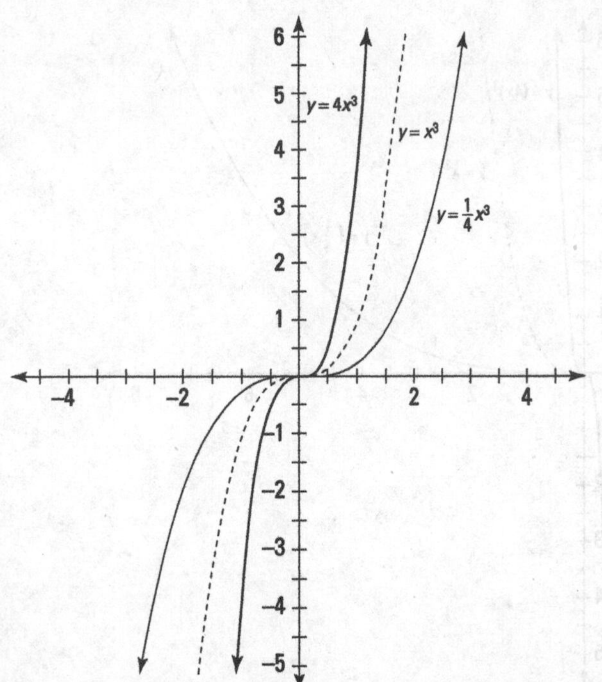

FIGURE 4-1: Graphing the transformations $y = 4x^3$ and $y = \frac{1}{4}x^3$.

EXAMPLE

Q. How does multiplying a number inside the function's operation affect the result?

A. **It flattens the curve.** Look at the graph of $y = (4x)^3$ (see Figure 4-2). The distance between any two consecutive values from the parent graph $y = x^3$ along the x-axis is always 1. If you set the inside of the new, transformed function equal to the distance between the x values, you get $4x = 1$. Solving the equation gives you $x = \frac{1}{4}$, which is how far you step along the x-axis. Beginning at the origin (0,0), you move right $\frac{1}{4}$, up 1; right $\frac{1}{2}$, up 2; right $\frac{3}{4}$, up 3; and so on. Also, in Figure 4-2, you see what happens when the multiplier within the function operation is between 0 and 1; in this case, the multiplier is $\frac{1}{4}$. This has the opposite effect and flattens out the curve.

YOUR TURN

With Questions 1 and 2, given the function $f(x) = x^2$, determine how the function changes with the multipliers.

1 $f(x) = \frac{1}{10}x^2$	**2** $f(x) = \left(\frac{8}{3}x\right)^2$

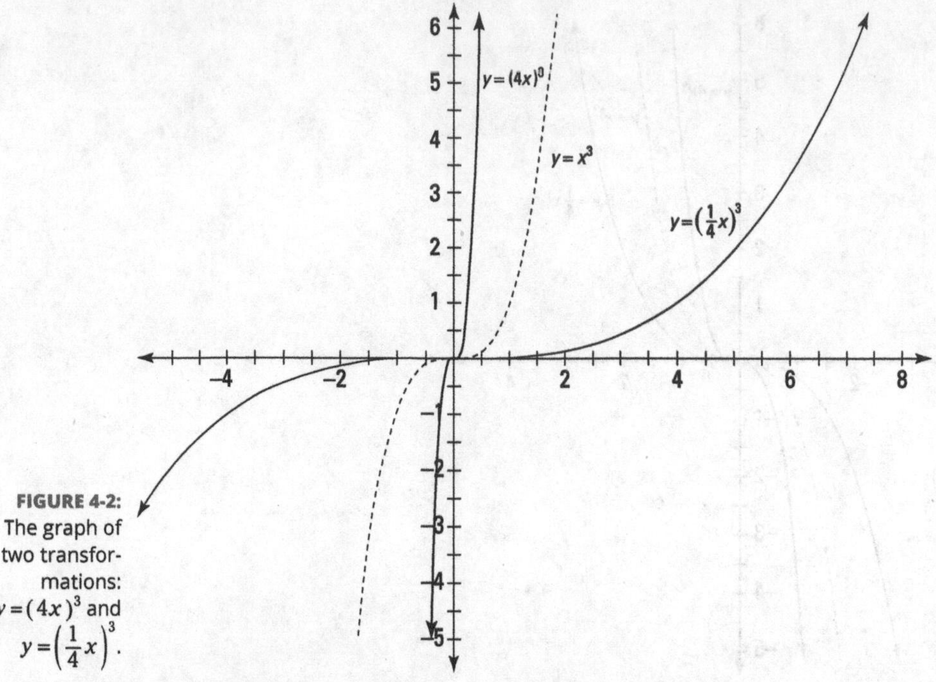

FIGURE 4-2: The graph of two transformations: $y = (4x)^3$ and $y = \left(\frac{1}{4}x\right)^3$.

With Questions 3 and 4, graph the function.

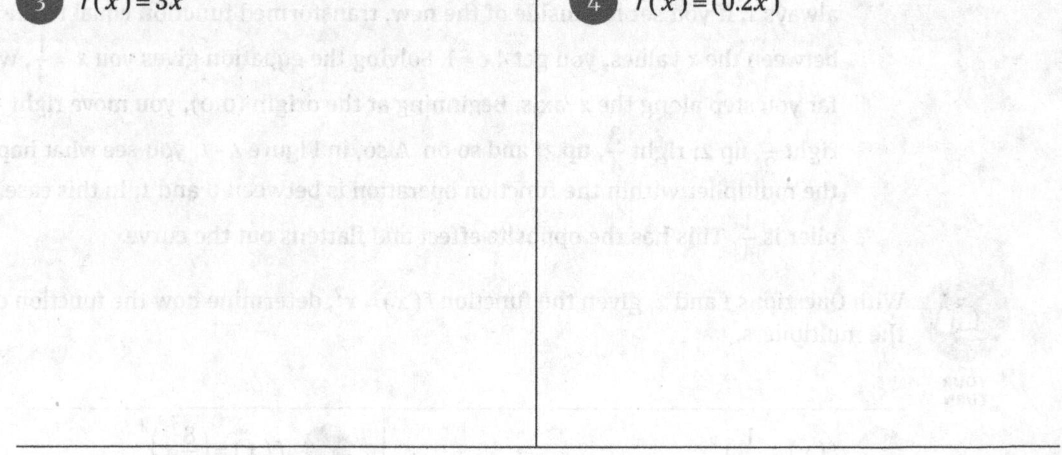

3 $f(x) = 3x^2$

4 $f(x) = (0.2x)^2$

Graphing translations

Shifting a graph horizontally or vertically is called a *translation.* In other words, every point on the parent graph is shifted left, right, up, or down. The graph doesn't change shape — it just moves around the coordinate plane. In this section, you find information on both kinds of translations: horizontal shifts and vertical shifts.

Horizontal shifts

A number added or subtracted inside the function operation, parentheses, or other grouping device of a function creates a *horizontal shift*. Such functions are written in the form $f(x \pm h)$, where h represents the horizontal shift.

REMEMBER

The number corresponding to the h in this function does the opposite of what it looks like it should do. For example, if you have the equation $g(x) = (x-h)^2$, the graph moves $f(x) = x^2$ to the right h units; in $k(x) = (x+h)^2$, the graph moves $f(x) = x^2$ to the left h units.

EXAMPLE

Q. Determine the horizontal shifts of the two functions $g(x) = (x-3)^2$ and $k(x) = (x+2)^2$.

A. **3 to the right and 2 to the left, respectively.** The parent function, $f(x) = x^2$, is translated to the right in $g(x) = (x-3)^2$ and translated to the left, $k(x) = (x+2)^2$.

Figure 4-3 shows the graphs for the parent function, $f(x) = x^2$, the translation to the right in $g(x) = (x-3)^2$, and the translation to the left, $k(x) = (x+2)^2$.

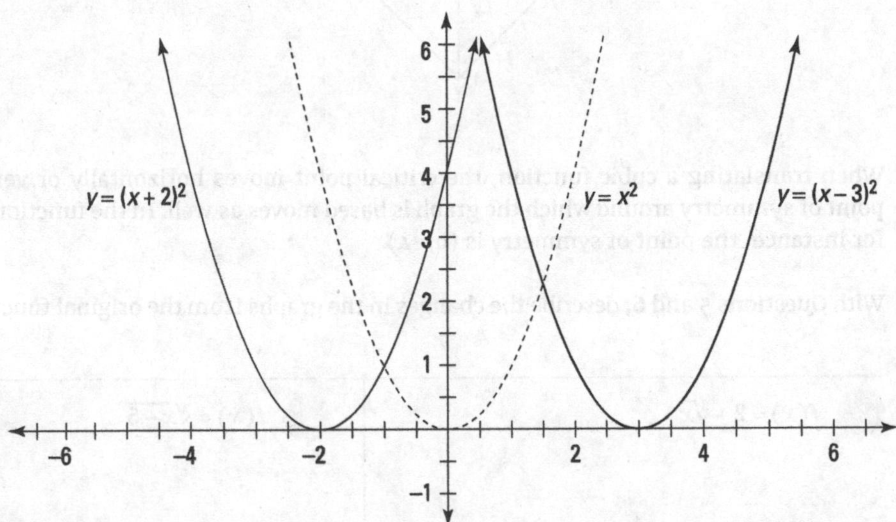

FIGURE 4-3: The graph of two horizontal shifts: 3 to the right and 2 to the left.

Vertical shifts

Adding or subtracting numbers completely separate from the function operation causes a *vertical shift* in the graph of the function. Consider the expression $f(x) + v$, where v represents the vertical shift. Notice that the addition of the variable exists outside the function operation.

Vertical shifts are less complicated than horizontal shifts (see the previous section), because reading them tells you exactly what to do. In the equation $h(x) = |x| - v$, where v is a positive integer, you can probably guess what the graph is going to do: It moves the function $f(x) = |x|$ down v units, whereas the graph of $g(x) = |x| + v$ moves the function $f(x)$ up v units.

Note: You see no vertical stretch or shrink for either $f(x)$ or $g(x)$, because the coefficient in front of x^2 for both functions is 1. If another number was multiplied with the functions, you'd have a vertical stretch or shrink.

Q. Determine the vertical shifts in the functions $h(x) = |x| - 4$ and $g(x) = |x| + 3$.

EXAMPLE **A.** **Down 4 units and up 3 units, respectively.** In the graph of the function $h(x) = |x| - 4$, the vertical shift is down 4 units; and in $g(x) = |x| + 3$, the shift is up 3 units. Figure 4-4 shows these translated graphs.

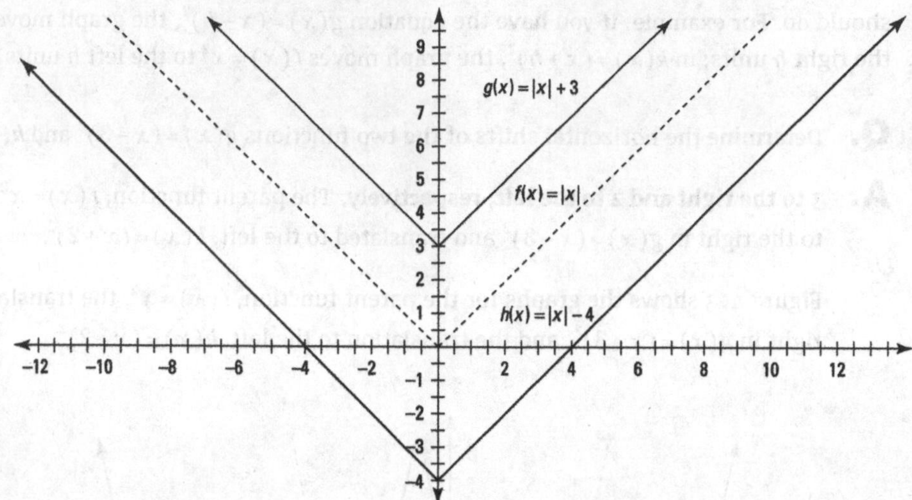

FIGURE 4-4: The graph of vertical shifts: $h(x) = |x| - 4$ and $g(x) = |x| + 3$.

REMEMBER

When translating a cubic function, the critical point moves horizontally or vertically, so the point of symmetry around which the graph is based moves as well. In the function $k(x) = x^3 - 4$, for instance, the point of symmetry is $(0, -4)$.

YOUR TURN

With Questions 5 and 6, describe the changes in the graphs from the original function $f(x) = \sqrt[3]{x}$.

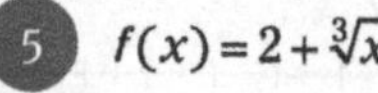

5 $f(x) = 2 + \sqrt[3]{x}$

6 $f(x) = \sqrt[3]{x - 5}$

With Questions 7 and 8, graph the functions.

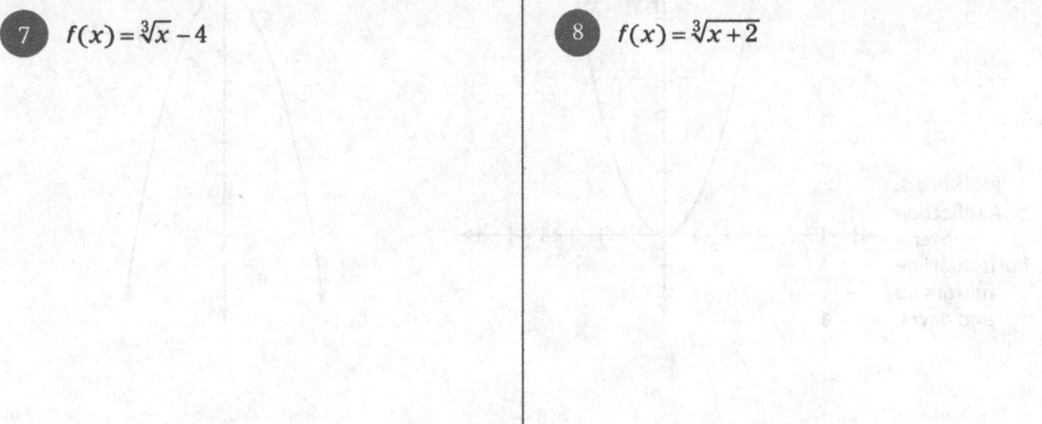

Seeing reflections

Reflections take the parent function and provide a mirror image of it over either a horizontal or vertical line. You'll come across two types of reflections.

- **Horizontal reflection:** the function is flipped over a horizontal line. If the reflection is over the *x*-axis, then positive function values become negative and negative function values become positive.
- **Vertical reflection:** the function is flipped over a vertical line. If the reflection is over the *y*-axis, then it's the input or *x* values that get reversed.

Q. Determine what happens when $f(x)=x^2$ is reflected over the horizontal axis.

EXAMPLE **A.** **Instead of $f(x)=x^2$, you have $f(x)=-1x^2$.** Look at Figure 4-5, which shows the parent function $f(x)=x^2$ and the reflection $g(x)=-1x^2$. If you find the value of both functions at the same number in the domain, you'll get opposite values in the range. For example, if $x=4$, $f(4)=16$ and $g(4)=-16$.

Q. Determine what happens when $f(x)=\sqrt{x}$ is reflected over the vertical axis.

EXAMPLE **A.** **Instead of $f(x)=\sqrt{x}$, you have $f(x)=\sqrt{-x}$.** Vertical reflections work the same as horizontal reflections, except the reflection occurs across a vertical line and reflects from side to side rather than up and down. You now have a negative inside the function. For this reflection, evaluating opposite inputs in both functions yields the same output. For example, if $f(x)=\sqrt{x}$, you can write its reflection over a vertical line as $g(x)=\sqrt{-x}$. When $f(4)=2$, $g(-4)=2$ as well (check out the graph in Figure 4-6).

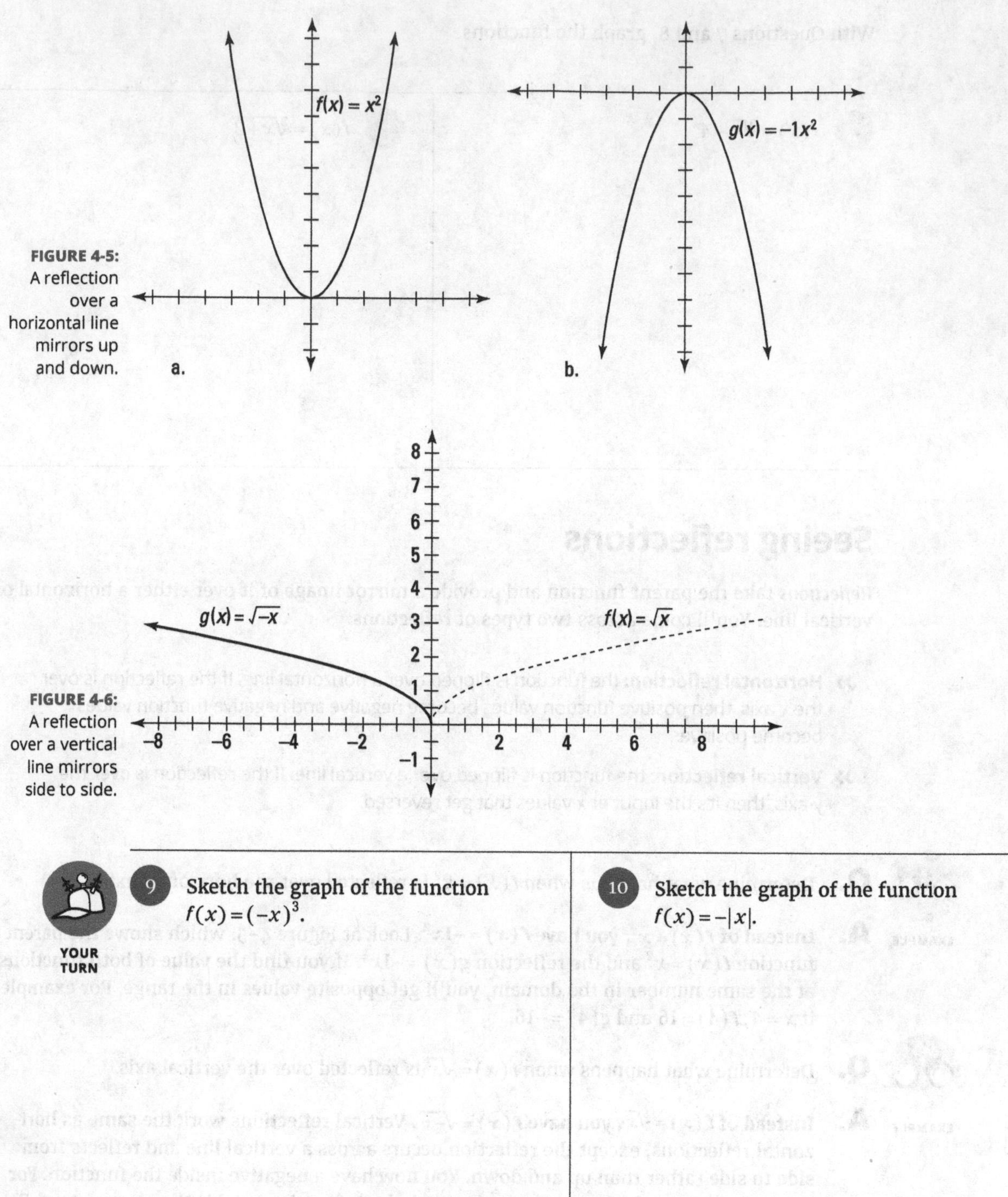

FIGURE 4-5: A reflection over a horizontal line mirrors up and down.

FIGURE 4-6: A reflection over a vertical line mirrors side to side.

YOUR TURN

9 **Sketch the graph of the function $f(x) = (-x)^3$.**

10 **Sketch the graph of the function $f(x) = -|x|$.**

Combining various transformations (a transformation in itself!)

Certain mathematical expressions allow you to combine stretching, shrinking, translating, and reflecting a function all into one graph. An expression that shows all the transformations in one is $a \cdot f[c(x-h)]+k$, where

a is the vertical transformation;

c is the horizontal transformation;

h is the horizontal shift; and

k is the vertical shift.

EXAMPLE

Q. Graph the function $f(x)=-2(x-1)^2+4$ by performing the transformations on a graph of the parent function $f(x)=x^2$.

A. The transformations are as follows:

- The 1 in the parentheses moves the graph to the right 1 unit.
- The 4 added at the end moves the graph upward 4 units.
- The 2 in the −2 stretches (steepens) the curve, and the negative sign reflects it horizontally over the x-axis.

Figure 4-7 shows each stage.

- Figure 4-7a is the parent graph: $k(x)=x^2$.
- Figure 4-7b is the horizontal shift to the right by 1: $h(x)=(x-1)^2$.
- Figure 4-7c is the vertical shift up by 4: $g(x)=(x-1)^2+4$.
- Figure 4-7d is the vertical stretch of 2 and the horizontal reflection: $f(x)=-2(x-1)^2+4$.

EXAMPLE

Q. Graph the function $q(x)=\sqrt{4-x}$ by performing the transformations on a graph of the parent function $f(x)=\sqrt{x}$.

A. First, rewrite the function equation to fit the function form $f(x)=a \cdot f[c(x-h)]+v$. Reverse the terms under the radical and factor out the negative sign.

$$q(x)=\sqrt{4-x}=\sqrt{-x+4}=\sqrt{-(x-4)}$$

The transformations are as follows:

- The 4 subtracted from *x* moves the graph to the right 4 units.
- The negative multiplier inside the radical reflects the function vertically over the *y*-axis.

Figure 4-8 shows the graph of q(x).

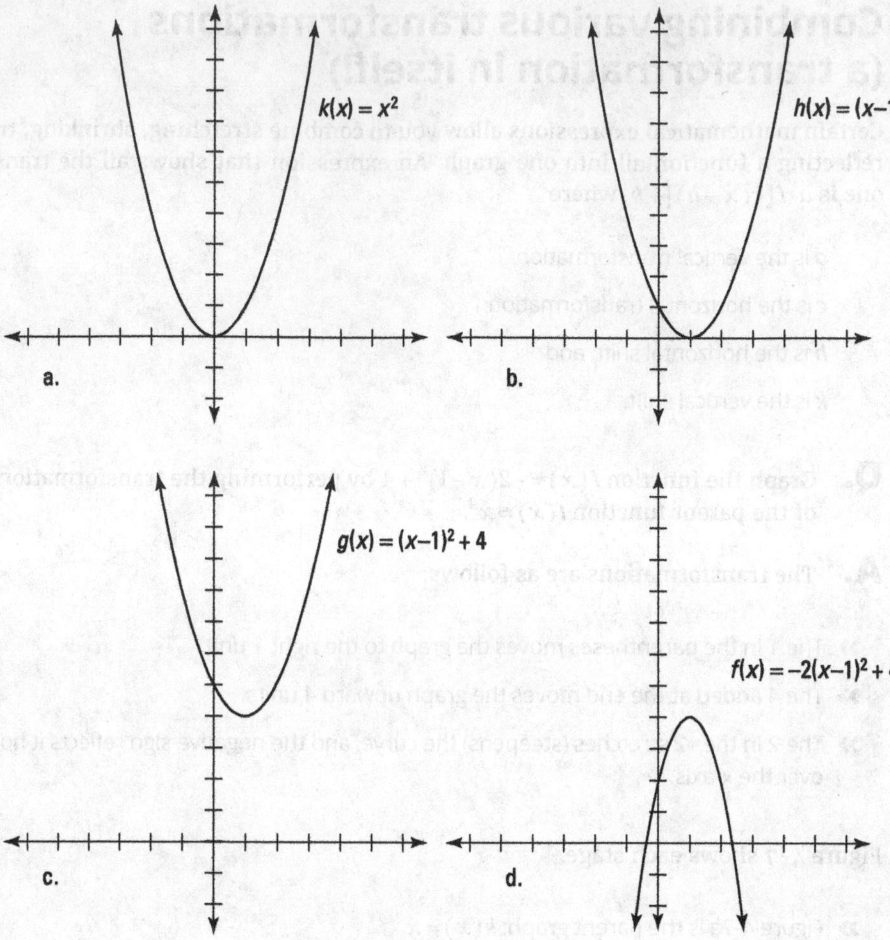

$k(x) = x^2$

a.

$h(x) = (x-1)^2$

b.

$g(x) = (x-1)^2 + 4$

c.

$f(x) = -2(x-1)^2 + 4$

d.

FIGURE 4-7: A view of multiple transformations.

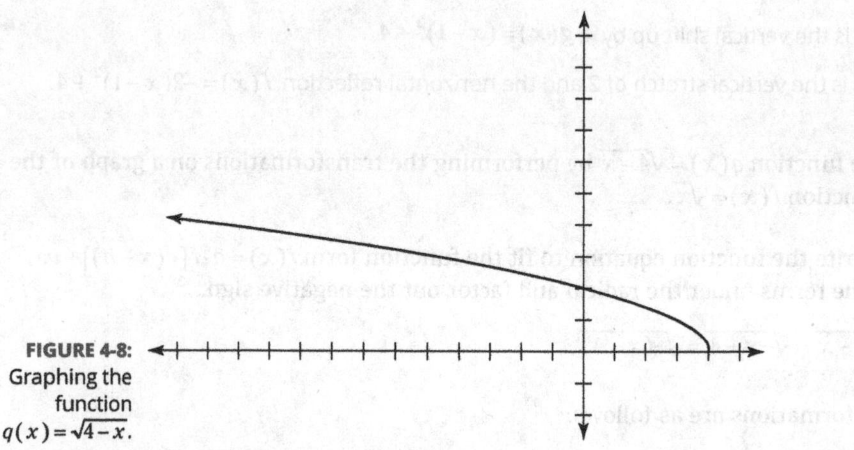

FIGURE 4-8: Graphing the function $q(x) = \sqrt{4-x}$.

YOUR TURN

 Sketch the graph of $f(x)=-\frac{1}{2}|x+1|+3$.

 Sketch the graph of $f(x)=2\sqrt[3]{5-x}-3$.

Transforming functions point by point

For some problems, you may be required to transform a function given only a set of random points on the coordinate plane. This is a way of making up some new kind of function that has never existed before. Just remember that *all* functions follow the same transformation rules, not just the common functions that have been explained so far in this chapter.

EXAMPLE

Q. Transform the graph of the parent function $y=f(x)$ using the transformation given in the function equation $y=\frac{1}{2}(f(x-4)-1)$. In Figure 4-9a, the dotted segments represent the parent function graph.

A. The transformations are as follows:

- The 4 moves each function value right 4 units.
- The -1 then drops the function value down by 1 unit.
- The $\frac{1}{2}$ shrinks the resulting function value by a factor of $\frac{1}{2}$.

For example, the first random point on the parent function is $(-5,3)$; shifting it to the right by 4 puts you at $(-1,3)$, and shifting it down 1 puts you at $(-1,2)$. Because the translated height is 2, you shrink the function by finding $\frac{1}{2}$ of the y-coordinate, 2. You end up at the final point, which is $(-1,1)$.

The graphs of $y=f(x)$ and $y=\frac{1}{2}(f(x-4)-1)$ are shown in Figure 4-9.

REMEMBER

You repeat that process for as many points as you see on the original graph to get the transformed one.

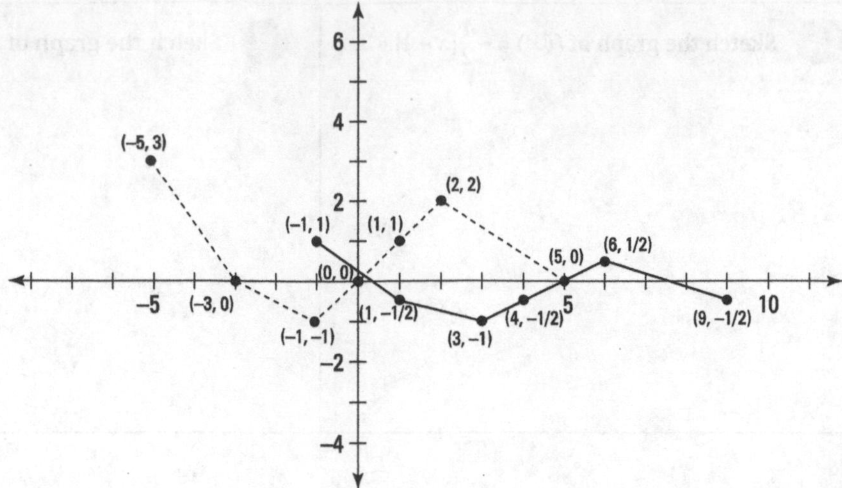

FIGURE 4-9: The graphs of (a) $y = f(x)$ and (b) $y = \frac{1}{2}(f(x-4)-1)$.

Sharpen Your Scalpel: Operating on Functions

Yes, graphing functions is fun, but what if you want more? Well, here's some good news: you also can perform mathematical operations with functions. That's right, you'll see how to add, subtract, multiply, or divide two or more functions. You'll also see an operation that's special to functions: composition.

REMEMBER

Operating on (sometimes called *combining*) functions is pretty easy, but the graphs of new, combined functions can be hard to create, because those combined functions don't always have the same parent functions and, therefore, no transformations of parent functions that allow you to graph easily. So you usually don't see multi-parent functions in pre-calculus . . . well, except for maybe a few. If you're asked to graph a combined function, you must resort to the old plug-and-chug method or perhaps use your graphing calculator.

This section walks you through various operations you may be asked to perform on functions, using the same three functions throughout the examples.

Adding and subtracting

When asked to add functions, you simply combine like terms, if the functions have any. For example, $(f+g)(x)$ is asking you to add the like terms in the $f(x)$ and the $g(x)$ functions. So, $(f+g)(x) = f(x) + g(x)$.

EXAMPLE

Q. Find $(f+g)(x)$ if $f(x) = x^2 - 6x + 1$ and $g(x) = 3x^2 - 10$.

A. $4x^2 - 6x - 9$. Adding the functions, you get $(f+g)(x) = (x^2 - 6x + 1) + (3x^2 - 10) = 4x^2 - 6x - 9$. The x^2 and $3x^2$ add up to $4x^2$; $-6x$ remains the same because it has no like terms; 1 and −10 add up to −9.

Q. Find $(g-h)(x)$ if $g(x)=3x^2-10$ and $h(x)=\sqrt{x^2-6x+1}$.

EXAMPLE **A.** $\mathbf{2x^2+6x-11.}$

When asked to subtract functions, you distribute the negative sign throughout the second function, using the distributive property (see Chapter 1), and then treat the process like an addition problem:

$$\begin{aligned}(g-f)(x)&=\left(3x^2-10\right)-\left(x^2-6x+1\right)\\&=\left(3x^2-10\right)+\left(-x^2+6x-1\right)\\&=2x^2+6x-11\end{aligned}$$

Q. Find $(f+h)(1)$ if $f(x)=x^2-6x+1$ and $h(x)=\sqrt{2x-1}$.

EXAMPLE **A.** **−3.** You have $(f+h)(x)=\left(x^2-6x+1\right)+\sqrt{2x-1}$. When you plug in 1, you get $(1)^2-6(1)+1)+\sqrt{2(1)-1}=1-6+1+\sqrt{2-1}=-4+\sqrt{1}=-4+1=-3.$

Multiplying and dividing

Multiplying and dividing functions is a similar concept to adding and subtracting them (see the previous section). When multiplying functions, you use the distributive property over and over and then add the like terms to simplify. Dividing functions is trickier, however. You take a look at multiplication first and save the trickier division for last. Here's the setup for multiplying $f(x)$ and $g(x)$: $(fg)(x)=[f(x)]\cdot[g(x)]$.

Q. Find $(f\cdot g)(x)$ if $f(x)=x^2-6x+1$ and $g(x)=3x^2-10$.

EXAMPLE **A.** $\mathbf{3x^4-18x^3-7x^2+60x-10}$. **Multiplying the functions, you get. Distribute each term of the polynomial on the left over each term of the polynomial on the right.**

$$x^2\left(3x^2\right)+x^2(-10)+(-6x)\left(3x^2\right)+(-6x)(-10)+1\left(3x^2\right)+1(-10)$$

$$=3x^4-10x^2-18x^3+60x+3x^2-10$$

Combine the like terms.

$$=3x^4-18x^3-7x^2+60x-10.$$

Operations that call for division of functions may involve factoring to cancel out terms and simplify the fraction. (If you're unfamiliar with this concept, check out Chapter 5.) If you're asked to divide $g(x)$ by $f(x)$, though, you write the following equation: $\left(\frac{g}{f}\right)(x)=\frac{g(x)}{f(x)}$

Q. Find $\left(\frac{g}{f}\right)(x)$ if $f(x)=x^2-6x+1$ and $g(x)=3x^2-10$.

EXAMPLE **A.** $\left(\frac{g}{f}\right)(x)=\frac{3x^2-10}{x^2-6x+1}$. **Because neither the denominator nor the numerator factor, the new, combined function is already simplified and you're done.**

YOUR TURN

Given the functions $f(x)=2x-3$, $g(x)=x^2+6x-5$, and $h(x)=\sqrt{6-x}$, perform the operations.

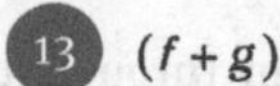

13 $(f+g)(x)$

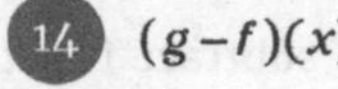

14 $(g-f)(x)$

15 $(f \cdot h)(x)$

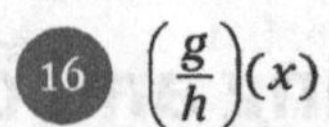

16 $\left(\frac{g}{h}\right)(x)$

Breaking down a composition of functions

A *composition* of functions is one function acting upon another. Think of it as putting one function inside of the other. The composition $f(g(x))$, also written $(f \circ g)(x)$, means that you plug the entire $g(x)$ function into $f(x)$. To solve such a problem, you work from the inside out.

EXAMPLE

Q. Given $f(x)=x^2-6x+1$ and $g(x)=3x^2-10$, find $(f \circ g)(x)$.

A. $9x^4-78x^2+161$. First, replace each x in the function $f(x)$ with the function rule for g:

$$f(g(x))=f(3x^2-10)=(3x^2-10)^2-6(3x^2-10)+1$$

Simplifying:

$$(3x^2-10)^2-6(3x^2-10)+1=9x^4-60x^2+100-18x^2+60+1=9x^4-78x^2+161$$

Q. Given $g(x) = 3x^2 - 10$ and $h(x) = \sqrt{2x-1}$, find $(g \circ h)(x)$.

EXAMPLE **A.** **$6x - 13$.** Begin by simplifying $g(h(x)) = (g \circ h)(x) = 3\left(\sqrt{2x-1}\right)^2 - 10$. You get the following: $(g \circ h)(x) = 3\left(\sqrt{2x-1}\right)^2 - 10 = 3(2x-1) - 10 = 6x - 3 - 10 = 6x - 13$.

You may also be asked to find one value of a composed function. You can always find the composition and then evaluate. But, if you don't need the actual function rule, you can just insert the value(s) into the functions. You work from left to right.

Q. Given $f(x) = x^2 - 6x + 1$ and $g(x) = 3x^2 - 10$, find $(g \circ f)(-3)$.

EXAMPLE **A.** **2,342.** It helps to rewrite this as $g(f(-3))$. You first find $f(-3)$ and then insert the result into the function g.

$$f(-3) = (-3)^2 - 6(-3) + 1 = 28$$
$$g(28) = 3(28)^2 - 10 = 2,342$$

YOUR TURN

Given the functions $f(x) = \sqrt{x+7}$ and $g(x) = 9 + x^2$, find the following:

17 $f(g(x))$

18 $(g \circ f)(x)$

19 $f(g(3))$

Adjusting the domain and range of combined functions (if applicable)

If you've looked over the previous sections that cover adding, subtracting, multiplying, and dividing functions, or putting one function inside of another, you may be wondering whether all these operations are messing with domain and range. Well, the answer depends on the operation performed and the original function or functions. But yes, the possibility *does* exist that the domain and range will change when you combine functions.

Following are the two main types of functions whose domains often are *not* all real numbers.

- **Rational functions:** The denominator of a fraction can never be zero, so at times rational functions are undefined and have numbers missing from their domain.
- **Square-root functions (and any root with an even index):** The *radicand* (the stuff underneath the root symbol) can't be negative. To find out how the domain is affected, set the radicand greater than or equal to zero and solve. This solution will tell you the effect.

When you begin combining functions (like adding a polynomial and a square root, for example), the domain of the new combined function is also affected. The same can be said for the range of a combined function; the new function will be based on the restriction(s) of the original functions.

REMEMBER

The domain is affected when you combine functions with division because variables end up in the denominator of the fraction. When this happens, you need to specify the values in the domain for which the quotient of the new function is undefined. The undefined values are also called the *excluded values* for the domain.

EXAMPLE

Q. Determine the range of the function $\left(\frac{g}{f}\right)(x)$ if $g(x)=3x^2-10$ and $f(x)=x^2-6x-7$.

A. $(-\infty,-1)\cup(-1,7)\cup(7,\infty)$. The new function is $h(x)=\frac{3x^2-10}{x^2-6x-7}$. The fraction has excluded values because *f*(*x*) is a quadratic equation with real roots. Factoring the denominator, $x^2-6x-7=(x-7)(x+1)$, you find that the denominator is 0 when $x=7$ or $x=-1$, so these values are excluded from the domain.

Unfortunately, there isn't one foolproof method for finding the domain and range of a combined function. The domain and range you find for a combined function depend on the domain and range of each of the original functions. The best way is to look at the functions visually, creating a graph by using the plug-and-chug method. This way, you can see the minimum and maximum of *x*, which may help determine your function domain, and the minimum and maximum of *y*, which may help determine your function range.

EXAMPLE

Q. Given two functions, $f(x)=\sqrt{x}$ and $g(x)=25-x^2$, find the domain of the new combined function $f(g(x))=\sqrt{(25-x^2)}$.

A. $-5\le x\le 5$ **or** $[-5,5]$. To do so, you need to find the domain of each individual function first. Here's how you find the domain of the composed function *f*(*g*(*x*)):

1. **Find the domain of *f*(*x*).**

   Because you can't find the square root of a negative number, the domain of *f* has to be all non-negative numbers. Mathematically, you write this as $x \geq 0$, or in interval notation, $[0, \infty)$.

2. **Find the domain of *g*(*x*).**

   Because the function *g*(*x*) is a polynomial, its domain is all real numbers, or $(-\infty, \infty)$.

3. **Find the domain of the combined function.**

   When specifically asked to look at the composed function *f*(*g*(*x*)), note that *g* is inside *f*. You're still dealing with a square root function, meaning that all the rules for square root functions still apply. So the new radicand of the composed function can't be smaller than –5 or larger than 5, which makes up the domain of the composed function: $-5 \leq x \leq 5$ or $[-5, 5]$.

To find the range of the same composed function, you must also consider the range of both original functions first:

1. **Find the range of *f*(*x*).**

   A square-root function always gives non-negative answers, so its range is $y \geq 0$.

2. **Find the range of *g*(*x*).**

   The function *g*(*x*) is a polynomial of even degree (specifically, a quadratic), and even-degree polynomials always have a minimum or a maximum value. The higher the degree on the polynomial, the harder it is to find the minimum or the maximum. Because this function is "just" a quadratic, you can find its min or max by locating the vertex.

   First, rewrite the function as $g(x) = -x^2 + 25$. This form tells you that the function is a transformed quadratic that has been shifted up 25 and turned upside down (see the earlier section, "Transforming the Parent Graphs"). Therefore, the function never gets higher than 25 in the *y* direction. The range is $y \leq 25$.

3. **Find the range of the composed function *f*(*g*(*x*)).**

   The function *g*(*x*) reaches its maximum (25) when $x = 0$. Therefore, the composed function also reaches its maximum of 5 at $x = 0$: $f(g(0)) = \sqrt{25 - 0^2} = 5$. The range of the composed function has to be less than or equal to that value, or $y \leq 5$.

TIP

The graph of this combined function also depends on the range of each individual function. Because the range of *g*(*x*) must be non-negative, so must be the combined function, which is written as $y \geq 0$. Therefore, the range of the combined function is $0 \leq y \leq 5$. If you graph this combined function on your graphing calculator, you get a half circle of radius 5 that's centered at the origin.

YOUR TURN

Find the domain and range of the composed function $f(g(x))$ if $f(x)=\frac{1}{x+2}$ and $g(x)=x^2-1$.

Turning Inside Out with Inverse Functions

Many operations in math have inverses: addition undoes subtraction and multiplication undoes division (and vice versa for both). Because functions are just more complicated forms of operations, functions also may have inverses. An *inverse function* simply undoes another function. Not all functions have inverses. Only one-to-one functions (see Chapter 3) can have inverses, so this is pretty special.

Perhaps the best reason to know whether functions are inverses of each other is that if you can graph the original function, you can *usually* graph the inverse as well. So that's where this section begins. At times in pre-calculus you'll be asked to show that two functions are inverses of one another or to find the inverse of a given function, so you find that info later in this section as well.

REMEMBER

If $f(x)$ is the original function, $f^{-1}(x)$ is the symbol for its inverse. This notation is used strictly to describe the inverse function and not $\frac{1}{f(x)}$.

The -1 exponent in $f^{-1}(x)$ is used in this way to represent the inverse, not the reciprocal. If you want to indicate the reciprocal of $f(x)$, use $[f(x)]^{-1}$.

Graphing an inverse

TIP

If you're asked to graph the inverse of a function, you can do it the long way and find the inverse first (see the next section), or you can remember one fact and get the graph. What's the one fact, you ask? Well, the fact is that a function and its inverse are reflections of one another over the line $y=x$. This line is a linear function that passes through the origin and has a slope of 1. When you're asked to draw a function and its inverse, you may choose to draw this line in as a dotted line; this way, it acts like a big mirror, and you can literally see the points of the function reflecting over the line to become the inverse function points. Reflecting over that line switches the x and the y and gives you a graphical way to find the inverse without plotting tons of points.

EXAMPLE

Q. Show that the functions $f(x)=2x-3$ and $g(x)=\frac{x+3}{2}$ are inverses of one another.

A. To see how x and y switch places, follow these steps:

1. **Take a number (any that you want) and plug it into the first given function.**

   A good choice is −4. With $f(-4)$, you get −11. As a point, this is written $(-4,-11)$.

2. **Take the result from Step 1 and plug it into the other function.**

   In this case, you need to find $g(-11)$. When you do, you get −4 back again. As a point, this is $(-11,-4)$. Whoa!

TIP

The entire domain and range swap places from a function to its inverse. For instance, knowing that just a few points from the given function $f(x)=2x-3$ include $(-4,-11)$, $(-2,-7)$, and $(0,-3)$, you automatically know that the points on the inverse $g(x)$ will be $(-11,-4)$, $(-7,-2)$, and $(-3,0)$.

So if you're asked to graph a function and its inverse, all you have to do is graph the function and then switch all x and y values in each point to graph the inverse. Just look at all those values switching places from the $f(x)$ function to its inverse $g(x)$ (and back again), reflected over the line $y=x$.

Figure 4-10 shows you the function $f(x)=2x-3$ and its inverse.

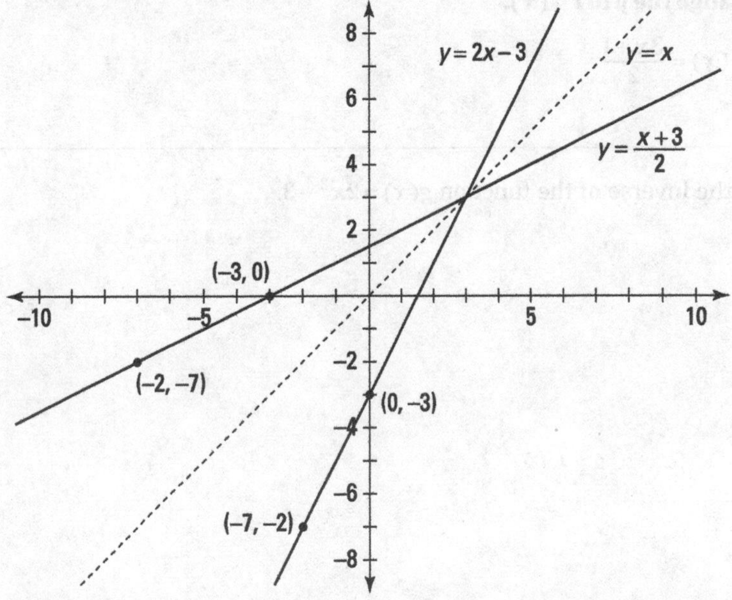

FIGURE 4-10: The graph of $f(x)=2x-3$ and its inverse.

Inverting a function to find its inverse

If you're given a function and must find its inverse, first remind yourself that domain and range swap places in the functions. Literally, you exchange $f(x)$ and x in the original equation. When you make that change, you call the new $f(x)$ by its true name, $f^{-1}(x)$, and solve for this function.

Q. Find the inverse of this function: $f(x)=\dfrac{2x-1}{3}$.

EXAMPLE **A.** $f^{-1}(x)=\dfrac{3x+1}{2}$. Follow these steps:

1. **Change the $f(x)$ to y.**

   $y=\dfrac{2x-1}{3}$

2. **Change the y to x and the x to y.**

   $x=\dfrac{2y-1}{3}$

   If there was more than one x in the equation, you would change them all to y.

3. **Solve for y.**

   This step has three parts:

   a. Multiply both sides by 3 to get $3x=2y-1$.

   b. Add 1 to both sides to get $3x+1=2y$.

   c. Divide both sides by 2 to get your inverse: $y=\dfrac{3x+1}{2}$.

4. **Change the y to $f^{-1}(x)$.**

   $f^{-1}(x)=\dfrac{3x+1}{2}$

YOUR TURN

21 Find the inverse of the function $g(x)=2x^3-3$.

Verifying an inverse

At times, you may be asked to verify that two given functions are actually inverses of each other. To do this, you need to show that both $f(g(x)) = x$ and $g(f(x)) = x$.

TIP

When you're asked to find an inverse of a function (like in the previous section), it's recommended that you verify on your own that what you did was correct, time permitting.

EXAMPLE

Q. Show that the functions $f(x) = 5x^3 + 4$ and $g(x) = \sqrt[3]{\frac{x-4}{5}}$ are inverses of each other.

A. Follow these steps:

1. **Show that** $f(g(x)) = x$.

   This step is a matter of plugging the function $g(x)$ in for every x in the equation:

   $$f(g(x)) = 5[g(x)]^3 + 4 = 5\left[\sqrt[3]{\frac{x-4}{5}}\right]^3 + 4$$

   And since cubing a cube root makes the exponent 1, $\cancel{5}\left[\frac{x-4}{\cancel{5}}\right] + 4 = x - 4 + 4 = x$.

2. **Show that** $g(f(x)) = x$.

   Again, plug in the numbers and start simplifying: $g(f(x)) = \sqrt[3]{\frac{f(x)-4}{5}} =$
   $\sqrt[3]{\frac{5x^3+4-4}{5}} = \sqrt[3]{\frac{\cancel{5}x^3}{\cancel{5}}} = \sqrt[3]{x^3} = x$

The functions are inverses of one another, because both compositions resulted in x.

YOUR TURN

22 Show that the functions $f(x) = \frac{7x-5}{2}$ and $g(x) = \frac{2x+5}{7}$ are inverses of each other.

Practice Questions Answers and Explanations

1. **Shrinks/flattens by $\frac{1}{10}$.** Since the multiplier is smaller than 1, the function values shrink.
2. **Stretches/steepens by $\frac{64}{9}$.** Since the multiplier is greater than 1, the values grow larger.
3. **Refer to the following figure.** The multiplier of 3 steepens the curve.

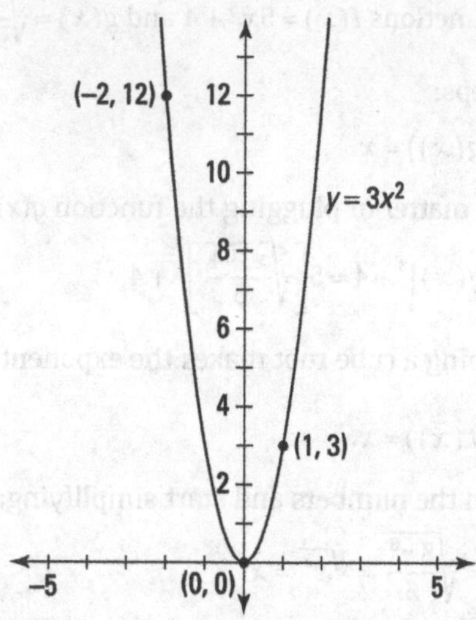

4. **Refer to the following figure.** The multiplier of 0.04 flattens the curve.

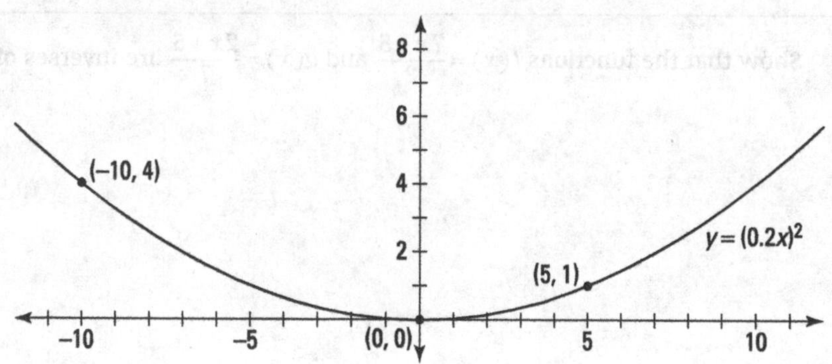

5. **Raises the graph by 2 units.** Adding 2 onto the parent function equation raises the graph.
6. **Moves the graph 5 units to the right.** Subtracting 5 from the variable before performing the function operation on the parent function moves the graph 5 units to the right.

7 **Refer to the following figure.** The parent function drops 4 units.

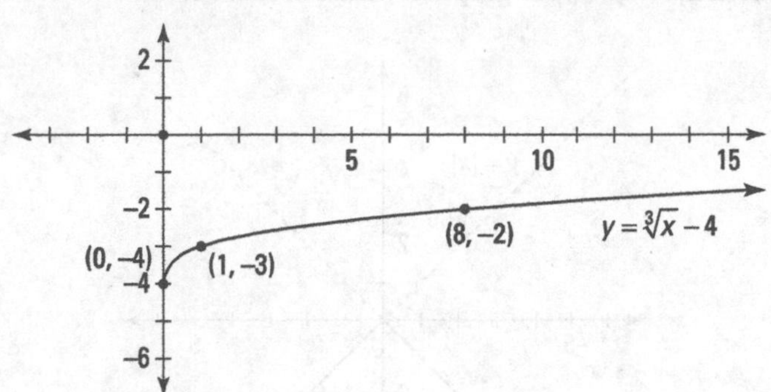

8 **Refer to the following figure.** The parent function moves 2 units to the left.

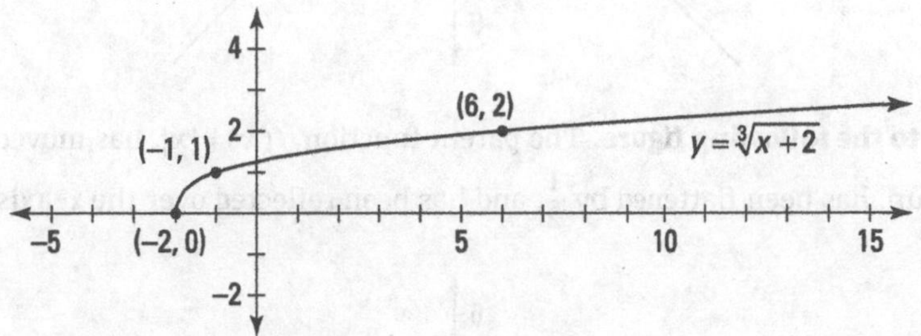

9 **Refer to the following figure.** The parent function reflects over the vertical y-axis. The parent function, $f(x)=x^3$, is graphed with a dotted curve. Because the exponent is odd, this reflection has the same effect as reflecting over the horizontal axis.

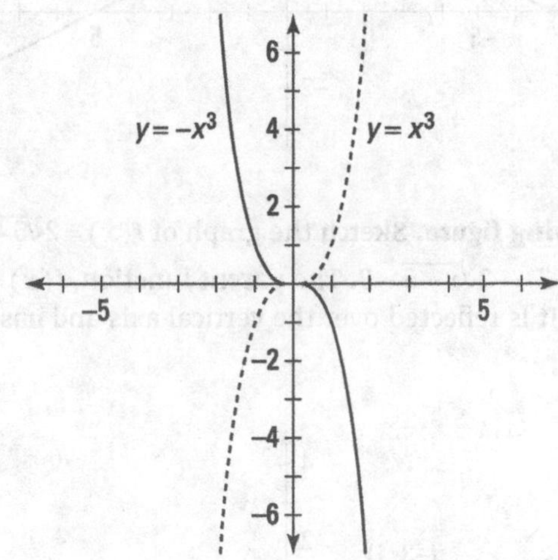

10 **Refer to the following figure.** The parent function reflects over the horizontal x-axis. The parent function, $f(x) = |x|$, is graphed with a dotted curve.

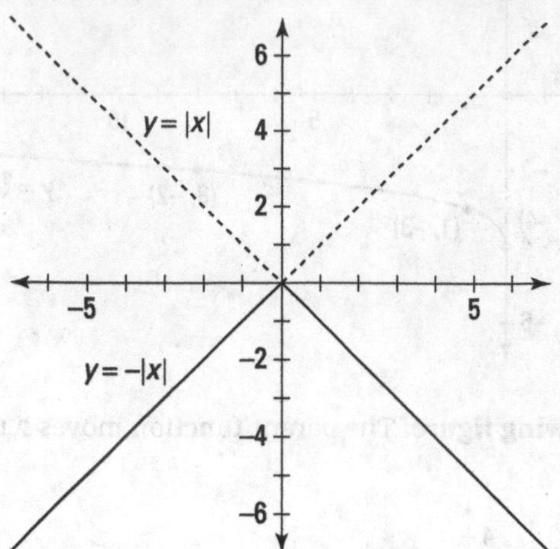

11 **Refer to the following figure.** The parent function, $f(x) = |x|$, has moved 1 unit left and 3 units up, has been flattened by $\frac{1}{2}$, and has been reflected over the x-axis.

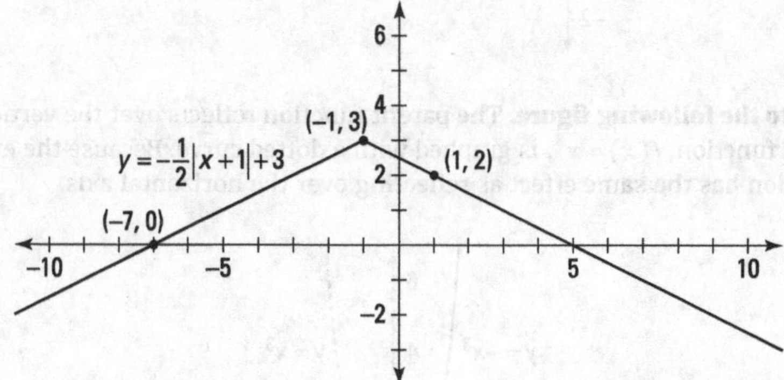

12 **Refer to the following figure.** Sketch the graph of $f(x) = 2\sqrt[3]{5-x} - 3$. Rewrite the function as $f(x) = 2\sqrt[3]{-(x-5)} - 3 = -2\sqrt[3]{x-5} - 3$. The parent function, $f(x) = \sqrt[3]{x}$, has moved 5 units right and 3 units down. It is reflected over the vertical axis and has stretched because of the multiplier of 2.

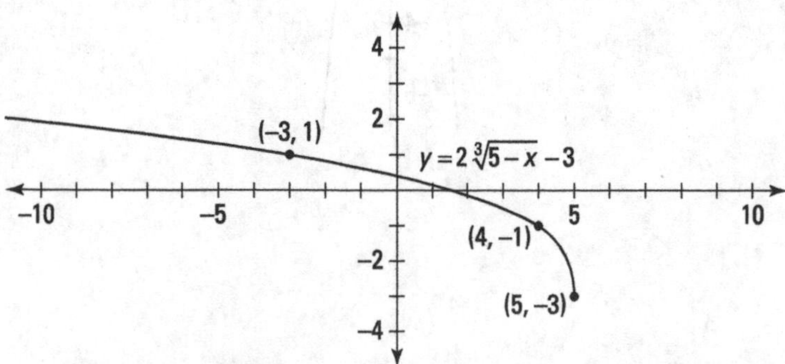

13 x^2+8x-8**.** You get $f(x)+g(x)=2x-3+x^2+6x-5=x^2+8x-8$.

14 x^2+4x-2**.** You get $g(x)-f(x)=x^2+6x-5-(2x-3)=x^2+6x-5-2x+3=x^2+4x-2$.

15 $2x\sqrt{6-x}-3\sqrt{6-x}$**.** You get $f(x)\cdot h(x)=(2x-3)\cdot\sqrt{6-x}=2x\sqrt{6-x}-3\sqrt{6-x}$.

16 $\dfrac{(x^2+6x-5)\sqrt{6-x}}{6-x}$**.** You get $\dfrac{g(x)}{h(x)}=\dfrac{x^2+6x-5}{\sqrt{6-x}}\cdot\dfrac{\sqrt{6-x}}{\sqrt{6-x}}=\dfrac{(x^2+6x-5)\sqrt{6-x}}{6-x}$.

17 $\sqrt{x^2+16}$**.** You get $f(g(x))=\sqrt{g(x)+7}=\sqrt{9+x^2+7}=\sqrt{x^2+16}$.

18 $x+16$**.** You get $(g\circ f)(x)=9+(f(x))^2=9+\left(\sqrt{x+7}\right)^2=9+x+7=x+16$.

19 **5.** First, finding $g(3)$, $g(3)=9+3^2=18$. Then, finding $f(18)$, $f(18)=\sqrt{18+7}=\sqrt{25}=5$. So $f(g(3))=f(18)=5$.

20 **Domain:** $(-\infty,-2)\cup(-2,\infty)$**; Range:** $(0,\infty)$. The domain of $f(x)$ is all real numbers except -2, which can be written $(-\infty,-2)\cup(-2,\infty)$. The domain of $g(x)$ is all real numbers, $(-\infty,\infty)$. The domain of the composition function, $f(g(x))=\dfrac{1}{g(x)+2}=\dfrac{1}{x^2-1+2}=\dfrac{1}{x^2+1}$, is all real numbers. So the domain of the composition is $(-\infty,-2)\cup(-2,\infty)$. The range of $f(x)=\dfrac{1}{x+2}$ is all real numbers except 0, or $(-\infty,0)\cup(0,\infty)$, and the range of $g(x)=x^2-1$ is $[-1,\infty)$. The range of the composition is all real numbers greater than 0, or $(0,\infty)$. So the range of the composition is $(0,\infty)$.

21 $g^{-1}(x)=\sqrt[3]{\dfrac{x+3}{2}}$**.** Solving the equation $x=2y^3-3$ for y, you have $x=2y^3-3\rightarrow x+3=2y^3\rightarrow\dfrac{x+3}{2}=y^3$. Then find the cube root of each side: $y=\sqrt[3]{\dfrac{x+3}{2}}$. Replace the y with $g^{-1}(x)$.

22 **Both compositions are equal to x.**

$$f(g(x))=\frac{7\left(\frac{2x+5}{7}\right)-5}{2}=\frac{2x+5-5}{2}=\frac{2x}{2}=x$$

$$g(f(x))=\frac{2\left(\frac{7x-5}{2}\right)+5}{7}=\frac{7x-5+5}{7}=\frac{7x}{7}=x$$

If you're ready to test your skills a bit more, take the following chapter quiz that incorporates all the chapter topics.

Whaddya Know? Chapter 4 Quiz

Quiz time! Complete each problem to test your knowledge on the various topics covered in this chapter. You can then find the solutions and explanations in the next section.

1 **Find the inverse of the function $f(x) = 4\sqrt[3]{x} - 11$.**

2 **Change the function equation $f(x) = x^2$ to one that drops by 2 units.**

3 **Change the function equation $f(x) = x^3$ to one that has moved left by 3 units.**

4 **Graph the function $f(x) = -4|x+1| + 3$.**

5 **Given $f(x) = \sqrt[3]{2x-2}$ and $g(x) = x^3 + 1$, find $(f \circ g)(x)$.**

6 **Describe the changes that occurred in $f(x) = |x|$ when the function equation became $f(x) = \frac{1}{4}|x+3| - 1$.**

7 **Given $f(x) = 3 - x$ and $g(x) = x^2 - 9$, find $\left(\frac{g}{f}\right)(x)$.**

8 **Given $f(x) = x^2 + 1$ and $g(x) = \sqrt{x+4}$, find $g(f(-2))$.**

9 **Describe the changes that occurred in $f(x) = x^2$ when the function equation became $f(x) = (x-5)^2 + 7$.**

10 **Graph the function $f(x) = 2\sqrt{x-3} + 1$.**

11 **Change the function equation $f(x) = \sqrt{x}$ to one that has reflected over the x-axis.**

12 **Given $f(x) = x^2 - 2$ and $g(x) = 3 - x^2$, find $(f + g)(x)$.**

13 **Given $f(x) = x^2 + 1$ and $g(x) = \sqrt{x}$, determine the domain and range of $(f \circ g)(x)$.**

Answers to Chapter 4 Quiz

1. $f^{-1}(x)=\frac{(x+11)^3}{64}$. Replace the $f(x)$ with y, and then reverse all the x's and y's in the equation: $f(x)=4\sqrt[3]{x}-11 \to y=4\sqrt[3]{x}-11 \to x=4\sqrt[3]{y}-11$. Now solve for y:. $x=4\sqrt[3]{y}-11 \to x+11=4\sqrt[3]{y} \to \frac{x+11}{4}=\sqrt[3]{y} \to \left(\frac{x+11}{4}\right)^3=\left(\sqrt[3]{y}\right)^3 \to \frac{(x+11)^3}{64}=y$. Replace the y with $f^{-1}(x)$.

2. $f(x)=x^2-2$. The number is subtracted from the function operation.

3. $f(x)=(x+3)^3$. Add 3 to the variable before the function operation is applied.

4. **See the following figure.** Graph the function $f(x)=-4|x+1|+3$. The parent function $f(x)=|x|$ is moved 1 unit to the left and 3 units up. The multiplier 4 steepens the graph, and the negative multiplier reflects the graph over the x-axis.

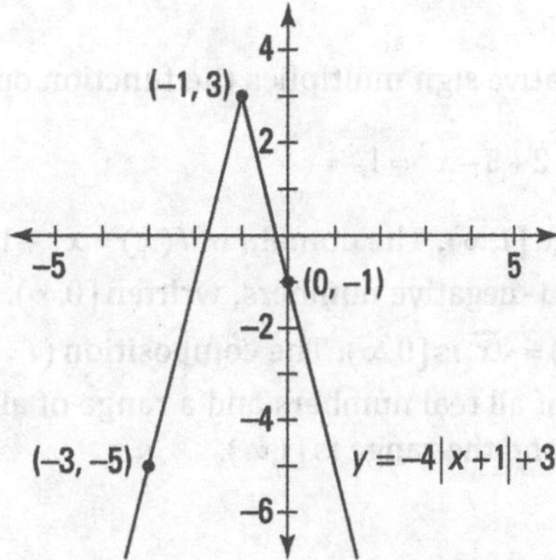

5. $x\sqrt[3]{2}$. You get $f(g(x))=\sqrt[3]{2g(x)-2}=\sqrt[3]{2(x^3+1)-2}=\sqrt[3]{2x^3+2-2}=\sqrt[3]{2x^3}=x\sqrt[3]{2}$.

6. **Moves 3 left and down 1; becomes flatter (shrinks).** The 3 is added before the function operation is applied, and the 1 is then subtracted. The multiplier $\frac{1}{4}$ is smaller than 1.

7. $-x-3$. Dividing: $\frac{g}{f}=\frac{x^2-9}{3-x}=\frac{\cancel{(x-3)}(x+3)}{-\cancel{(x-3)}}=\frac{x+3}{-1}=-x-3$

8. **3.** You get $f(-2)=(-2)^2+1=5$ and $g(5)=\sqrt{5+4}=\sqrt{9}=3$.

9. **Moves 5 right and up 7.** The 5 is subtracted before the function operation is applied, and the 7 is added after.

(10) **Refer to the following figure.** Graph the function $f(x)=2\sqrt{x-3}+1$. The parent function $f(x)=\sqrt{x}$ is moved 3 units to the right and 1 unit up. The multiplier 2 steepens (stretches) the curve.

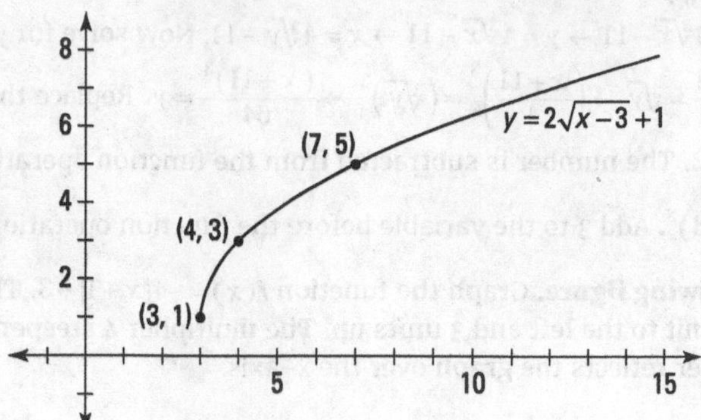

(11) $f(x)=-\sqrt{x}$**.** The negative sign multiplies the function operation.

(12) **1.** Adding: $f+g=x^2-2+3-x^2=1$

(13) **Domain:** $[0,\infty)$**; Range** $[1,\infty)$**.** The domain of $f(x)=x^2+1$ is all real numbers, and the domain of $g(x)=\sqrt{x}$ is all non-negative numbers, written $[0,\infty)$. The range of $f(x)=x^2+1$ is $[1,\infty)$, and the range of $g(x)=\sqrt{x}$ is $[0,\infty)$. The composition $(f\circ g)(x)=f\left(\sqrt{x}\right)=\left(\sqrt{x}\right)^2+1=x+1$, which has a domain of all real numbers and a range of all real numbers. So the domain of the composition is $[0,\infty)$ and the range is $[1,\infty)$.

2

Getting the Grip on Graphing

IN THIS UNIT . . .

Applying your graphing skills to polynomial functions.

Dealing with exponential and logarithmic functions.

Piecing together piece-wise functions.

Grappling with greatest integer functions.

IN THIS CHAPTER

» Exploring the factoring of quadratic equations

» Solving non-factorable quadratic equations

» Deciphering and counting a polynomial's roots

» Employing solutions to find factors

» Graphing polynomials on the coordinate plane

Chapter 5

Graphing Polynomial Functions

Ever since those bygone days of algebra, variables have been standing in for unknowns in equations. You're probably very comfortable with using variables by now, so you're ready to move on and find out how to deal with equations that use multiple terms and figure out how to graph them.

When variables and constants start multiplying, the result of a variable times a constant is a *monomial*, which means "one term." Examples of monomials include $3y$, x^2, and $4ab^3c^2$. When you start creating expressions by adding and subtracting distinct monomials, you get *polynomials*, because you create something with one or more terms. Usually, *monomial* refers to a polynomial with one term only, *binomial* refers to two terms, *trinomial* refers to three, and the word *polynomial* is reserved for four or more. Think of a polynomial as the umbrella under which are binomials and trinomials. Each part of a polynomial that's added or subtracted is a term; so, for example, the polynomial $2x + 3$ has two terms: $2x$ and 3.

REMEMBER

Part of the official definition of a polynomial is that it can never have a variable in the denominator of a fraction, it can't have negative exponents, and it can't have fractional exponents. The exponents on the variables have to be non-negative integers. A general format of a polynomial is $a_nx^n + a_{n-1}x^{n-1} + a_{n-2}x^{n-2} + \cdots + a_1x^1 + a_0$. The a_n factors represent the coefficients, and the n's are always non-negative integers.

In this chapter, you go searching for the *solution(s)* of the given polynomial equation — the value(s) that make it true. When the given equation is set equal to zero, these solutions are called *roots* or *zeros*. These words are used interchangeably because they represent the same idea: where the graph crosses the x-axis (a point called the x-intercept). You see how to find the roots of polynomial functions and how to represent the functions with graphs.

Understanding Degrees and Roots

The *degree* of a polynomial is closely related to its exponents, and it determines how you work with the polynomial to find the roots. To find the degree of a polynomial, you simply find the degree of each term by adding the exponents of variables (remembering that when no exponent is given, 1 is implied). The greatest of these sums is the degree of the whole polynomial. For example, consider the expression $3x^4y^6 - 2x^4y - 5xy + 2$:

The degree of $3x^4y^6$ is $4+6$, or 10.

The degree of $2x^4y$ is $4+1$, or 5.

The degree of $5xy$ is $1+1$, or 2.

The degree of 2 is 0, because it has no variables.

Therefore, this polynomial has a degree of 10.

A *quadratic expression* is a polynomial in which the highest degree is two. One example of a quadratic polynomial is $3x^2 - 10x + 5$. The x^2 term in the polynomial is called the *quadratic term* because it's the one that makes the whole expression quadratic. The number in front of x^2 is called the *leading coefficient* (in the example here, it's the 3). The x term is called the *linear term* ($-10x$), and the number by itself is called the *constant* (5).

Without taking calculus, getting a perfectly accurate graph of a polynomial function by plotting points is nearly impossible. However, in pre-calculus you can find the roots of a polynomial (if it has any), and you can use those roots as a guide to get an idea of what the graph of that polynomial looks like. You simply plug in an x value between the two roots (the x-intercepts) to see if the function is positive or negative between those roots. For example, you may be asked to graph the equation $y = 3x^2 - 10x + 5$. You now know that this equation is a second-degree polynomial, so it may have two roots and, therefore, may cross the x-axis up to two times (more on why later).

This chapter begins with solving quadratics because the techniques required to solve them are specific. Factoring, completing the square, and using the quadratic formula are excellent methods to solve quadratics; however, they often don't work for polynomials of higher degrees. Then you move on to higher-degree polynomials (like x^3 or x^5, for example) because the steps required to solve them are often longer and more complicated.

TIP

You can solve many types of polynomial equations (including quadratics) using the steps described near the end of this chapter. However, solving quadratics using the techniques specifically reserved for them saves you time and effort.

Factoring a Polynomial Expression

Recall that when two or more variables or constants are multiplied to get a product, each multiplier of the product is called a *factor*. You first ran into factors when multiplication was introduced (remember factor trees, prime factorization, and so on?). For example, one set of factors of 24 is 6 and 4 because $6 \cdot 4 = 24$.

In mathematics, *factorization* or *factoring* a polynomial is the rearrangement of the polynomial into a product of other smaller polynomials. If you choose, you can then multiply these factors together to get the original polynomial (which is a great way to check yourself on your factoring skills). For example, the polynomial $f(x) = x^5 - 4x^4 - 8x^3 + 32x^2 - 9x + 36$ consists of six terms that can be rewritten as four factors, $f(x) = (x+3)(x-3)(x-4)(x^2+1)$.

REMEMBER

One way of solving a polynomial equation is to factor the polynomial into the product of two or more binomials (two-term expressions). After the polynomial is fully factored, you can use the zero-product property to solve the equation. This idea is discussed later in the section, "Finding the Roots of a Factored Equation."

You have multiple options to choose from when factoring:

- For a polynomial, no matter how many terms it has, always check for a greatest common factor (GCF) first. The *greatest common factor* is the biggest expression that goes into all of the terms evenly. Using the GCF is like doing the distributive property backward (see Chapter 1).
- If the expression is a *trinomial* — it has three terms — you can use the FOIL method (for multiplying binomials) backward or the Box method.
- If the expression is a binomial, look for difference of squares, difference of cubes, or sum of cubes after factoring out the GCF, if there is one.

The following sections show each of these methods in detail.

REMEMBER

If a quadratic doesn't factor, it's called *prime* because its only factors are 1 and itself. When you have tried all the factoring tricks in your bag (GCF, backward FOIL, difference of squares, and so on), and the quadratic equation doesn't factor, then you can either complete the square or use the quadratic formula to solve the equation. The choice is yours. You could even potentially choose to *always* use either completing the square or quadratic formula (and skip the factoring) to solve an equation. Factoring can sometimes be quicker, which is why this is recommended to try first.

TIP

The standard form for a quadratic expression is $ax^2 + bx + c$. If you're given a quadratic expression that isn't in standard form, rewrite it in standard form by putting the terms in descending order of the degrees. This step makes factoring easier (and is sometimes even necessary to factor).

Always the first step: Looking for a GCF

No matter how many terms a polynomial has, you always want to check for a greatest common factor first. If the polynomial has a GCF, factoring the polynomial makes the rest much easier because each term now has fewer factors and smaller numbers. If the GCF includes a variable, your job becomes even easier, because the degrees of the results are smaller.

EXAMPLE

Q. Factor the polynomial $6x^4 - 12x^3 + 4x^2$.

A. $2x^2(3x^2 - 6x + 2)$. Follow these steps:

1. **Determine the smallest power of the variable appearing in the terms.**

   The smallest power is x^2 in the $4x^2$ term.

2. **Look for the GCF of the constant terms.**

   You can see one 2 in every term: $3 \cdot 2 \cdot x^4 - 3 \cdot 2 \cdot 2 \cdot x^3 + 2 \cdot 2 \cdot x^2$. The GCF of the constant terms is 2.

3. **Create the GCF from the two factors.**

   The GCF is $2x^2$.

4. **Divide each term by the GCF and write the results in parentheses.**

   $2x^2(3x^2 - 6x + 2)$

5. **Distribute to make sure the GCF is correct.**

   If you multiply the $2x^2$ times each term in the parentheses, you get $6x^4 - 12x^3 + 4x^2$. You can now say with confidence that $2x^2$ is the GCF.

YOUR TURN

Find the GCF of the polynomials and factor them.

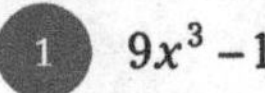

1 $9x^3 - 12x^5 + 24x$

2 $20x^4 + 60x^3 - 30x^2 - 10x$

Unwrapping the box containing a trinomial

After you check a polynomial for a GCF (regardless of whether it had one or not), try to factor again. You may find that it is easier to factor after the GCF has been factored out. Many people prefer to use the guess-and-check method of factoring, where you write down two sets of parentheses — ( )( ) — and literally plug in guesses for the factors to see if anything works. A second method that's used is called the *FOIL method* of factoring (sometimes called the *British method*). The FOIL method of factoring calls for you to follow the steps required to FOIL binomials, only backward. And a third method, which has become very popular, is the *Box method*, which is shown here.

EXAMPLE

Q. Factor $3x^2+19x-14$ using the *Box method.*

A. Follow these steps.

1. **Draw a two-by-two square, and put the first term of the binomial in the upper left and the last term in the lower right.**

$3x^2$	
	-14

2. **Multiply the two terms in the box together:** $3x^2(-14)=-42x^2$.
3. **Write down all the linear factors of the resulting product, in pairs.**

   The factors of $-42x^2$ are $-42x$ and $1x$, $42x$ and $-1x$, $-21x$ and $2x$, $21x$ and $-2x$, $-14x$ and $3x$, $14x$ and $-3x$,$-7x$ and $6x$, $7x$ and $-6x$. You've probably noticed that there are really just four sets of factors — with changes in the signs of the two terms.
4. **From this list, find the pair that adds up to produce the linear term.**

   You want the pair whose sum is $+19x$. For this problem, the answer is $21x$ and $-2x$.
5. **Put those two terms in the box; it doesn't matter which goes where.**

$3x^2$	$21x$
$-2x$	-14

6. **Find the GCF of each pair of terms, horizontally and vertically; write the GCFs on the top and side of the box.**

x	7	
$3x^2$	$21x$	$3x$
$-2x$	-14	-2

7. **Write the product of two binomials formed by the terms on the top and side of the square, respectively:** $(x+7)(3x-2)$.

   These are the two binomial factors of the trinomial.

   Depending on where you put the two factors in Step 5, the order of the two binomials may change. But that doesn't affect the final answer.

REMEMBER

If none of the pairs of factors listed in Step 3 give you the middle term, then you're done. The factors don't have to be pretty — even radicals can work. But, if it doesn't factor, then the trinomial is prime — it doesn't factor.

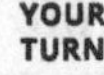

YOUR TURN

Use the *Box method* to factor the trinomials.

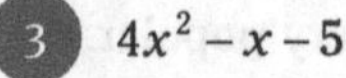

3 $4x^2 - x - 5$

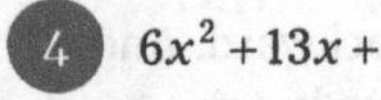

4 $6x^2 + 13x + 6$

Recognizing and factoring special polynomials

The whole point of factoring is to discover the original polynomial factors that give you a particular end product. You can spend a long time in algebra factoring polynomials, and factoring just undoes the original process of multiplying. It's a little like *Jeopardy!* — you know the answer, but you are looking for the question.

Special cases can occur when factoring trinomials and binomials; you should recognize them quickly so that you can save time.

- **Perfect-square trinomials:** When you multiply a binomial times itself, the product is called a *perfect square*. For example, $(a+b)^2$ gives you the perfect-square trinomial $a^2 + 2ab + b^2$. Notice that the middle term is twice the product of the roots of the first and third terms.
- **Difference of squares binomials:** When you multiply a binomial times its conjugate, the product is called a *difference of squares*. The product of $(a-b)(a+b)$ is $a^2 - b^2$. Factoring a difference of squares also requires its own set of steps, which is explained for you in this section.

Two other special types of factoring didn't come up when you were learning how to FOIL, because they aren't the product of two binomials.

- **Sum of cubes:** One factor is a binomial and the other is a trinomial. The binomial $a^3 + b^3$ can be factored to $(a+b)(a^2 - ab + b^2)$.
- **Difference of cubes:** These expressions factor almost like a sum of cubes, except that some signs are different in the factors: $a^3 - b^3 = (a-b)(a^2 + ab + b^2)$.

REMEMBER

No matter what type of problem you face, you should always check for the GCF first; however, none of the following examples has a GCF, so skip over that step in the directions. In another section, you find out how to factor more than once when you can.

Seeing double with perfect squares

Because a perfect-square trinomial is still a trinomial, you can always use the FOIL method or the Box method, but it's much quicker and nicer if you recognize that you have a perfect-square trinomial — it saves time.

EXAMPLE

Q. Factor the polynomial $4x^2 - 12x + 9$.

A. $(\mathbf{2x-3})(\mathbf{2x-3})$. You suspect it's a perfect square:

1. **Check to see that the first and last terms are perfect squares.**

   The term $4x^2$ is $2x \cdot 2x$ and the constant 9 is $3 \cdot 3$, so both are perfect squares.

2. **The middle term has to be twice the product of the square roots of the first and last terms.**

   Since the root of $4x^2$ is 2x and the root of 9 is 3, twice their product is $2 \cdot 2x \cdot 3 = 12x$. You have a winner!

3. **Write the product of the two identical binomials, inserting the correct sign.**

   Starting with $(2x \quad 3)(2x \quad 3)$, you see that the middle term of the trinomial you're factoring is $-12x$, so the two signs in the binomials must be minus: $(2x-3)(2x-3) = 4x^2 - 12x + 9$.

Working with differences of squares

You can recognize a *difference of squares* because it's always a binomial where both terms are perfect squares and a subtraction sign appears between them. It *always* appears as $a^2 - b^2$, or $(\text{something})^2 - (\text{something else})^2$. When you do have a difference of squares on your hands — after checking it for a GCF in both terms — you follow a simple procedure: $a^2 - b^2 = (a-b)(a+b)$.

EXAMPLE

Q. Factor $25y^4 - 9$.

A. $(\mathbf{5y^2-3})(\mathbf{5y^2+3})$. Use these steps:

1. **Rewrite each term as (something)².**

   This example becomes $(5y^2)^2 - (3)^2$, which clearly shows the difference of squares ("difference of" meaning subtraction).

2. **Factor the difference of squares $(a)^2 - (b)^2$ to $(a-b)(a+b)$.**

   Each difference of squares $(a)^2 - (b)^2$ always factors to $(a-b)(a+b)$. This example factors to $(5y^2-3)(5y^2+3)$.

Factor the trinomials.

YOUR TURN

5 $4x^2-49$	**6** $9y^2-30y+25$

Breaking down a cubic difference or sum

After you've checked to see if there's a GCF in the given polynomial and discovered it's a binomial that isn't a difference of squares, consider that it may be a sum or difference of cubes.

A *sum of cubes* is always a binomial with a plus sign in between — the only one where that happens: $(\text{something})^3+(\text{something else})^3$. When you recognize a sum of cubes a^3+b^3, it factors as $(a+b)(a^2-ab+b^2)$.

Q. Factor $8x^3+27$.

EXAMPLE **A.** $(2x+3)(4x^2-6x+9)$. You first look for the GCF. You find none, so now you use the following steps:

1. **Determine that you have a sum of cubes.**

   The plus sign tells you that it may be a sum of cubes, but that clue isn't foolproof. Time for some trial and error: Try to rewrite the expression as the sum of cubes; if you select $(2x)^3+(3)^3$, you've found a winner.

2. **Use the factoring format, substituting the two cube roots for *a* and *b*.**

   Replace a with $2x$ and b with 3. Using the formula $a^3+b^3=(a+b)(a^2-ab+b^2)$, you get $(2x+3)\left[(2x)^2-(2x)(3)+3^2\right]$.

3. **Simplify the expression.**

   This example simplifies to $(2x+3)(4x^2-6x+9)$.

4. **Check the factored polynomial to see if it will factor again.**

REMEMBER

You're not done factoring until you're done. Always look at the "leftovers" to see if they'll factor again. Sometimes the binomial term may factor again as the difference of squares. However, the trinomial factor *never* factors again.

In the previous example, the binomial term $2x+3$ is a first-degree binomial (the exponent on the variable is 1) without a GCF, so it won't factor again. Therefore, $(2x+3)(4x^2-6x+9)$ is your final answer.

A *difference of cubes* sounds an awful lot like a difference of squares (see the last section), but it factors quite differently — it factors very much like the sum of cubes. A difference of cubes always starts off as a binomial with a subtraction sign in between, but it's written as $(\text{something})^3 - (\text{something else})^3$. To factor any difference of cubes, you use the formula $a^3 - b^3 = (a-b)(a^2 + ab + b^2)$. Use the same steps as when factoring the sum of cubes; just change the two signs in the formula.

YOUR TURN

Factor the binomials.

 7 $27x^3 - 1$	 8 $1000y^2 + 343$

Grouping to factor four or more terms

When a polynomial has four or more terms, the easiest way to factor it (if possible in this case) is with *grouping*. In this method, you look at only two terms at a time to see if any of the previous factoring techniques becomes apparent (you may see a GCF in two terms, or you may recognize a trinomial as a perfect square). Here, you see how to group when the given polynomial *starts off* with four (or more) terms.

REMEMBER

Sometimes you can group a polynomial into sets with two terms each to find a GCF in each set. You should try this method first when faced with a polynomial with four or more terms. This type of grouping is the most common method in a pre-calculus text.

EXAMPLE

Q. Factor $x^3 + x^2 - x - 1$ by using grouping.

A. $(x+1)^2(x-1)$. Just follow these steps:

1. **Break up the polynomial into sets of two.**

   You can go with $(x^3 + x^2) + (-x - 1)$. Put the plus sign between the sets, just like when you factor trinomials.

2. **Find the GCF of each set and factor it out.**

   The square x^2 is the GCF of the first set, and −1 is the GCF of the second set. Factoring out both of them, you get $x^2(x+1) - 1(x+1)$.

3. Factor again as many times as you can.

The two terms you've created have a GCF of $(x+1)$. When factored out, you get $(x+1)(x^2-1)$.

However, x^2-1 is a difference of squares and factors again. In the end, you get the following factors after grouping: $(x+1)(x+1)(x-1)$, or $(x+1)^2(x-1)$.

If the previous method doesn't work, you may have to group the polynomial some other way. Of course, after all your effort, the polynomial may end up being prime, which is okay.

EXAMPLE

Q. Factor $x^2-4xy+4y^2-16$.

A. $[(x-2y)+4][(x-2y)-4]$. You can group it into sets of two, and it becomes $x(x-4y)+4(y^2-4)$. This expression, however, doesn't factor again. The grouping should result in a common factor in the two terms. So look again at the original expression. You can try grouping it in some other way. In this case, if you look at the first three terms, you'll discover a perfect-square trinomial, $x^2-4xy+4y^2$, which factors to $(x-2y)^2$.

Rewrite the expression as $(x-2y)^2-16$. Now you have a difference of squares, which factors again to $[(x-2y)+4][(x-2y)-4]$.

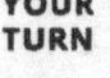

YOUR TURN

Factor the polynomials.

9 $2x^3+7x^2-8x-28$	10 $4x^3-12x^2-9x+27$

Finding the Roots of a Factored Equation

Sometimes after you've factored, the two factors can be factored again, in which case you should continue factoring. In other cases, you may have a trinomial factor that is not factorable. In this case, you can solve the equation only by using the quadratic formula. For example, $6x^4-12x^3+4x^2=0$ factors to $2x^2(3x^2-6x+2)=0$. The first term, $2x^2=0$, is solvable using algebra, but the second factor, $3x^2-6x+2=0$, is not factorable and requires the quadratic formula (see the following section).

REMEMBER

After you factor a polynomial into its different pieces, you can set each piece equal to zero to solve for the roots with the zero-product property. The *zero-product property* says that if several factors are multiplying to give you zero, at least one of them has to be zero. Your job is to find all the values of *x* that make the polynomial equal to zero. This task is much easier if the polynomial is factored because you can set each factor equal to zero and solve for *x*.

EXAMPLE

Q. Use the *zero-product property* to find the roots of $x^2+3x-10=0$.

A. **−5,2.** Factoring the trinomial gives you $(x+5)(x-2)=0$. Each factor is linear (first degree). The term $x+5=0$ gives you one solution, $x=-5$, and $x-2=0$ gives you the other solution, $x=2$.

These solutions each become an *x*-intercept on the graph of the polynomial (see the section, "Graphing Polynomials").

Find the solutions of the equations.

YOUR TURN

 **11** $x^2+4x-21=0$	 **12** $2x^2+7x-4=0$

Cracking a Quadratic Equation When It Won't Factor

When asked to solve a quadratic equation that just doesn't factor, you have to employ other methods to do the solving. The inability to factor means that the equation has solutions that you can't find by using normal techniques. Perhaps the solutions involve square roots of non-perfect squares; they can even be complex numbers involving imaginary numbers (see Chapter 15).

One such method is to use the *quadratic formula*, which is used to solve for the variable in a quadratic equation in standard form. Another is to *complete the square*, which means to manipulate an expression to create a perfect-square trinomial that you can then factor. The following sections present these methods in detail.

Using the quadratic formula

When a quadratic equation just won't factor, or if you just can't figure out the factorization, the quadratic formula is available to solve the equation. Given a quadratic equation in standard form $ax^2+bx+c=0$, $x=\dfrac{-b\pm\sqrt{b^2-4ac}}{2a}$.

TIP

Before you apply the formula, rewrite the equation in standard form (if it isn't already) and determine the *a*, *b*, and *c* values.

EXAMPLE

Q. Solve $x^2-3x+1=0$.

A. $\left(\dfrac{3+\sqrt{5}}{2},0\right)$ **and** $\left(\dfrac{3-\sqrt{5}}{2},0\right)$. **First, determine that $a=1$, $b=-3$, and $c=1$. The *a*, *b*, and *c* terms simply plug into the formula to give you the values for *x*: $x=\dfrac{-(-3)\pm\sqrt{(-3)^2-4(1)(1)}}{2(1)}$.**

Simplify this formula to get $x=\dfrac{3\pm\sqrt{9-4}}{2}=\dfrac{3\pm\sqrt{5}}{2}$.

If you're graphing the quadratic $y=x^2-3x+1$, this gives you the two *x*-intercepts, $\left(\dfrac{3+\sqrt{5}}{2},0\right)$ and $\left(\dfrac{3-\sqrt{5}}{2},0\right)$.

Completing the square

Completing the square comes in handy when you're asked to solve an uncooperative quadratic equation or when you need a particular equation format to graph a conic section (circle, ellipse, parabola, and hyperbola), which is explained in Chapter 17. For now, it's recommended that you find the roots of a quadratic using completing the square only when you're specifically asked to do so, because factoring a quadratic and using the quadratic formula work just as well (if not better). Those methods are less complicated than completing the square.

Q. Solve the equation $2x^2-4x-5=0$ using the *completing the square* method.

EXAMPLE

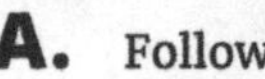

A. Follow these steps:

1. **Divide every term by the leading coefficient so that $a=1$.** If the equation already has a plain x^2 term, you can skip to Step 2.

   Be prepared to deal with fractions in this step. Dividing each term by 2, the equation now becomes $x^2-2x-\dfrac{5}{2}=0$.

2. **Move the constant term to the other side of the equal sign by adding its opposite to both sides. You want to have only terms with the variable on the left side of the equation.**

   You can add $\dfrac{5}{2}$ to both sides to get $x^2-2x=\dfrac{5}{2}$.

3. **Now to complete the square on the left: Divide the linear coefficient by 2. Square this number, and then add the square to both sides of the equation.**

   Divide −2 by 2 to get −1. Square this answer to get 1, and add it to both sides: $x^2 - 2x + 1 = \frac{5}{2} + 1$.

4. **Simplify the equation.**

   The equation becomes $x^2 - 2x + 1 = \frac{7}{2}$.

5. **Factor the newly created quadratic equation.**

   The new equation should be a perfect-square trinomial: $(x-1)(x-1) = \frac{7}{2}$ or $(x-1)^2 = \frac{7}{2}$.

6. **Get rid of the square exponent by taking the square root of both sides. Remember that the positive and negative roots could both be squared to get the answer!**

   This step gives you $x - 1 = \pm\sqrt{\frac{7}{2}}$.

7. **Simplify any square roots if possible.**

   The example equation doesn't simplify, but the denominator needs to be rationalized (see Chapter 2). Do the rationalization to get $x - 1 = \pm\sqrt{\frac{7}{2}} = \pm\frac{\sqrt{7}}{\sqrt{2}} \cdot \frac{\sqrt{2}}{\sqrt{2}} = \pm\frac{\sqrt{14}}{2}$.

8. **Solve for the variable by isolating it.**

   You add 1 to both sides to get $x = 1 \pm \frac{\sqrt{14}}{2}$. ***Note:*** You may be asked to express your answer as one fraction; in this case, find the common denominator and add to get $x = \frac{2 \pm \sqrt{14}}{2}$.

YOUR TURN

13 Solve the equation using the *quadratic formula*: $3x^2 - x - 1 = 0$.

14 Solve the equation using *completing the square*: $4x^2 + 24x - 5 = 0$.

Solving Polynomial Equations with a Degree Higher Than Two

By now you're a professional at solving second-degree polynomial equations (quadratics), and you have various tools at your disposal for solving these types of problems. You may have noticed while solving quadratics that when a quadratic equation has real solutions, it always has two solutions. Note that sometimes both solutions are the same (as in perfect-square trinomials). Even though you get the same solution twice, they still count as two solutions (how many times a solution is a root is called the *multiplicity* of the solution).

When the polynomial degree is higher than two and the polynomial won't factor using any of the techniques that are discussed earlier in this chapter, finding the roots gets more challenging. For example, you may be asked to solve a cubic polynomial that is *not* a sum or difference of cubes or a polynomial that is a fourth degree or greater that can't be factored by grouping. The higher the degree, the more possible roots exist, and the harder it is to find them. To find the roots, many different scenarios can guide you in the right direction. You can make very educated guesses about how many roots a polynomial has, as well as how many of them are positive or negative and how many are real or imaginary.

Counting a polynomial's total roots

REMEMBER

Usually, the first step you take before solving a polynomial equation is to find its *degree*, which helps you determine the number of solutions you'll find later. The degree tells you the maximum number of possible roots.

When you're being asked to solve a polynomial equation, finding its degree is relatively easy, because only one variable is in any term. Therefore, the highest exponent indicates the degree. For example, $f(x) = 2x^4 - 9x^3 - 21x^2 + 88x + 48$ is a fourth-degree polynomial with up to, but no more than, four possible solutions. In the next two sections, you see how to figure out how many of those roots may be real roots and how many may be imaginary.

Tallying the real roots: Descartes's rule of signs

The terms *solutions/zeros/roots* are synonymous because they all represent where the graph of the polynomial intersects the x-axis. The roots that are found when the graph meets with the x-axis are called *real roots;* you can see them and deal with them as real numbers in the real world. Also, because they cross or touch the x-axis, some roots may be *negative roots* (which means they intersect the negative x-axis), and some may be *positive roots* (which intersect the positive x-axis).

If you know how many total roots you have (see the last section), you can use a pretty cool theorem called *Descartes's rule of signs* to count how many roots are possibly real numbers (both positive *and* negative) and how many are imaginary (see Chapter 15). You see, the same man who pretty much invented graphing, Descartes, also came up with a way to figure out how many times a polynomial intersects the x-axis — in other words, how many real roots it has. All you have to be able to do is count!

Here's how Descartes's rule of signs can give you the numbers of possible real roots, both positive and negative.

- **Positive real roots:** For the number of positive real roots, look at the polynomial, written in descending order of powers, and count how many times the sign changes from term to term. This value represents the maximum number of positive roots in the polynomial.

  Descartes's rule of signs says the number of positive roots is equal to changes in sign of $f(x)$, or is less than that by an even number (so you keep subtracting 2 until you get either 1 or 0). Therefore, the previous $f(x)$ may have 2 or 0 positive roots.

- **Negative real roots:** For the number of negative real roots, find $f(-x)$ and count again. Because negative numbers raised to even powers are positive and negative numbers raised to odd powers are negative, this change affects only terms with odd powers. This step is the same as changing each term with an odd degree to its opposite sign and counting the sign changes again, which gives you the maximum number of negative roots.

Q. **Determine the possible number of positive real roots in the polynomial $f(x)=2x^4-9x^3-21x^2+88x+48$.**

A. **2 or 0. You see two changes in sign (don't forget to include the sign of the first term!) — from the first term to the second and from the third term to the fourth. That means this equation can have up to two positive solutions.**

Q. **Determine the possible number of negative real roots in the polynomial $f(x)=2x^4-9x^3-21x^2+88x+48$.**

A. **2 or 0. Replacing each x with $-x$, the example equation becomes $f(-x)=2x^4+9x^3-21x^2-88x+48$, in which the terms change signs twice. There can be, at most, two negative roots. However, similar to the rule for positive roots, the number of negative roots is equal to the changes in sign for $f(-x)$, or must be less than that by an even number. Therefore, this example can have either 2 or 0 negative roots.**

Are you wondering how many are real and how many are imaginary? Read on.

15 **Determine the possible number of positive and negative real roots in the polynomial $f(x)=3x^4+4x^3-2x^2-8x-4$.**

16 **Determine the possible number of positive and negative real roots in the polynomial $f(x)=x^6-2x^5-4x^4+5x^3+x^2$.**

Accounting for imaginary roots: The fundamental theorem of algebra

Imaginary roots appear in a quadratic equation when the discriminant of the quadratic equation — the part under the square root sign $(b^2 - 4ac)$ — is negative. If this value is negative, you can't actually take the square root, and the answers are not real. In other words, there is no solution; therefore, the graph won't cross the x-axis.

Using the quadratic formula always gives you two solutions, because the ± sign means you're both adding and subtracting and getting two completely different answers. When the number underneath the square-root sign in the quadratic formula is negative, the answers are called *complex conjugates*. One is $r + si$ and the other is $r - si$. These numbers have both real (the r) and imaginary (the si) parts. The s represents the real number multiplier of the imaginary number $i = \sqrt{-1}$.

REMEMBER

The fundamental theorem of algebra says that every non-constant polynomial function has at least one root in the complex number system. Recall, from Chapter 1, that a complex number can be either real or have an imaginary part. This concept is one you may remember from Algebra II. (For reference, flip to Chapter 15 to read the parts on imaginary and complex numbers first.)

The highest degree of a polynomial gives you the highest possible number of *complex* roots for the polynomial. Between this fact and Descartes's rule of signs, you can figure out how many complex roots a polynomial has. Pair up every possible number of positive real roots with every possible number of negative real roots (see the previous section); the remaining number of roots for each situation represents the number of roots that are not real.

EXAMPLE

Q. Determine the possible number of complex roots in the polynomial $f(x) = 2x^4 - 9x^3 - 21x^2 + 88x + 48$.

A. The polynomial has a degree of 4, with two or zero positive real roots, and two or zero negative real roots. Pair up the possible situations:

- Two positive and two negative real roots, with zero non-real roots
- Two positive and zero negative real roots, with two imaginary complex roots
- Zero positive and two negative real roots, with two imaginary complex roots
- Zero positive and zero negative real roots, with four imaginary complex roots

Complex numbers are covered in more detail in Chapter 15. For now, when dealing with roots of polynomials, you're mainly trying to be sure you found all the possible places where the curve is either crossing or touching the x-axis. If you can account for all the possible roots — how many are real and how many are imaginary — then you know you haven't missed anything.

For example, if you have a polynomial of degree 7 where $x = 5$, $x = 1$, and $x = -2$ are all real roots, then these three are all considered to be both real and complex because they can be rewritten as $x = 5 + 0i$, $x = 1 + 0i$, and $x = -2 + 0i$ (the imaginary part is 0). The other four, non-real roots must then come in pairs of the form $x = r + si$, where s is not equal to 0.

The fundamental theorem of algebra gives the total number of complex roots (in this example, there are seven); Descartes's rule of signs tells you how many possible real roots exist and how many of them are positive and negative.

Guessing and checking the real roots

Using Descartes's rule of signs from the preceding section, you can determine exactly how many roots (and what type of roots) exist. Now, the rational root theorem is another method you can use to narrow down the search for roots of polynomials. Descartes's rule of signs only narrows down the real roots into possible positive and negative. The rational root theorem says that some real roots are rational (they can be expressed as a fraction). It also helps you create a list of the *possible* rational roots of any polynomial.

The problem? Not every real root is rational; some may be irrational. A polynomial may even have *only* irrational roots. But this theorem is always the next place to start in your search for roots; it will at least give you a diving-off point. Besides, the problems you're presented with in pre-calculus are more than likely to have at least one rational root, so the information in this section greatly improves your odds of finding more!

Follow these general steps to ensure that you find every root (this is discussed in detail later in this section):

1. **Use the rational root theorem to list all possible rational roots.**

   The rational root theorem has you create a list of possible rational roots using the lead coefficient and constant in the polynomial equation. You see how to use this theorem next.

2. **Pick one root from the list in Step 1 and use long division or synthetic division to find out if it is, in fact, a root.**

   - If the root doesn't work, try another guess.
   - If the root works, proceed to Step 3.

3. **Using the depressed polynomial (the one you get after doing the division in Step 2), test the root that worked to see if it works again.**

   - If it works, repeat Step 3 *again.*
   - If it doesn't work, return to Step 2 and try a different root from the list in Step 1.

4. **List all the roots you find that work; you should have as many roots as the degree of the polynomial.**

   Don't stop until you've found them all. And keep in mind that some will be real and some will be imaginary.

Listing the possibilities with the rational root theorem

REMEMBER

The rational root theorem says that if you take all the factors of the constant term in a polynomial and divide by all the factors of the leading coefficient, you produce a list of all the possible rational roots of the polynomial. However, keep in mind that you're finding only the *rational* ones, and sometimes the roots of a polynomial are irrational. Some of your roots can also be non-real, but save those until the end of your search.

EXAMPLE

Q. List all the possible rational roots of the polynomial $f(x) = 2x^4 - 9x^3 - 21x^2 + 88x + 48$. The constant term is 48, and its factors follow: $\pm 1, \pm 2, \pm 3, \pm 4, \pm 6, \pm 8, \pm 12, \pm 16, \pm 24, \pm 48$.

A. The leading coefficient is 2, and its factors are $\pm 1, \pm 2$. So the list of possible real roots includes all the factors of 48 divided by all the factors of 2. There are many repeats after performing those divisions, so the final list boils down to $\pm\frac{1}{2}, \pm 1, \pm\frac{3}{2}, \pm 2, \pm 3, \pm 4, \pm 6, \pm 8, \pm 12, \pm 16, \pm 24, \pm 48$.

YOUR TURN

17 List all the possible rational roots of the polynomial $f(x) = 3x^4 - 8x^3 - 20x^2 + 11x + 10$.

18 List all the possible rational roots of the polynomial $f(x) = -8x^3 - 20x^2 + 11x + 8$.

Testing roots by dividing polynomials

Dividing polynomials follows the same algorithm as long division with real numbers. The polynomial you're dividing by is called the *divisor.* The polynomial being divided is called the *dividend.* The answer is called the *quotient*, and the leftover polynomial is called the *remainder.*

One way, other than synthetic division, that you can test possible roots from the rational root theorem is to use long division of polynomials and hope that when you divide, you get a remainder of 0. For example, when you have your list of possible rational roots (as found in the last section), pick one and assume that it's a root. If $x = c$ is a root, then $x - c$ is a factor. So if you pick $x = 2$ as your guess for the root, $x - 2$ should be a factor. You see in this section how to use long division to test if $x - 2$ is actually a factor and, therefore, $x = 2$ is a root.

Dividing polynomials to get a specific answer isn't something you do every day, but the idea of a function or expression that's written as the quotient of two polynomials is important for pre-calculus. If you divide a polynomial by another and get a remainder of 0, the divisor is a factor, which in turn gives a root. The following sections review two methods of checking your real roots: long division and synthetic division.

REMEMBER

In math lingo, the division algorithm states the following: If $f(x)$ and $d(x)$ are polynomials such that $d(x)$ isn't equal to 0, and the degree of $d(x)$ isn't larger than the degree of $f(x)$, there are unique polynomials $q(x)$ and $r(x)$ such that $f(x)=d(x)\cdot q(x)+r(x)$. In plain English, the dividend equals the divisor times the quotient plus the remainder. You can always check your results by remembering this information.

LONG DIVISION

You can use long division to find out if your possible rational roots are actual roots or not. It isn't recommended to use long division, but you can do it. Instead, it is suggested that you use synthetic division, which is covered later. However, in case you need to do long division for this process or for some other application, you see long division explained in the following steps — while trying to figure out a root at the same time.

TIP

Remember the mnemonic device Dirty Monkeys Smell Bad when doing long division to check your roots. Make sure all terms in the polynomial are listed in descending order and that every degree is represented. In other words, if x^2 is missing, put in a placeholder of $0x^2$ and then do the division. (This step is just to make the division process easier by lining up the like terms.)

To divide two polynomials, follow these steps:

1. **Divide.**

   Divide the leading term of the dividend by the leading term of the divisor. Write this quotient directly above the term you just divided into.

2. **Multiply.**

   Multiply the quotient term from Step 1 by the entire divisor. Write this polynomial under the dividend so that like terms are lined up.

3. **Subtract.**

   Subtract the whole line you just wrote from the dividend.

   TIP

   You can change all the signs and add if it makes you feel more comfortable. This way, you won't forget signs.

4. **Bring down the next term.**

   Do exactly what this says; bring down the next term in the dividend.

5. **Repeat Steps 1 to 4 over and over until the remainder polynomial has a degree that's less than the dividend's.**

Q. Use long division to divide $2x^4 - 9x^3 - 21x^2 + 88x + 48$ by $x - 2$.

EXAMPLE **A.** Here are the results using the steps:

$$\begin{array}{r} 2x^3 - 5x^2 - 31x + 26 \\ x-2 \overline{) 2x^4 - 9x^3 - 21x^2 + 88x + 48} \\ \underline{2x^4 - 4x^3} \qquad\qquad\qquad\qquad \\ -5x^3 - 21x^2 \qquad\qquad\quad \\ \underline{-5x^3 + 10x^2} \qquad\qquad\quad \\ -31x^2 + 88x \qquad \\ \underline{-31x^2 + 62x} \qquad \\ 26x + 48 \\ \underline{26x - 52} \\ 100 \end{array}$$

After all that work, you find that there's a remainder of 100, so $x - 2$ isn't a factor of the polynomial. Before you panic and wonder how many times you have to do this, read on about synthetic division.

SYNTHETIC DIVISION

The good news: A shortcut exists for long division of polynomials, and that shortcut is synthetic division. It's a special case of division that can be used when the divisor is a linear factor in the form $x + c$, where c is a constant.

The bad news, however, is that the shortcut only works if the divisor $(x + c)$ is a first-degree binomial with a leading coefficient of 1 (you can always make it 1 by dividing everything by the leading coefficient first). The *great* news — yep, more news — is that you can always use synthetic division to figure out if a possible root is actually a root.

Here are the general steps for synthetic division:

1. **Make sure the polynomial is written in descending order of the powers.**

   The term with the highest exponent comes first.

2. **Write down the coefficients and the constant of the polynomial from left to right, filling in a zero if any degree terms are missing; place the root you're testing outside the synthetic division sign.**

   The division sign looks like the right and bottom sides of a rectangle. Leave room below the coefficients to write another row of numbers.

3. **Drop down the first coefficient below the division sign.**

4. **Multiply the root you're testing by the number you just dropped down and write the answer below the next coefficient.**

5. **Add the coefficient and product from Step 4 and put the answer below the line.**
6. **Multiply the root you're testing by the answer from Step 5 and put the product below the next coefficient.**
7. **Continue multiplying and adding until you use the last number inside the synthetic division sign.**

If the last number you get isn't a 0, the number you tested isn't a root.

If the answer is 0, congratulations! You've found a root. The numbers below the synthetic division sign are the coefficients of the quotient polynomial. The degree of this polynomial is one less than the original (the dividend), so the exponent on the first x term should be one less than what you started with.

Q. Use synthetic division to see if $x = 4$ is a solution of $2x^4 - 9x^3 - 21x^2 + 88x + 48$.

EXAMPLE **A.** The number 4 is the root you're testing. Write it outside, on the left. The numbers on the inside are the coefficients of the polynomial. Here's the synthetic process, step by step, for this root:

$$\begin{array}{r|rrrrr} 4 & 2 & -9 & -21 & 88 & 48 \\ & & 8 & -4 & -100 & -48 \\ \hline & 2 & -1 & -25 & -12 & 0 \end{array}$$

1. The 2 below the line just drops down from the line above.
2. Multiply 4 with 2 to get 8 and write that under the next term, −9.
3. Add $-9+8$ to get −1.
4. Multiply 4 with −1 to get −4, and write that under the −21.
5. Add $-21+-4$ to get −25.
6. Multiply 4 with −25 to get −100, and write that under 88.
7. Add 88 to −100 to get −12.
8. Multiply 4 with −12 to get −48, and write that under 48.
9. Add 48 to −48 to get 0.

All you do is multiply and add, which is why synthetic division is the shortcut. The last number, 0, is your remainder. Because you get a remainder of 0, $x = 4$ is a root. The other numbers are the coefficients of the quotient, in order from the greatest degree to the least; however, your answer is always one degree lower than the original. So the quotient in this division is $2x^3 - x^2 - 25x - 12$.

REMEMBER

If you divide by c and the remainder is 0, then the linear expression $(x - c)$ is a factor and c is a root. A remainder other than 0 implies that $(x - c)$ isn't a factor and that c isn't a root.

TECHNICAL STUFF

Whenever a root works, you should always automatically test it again in the answer quotient to see if it's a double root, using the same process. A *double root* occurs when a factor has a multiplicity of two. A double root is one example of multiplicity (as described earlier in the section, "Accounting for imaginary roots: The fundamental theorem of algebra"). But a root can occur more than twice — you just need to keep checking.

EXAMPLE

Q. Test $x = 4$ again in the previous example with synthetic division.

A. Perform the synthetic division.

$$\begin{array}{r|rrrr} 4 & 2 & -1 & -25 & -12 \\ & & 8 & 28 & 12 \\ \hline & 2 & 7 & 3 & 0 \end{array}$$

You get a remainder of 0 again, so $x = 4$ is a double root. (In math terms, you say that $x = 4$ is a root with multiplicity of two.) You have to check it again, though, to see if it has a higher multiplicity. When you synthetically divide $x = 4$ one more time, it doesn't work. Before you tested $x = 4$ for a final time, the polynomial (called a *depressed polynomial*) was down to a quadratic: $2x^2 + 7x + 3$. If you factor this expression, you get $(2x + 1)(x + 3)$. This gives you two more roots of $-\frac{1}{2}$ and -3. To sum it all up, you've found $x = 4$ (multiplicity two), $x = -\frac{1}{2}$, and $x = -3$. You found four complex roots — two of them are negative real numbers, and two of them are positive real numbers.

REMEMBER

Always work off the newest quotient when using synthetic division. This way, the degree gets lower and lower until you end up with a quadratic expression. At that point, you can solve the quadratic by using any of the techniques covered earlier in this chapter: factoring, completing the square, or applying the quadratic formula.

TECHNICAL STUFF

The *remainder theorem* says that the remainder you get when you divide a polynomial by a binomial is the same as the result you get from plugging the number into the polynomial. Given $f(x) = a_n x^n + a_{n-1}x^{n-1} + a_{n-2}x^{n-2} + \cdots + a_1 x^1 + a_0$, the result when evaluating $f(h)$ is the same number as the remainder when dividing $f(x)$ by $x - h$.

EXAMPLE

Q. Refer to the example where $f(x) = 2x^4 - 9x^3 - 21x^2 + 88x + 48$ is divided by $x - 2$. What is $f(2)$?

A. $f(2) = 100$, because that's the remainder resulting from the division.

YOUR TURN

19 Determine the remainder when $x^4 + 10x^3 + 3x^2 + 66x + 45$ is divided by $x + 1$.

20 Find the four roots of $x^4 - 5x^3 - 20x^2 + 60x + 144 = 0$.

Put It in Reverse: Using Solutions to Find Factors

The *factor theorem* states that you can go back and forth between the roots of a polynomial and the factors of a polynomial. In other words, if you know one, you know the other. At times, you may need to factor a polynomial with a degree higher than two. If you can find its roots, you can find its factors. You see how in this section.

TECHNICAL STUFF

In symbols, the factor theorem states that if $x-c$ is a factor of the polynomial $f(x)$, then $f(c)=0$. The variable c is a zero or a root or a solution — whatever you want to call it (the terms all mean the same thing).

In the previous sections of this chapter, you employ many different techniques to find the roots of the polynomial $f(x)=2x^4-9x^3-21x^2+88x+48$. You find that they are $x=-\frac{1}{2}$, $x=-3$, and $x=4$ (multiplicity two). How do you use those roots to find the factors of the polynomial?

The factor theorem states that if $x=c$ is a root, $(x-c)$ is a factor. For example, look at the following roots:

- If $x=-\frac{1}{2}$, $\left(x-\left(-\frac{1}{2}\right)\right)$ is your factor, which is the same thing as saying $\left(x+\frac{1}{2}\right)$.
- If $x=-3$ is a root, $(x-(-3))$ is a factor, which is also $(x+3)$.
- If $x=4$ is a root with multiplicity two, $(x-4)$ is a factor with multiplicity two.

You can now factor $f(x)=2x^4-9x^3-21x^2+88x+48$ to get $f(x)=2\left(x+\frac{1}{2}\right)(x+3)(x-4)^2$.

YOUR TURN

21 Write the equation of a polynomial with roots $x=3$, $x=-1$, $x=-2$.

22 Write the equation of a polynomial with roots $x=-3$, $x=-3$, $x=\frac{1}{4}$, $x=0$.

Graphing Polynomials

The hard graphing work is over after you find the zeros of a polynomial function (using the techniques presented earlier in this chapter). Finding the zeros is very important to graphing the polynomial, because they give you a general template for what your graph should look like. Remember that zeros are x-intercepts, and knowing where the graph crosses or touches the x-axis is half the battle. The other half is knowing what the graph does in between these points. This section shows you how to figure that out.

TIP

If you're lucky enough to own a graphing calculator *and* are in a situation where you're allowed to use it, you can enter any polynomial equation into the calculator's graphing utility and graph the equation. The calculator will not only identify the zeros, but also show you the maximum and minimum values of the graph so that you can draw the best possible representation.

Graphing when all the roots are real numbers

When graphing a polynomial function, you want to use the following steps:

1. Plot the zeros (x-intercepts) on the coordinate plane and the y-intercept.
2. Determine which way the ends of the graph point.
3. Figure out what happens between the zeros by picking any value to the left and right of each intercept and plugging it into the function.
4. Plot the graph.

EXAMPLE

Q. Plot the graph of the polynomial $f(x) = 2x^4 - 9x^3 - 21x^2 + 88x + 48$. This is an example that appears earlier. The zeros are $x = -3$, $x = -\frac{1}{2}$, and $x = 4$.

A. Use the following steps.

1. Plot the zeros (x-intercepts) on the coordinate plane.

Mark the zeros that you found previously: $x = -3$, $x = -\frac{1}{2}$, and $x = 4$.

REMEMBER

Now plot the y-intercept of the polynomial. The y-intercept is *always* the constant term of the polynomial — in this case, $y = 48$. If no constant term is written, the y-intercept is 0.

2. Determine which way the ends of the graph point.

You can use a handy test called the *leading coefficient test*, which helps you figure out how the polynomial begins and ends. The degree and leading coefficient of a polynomial always explain the end behavior of its graph (see the section, "Understanding Degrees and Roots," for more on finding degree):

- **If the degree of the polynomial is even and the leading coefficient is positive, both ends of the graph point up.**
- **If the degree is even and the leading coefficient is negative, both ends of the graph point down.**
- **If the degree is odd and the leading coefficient is positive, the left side of the graph points down and the right side points up.**
- **If the degree is odd and the leading coefficient is negative, the left side of the graph points up and the right side points down.**

Figure 5-1 displays this concept in correct mathematical terms.

The function $f(x) = 2x^4 - 9x^3 - 21x^2 + 88x + 48$ is even in degree and has a positive leading coefficient, so both ends of its graph point up (they go to positive infinity).

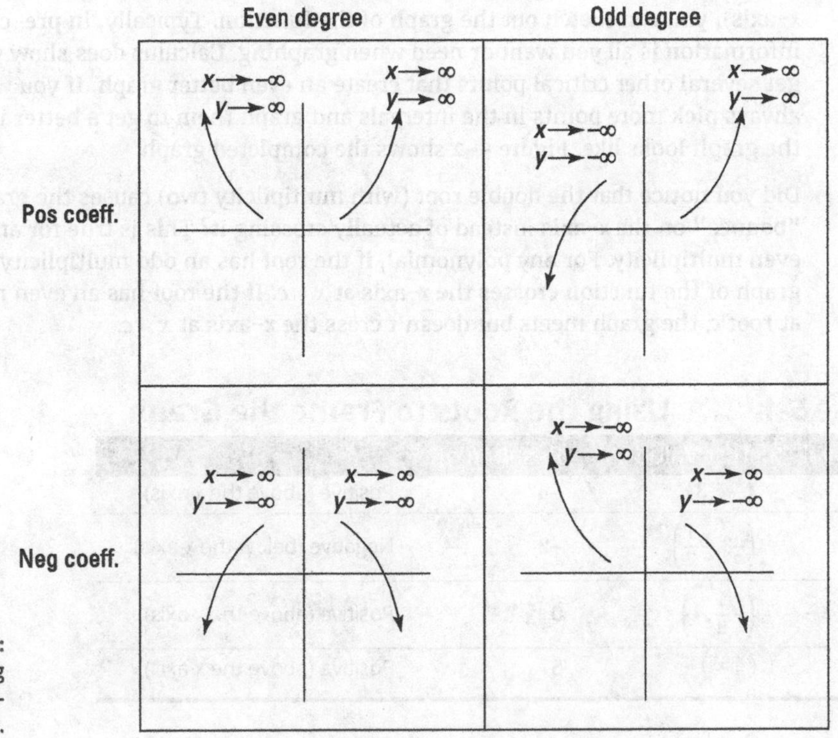

FIGURE 5-1: Illustrating the leading-coefficient test.

3. **Figure out what happens between the zeros by picking any value to the left and right of each intercept and plugging it into the function.**

   Just pick some convenient values, and see if the results are positive or negative and just how large or small the values are becoming in the interval.

A graphing calculator gives a very accurate picture of the graph. Calculus allows you to find high points, low points, and intercepts exactly, using an algebraic process, but you can easily use the calculator to find them. You can use your graphing calculator to check your work, and make sure the graph you've created looks like the one the calculator gives you.

Using the zeros for the function, set up a table to help you figure out whether the graph is above or below the x-axis between the zeros. See Table 5-1 for some values.

The first interval, $(-\infty, 3)$, and the last interval, $(4, \infty)$, both confirm the leading coefficient test from Step 2 — this graph points up (to positive infinity) in both directions.

4. **Plot the graph.**

   Now that you know where the graph crosses the x-axis, how the graph begins and ends, and whether the graph is positive (above the x-axis) or negative (below the x-axis), you can sketch out the graph of the function. Typically, in pre-calculus, this information is all you want or need when graphing. Calculus does show you how to get several other critical points that create an even better graph. If you want, you can always pick more points in the intervals and graph them to get a better idea of what the graph looks like. Figure 5-2 shows the completed graph.

Did you notice that the double root (with multiplicity two) causes the graph to "bounce" on the x-axis instead of actually crossing it? This is true for any root with even multiplicity. For any polynomial, if the root has an odd multiplicity at root c, the graph of the function crosses the x-axis at $x = c$. If the root has an even multiplicity at root c, the graph meets but doesn't cross the x-axis at $x = c$.

Table 5-1 Using the Roots to Frame the Graph

Interval	Test Value (x)	Result [$f(x)$]
$(-\infty, 3)$	-4	Positive (above the x-axis)
$\left(-3, -\frac{1}{2}\right)$	-2	Negative (below the x-axis)
$\left(-\frac{1}{2}, 4\right)$	0	Positive (above the x-axis)
$(4, \infty)$	5	Positive (above the x-axis)

Graphing when roots are imaginary numbers: Combining all techniques

In pre-calculus and calculus, certain polynomial functions have non-real roots in addition to real roots (and some of the more complicated functions have *all* imaginary roots). When you must find both, start off by finding the real roots, using all the techniques described earlier in this chapter (such as synthetic division). Then you're left with a depressed quadratic polynomial to solve that's unsolvable using real-number answers. No fear! You just have to use the quadratic formula, through which you'll end up with a negative number under the square-root sign. Therefore, you express the answer as a complex number (for more, see Chapter 15).

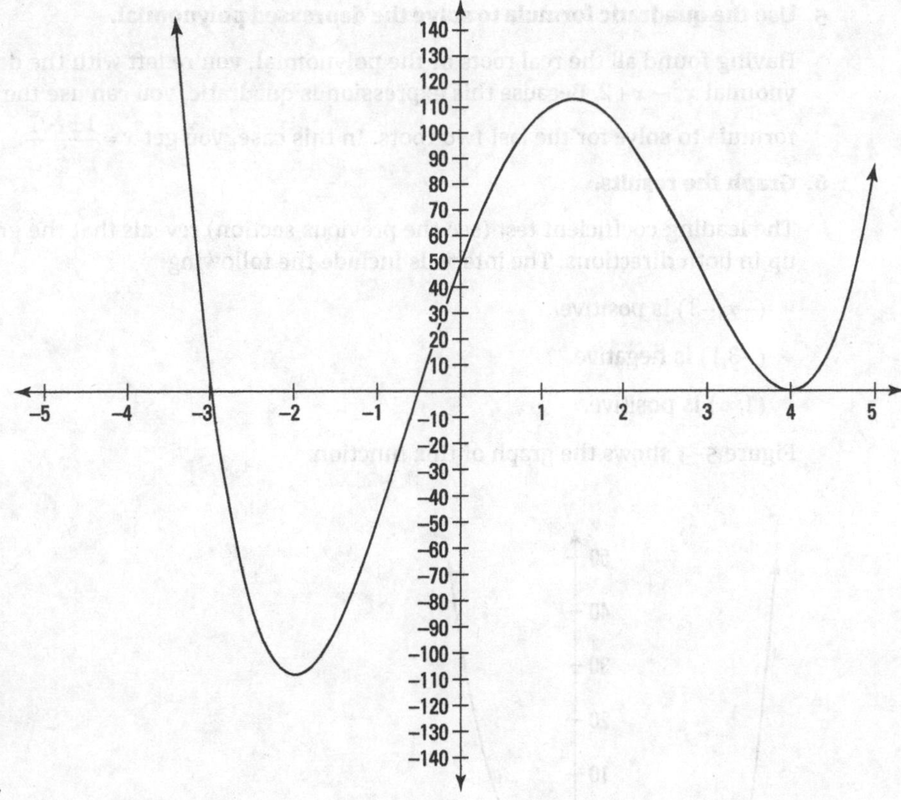

FIGURE 5-2: Graphing the polynomial $f(x)=2x^4-9x^3-21x^2+88x+48$.

EXAMPLE

Q. The polynomial $g(x)=x^4+x^3-3x^2+7x-6$ has non-real roots. Find these roots.

A. Follow these steps to find *all* the roots for this (or any) polynomial; each step involves a major section of this chapter:

1. **Classify the real roots as positive and negative by using Descartes's rule of signs.**

   Three changes of sign in the $g(x)$ function reveals that you could have three or one positive real roots. One change in sign in the $g(-x)$ function reveals that you have one negative real root.

2. **Find how many roots are possibly imaginary by using the fundamental theorem of algebra.**

   The theorem reveals that, in this case, up to four complex roots exist. Combining this fact with Descartes's rule of signs gives you several possibilities:

   - One real positive root and one real negative root means that two roots aren't real.
   - Three real positive roots and one real negative root means that all roots are real.

3. **List the possible rational roots, using the rational root theorem.**

   The possible rational roots include ± 1, ± 2, ± 3, and ± 6.

4. **Determine the rational roots (if any), using synthetic division.**

   Utilizing the rules of synthetic division, you find that $x=1$ is a root and that $x=-3$ is another root. These roots are the only real ones.

5. **Use the quadratic formula to solve the depressed polynomial.**

   Having found all the real roots of the polynomial, you're left with the depressed polynomial $x^2 - x + 2$. Because this expression is quadratic, you can use the quadratic formula to solve for the last two roots. In this case, you get $x = \frac{1 \pm i\sqrt{7}}{2}$.

6. **Graph the results.**

   The leading coefficient test (see the previous section) reveals that the graph points up in both directions. The intervals include the following:

   - $(-\infty,-3)$ is positive.
   - $(-3,1)$ is negative.
   - $(1,\infty)$ is positive.

   Figure 5-3 shows the graph of this function.

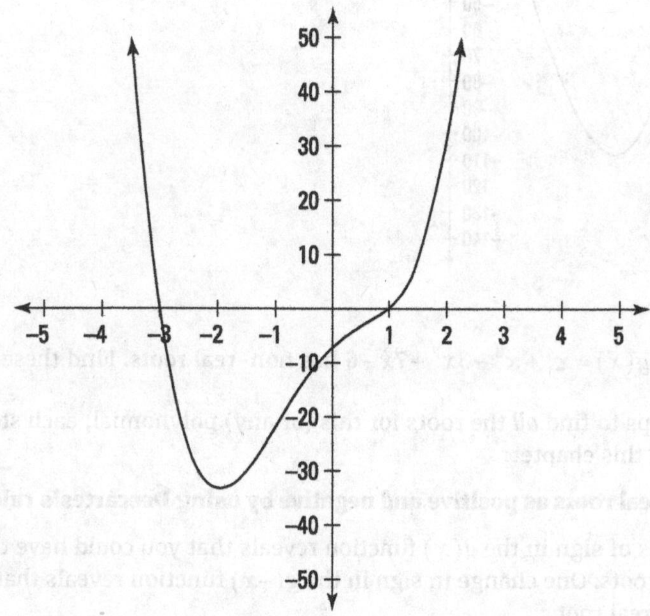

FIGURE 5-3: Graphing the polynomial $g(x) = x^4 + x^3 - 3x^2 + 7x - 6$.

You probably noticed that there's a slight bump or plateau in the graph, under the x-axis. This is an indication that there are imaginary roots. These values show up in calculus with some of its techniques.

YOUR TURN

23 Graph the function $f(x) = x^4 + 2x^3 - 13x^2 - 14x + 24$.

24 Graph the function $f(x) = x^4 - 8x^2 - 9$.

Practice Questions Answers and Explanations

1. **$3x$.** Start by finding the GCF: $9x^3 - 12x^5 + 24x = 3x(3x^2 - 4x^4 + 8) = 3x(-4x^4 + 3x^2 + 8)$. You usually want to write the polynomial terms in order of their exponents — highest to lowest or lowest to highest.

2. **$10x$.** You get $20x^4 + 60x^3 - 30x^2 - 10x = 10x(2x^3 + 6x^2 - 3x - 1)$.

3. **$(4x-5)(x+1)$.** Place the first and third terms diagonally. Their product is $-20x^2$. Find two factors of that product whose sum is the middle term of the quadratic. Place them in the other cells. Write the GCF of each row and column outside the box. These are the terms needed to create the binomial products.

$4x^2$	
	-5

→

$4x^2$	$-5x$
$4x$	-5

→

$4x$	-5	
$4x^2$	$-5x$	x
$4x$	-5	1

4. **$(2x+3)(3x+2)$.** Place the first and third terms diagonally. Their product is $36x^2$. Find two factors of that product whose sum is the middle term of the quadratic. Place them in the other cells. Write the GCF of each row and column outside the box. These are the terms needed to create the binomial products.

$6x^2$	
	6

→

$6x^2$	$9x$
$4x$	6

→

$2x$	3	
$6x^2$	$9x$	$3x$
$4x$	6	2

5. **$(2x-7)(2x+7)$.** $4x^2 - 49$ is the difference of squares.

6. **$(3y-5)^2$.** $9y^2 - 30y + 25$ is a perfect-square trinomial.

7. **$(3x-1)(9x^2+3x+1)$.** $27x^3 - 1$ is the difference of cubes.

8. **$(10y+7)(100y^2-70y+49)$.** $1000y^2 + 343$ is the sum of cubes.

9. **$(2x+7)(x-2)(x+2)$.** Using grouping, factor the GCF x^2 from the first two terms and the GCF -4 from the second two terms: $2x^3 + 7x^2 - 8x - 28 = x^2(2x+7) - 4(2x+7)$. Now factor out the common factor $2x+7$ and then factor the binomial that is the difference of squares: $x^2(2x+7) - 4(2x+7) = (2x+7)(x^2-4) = (2x+7)(x-2)(x+2)$.

10. **$(x-3)(2x-3)(2x+3)$.** Using grouping, factor the GCF $4x^2$ from the first two terms and the GCF -9 from the second two terms: $4x^3 - 12x^2 - 9x + 27 = 4x^2(x-3) - 9(x-3)$. Now factor out the common factor $x-3$ and then factor the binomial that is the difference of squares: $4x^2(x-3) - 9(x-3) = (x-3)(4x^2-9) = (x-3)(2x-3)(2x+3)$.

11. **$-7, 3$.** Factor the quadratic. Then set each of the factors equal to 0 and solve for x: $x^2 + 4x - 21 = (x+7)(x-3)$; $x+7=0 \rightarrow x=-7$; $x-3=0 \rightarrow x=3$.

12. **$\frac{1}{2}, -4$.** Factor the quadratic. Then set each of the factors equal to 0 and solve for x: $2x^2 + 7x - 4 = (2x-1)(x+4)$; $2x-1=0 \rightarrow x=\frac{1}{2}$; $x+4=0 \rightarrow x=-4$.

13 $\frac{1\pm\sqrt{13}}{6}$. You get $x=\frac{-(-1)\pm\sqrt{(-1)^2-4(3)(-1)}}{2(3)}=\frac{1\pm\sqrt{1+12}}{6}=\frac{1\pm\sqrt{13}}{6}$.

14 $-3\pm\frac{\sqrt{41}}{2}$. Add 5 to each side, and then divide each term by 4.

$$4x^2+24x-5=0\rightarrow 4x^2+24x=5\rightarrow x^2+6x=\frac{5}{4}$$

Find half of the coefficient of the x term, square it, and add that result to each side.

$$x^2+6x+\left(\frac{6}{2}\right)^2=\frac{5}{4}+\left(\frac{6}{2}\right)^2\rightarrow x^2+6x+9=\frac{5}{4}+9=\frac{41}{4}$$

Factor the perfect-square trinomial on the left: $(x+3)^2=\frac{41}{4}$. Now find the square root of each side: $(x+3)^2=\frac{41}{4}\rightarrow x+3=\pm\sqrt{\frac{41}{4}}=\pm\frac{\sqrt{41}}{2}$. Subtract 3 from each side to get $x=-3\pm\frac{\sqrt{41}}{2}$.

15 **1 positive; 3 or 1 negative.** The polynomial $f(x)=3x^4+4x^3-2x^2-8x-4$ has 1 change in sign, and $f(-x)=3x^4-4x^3-2x^2+8x-4$ has 3 changes in sign.

16 **2 or 0 positive; 2 or 0 negative.** The polynomial $f(x)=x^6-2x^5-4x^4+5x^3+x^2$ has 2 changes in sign, and $f(-x)=x^6+2x^5-4x^4-5x^3+x^2$ has 2 changes in sign.

17 $\pm1, \pm2, \pm5, \pm10, \pm\frac{1}{3}, \pm\frac{2}{3}, \pm\frac{5}{3}, \pm\frac{10}{3}$. Divide all the possible factors of 10 by all the possible factors of 3.

18 $\pm1, \pm2, \pm4, \pm8, \pm\frac{1}{2}, \pm\frac{1}{4}, \pm\frac{1}{8}$. Divide all the possible factors of –8 by all the possible factors of 8. Several reduce to a number already listed.

19 **–27.** Using synthetic division, change the +1 to –1 and divide.

$$\begin{array}{r|rrrrr} -1 & 1 & 10 & 3 & 66 & 45 \\ & & -1 & -9 & 6 & -72 \\ \hline & 1 & 9 & -6 & 72 & -27 \end{array}$$

The remainder is –27.

20 $x=4, x=6, x=-2, x=-3$. From the rule of signs, there are 2 or 0 positive real roots and 2 or 0 negative real roots. From the rational root theorem, the possible roots are as follows: $\pm1, \pm2, \pm3, \pm4, \pm6, \pm8, \pm9, \pm12, \pm18, \pm24, \pm36, \pm48, \pm72, \pm144$.

Using synthetic division and trying 4 and then 6:

$$\begin{array}{r|rrrrr} 4 & 1 & -5 & -20 & 60 & 144 \\ & & 4 & -4 & -96 & -144 \\ \hline & 1 & -1 & -24 & -36 & 0 \end{array} \qquad \begin{array}{r|rrrr} 6 & 1 & -1 & -24 & -36 \\ & & 6 & 30 & 36 \\ \hline & 1 & 5 & 6 & 0 \end{array}$$

Factoring the quadratic, $x^2+5x+6=(x+2)(x+3)$. Setting the two factors equal to 0, you get $x=-2, x=-3$. So the four roots are $x=4, x=6, x=-2, x=-3$.

21 $f(x)=x^3-7x-6$. You get $(x-3)(x+1)(x+2)=x^3-7x-6$.

22 $f(x)=x^4+\frac{17}{3}x^3+7x^2-3x$. You get $(x+3)(x+3)\left(x-\frac{1}{4}\right)(x-0)=x^4+\frac{23}{4}x^3+\frac{15}{2}x^2-\frac{9}{4}x$.

23. **Refer to the following graph.** The intercepts are (–4,0), (–2,0), (1,0), (3,0), (0,24).

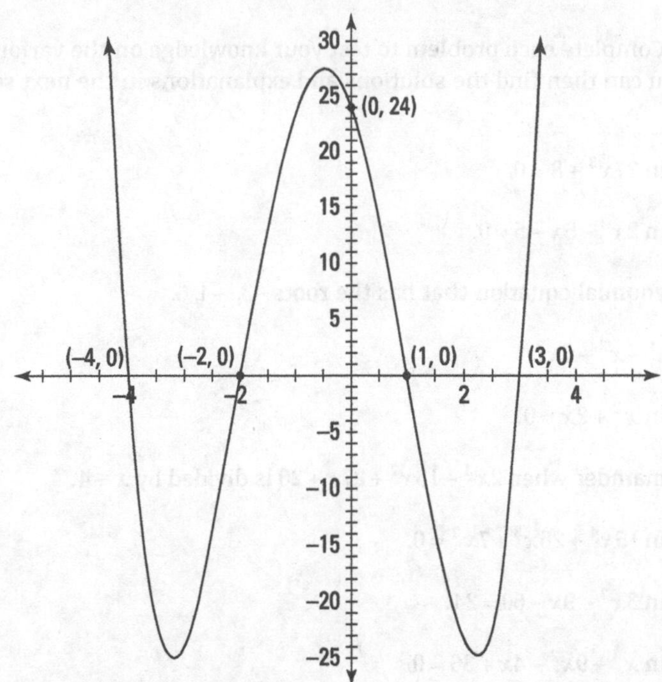

24. **Refer to the following graph.** The intercepts are (–3,0), (3,0), (0,–9).

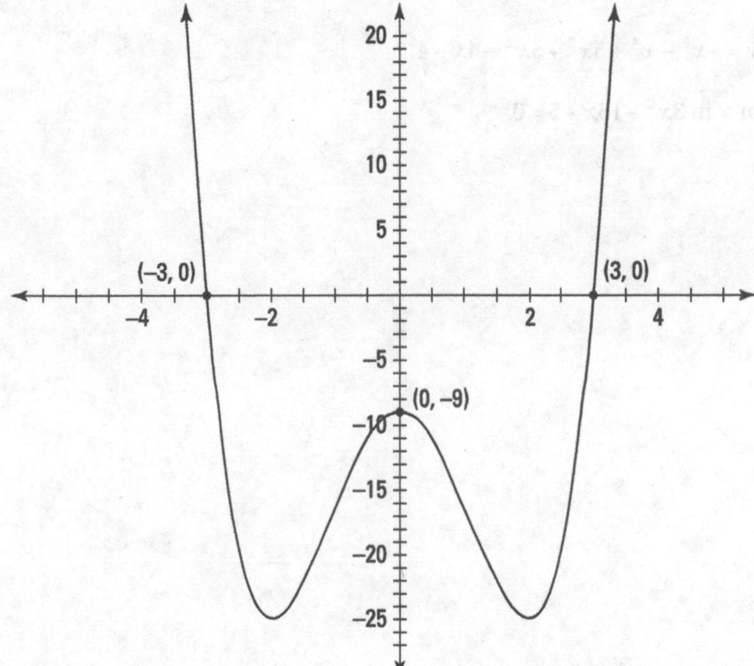

If you're ready to test your skills a bit more, take the following chapter quiz that incorporates all the chapter topics.

Whaddya Know? Chapter 5 Quiz

Quiz time! Complete each problem to test your knowledge on the various topics covered in this chapter. You can then find the solutions and explanations in the next section.

1. Solve for x in $27x^3+8=0$.

2. Solve for x in $2x^2-5x-5=0$.

3. Write a polynomial equation that has the roots –3, –1, 6.

4. Graph $y=x^4-x^3-6x^2$.

5. Solve for x in $x^2+2x=0$.

6. Find the remainder when $2x^3-13x^2+27x+20$ is divided by $x-4$.

7. Solve for x in $15x^4+26x^3+7x^2=0$.

8. Solve for x in $3x^2-9x-60=24$.

9. Solve for x in $x^3-9x^2-4x+36=0$.

10. Solve by completing the square: $x^2-12x-60=0$.

11. List all the possible real roots of $f(x)=5x^3+2x^2-10x-14$.

12. Graph $y=-x^5-x^4+5x^3+5x^2-4x-4$.

13. Solve for x in $3x^2-16x+5=0$.

Answers to Chapter 5 Quiz

1. $-\frac{2}{3}$**.** Factor the sum of cubes: $(3x+2)(9x^2-6x+4)=0$. Setting the first factor equal to 0 and solving for x, you get $x=-\frac{2}{3}$. The second factor has no real solution.

2. $\frac{5\pm\sqrt{65}}{4}$**.** Using the quadratic formula, $x=\frac{-(-5)\pm\sqrt{(-5)^2-4(2)(-5)}}{2(2)}=\frac{5\pm\sqrt{25+40}}{4}=\frac{5\pm\sqrt{65}}{4}$.

3. $f(x)=x^3-2x^2-21x-18$**.** Writing the roots in a factored polynomial form, $f(x)=(x+3)(x+1)(x-6)$. Multiplying, you get $f(x)=x^3-2x^2-21x-18$. This is not the only polynomial with these roots, but it's the simplest.

4. **Refer to the following figure.**

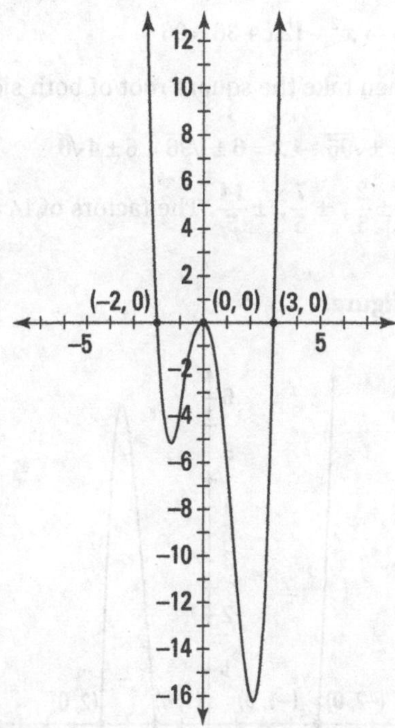

Factoring the equation, $y=x^4-x^3-6x^2=x^2(x^2-x-6)=x^2(x-3)(x+2)$. Setting the factors equal to 0 and solving for x gives you the x-intercepts (0,0), (3,0), (−2,0). The y-intercept is (0,0).

5. **0, −2.** Factor, set the factors equal to 0, and solve for x: $x^2+2x=x(x+2)=0$.

6. **48.** Using synthetic division, change the constant to its opposite:

4	2	−13	27	20
		8	−20	28
	2	−5	7	48

7) **$0, -\frac{7}{5}, -\frac{1}{3}$. Factor, then set the factors equal to 0 and solve for x:**

$15x^4 + 26x^3 + 7x^2 = x^2(15x^2 + 26x + 7) = x^2(5x+7)(3x+1) = 0.$

8) **$7, -4$. Subtract 24 from each side. Then factor the quadratic, set the factors equal to 0, and solve for x:** $3x^2 - 9x - 60 = 24 \rightarrow 3x^2 - 9x - 84 = 3(x^2 - 3x - 28) = 3(x-7)(x+4)$.

9) **$9, 2, -2$. Factor by grouping. Then set the factors equal to 0 and solve for x:**

$x^3 - 9x^2 - 4x + 36 = x^2(x-9) - 4(x-9) = (x-9)(x^2-4) = (x-9)(x-2)(x+2) = 0.$

10) **$6 \pm 4\sqrt{6}$. Add 60 to each side. Then square half of the coefficient on the x term and add it to each side.**

$$x^2 - 12x - 60 = 0 \rightarrow x^2 - 12x = 60 \rightarrow x^2 - 12x + \left(\frac{12}{2}\right)^2 = 60 + \left(\frac{12}{2}\right)^2$$

$$\rightarrow x^2 - 12x + 36 = 60 + 36 \rightarrow x^2 - 12x + 36 = 96$$

Factor the quadratic. Then take the square root of both sides and solve for x.

$$\rightarrow (x-6)^2 = 96 \rightarrow x - 6 = \pm\sqrt{96} \rightarrow x = 6 \pm \sqrt{96} = 6 \pm 4\sqrt{6}$$

11) **$\pm 1, \pm 2, \pm 7, \pm 14, \pm\frac{1}{5}, \pm\frac{2}{5}, \pm\frac{7}{5}, \pm\frac{14}{5}$. The factors of 14 are $\pm 1, \pm 2, \pm 7, \pm 14$, and the factors of 5 are $\pm 1, \pm 5$.**

12) **Refer to the following figure.**

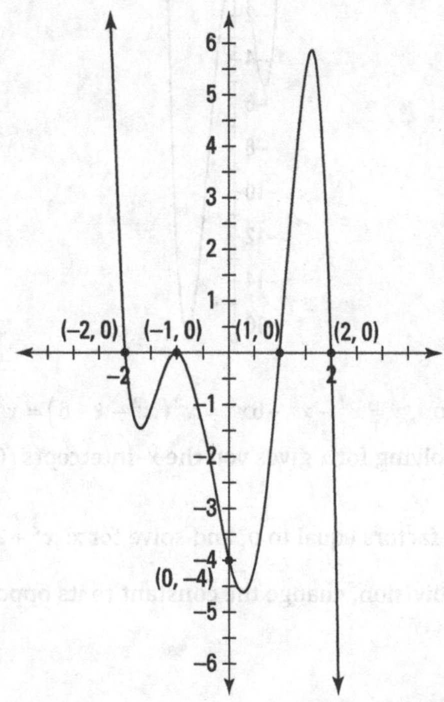

Factor the polynomial and set the factors equal to 0 to determine the x-intercepts. Factoring here works best when you start with grouping.

$$y=-x^5-x^4+5x^3+5x^2-4x-4=-x^4(x+1)+5x^2(x+1)-4(x+1)$$
$$=(x+1)(-x^4+5x^2-4)=-(x+1)(x^4-5x^2+4)=-(x+1)(x^2-4)(x^2-1)$$
$$=-(x+1)(x-2)(x+2)(x-1)(x+1)=-(x+1)^2(x-2)(x+2)(x-1)$$

The x-intercepts are $(-1,0)$, $(2,0)$, $(-2,0)$, $(1,0)$.

The y-intercept is $(0,-4)$.

13 $\mathbf{\frac{1}{3}, 5}$. Factor the quadratic, set the factors equal to 0, and solve for x.

$$3x^2-16x+5=(3x-1)(x-5)=0$$

Setting the two factors equal to 0, you have $3x-1=0$ giving you $x=\frac{1}{3}$ and $x-5=0$ giving you $x=5$.

IN THIS CHAPTER

» Simplifying, solving, and graphing exponential functions

» Checking all the ins and outs of logarithms

» Working through equations with exponents and logs

» Conquering a growth-and-decay example problem

Chapter 6 Exponential and Logarithmic Functions

If someone presented you with the choice of taking $1 million right now or taking one penny with the stipulation that the amount would double every day for 30 days, which would you choose? Most people would take the million without even thinking about it, and it would surely surprise them that the other plan is the better offer. Take a look: On the first day, you have only a penny, and you feel like you've been duped. But by the last day, you have $5,368,709.12! As you can see, doubling something (in this case, your money) makes it get big pretty fast. This idea is the basic concept behind an exponential function. Sound interesting?

In this chapter, you see two unique types of functions from pre-calculus: exponential and logarithmic. These functions can be graphed, solved, or simplified just like any other function discussed in this book. You see all the new rules you need in order to work with these functions; they may take some getting used to, but you'll find them broken down into the simplest terms.

That's great and all, you may be saying, but when will I ever use this complex stuff? (No one in their right mind would offer you the money, anyway.) Well, this chapter's info on exponential and logarithmic functions will come in handy when working with numbers that grow or shrink (usually with respect to time). Populations usually grow (get larger), whereas the monetary value of objects usually shrinks (gets smaller). You can describe these ideas with exponential functions. In the real world, you can also figure out compounded interest, carbon dating, inflation, and so much more!

Exploring Exponential Functions

REMEMBER

An *exponential function* is a function with a variable in the exponent. In math terms, you write $f(x) = b^x$, where b is the base and is a positive number. Because $1^x = 1$ for all x, for the purposes of this chapter, you assume b is not 1. If you've read Chapter 2, you know all about exponents and their place in math. So what's the difference between exponents and exponential functions? Prior to now, the variable was always the base — as in $g(x) = x^2$, for example. The exponent always stayed the same. In an exponential function, however, the variable is the exponent and the base stays the same — as in the function $f(x) = b^x$.

The concepts of exponential growth and exponential decay play an important role in biology. Bacteria and viruses especially love to grow exponentially. If one cell of a cold virus gets into your body and then the number of viruses doubles every hour, at the end of one day you'll have $2^{24} = 16{,}777{,}216$ of the little bugs moving around inside your body. So the next time you get a cold, just remember to thank (or curse) your old friend, the exponential function.

In this section, you dig deeper to uncover what an exponential function really is and how you can use one to describe the growth or decay of anything that gets bigger or smaller.

Searching the ins and outs of exponential functions

Exponential functions follow all the rules of functions, which are discussed in Chapter 3. But because they also make up their own unique family, they have their own subset of rules. The following list outlines some basic rules that apply to exponential functions:

- **The parent exponential function $f(x) = b^x$ always has a horizontal asymptote at** $y = 0$ **(except when $b = 1$).** You can't raise a positive number to any power and get 0 (it also will never become negative). For more on asymptotes, refer to Chapter 3.
- **The domain of any exponential function is $(-\infty, \infty)$.** This rule is true because you can raise a positive number to any power. However, the range of exponential functions reflects that all exponential functions have horizontal asymptotes. All parent exponential functions (except when $b = 1$) have ranges that consist of all real numbers greater than 0, or $(0, \infty)$.
- **The order of operations still governs how you act on the function.** When the idea of a vertical transformation (see Chapter 4) applies to an exponential function, most people take the order of operations and throw it out the window. Avoid this mistake. For example, $y = 2 \cdot 3^x$ doesn't become $y = 6^x$. You can't multiply before you deal with the exponent.
- **An exponential function can't have a base that's negative.** You may be asked to graph $y = -2^x$, which is fine. You read this as "the opposite of 2 to the x," which means that (remember the order of operations) you raise 2 to the power first and then multiply by –1. But you won't be graphing $y = (-2)^x$, because the base here is negative. The graph of $y = -2^x$ is a simple reflection that flips the graph upside down and changes its range to $(-\infty, 0)$.
- **Negative exponents take the reciprocal of the number to the positive power.** For instance, $y = 2^{-3}$ doesn't equal $(-2)^3$ or -2^3. Raising any number to a negative power takes the reciprocal of the number to the positive power: $2^{-3} = \frac{1}{2^3}$ or $\frac{1}{8}$.

» **When you multiply monomials with exponents, you add the exponents.** For instance, $2^{x+1} \cdot 2^{2x+5} = 2^{(x+1)+(2x+5)} = 2^{3x+6}$. And, of course, the bases have to be the same.

» **When you have multiple factors inside parentheses raised to a power, you raise every single term to that power.** For instance, $\left(2^x \cdot 3^{4y}\right)^7 = 2^{x(7)} \cdot 3^{4y(7)} = 2^{7x}3^{28y}$.

» **When graphing an exponential function, remember that base numbers greater than 1 always get bigger (or *rise*) as they move to the right; as they move to the left, they always approach 0 but never actually get there. If the base is between 0 and 1, then the graph falls as it moves to the right.** For example, $f(x) = 2^x$ is an exponential function, as is $g(x) = \left(\frac{1}{3}\right)^x$.

Table 6-1 shows the *x* and *y* values of two exponential functions. These parent functions illustrate that when a base is greater than 1, you have exponential growth, and when the base is between 0 and 1, you have exponential decay.

» **Exponential functions with base numbers that are fractions between 0 and 1 always fall from the left and approach 0 to the right.** This rule holds true until you start to transform the parent graphs, which you get to in the next section.

Table 6-1 The Values of *x* in Two Exponential Functions

x	$f(x) = 2^x$	$g(x) = \left(\frac{1}{3}\right)^x$
–3	$\frac{1}{8}$	27
–2	$\frac{1}{4}$	9
–1	$\frac{1}{2}$	3
0	1	1
1	2	$\frac{1}{3}$
2	4	$\frac{1}{9}$
3	8	$\frac{1}{27}$

YOUR TURN

 $\left(\frac{1}{4}\right)^{-1} =$ and $\left(\frac{1}{4}\right)^{-3} =$

 $\left(\frac{1}{4}\right)^{0} =$ and $\left(\frac{1}{4}\right)^{-7} \cdot \left(\frac{1}{4}\right)^{8} =$

Graphing and transforming exponential functions

Graphing an exponential function is helpful when you want to visually analyze the function. Doing so allows you to really see the growth or decay of what you're dealing with. The basic parent graph of any exponential function is $f(x)=b^x$, where b is the base. Figure 6-1a, for instance, shows the graph of $f(x)=2^x$, and Figure 6-1b shows $g(x)=\left(\frac{1}{3}\right)^x$. Using the x and y values from Table 6-1, you simply plot the coordinates to get the graphs.

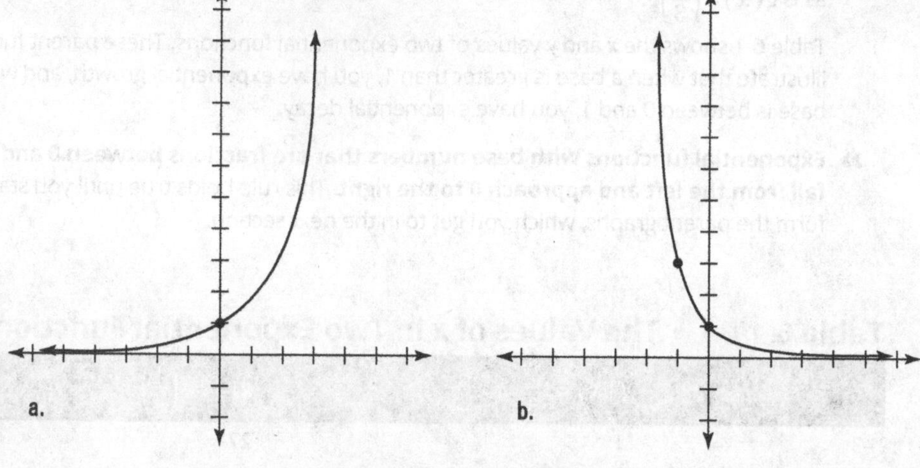

FIGURE 6-1: The graphs of the exponential functions $f(x)=2^x$ and $g(x)=\left(\frac{1}{3}\right)^x$.

REMEMBER

The parent graph of any exponential function crosses the y-axis at (0, 1), because anything raised to the 0 power is always 1. This intercept is referred to as the *key point* because it's shared among all exponential parent functions.

REMEMBER

Because an exponential function is simply a function, you can transform the parent graph of an exponential function in the same way as any other function (see Chapter 4 for the rules): $y=a\cdot b^{x-h}+k$, where a changes steepness, h is the horizontal shift, and k is the vertical shift.

For example, you can graph $y=2^{x+3}+1$ by transforming the parent graph of $f(x)=2^x$. Based on the general equation $y=a\cdot b^{x-h}+k$, y has been shifted 3 to the left and 1 up ($k=1$). Figure 6-2 shows each of these shifts as steps: the parent function $y=2^x$ appears as a heavy solid line, the lighter solid line illustrates the horizontal shift, and the dotted line is the vertical shift following the horizontal shift.

REMEMBER

Moving an exponential function up or down moves the horizontal asymptote. The function $y=2^{x+3}+1$ has a horizontal asymptote at $y=1$ (for more info on horizontal asymptotes, see Chapter 3). This change also shifts the range up 1 to $(1,\infty)$.

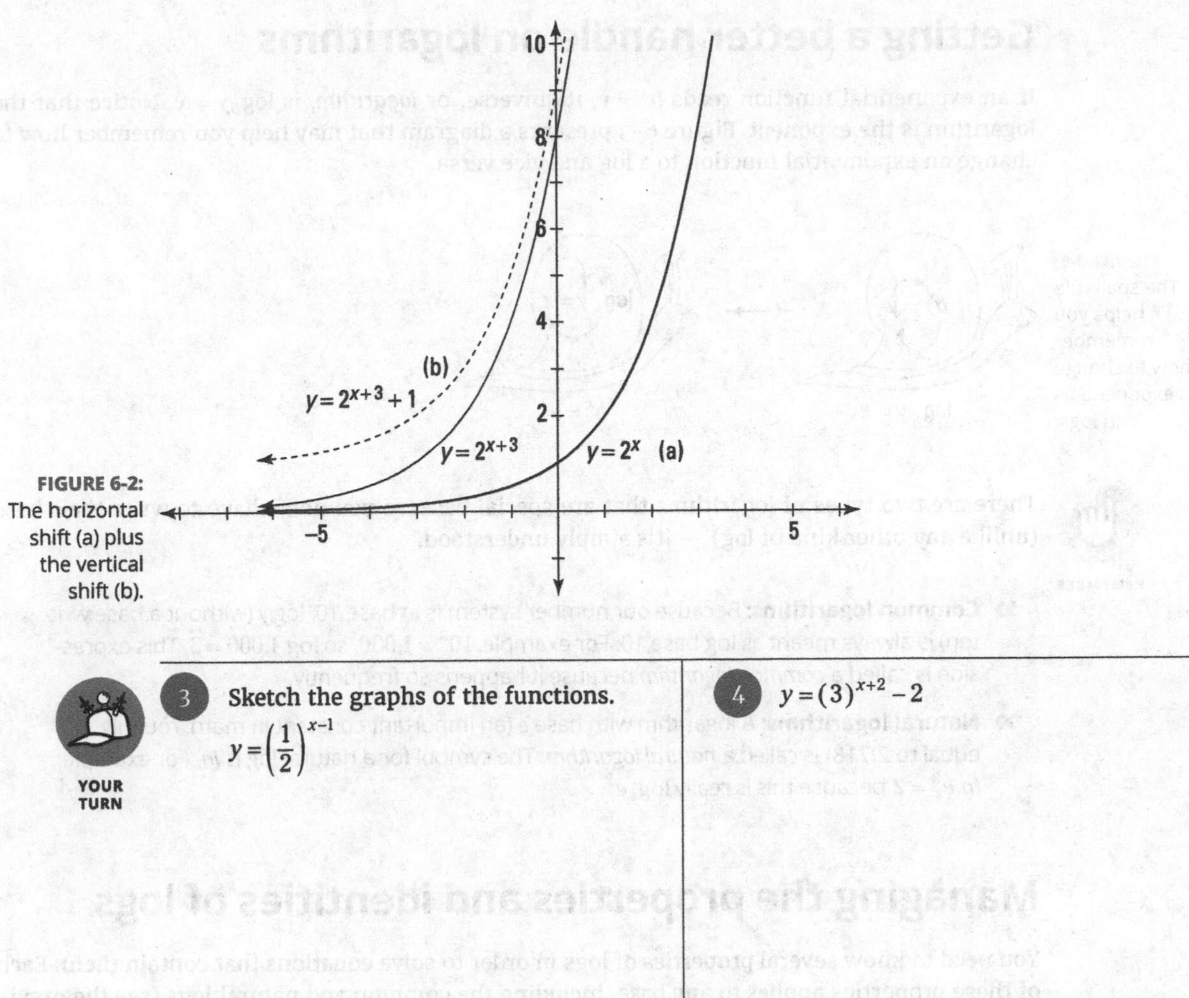

FIGURE 6-2: The horizontal shift (a) plus the vertical shift (b).

YOUR TURN

3 **Sketch the graphs of the functions.**

$$y = \left(\frac{1}{2}\right)^{x-1}$$

4 $y = (3)^{x+2} - 2$

Logarithms: The Inverse of Exponential Functions

Almost every function has an inverse. (Chapter 4 discusses what an inverse function is and how to find one.) But this question once stumped mathematicians: What could possibly be the inverse for an exponential function? They couldn't find one, so they invented one! They defined the inverse of an exponential function to be a *logarithmic* function (or *log* function).

A logarithm *is* an exponent, plain and simple. Recall, for example, that $4^2 = 16$; 4 is called the *base* and 2 is called the *exponent*. The corresponding logarithm is $\log_4 16 = 2$, where 2 is called the logarithm of 16 with base 4. In math, a logarithm is written $\log_b y = x$, where *b* is the base of the log, *y* is the number you're taking the log of, and *x* is the logarithm. So really, the logarithm and exponential forms are saying the same thing in different ways. The base, *b*, must always be a positive number (and not 1).

Getting a better handle on logarithms

If an exponential function reads $b^x = y$, its inverse, or *logarithm,* is $\log_b y = x$. Notice that the logarithm is the exponent. Figure 6-3 presents a diagram that may help you remember how to change an exponential function to a log and vice versa.

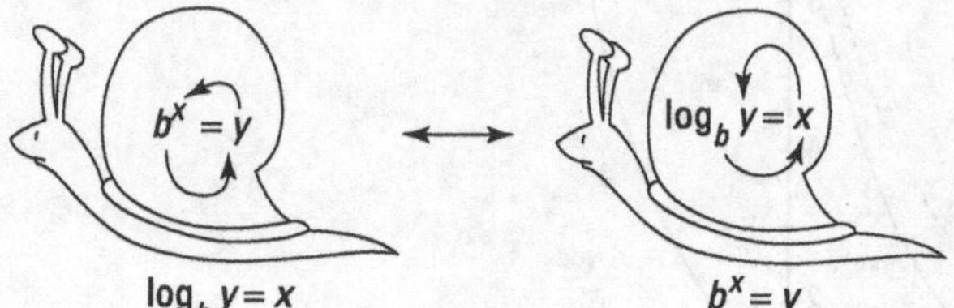

FIGURE 6-3: The snail rule helps you remember how to change exponentials and logs.

REMEMBER

There are two types of logarithms that are special because you don't have to write their base (unlike any other kind of log) — it's simply understood.

- **Common logarithms:** Because our number system is in base 10, $\log y$ (without a base written) is always meant as log base 10. For example, $10^3 = 1{,}000$, so $\log 1{,}000 = 3$. This expression is called a *common logarithm* because it happens so frequently.
- **Natural logarithms:** A logarithm with base *e* (an important constant in math, roughly equal to 2.718) is called a *natural logarithm.* The symbol for a natural log is *ln.* For example: $\ln e^2 = 2$ because this is really $\log_e e^2$.

Managing the properties and identities of logs

You need to know several properties of logs in order to solve equations that contain them. Each of these properties applies to any base, including the common and natural logs (see the previous section):

- $\log_b 1 = 0$

  If you change back to an exponential function, $b^0 = 1$ no matter what the base is. So, it makes sense that $\log_b 1 = 0$. Remember, though, that *b* must be a positive number.

- $\log_b b = 1$

  As an exponential function, $b^1 = b$.

- $\log_b x$ exists only when $x > 0$.

  The domain $(-\infty, \infty)$ and range $(0, \infty)$ of the original exponential parent function switch places in the inverse function. Therefore, any logarithmic function has the domain of $(0, \infty)$ and range of $(-\infty, \infty)$.

- $\log_b b^x = x$

  You can change this logarithmic property into an exponential property by using the snail rule: $b^x = b^x$. (Refer to Figure 6-3 for an illustration.) No matter what value you put in for *b*, this equation always works. So, $\log_b b = 1$ no matter what the base is (because it's really just $\log_b b^1$).

REMEMBER

The fact that you can use any base you want in this equation illustrates how this property works for common and natural logs: $\log 10^x = x$ and $\ln e^x = x$.

- $b^{\log_b x} = x$

  When the base is raised to a log of the same base, the number you're taking the log of is the answer. You can change this equation back to a log to confirm that it works: $\log_b x = \log_b x$.

- $\log_b x + \log_b y = \log_b (x \cdot y)$

  This is called the *product rule,* rewriting the sum of two logs as a product.

- $\log_b x - \log_b y = \log_b \left(\frac{x}{y}\right)$

  This is called the *quotient rule,* rewriting the difference of two logs as a quotient.

- $\log_b x^y = y \cdot \log_b x$

  This is called the *power rule,* rewriting a power as a product.

Q. **Rewrite as the log of a single number: $\log_4 10 + \log_4 2$.**

EXAMPLE **A.** **$\log_4 20$. Using the product rule, you can write the sum of two logs with the same base as the log of the product of the two logs: $\log_4 10 + \log_4 2 = \log_4 20$.**

Q. **Rewrite as the log of a single expression: $\log 4 - \log(x-3)$.**

EXAMPLE **A.** **$\log\left(\frac{4}{x-3}\right)$. Using the quotient rule, you can write the difference of two logs with the same base as the log of the quotient of the two logs: $\log 4 - \log(x-3) = \log\left(\frac{4}{x-3}\right)$.**

Q. **Rewrite as the log of a power: $4 \cdot \log_3 x$.**

EXAMPLE **A.** **$\log_3 x^4$. The multiplier of the log expression is the exponent of the log: $4 \cdot \log_3 x = \log_3 x^4$.**

Q. **Rewrite as a single number: $\log_6 6$.**

EXAMPLE **A.** **1. When the base and the log are the same number, the exponent of the base must be 1: $\log_6 6 = 1$.**

Q. **Rewrite as a single number: $\log_{10} 100 + \log_{30} 1$.**

EXAMPLE **A.** **2. First, write 100 as 10^2, and then move the 2 as a multiplier of the log.**

$\log_{10} 100 + \log_{30} 1 \rightarrow \log_{10} 10^2 + \log_{30} 1 \rightarrow 2\log_{10} 10 + \log_{30} 1$

Since $\log_{10} 10 = 1$ and $\log_{30} 1 = 0$, $2\log_{10} 10 + \log_{30} 1 \rightarrow 2(1) + 0 = 2$.

YOUR TURN

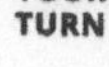

 5 Write as a single term: $\log_2 x + \log_2 y$.	 6 Write as a single term: $3\log_5 z - 4\log_5 w$.

WARNING

Keep the properties of logs straight so you don't get confused and make a critical mistake. The following list highlights many of the mistakes that people make when it comes to working with logs.

- **Misusing the product rule:** $\log_b x + \log_b y \neq \log_b(x+y)$; $\log_b x + \log_b y = \log_b(x \cdot y)$. You can't add two logs inside of one. Similarly, $\log_b x \cdot \log_b y \neq \log_b(x \cdot y)$.
- **Misusing the quotient rule:** $\log_b x - \log_b y \neq \log_b(x-y)$; $\log_b x - \log_b y = \log_b\left(\frac{x}{y}\right)$. Also, $\frac{\log_b x}{\log_b y} \neq \log_b\left(\frac{x}{y}\right)$.

  This error messes up the change-of-base formula (see the following section).
- **Misusing the power rule:** $\log_b(xy^p) \neq p\log_b(xy)$; the power is on the second variable only. If the formula was written as $\log_b(xy)^p$, it would equal $p\log_b(xy)$.

  Note: Watch what those exponents are doing. You should split up the multiplication from $\log_b(xy^p)$ first by using the product rule: $\log_b x + \log_b y^p$. Only then can you apply the power rule to get $\log_b(xy^p) = \log_b x + p\log_b y$.

Changing a log's base

Calculators usually come equipped with only common log and/or natural log buttons, so you must know what to do when a log has a base your calculator can't recognize, such as $\log_5 2$; the base is 5 in this case. In these situations, you must use the *change-of-base formula* to change the base to either base 10 or base *e* (the decision depends on your personal preference) in order to perform a calculation using the buttons that your calculator does have.

TECHNICAL STUFF

The following is the change-of-base formula:

$$\log_m n = \frac{\log_b n}{\log_b m},$$ **where *m* and *n* are real numbers.**

You can make the new base anything you want (5, 30, or even 3,000) by using the change-of-base formula, but remember that your goal is to be able to utilize your calculator by using either base 10 or base *e* to simplify the process.

Q. Use the change-of-base formula to find $\log_3 5$.

EXAMPLE

A. **1.465.** Choosing the common log in the change-of-base formula, you find that $\log_3 5 = \frac{\log 5}{\log 3} \approx 1.465$. However, if you're a fan of natural logs, you can go this route: $\log_3 5 = \frac{\ln 5}{\ln 3}$, which is still 1.465.

Calculating a number when you know its log: Inverse logs

If you know the logarithm of a number but need to find out what the original number actually was, you must use an *inverse logarithm*, which is also known as an *antilogarithm*. If $\log_b y = x$, y is the antilogarithm. An inverse logarithm undoes a log (makes it go away) so that you can solve certain log equations.

Q. Solve for x in $\log x = 0.699$ in exponential form.

EXAMPLE

A. **5.** Change it back to an exponential (take the inverse log) to solve it. First, write an equation with 10 as the base and the two sides of the equation as exponents: $10^{\log x} = 10^{0.699}$. Using the rule $b^{\log_b x} = x$, you get that the left side is equal to x, so $x = 10^{0.699}$. Using your calculator, you get $x = 5.00034535$ or x is about 5.

Q. Solve for x in $\ln x = 1.099$ in exponential form.

EXAMPLE

A. **3.** Write an equation with e as the base and the two sides of the equation as exponents. You have $e^{\ln x} = e^{1.099}$, and you get $x = e^{1.009}$, which, with your calculator, gives you $x = 3.00116336$ or x is about 3.

REMEMBER

The base you use in an antilogarithm depends on the base of the given log. For example, if you're asked to solve the equation $\log_5 x = 3$, you must use base 5 on both sides to get $5^{\log_5 x} = 5^3$, which simplifies to $x = 5^3$, or $x = 125$.

YOUR TURN

7 Solve for x: $\log x = 1.301$.

8 Solve for x: $\log_8 x = -3$.

Graphing logs

Want some good news, free of charge? Graphing logs is a snap! You can change any log into an exponential expression, so this step comes first. You then graph the exponential (or its inverse), remembering the rules for transforming (see Chapter 4), and then use the fact that exponentials and logs are inverses to get the graph of the log. The following sections explain these steps for both parent functions and transformed logs.

A parent function

Exponential functions each have a parent function that depends on the base; logarithmic functions also have parent functions for each different base. The parent function for any log is written $f(x) = \log_b x$. For example, $g(x) = \log_4 x$ is a different family than $h(x) = \log_8 x$ (although they are related).

EXAMPLE

Q. Graph the common log function: $f(x) = \log x$.

A. Use the following steps.

1. **Change the log to an exponential.**

   Because $f(x)$ and y represent the same thing mathematically, and because dealing with y is easier in this case, you can rewrite the equation as $y = \log x$ or $y = \log_{10} x$. The exponential equation of this log is $10^y = x$.

2. **Find the inverse function by switching x and y.**

   As you discover in Chapter 4, you write the inverse function as $y = 10^x$.

3. **Graph the inverse function.**

   Because you're now graphing an exponential function, you can plug and chug a few x values to find y values and get points. The graph of $y = 10^x$ gets really big, really fast. You can see its graph in Figure 6-4.

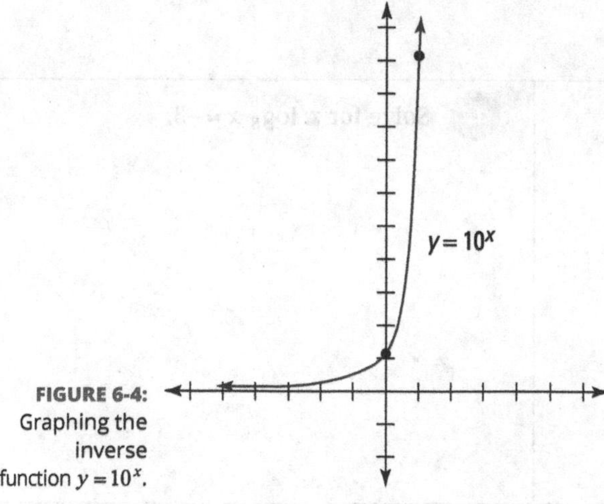

FIGURE 6-4: Graphing the inverse function $y = 10^x$.

4. **Reflect every point on the inverse function graph over the line $y = x$.**

   Figure 6-5 illustrates this last step, which yields the parent log's graph.

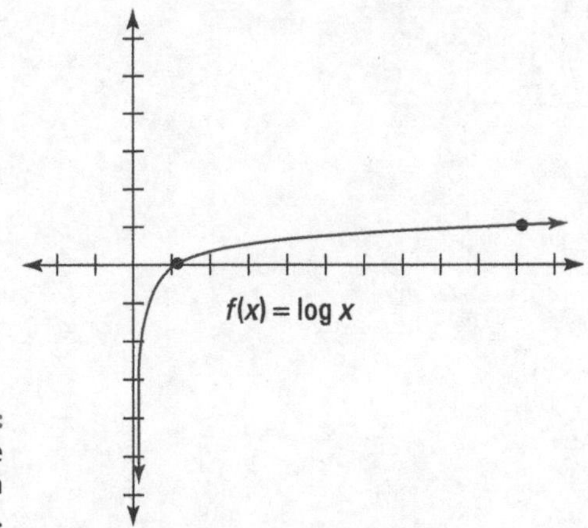

FIGURE 6-5: Graphing the logarithm $f(x) = \log x$.

A transformed log

The most general equation for logs can be written as $f(x) = a \cdot \log_b(x - h) + k$, where a is the vertical stretch or shrink, h is the horizontal shift, and k is the vertical shift.

So if you can find the graph of the parent function $y = \log_b x$, you can transform it. However, you'll find that many still prefer to change the log function to an exponential one and then graph.

EXAMPLE

Q. Graph the function $f(x) = \log_3(x - 1) + 2$.

A. Use the following steps:

1. **Get the logarithm by itself.**

   First, rewrite the equation as $y = \log_3(x - 1) + 2$. Then subtract 2 from both sides to get $y - 2 = \log_3(x - 1)$.

2. **Change the log to an exponential expression and find the inverse function.**

   If $y - 2 = \log_3(x - 1)$ is the logarithmic function, then $3^{y-2} = x - 1$ is the corresponding exponential; the inverse function is $3^{x-2} = y - 1$ because x and y switch places in the inverse.

3. **Solve for the variable *not* in the exponential of the inverse.**

   To solve for y in this case, add 1 to both sides to get $3^{x-2} + 1 = y$.

4. **Graph the exponential function.**

   The parent graph of $y = 3^x$ transforms right 2 $(x - 2)$ and up 1 $(+1)$, as shown in Figure 6-6. Its horizontal asymptote is at $y = 1$ (for more on graphing exponentials, refer to Chapter 3).

5. **Swap the domain and range values to get the inverse function.**

   Switch every x and y value in each point to get the graph of the inverse function. Figure 6-7 shows the graph of the logarithm.

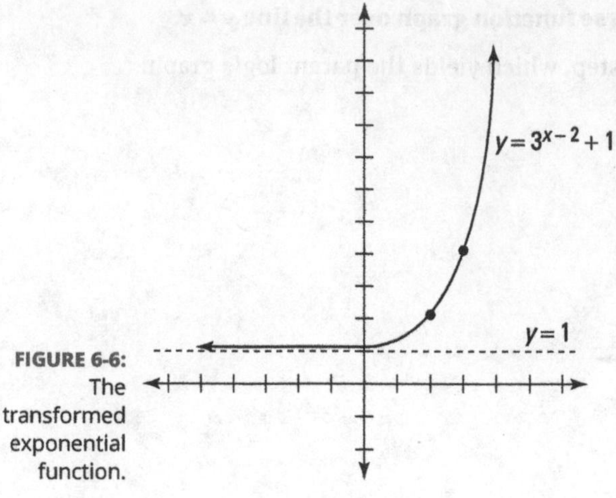

FIGURE 6-6: The transformed exponential function.

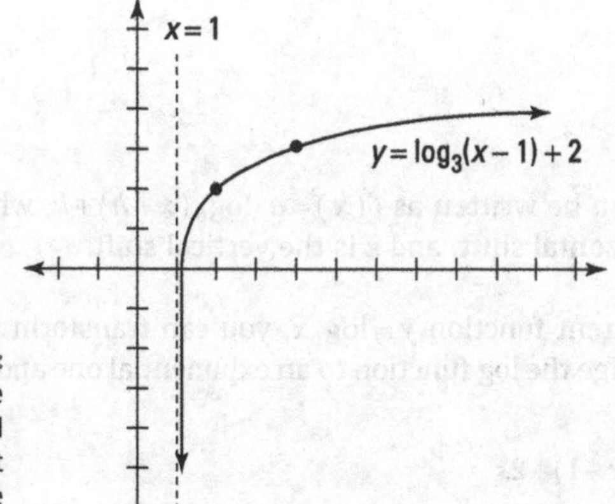

FIGURE 6-7: You change the domain and range to get the inverse function (log).

REMEMBER

Did you notice that the asymptote for the log changed as well? You now have a vertical asymptote at $x = 1$. The parent function for any log has a vertical asymptote at $x = 0$. The function $f(x) = \log_3(x-1) + 2$ is shifted to the right 1 and up 2 from its parent function $p(x) = \log_3 x$ (using transformation rules; see Chapter 4), so the vertical asymptote is now $x = 1$.

YOUR TURN

9 Graph the function $y = \log_2 x$.

10 Graph the function $f(x) = 3\log(x+2) - 4$.

Base Jumping to Simplify and Solve Equations

At some point, you will want to solve an equation with an exponent or a logarithm in it. Have no fear, *Pre-Calculus For Dummies* is here! You must remember one simple rule, and it's all about the base: If you can make the base on one side the same as the base on the other, you can use the properties of exponents or logs (see the corresponding sections earlier in this chapter) to simplify the equation. Now you have it made in the shade, because this simplification makes ultimately solving the problem a heck of a lot easier!

In the following sections, you discover how to solve exponential equations with the same base. You also find out how to deal with exponential equations that don't have the same base. And to round things out, you see the process of solving logarithmic equations.

Stepping through the process of solving exponential equations

The type of exponential equation you're asked to solve determines the steps you take to solve it. The following sections break down the types of equations you'll see, along with the steps you follow to solve them.

The basics: Solving an equation with a variable on one side

The basic type of exponential equation has a variable on only one side and can be written with the same base for each side.

EXAMPLE

Q. Solve for x: $4^{x-2} = 64$.

A. 5. Follow these steps:

1. **Rewrite both sides of the equation so that the bases match.**

   You know that $64 = 4^3$, so you can say $4^{x-2} = 4^3$.

2. **Drop the base on both sides and just look at the exponents.**

   When the bases are equal, the exponents have to be equal. This step gives you the equation $x - 2 = 3$.

3. **Solve the equation.**

   The equation $4^{x-2} = 64$ has the solution $x = 5$.

Getting fancy: Solving when variables appear on both sides

If you must solve an equation with variables on both sides, you still want to look for a common base.

Q. Solve for x: $2^{x-5} = 8^{x-3}$.

EXAMPLE **A.** 2. Follow these steps:

1. **Rewrite all exponential equations so that they have the same base.**

   The number 8 is a power of 2, so you can write $2^{x-5} = \left(2^3\right)^{x-3}$.

2. **Use the properties of exponents to simplify.**

   A power to a power signifies that you multiply the exponents. Distributing the right-hand exponent inside the parentheses, you get $3(x-3) = 3x-9$, so you have $2^{x-5} = 2^{3x-9}$.

3. **Drop the base on both sides.**

   The result is $x-5 = 3x-9$.

4. **Solve the equation.**

   Subtract x from both sides and add 9 to each side to get $4 = 2x$. Lastly, divide both sides by 2 to get $2 = x$. The equation $2^{x-5} = 8^{x-3}$ has the solution $x = 2$.

Solving when you can't simplify: Taking the log of both sides

Sometimes you just can't express both sides as powers of the same base. When facing that problem, you can make the exponent go away by taking the log of both sides.

Q. Solve for x: $4^{3x-1} = 11$.

EXAMPLE **A.** **About 0.91.** No integer with the power of 4 gives you 11, so you use the following technique:

1. **Take the log of both sides.**

   You can take any log you want, but remember that you actually need to solve the equation with this log, so it is suggested that you stick with common or natural logs only (see the section, "Getting a better handle on logarithms," earlier in this chapter for more info).

   Using the common log on both sides gives you $\log 4^{3x-1} = \log 11$.

2. **Use the power rule to drop down the exponent.**

   This step gives you $(3x-1)\log 4 = \log 11$.

3. **Divide the log away to isolate the variable.**

   You get $3x-1 = \frac{\log 11}{\log 4}$.

4. **Solve for the variable.**

   First, find the value of the logs to get $3x-1 \approx \frac{1.04139}{0.60206} \approx 1.72972$. Adding 1 to both sides of the equation and dividing by 3, you get that $x \approx 0.90991$ or about 0.91.

In the previous problem, you have to use the power rule on only one side of the equation because the variable appeared on only one side. When you have to use the power rule on both sides, the equations can get a little messy. But with persistence, you can figure it out.

EXAMPLE

Q. Solve for x: $5^{2-x}=3^{3x+1}$.

A. **About 0.4322.** Follow these steps:

1. **Take the log of both sides.**

   As with the previous problem, it's suggested that you use either a common log or a natural log. This time, using a natural log, you get $\ln 5^{2-x}=\ln 3^{3x+1}$.

2. **Use the power rule to drop down both exponents.**

   Don't forget to include your parentheses! You get $(2-x)\ln 5=(3x+1)\ln 3$.

3. **Distribute the lns over the inside of the parentheses.**

   This step gives you $2\ln 5-x\ln 5=3x\ln 3+\ln 3$.

4. **Isolate the variables on one side and move everything else to the other by adding or subtracting.**

   You now have $2\ln 5-\ln 3=3x\ln 3+x\ln 5$.

5. **Factor out the *x* variable from the two terms on the left.**

   That leaves you with $2\ln 5-\ln 3=x(3\ln 3+\ln 5)$.

6. **Divide both sides by the quantity in parentheses and solve for *x*:** $x=\dfrac{2\ln 5-\ln 3}{3\ln 3+\ln 5}$.

   Before entering the logarithms into your calculator, use the laws of logarithms to make two of the terms simpler: $x=\dfrac{2\ln 5-\ln 3}{3\ln 3+\ln 5}=\dfrac{\ln 5^2-\ln 3}{\ln 3^3+\ln 5}=\dfrac{\ln 25-\ln 3}{\ln 27+\ln 5}\approx 0.4322$.

YOUR TURN

 Solve for x: $5^{3x-1}=25$

12 $9^{x+2}=27^{3x-1}$

 $2^{x+7}=9$

 $4^{2x-3}=5^{3x-2}$

Solving logarithmic equations

Before solving equations with logs in them, you need to know the following four types of log equations.

- **Type 1:** The variable you need to solve for is inside the log operation, with one log on one side of the equation and a constant on the other.
- **Type 2:** The variable you need to solve for is the base.
- **Type 3:** The variable you need to solve for is inside the log, but the equation has more than one log and a constant.
- **Type 4:** The variable you need to solve for is inside the log, and all the terms in the equation involve logs.

EXAMPLE

Q. Solve for *x*: $\log_3 x = -4$.

A. $\frac{1}{81}$. The variable you need to solve for is inside the log operation, with one log on one side of the equation and a constant on the other. Rewrite as an exponential equation and solve for *x*: $3^{-4} = x$ or $x = \frac{1}{3^4} = \frac{1}{81}$.

EXAMPLE

Q. Solve for *x*: $\log_x 16 = 2$.

A. 4. The variable you need to solve for is the base. Again, rewrite as an exponential equation and solve for *x*: $x^2 = 16$ or $x = \pm 4$.

REMEMBER

Because logs don't have negative bases, you throw the negative one out the window and say $x = 4$ only.

EXAMPLE

Q. Solve for *x*: $\log_2(x-1) + \log_2 3 = 5$.

A. $\frac{35}{3}$. The variable you need to solve for is inside the log, but the equation has more than one log and a constant. Using the rules found in the section, "Managing the properties and identities of logs," you can solve equations with more than one log. To solve $\log_2(x-1) + \log_2 3 = 5$, first combine the two logs into one log by using the product rule: $\log_2[(x-1)\cdot 3] = 5$.

Now, rewrite as an exponential equation and solve for *x*.

$$2^5 = 3(x-1) = 3x-3$$
$$32 = 3x-3$$

And, finally, $3x = 35$ or $x = \frac{35}{3}$.

Q. Solve for x: $\log_3(x-1)-\log_3(x+4)=\log_3 5$.

EXAMPLE

A. $x=-\frac{21}{4}$. The variable you need to solve for is inside the log, and all the terms in the equation involve logs.

If all the terms in a problem are logs, they have to have the same base in order for you to solve the equation algebraically. You can combine all the logs so that you have one log on the left and one log on the right, and then you can drop the log from both sides:

$$\log_3(x-1)-\log_3(x+4)=\log_3 5$$

$$\log_3\frac{x-1}{x+4}=\log_3 5.$$

You can drop the log base 3 from both sides to get $\frac{x-1}{x+4}=5$, which you can solve relatively easily.

$$x-1=5(x+4)$$

$$x-1=5x+20$$

$$-4x=21$$

$$x=-\frac{21}{4}$$

WARNING

The number inside a log can never be negative. Plugging this answer back into part of the original equation gives you $\log_3\left(-\frac{21}{4}-1\right)-\log_3\left(-\frac{21}{4}+4\right)=\log_3 5$. The solution to this equation, therefore, is actually the empty set: no solution.

TIP

Always plug your answer to a logarithm equation back into the equation to make sure you get a positive number inside the log (not 0 or a negative number). Sometimes, even though x is negative, it's still a solution. Just be sure to check.

YOUR TURN

15 Solve for x: $\log_5 x=2$

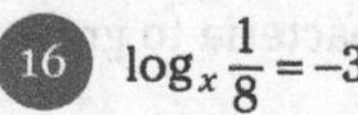

16 $\log_x\frac{1}{8}=-3$

17 $\log(x+3)-\log x=2$

18 $\log_7(3x+1)+\log_7 x=\log_7 2$

Growing Exponentially: Word Problems in the Kitchen

You can use exponential equations in many real-world applications: to predict populations in people or bacteria, to estimate financial values, and even to solve mysteries! And here's an example of a situation involving an exponential word problem.

Exponential word problems come in many different varieties, but they all follow one simple formula: $B(t) = Pe^{rt}$, where

P stands for the initial value of the function — usually referred to as the number of objects whenever $t = 0$.

t is the time (measured in many different units, so be careful!).

B(*t*) is the value of how many people, bacteria, dollars, and so on you have after time *t*.

r is a constant that describes the rate at which the population is changing. If *r* is positive, it's called the growth constant; if *r* is negative, it's called the decay constant.

e is the base of the natural logarithm, used for continuous growth or decay.

When solving word problems, remember that if the object grows continuously, then the base of the exponential function can be *e*.

Take a look at the following word problem, which this formula enables you to solve:

Q. Exponential growth exists in your kitchen on a daily basis in the form of bacteria. Suppose that you leave your leftover breakfast on the kitchen counter when leaving for work. Assume that 5 bacteria are present on the breakfast at 8:00 a.m., and 50 bacteria are present at 10:00 a.m. Use $B(t) = Pe^{rt}$ to find out how long it will take for the population of bacteria to grow to 1 million if the growth is continuous.

A. **About 10.602 hours.** You need to solve two parts of this problem: First, you need to know the rate at which the bacteria are growing, and then you can use that rate to find the time at which the population of bacteria will reach 1 million. Here are the steps for solving this word problem:

1. **Calculate the time that elapsed between the initial reading and the reading at time *t*.**

   Two hours elapsed between 8:00 a.m. and 10:00 a.m.

2. **Identify the population at time *t*, the initial population, and the time, and plug these values into the formula: $50 = 5 \cdot e^{r \cdot 2}$.**

3. **Divide both sides by the initial population to isolate the exponential: $10 = e^{2r}$.**

4. **Take the appropriate logarithm of both sides, depending on the base.**

   In the case of continuous growth, the base is always *e*: $\ln 10 = \ln e^{2r}$.

5. **Using the power rule (see the section, "Managing the properties and identities of logs"), simplify the equation: $\ln\ 10 = \ln e^{2r} \rightarrow \ln 10 = 2r$.**

6. **Divide by the time to find the rate; use your calculator to find the decimal approximation.**

   You find that $r = \frac{\ln 10}{2} \approx 1.1513$. This rate means that the population is growing by more than 115 percent per hour.

7. **Plug r back into the original equation and leave t as the variable: $B(t) = 5e^{1.1513t}$.**

8. **Plug the final amount in B(t) and solve for t, leaving the initial population the same: $1{,}000{,}000 = 5e^{1.1513t}$.**

9. **Divide by the initial population to isolate the exponential: $200{,}000 = e^{1.1513t}$.**

10. **Take the log (or ln) of both sides: $\ln 200{,}000 = \ln e^{1.1513t}$, giving you $\ln 200{,}000 = 1.1513t$.**

11. **Solve for t and use your calculator to do the computations: $t = \frac{\ln 200{,}000}{1.1513} \approx 10.602$ hours.**

    Phew, that was quite a workout! One million bacteria in a little more than ten hours is a good reason to refrigerate promptly.

YOUR TURN

19 You invest $10,000 in an account that has an interest rate of 1.5% compounded quarterly. Find the amount in the account after 8 years using the compound interest formula $A = P\left(1+\frac{r}{n}\right)^{nt}$, where A is the total amount after t years, r is the rate, n is the number of times each year the money is compounded, and P is the principal or starting amount.

20 Ammonia water has a pH of 11.6. What is the hydrogen ion concentration (in moles/liter)? Use the pH formula $pH = -\log\left(H^+\right)$, where H^+ is the hydrogen ion concentration in moles/liter.

Practice Questions Answers and Explanations

1. **4 and 64.** The –1 exponent says to flip the fraction, so $\left(\frac{1}{4}\right)^{-1} = \frac{4}{1} = 4$, and the –3 exponent says to flip the fraction and raise the result to the third power, so $\left(\frac{1}{4}\right)^{-3} = \left(\frac{4}{1}\right)^{3} = 64$.

2. **1 and $\frac{1}{4}$.** The zero exponent gives you 1, so $\left(\frac{1}{4}\right)^{0} = 1$. Multiplying numbers with the same base, you add the exponents: $\left(\frac{1}{4}\right)^{-7+8} = \left(\frac{1}{4}\right)^{1} = \frac{1}{4}$.

3. **Refer to the following graph.** The parent graph, $y = b^x$ moves 1 unit to the right.

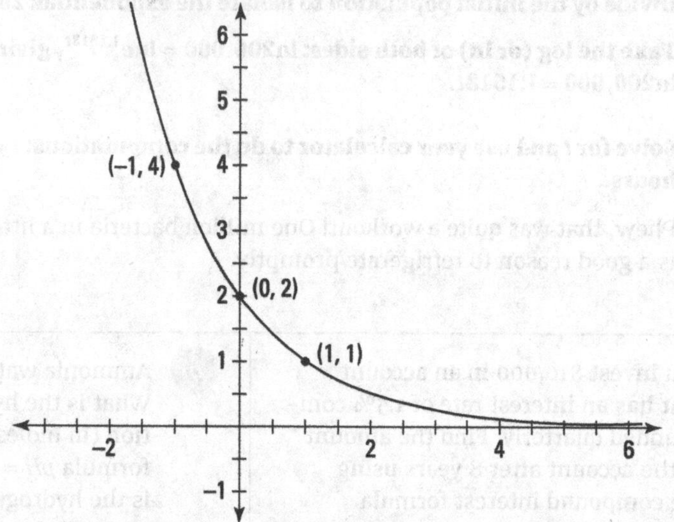

4. **Refer to the following graph.** The parent graph, $y = b^x$ moves 2 units to the left and 2 units down. The horizontal asymptote is $y = -2$.

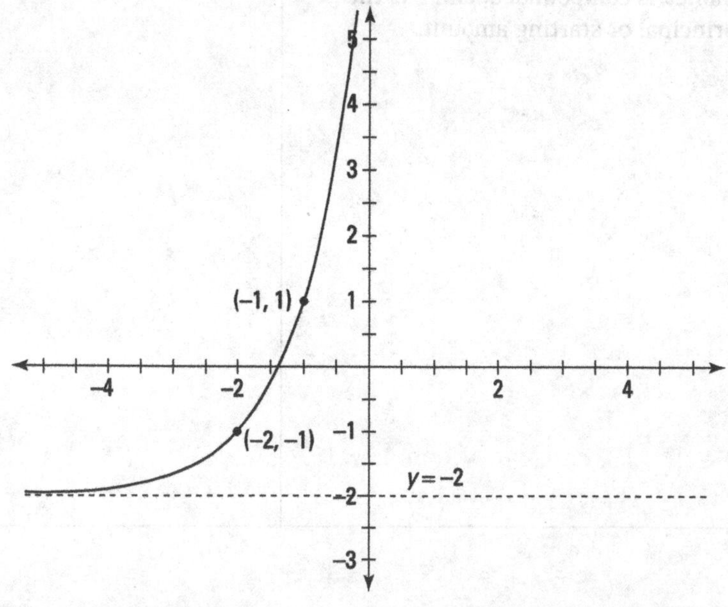

5. $\log_2 xy$. Apply the rule for the log of a product: $\log_2 x + \log_2 y = \log_2 xy$.

6. $\log_5 \frac{z^3}{w^4}$. First, apply the rule for the log involving an exponent and then the rule for the log of a quotient: $3\log_5 z - 4\log_5 w = \log_5 z^3 - \log_5 w^4 = \log_5 \frac{z^3}{w^4}$.

7. **About 19.999.** Rewrite in the exponential form: $\log x = 1.301 \rightarrow 10^{1.301} = x$. Using your calculator, determine the value correct to three decimal places: $10^{1.301} \approx 19.999$.

8. $\frac{1}{512}$. Rewrite in the exponential form: $\log_8 x = -3 \rightarrow 8^{-3} = \frac{1}{8^3} = \frac{1}{512} = x$.

9. **Refer to the following graph.** Sketch the graph of $y = 2^x$ and reflect it over the line $y = x$: $y = \log_2 x$.

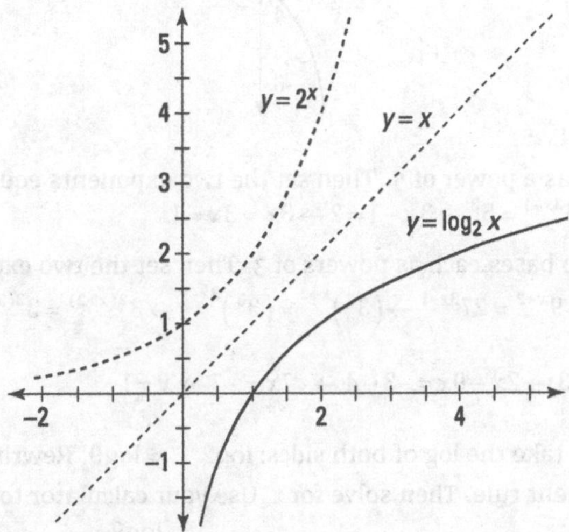

10. **Refer to the following graph.** First, rewrite the log function in its related exponential form. To do this, first add 4 to each side, then divide each side by 3, and then change the log form to the exponential form.

$$y = 3\log(x+2) - 4 \rightarrow y + 4 = 3\log(x+2) \rightarrow \frac{y+4}{3} = \log(x+2)$$

Reverse the x and y before writing as a power of 10. Then solve for y by subtracting 2 from each side.

$$\frac{x+4}{3} = \log(y+2) \rightarrow 10^{\left(\frac{x+4}{3}\right)} = y + 2 \rightarrow y = 10^{\left(\frac{x+4}{3}\right)} - 2$$

Graph the exponential $y = 10^{\left(\frac{x+4}{3}\right)} - 2$ and reflect it over the line $y = x$ to get the graph of the log function. The vertical asymptote is $x = -2$.

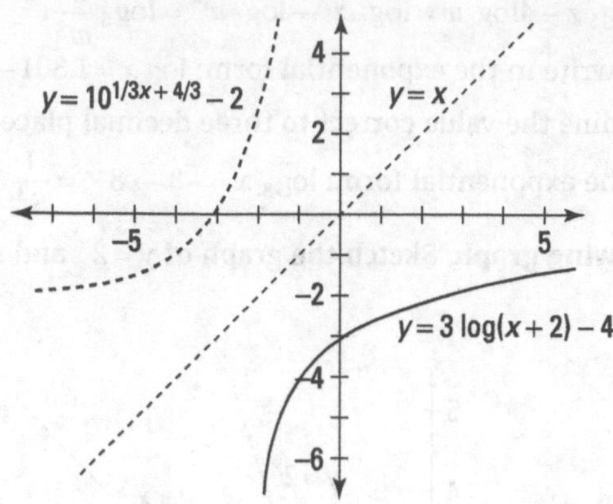

(11) **1.** Write the 25 as a power of 5. Then set the two exponents equal to one another and solve for x: $5^{3x-1} = 25 \rightarrow 5^{3x-1} = 5^2 \rightarrow 3x - 1 = 2 \rightarrow 3x = 3x = 1$.

(12) **1.** Write the two bases each as powers of 3. Then set the two exponents equal to one another and solve for x: $9^{x+2} = 27^{3x-1} \rightarrow \left(3^2\right)^{x+2} = \left(3^3\right)^{3x-1} \rightarrow 3^{2(x+2)} = 3^{3(3x-1)} \rightarrow 2(x+2) = 3(3x-1)$.

$\rightarrow 2x + 4 = 9x - 3 \rightarrow 2x - 9x = -3 - 4 \rightarrow -7x = -7 \rightarrow x = 1$

(13) **≈ –3.830.** First, take the log of both sides: $\log 2^{x+7} = \log 9$. Rewrite the left expression using the log of an exponent rule. Then solve for x. Use your calculator to estimate the value of the logs: $\log 2^{x+7} = \log 9 \rightarrow (x+7)\log 2 = \log 9 \rightarrow x + 7 = \frac{\log 9}{\log 2} \rightarrow x = -7 + \frac{\log 9}{\log 2}$.

(14) **≈ –0.457.** First, take the ln of both sides. Apply the rule for the log of a power and solve for x: $4^{2x-3} = 5^{3x-2} \rightarrow \ln 4^{2x-3} = \ln 5^{3x-2} \rightarrow (2x-3)\ln 4 = (3x-2)\ln 5 \rightarrow 2x\ln 4 - 3\ln 4 = 3x\ln 5 - 2\ln 5$.

$\rightarrow 2x\ln 4 - 3x\ln 5 = 3\ln 4 - 2\ln 5 \rightarrow x\left(\ln 4^2 - \ln 5^3\right) = \ln 4^3 - \ln 5^2$

$\rightarrow x = \frac{\ln 4^3 - \ln 5^2}{\ln 4^2 - \ln 5^3} \rightarrow x = \frac{\ln 64 - \ln 25}{\ln 16 - \ln 125} \approx -0.457$. The numerator and denominator in the last step could have been written as logs of quotients: $x = \frac{\ln 64 - \ln 25}{\ln 16 - \ln 125} = \frac{\ln \frac{64}{25}}{\ln \frac{16}{125}} = \frac{\ln 2.56}{\ln 0.128}$. This makes entering the values in the calculator easier. And, of course, the answers come out the same.

(15) **25.** Rewrite the log as an exponential equation: $\log_5 x = 2 \rightarrow 5^2 = x = 25$.

(16) **2.** Rewrite the log as an exponential equation: $\log_x \frac{1}{8} = -3 \rightarrow x^{-3} = \frac{1}{8}$. Now write the $\frac{1}{8}$ as a power of 2: $x^{-3} = \frac{1}{8} \rightarrow x^{-3} = 2^{-3}$. The bases are equal, so $x = 2$.

17 $\frac{1}{33}$. **Apply the rule for the log of a quotient. Then rewrite the log equation in its exponential form and solve for x:** $\log(x+3)-\log x=2 \rightarrow \log\frac{x+3}{x}=2 \rightarrow 10^2=\frac{x+3}{x}$.

$\rightarrow 100=\frac{x+3}{x} \rightarrow 100x=x+3 \rightarrow 99x=3 \rightarrow x=\frac{3}{99}=\frac{1}{33}$

18 $\frac{2}{3}$. **Apply the rule for the log of a product. The bases are the same, so set the logs equal to one another and solve for x:** $\log_7(3x+1)+\log_7 x=\log_7 2 \rightarrow \log_7(3x+1)x=\log_7 2 \rightarrow (3x+1)x=2.$

$\rightarrow 3x^2+x=2 \rightarrow 3x^2+x-2=0 \rightarrow (3x-2)(x+1)=0$

The two solutions are $x=\frac{2}{3}$ **and** -1**. The negative value can't be used, because the log has to be a positive number.**

19 **\$11,272.44. Using the formula,** $A=P\left(1+\frac{r}{n}\right)^{nt}=10{,}000\left(1+\frac{0.015}{4}\right)^{4\cdot 8}=10{,}000(1.00375)^{32}\approx$ $11{,}272.44$.

20 2.512×10^{-12}. **Replace the pH in the formula:** $11.6=-\log\left[H^+\right]$**. Then multiply each side of the equation by −1, rewrite the equation in exponential form, and then solve for** H^+**:** $11.6=-\log\left[H^+\right] \rightarrow -11.6=\log\left[H^+\right] \rightarrow 10^{-11.6}=H^+ \approx 2.512\times 10^{-12}$ **moles per liter.**

If you're ready to test your skills a bit more, take the following chapter quiz that incorporates all the chapter topics.

Whaddya Know? Chapter 6 Quiz

Quiz time! Complete each problem to test your knowledge on the various topics covered in this chapter. You can then find the solutions and explanations in the next section.

1. **Solve for x: $\log(x-1)+\log(x+2)=1$.**

2. **Solve for x: $3^x = 27$.**

3. **Simplify: $(40)^6 \cdot \left(\frac{1}{40}\right)^5$.**

4. **Graph: $y = 3 + 2^{x-4}$.**

5. **Solve for x: $4^{x-1} = 9$.**

6. **Write as a single term: $\log_3(x-4) + \log_3 6$.**

7. **Solve for x: $\log_5 x + \log_5(x-3) = \log_5 40$.**

8. **The magnitude of an earthquake is measured with $M = \log\frac{I}{S}$, where I is the intensity of the earthquake and S is the standard intensity. In San Francisco, an earthquake had a magnitude of 8.3. Shortly after that, there was an earthquake in the Pacific whose intensity was 4 times the one in San Francisco. What was the magnitude of that earthquake?**

9. **Solve for x: $3^x = 7^{2x+3}$.**

10. **Graph: $y = \log(x-5) + 2$.**

11. **Solve for x: $\log_4(x+1) = 3$.**

12. **Simplify: $\left(\frac{1}{3}\right)^{-4}(9)^2\left(\frac{2}{27}\right)$.**

13. **Write as a single term: $2\ln x - 5\ln(x-3)$.**

14. **Solve for x: $\log_x 3 = -1$.**

15. **Solve for x: $5^{x+1} = 125^x$.**

Answers to Chapter 6 Quiz

1. **3.** First, use the sum of logs rule, and then write the expression in exponential form: $\log(x-1)+\log(x+2)=1 \rightarrow \log(x-1)(x+2)=1 \rightarrow 10^1=(x-1)(x+2)$.

   Multiply the binomials and rewrite as a quadratic equation. Solve the equation for x.

   $10^1=(x-1)(x+2) \rightarrow 10=x^2+x-2 \rightarrow 0=x^2+x-12 \rightarrow 0=(x+4)(x-3)$

   The two solutions are $x=-4$ and $x=3$. The negative solution cannot be a solution of the log equation.

2. **3.** Write the 27 as a power of 3. With the bases the same, the exponents must be equal: $3^x=27 \rightarrow 3^x=3^3$ so $x=3$.

3. **40.** Rewrite the fraction using a negative exponent. Then multiply the two powers by adding the exponents: $(40)^6 \cdot \left(\frac{1}{40}\right)^5=(40)^6 \cdot (40)^{-5}=40^1$

4. **Refer to the following graph.** The parent graph $y=2^x$ is shifted 4 units to the right and 3 units up. The horizontal asymptote is the line $y=3$.

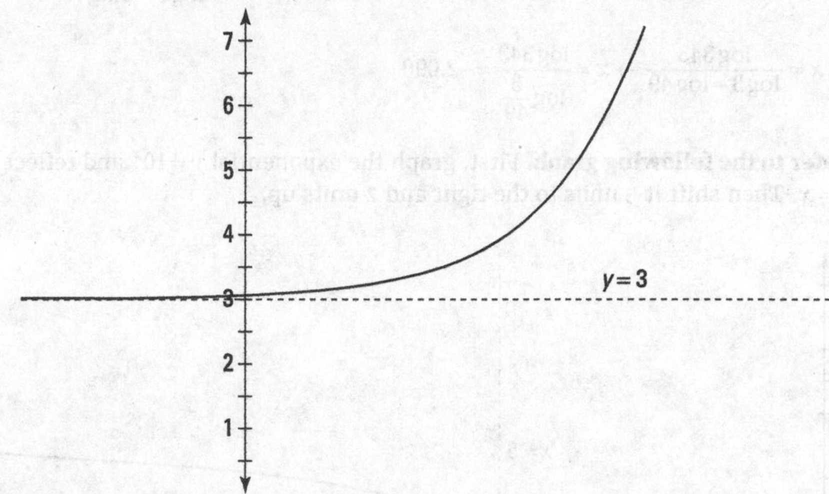

5. **About 2.585.** Find the ln of each side of the equation. (You can also use log.) Apply the rule for the log of an exponent and solve for x: $4^{x-1}=9 \rightarrow \ln 4^{x-1}=\ln 9 \rightarrow (x-1)\ln 4=\ln 9 \rightarrow x-1=\frac{\ln 9}{\ln 4} \rightarrow x=1+\frac{\ln 9}{\ln 4} \approx 2.585$.

6. **$\log_3 6(x-4)$.** Apply the rule for the sum of logs: $\log_3(x-4)+\log_3 6=\log_3 6(x-4)$.

7. **8.** First, apply the rule for the sum of logs. The bases are the same, so the logs are equal. Solve the resulting equation for x: $\log_5 x + \log_5(x-3) = \log_5 40 \rightarrow \log_5 x(x-3) = \log_5 40 \rightarrow x(x-3) = 40$.

$$x(x-3) = 40 \rightarrow x^2 - 3x - 40 = 0 \rightarrow (x-8)(x+5) = 0$$

The two solutions are $x = 8$ and $x = -5$, but only the positive number is a solution of the log equation.

8. **About 8.9.** The magnitude of the San Francisco earthquake was 8.3, so $8.3 = \log\frac{I}{S}$. The Pacific earthquake had an intensity 4 times that, so its magnitude was $M_P = \log\frac{4I}{S}$. Use the rules for logs and write $M_P = \log\frac{4I}{S} = \log 4 + \log\frac{I}{S}$. Replace the $\log\frac{I}{S}$, the magnitude of the San Francisco earthquake, with 8.3, and $M_P = \log 4 + 8.3 = 0.602 + 8.3 = 8.902$. The magnitude of the Pacific earthquake was about 8.9.

9. **About –2.090.** Take the log of each side of the equation. (You can also use ln.) Then apply the rule for the log of a power and solve for x: $3^x = 7^{2x+3} \rightarrow \log 3^x = \log 7^{2x+3} \rightarrow x\log 3 = (2x+3)\log 7 \rightarrow x\log 3 = 2x\log 7 + 3\log 7$.

$$\rightarrow x\log 3 - 2x\log 7 = 3\log 7 \rightarrow x(\log 3 - 2\log 7) = 3\log 7 \rightarrow x = \frac{3\log 7}{\log 3 - 2\log 7}$$

$$\rightarrow x = \frac{\log 343}{\log 3 - \log 49} \rightarrow x = \frac{\log 343}{\log\frac{3}{49}} \approx -2.090$$

10. **Refer to the following graph.** First, graph the exponential $y = 10^x$ and reflect it over the line $y = x$. Then shift it 5 units to the right and 2 units up.

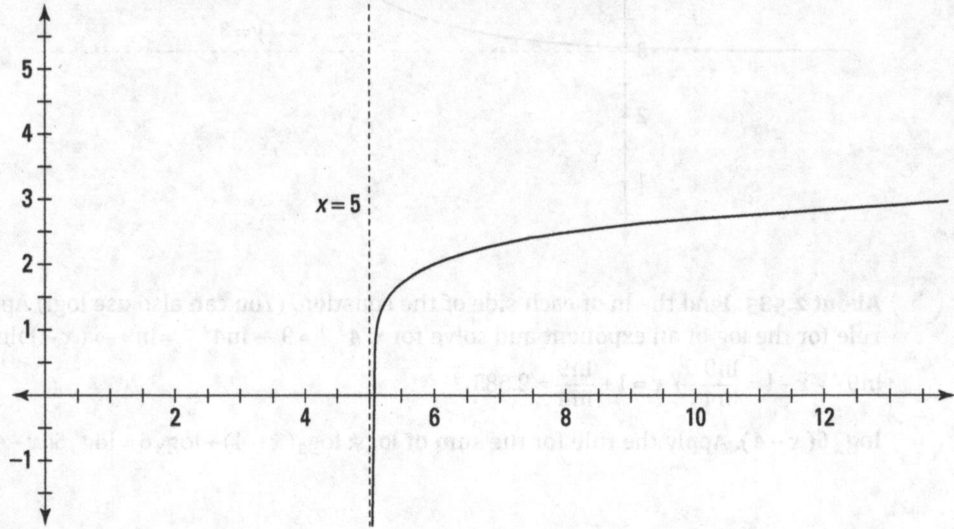

11 **63.** Rewrite the log function in its exponential form and solve for x.

$\log_4(x+1)=3 \rightarrow 4^3 = x+1 \rightarrow 64-1=x$

12 **486.** Write each factor using powers of 3: $\left(\frac{1}{3}\right)^{-4}(9)^2\left(\frac{2}{27}\right)=(3)^4(3^2)^2\left(\frac{2}{3^3}\right)=$ $(3)^4(3)^4 2\cdot(3)^{-3}=2\cdot 3^5=486.$

13 $\ln\frac{x^2}{(x-3)^5}$. Use the rules for powers and the difference of logs: $2\ln x-5\ln(x-3)=$ $\ln x^2-\ln(x-3)^5=\ln\frac{x^2}{(x-3)^5}.$

14 $\frac{1}{3}$. Write the equation in its exponential form and solve for x.

$\log_x 3=-1 \rightarrow x^{-1}=3 \rightarrow x^{-1}=\left(\frac{1}{3}\right)^{-1} \rightarrow x=\frac{1}{3}$

15 $\frac{1}{2}$. Write the 125 as a power of 5 and simplify. Set the two exponents equal to one another and solve for x.

$5^{x+1}=125^x \rightarrow 5^{x+1}=\left(5^3\right)^x \rightarrow 5^{x+1}=5^{3x} \rightarrow x+1=3x \rightarrow 1=2x \rightarrow x=\frac{1}{2}$

IN THIS CHAPTER

» Recognizing and graphing piece-wise functions

» Getting the greatest thrill out of greatest-integer functions

» Putting special functions to work

Chapter 7
Piece-Wise and Greatest-Integer Functions

Just like pieces of a puzzle, you can put function equations together, each sharing their own space, and each providing their own picture. Piece-wise functions are valuable for instances when the rules have to change depending on the place and the situation.

Greatest-integer functions do just that: find the greatest integer that fits the situation. But wait! There's also a partner of the greatest integer called the least integer. They all have their place in the picture. And remember, a *function* is a set of ordered pairs where every *x* value gives one and only one *y* value, so these specialty functions still follow the rules!

This chapter shows you how to work with the special functions and find their place in the world.

Looking at Functions with More Than One Rule: Piece-Wise Functions

Piece-wise functions are functions broken into pieces; they consist of more than one rule assigned to different parts of the domain. Each rule in a piece-wise function is defined only on a specific interval. Basically, the output depends on the input, and the graph of the function sometimes looks like it's literally been broken into pieces.

EXAMPLE

Q. Determine the domain and range of each function in the piece-wise function:

$$f(x)=\begin{cases} x^2 & \text{if} \quad x\le -2 \\ |x| & \text{if} \quad -2<x\le 3 \\ \sqrt{x} & \text{if} \quad x>3 \end{cases}$$

A. This function is broken into three pieces, depending on the domain values for each piece:

- The first piece is the quadratic function $f(x)=x^2$, and it exists only on the interval $(-\infty,-2]$. As long as the input for this function is less than or equal to –2, the output comes from inserting the input value into the quadratic equation. And the output, or range, in this case is $[4,\infty)$.
- The second piece is the absolute-value function $f(x)=|x|$, and it exists only on the interval $(-2,3]$. And the range in this case is $[0,3]$.
- The third piece is the square-root function $f(x)=\sqrt{x}$, and it exists only on the interval $(3,\infty)$. The range here is $\left[\sqrt{3},\infty\right)$.

You can see from the different range descriptions that this piece-wise function has no values lower than 0. Check out Figure 7-1 for a picture of what is happening.

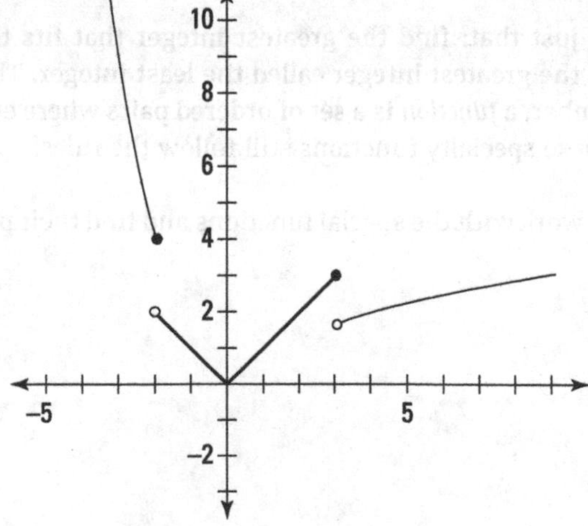

FIGURE 7-1: This piece-wise function is discontinuous.

YOUR TURN

Determine the domain and range of the functions.

1 $g(x)=\begin{cases}\sqrt{x+10} & \text{if } x\le -1\\ (x+10)^2 & \text{if } x>-1\end{cases}$	2 $h(x)=\begin{cases}x^2+2 & \text{if } x\le 0\\ 3x-1 & \text{if } x>0\end{cases}$

Graphing piece-wise functions

A *piece-wise function* is broken into two or more parts. In other words, it actually contains several functions, each of which is defined on a restricted interval. The output depends on what the input is. The graphs of these functions may look like they've been broken into pieces. Because of this broken quality, a piece-wise function that jumps is called *discontinuous*.

EXAMPLE

Q. Graph the piece-wise function $f(x)=\begin{cases}x^2 & \text{if} \quad x\le -2\\ |x| & \text{if} \quad -2<x\le 3\\ \sqrt{x} & \text{if} \quad x>3\end{cases}$

A. To graph this function, follow these steps:

1. **Lightly sketch out a quadratic function and darken just the values in the interval.**

   Because of the interval of the quadratic function of the first piece, you darken all points to the left of -2. And because $x=-2$ is included (the interval is $x\le -2$), the circle at $(-2,4)$ is filled in.

2. **Between -2 and 3, the graph moves to the second function, $|x|$ if $-2<x\le 3$; sketch the absolute-value graph, but pay attention only to the x values between -2 and 3.**

   You don't include -2 (use an open circle), but you include the 3 (closed circle).

3. **For x values bigger than 3, the graph follows the third function: $\sqrt{x}$ if $x>3$.**

   You sketch in this linear function, but only to the right of $x=3$ (that point is an open circle). The finished product is shown in Figure 7-1.

TECHNICAL STUFF

Notice that you can't draw the graph of this piece-wise function without lifting your pencil from the paper. Mathematically speaking, this is called a *discontinuous function*. There are more examples of discontinuities in Chapter 3, in the discussion of rational functions.

Absolute-value functions can also be described or defined as piece-wise functions. The absolute function $f(x)=|x|$ can be written $f(x)=\begin{cases} x & \text{if} \quad x\geq 0 \\ -x & \text{if} \quad x<0 \end{cases}$.

This just says that the input x is the same as its output, as long as x isn't negative. If x is a negative number, then the output is the opposite of x.

EXAMPLE

Q. Graph $f(x)=\begin{cases} x^2+2 & \text{if } x\leq 1 \\ 3x-1 & \text{if } x>-1 \end{cases}$.

A. Look at the following graph. This function has been broken into two pieces: When $x\leq -1$, the function follows the graph of the quadratic function, and when x > 1, the function follows the graph of the linear function. (Notice the hole in this second piece of the graph, which indicates that the point isn't actually there.)

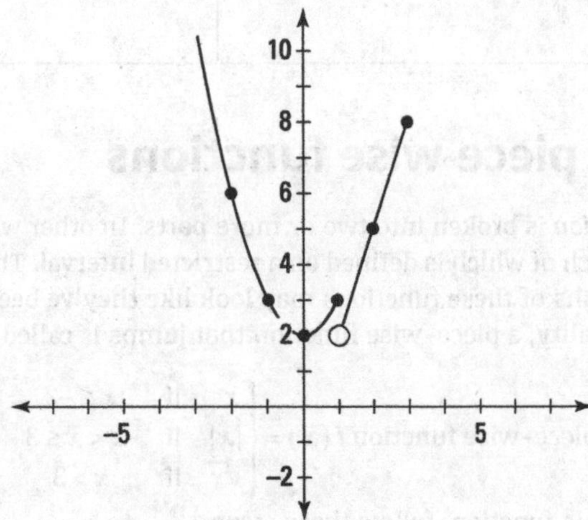

YOUR TURN

Graph the piece-wise functions.

3 $m(x)=\begin{cases} x^3+2 & \text{if} \quad x<0 \\ x^2+2 & \text{if} \quad 0\leq x<2 \\ x+2 & \text{if} \quad x\geq 2 \end{cases}$	**4** $n(x)=\begin{cases} \|x-1\| & \text{if} \quad x<-3 \\ 3 & \text{if } x=-3 \\ \|x\|-1 & \text{if} \quad x>-3 \end{cases}$

Using piece-wise functions in real-world applications

Piece-wise functions occur in all walks of life. Sunscreens have different SPF measures depending on the skin of the user, medicines have different dosages depending on the weight of the patient, taxes have different rates depending on the income of the worker, and so on.

EXAMPLE

Q. **You just got a summer job that pays $15 per hour, and you are required to work at least 30 hours each week. Any hours more than the 30 hours earn you twice that amount! Write a piece-wise function that determines your weekly pay when working *h* hours. How much would you make if you worked 40 hours in one week?**

A. When you work up to 30 hours, your pay is $\$15h$. Since you can't work a negative number of hours, the domain here is $[0,30]$. Working more than 30 hours, you multiply by $30, but you have to subtract that first 30 hours, so you multiply $30 times $(h-30)$ and then add on the amount you earned for the first 30 hours: $\$15\cdot 30=\$450$.

$$p(h)=\begin{cases}\$15h & \text{if } h\le 30\\ \$30(h-30)+\$450 & \text{if } h>30\end{cases}$$

When you work 40 hours, your pay is $p(40)=\$30(40-30)+\$450=\$30\cdot 10+\$450=\$750$.

YOUR TURN

5 **The velocity of a rocket can be described with the following piece-wise function, which gives you the velocity in miles per hour *t* seconds after the rocket is launched. What is the velocity 10 seconds after launch?**

$$v(t)=\begin{cases}10t^2+60 & \text{if } 0\le t<10\\ 1100-4t & \text{if } 10\le t<20\\ 53t+2(t-20)^2 & \text{if } 20\le t<40\\ 2920e^{-0.2(t-40)} & \text{if } t\ge 40\end{cases}$$

 Write a piece-wise function for the income tax paid in a certain country where there is no tax up to 15,000 euros, 12% tax over 15,000 euros and up to 35,000 euros, and 18% tax over 35,000 euros.

Grappling with the Greatest-Integer Function

You've probably had lots of experience with rounding numbers. You round the number 161 to the nearest hundred by raising it to 200. You round the same number 161 to the nearest ten by lowering it to 160. Now you get to use the *greatest-integer* function (sometimes referred to as the *step* function) with its special notation. And, as a bonus, you find out about the *ceiling* and the *floor*.

» *Greatest integer,* given x: $[x]$ = the largest integer less than or equal to x

» *Ceiling,* given x: $\lceil x \rceil$ = the smallest integer larger than or equal to x

» *Floor,* given x: $\lfloor x \rfloor$ = the largest integer less than or equal to x

You probably noticed that the definitions for *greatest integer* and *floor* are the same. You'll run across the symbols in different situations and just need to recognize which is which. The ceiling and floor operation symbols are really more descriptive of what is going on with the function.

EXAMPLE

Q. Evaluate the following function values.

a. $[4.8] =$

b. $\lceil 4.8 \rceil =$

c. $[-3.2] =$

d. $\lceil -3.2 \rceil =$

e. $\lfloor 6 \rfloor =$

A. Here are the results.

a. $[4.8] = 4$. The integer 4 is the largest integer less than 4.8.

b. $\lceil 4.8 \rceil = 5$. The integer 5 is the smallest integer greater than 4.8.

c. $[-3.2] = -4$. Think about the number line. The integer –4 is just to the left of –3.2, making it the largest integer smaller than that number.

d. $\lceil -3.2 \rceil = -3$. The integer –3 is greater than –3.2. Think in terms of temperatures.

e. $\lfloor 6 \rfloor = 6$. The number 6 is the *equal to* part of the definition of less than or equal to.

YOUR TURN

Evaluate the following function values.

7 $[-1.9] =$ and $\lceil -1.9 \rceil =$	 **8** $\lfloor 2.203 \rfloor =$ and $\lceil 2.203 \rceil =$

Several interesting properties are associated with the *greatest-integer* function:

- $[x] = x$, if x is an integer.
- $[x + n] = [x] + n$, if x is an integer.
- $[-x] = -x$, if x is an integer.
- $[-x] = -[x] - 1$, if x is **not** an integer.

Observing the greatest-integer function

This is where the alternate name *step* function for the *greatest-integer* function comes in handy. When you graph the greatest-integer function, you have something that looks like steps going upward. At the left end of each step, you have a solid dot for that point, and on the right end, you have a hollow dot. This function is definitely discontinuous — repeatedly! Just look at Figure 7-2.

Making use of the greatest-integer function

The greatest-integer function is used all over the world for all sorts of purposes. Money amounts are often subject to this function in utility and phone bills — and this would be with the *ceiling* function, of course. When paying interest on your savings, the *floor* function would be applied.

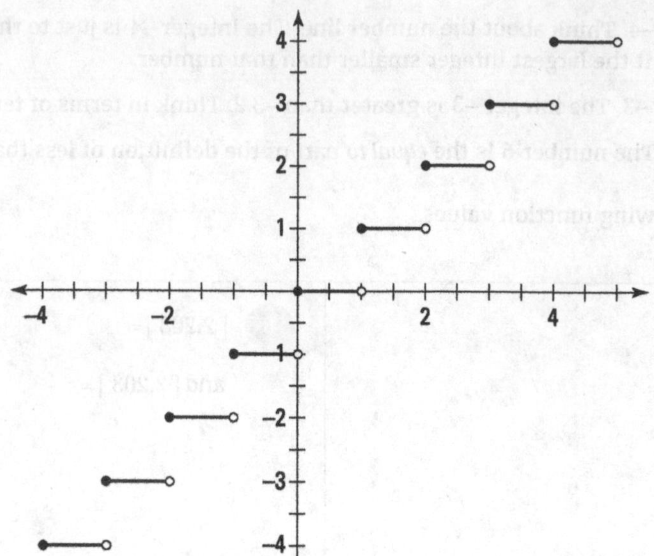

FIGURE 7-2: The greatest-integer (floor) function.

EXAMPLE

Q. **The cost to mail a particular parcel is computed with the function $P(x) = 0.60 + 0.15[x]$, where P is in cents, x is the weight in ounces (and $0 < x \leq 16$), and $[x]$ indicates the greatest-integer function. What is the cost of mailing something weighing 8.47 ounces?**

A. **\$1.80. Using the function: $P(8.47) = 0.60 + 0.15[8.47] = 0.60 + 0.15(8) = 0.60 + 1.20 = 1.80$. The greatest-integer function changes the 8.47 to the greatest integer lower than 8.47, which is 8.**

YOUR TURN

9 **A taxi company charges the following: $R(m) = 8 + 5\lceil m \rceil$, where R is in dollars, m is the number of miles traveled, and $\lceil m \rceil$ is the ceiling function. What is the charge for someone traveling 8.1 miles?**

10 **You've determined that it will take 9.5 cans of paint to decorate your house. If the paint costs \$20 per can (and you can't buy half-cans), what is this job going to cost you?**

Practice Questions Answers and Explanations

1. **Domain:** $[-10,\infty)$**; Range:** $[0,3]\cup(81,\infty)$**.** The first piece, $g(x)=\sqrt{x+10}$, has the domain $[-10,-1]$, and the second piece, $g(x)=(x+10)^2$, has a domain $(-1,\infty)$. The range of $g(x)=\sqrt{x+10}$ is $[0,3]$, and the range of $g(x)=(x+10)^2$ is $(81,\infty)$.

2. **Domain:** $(-\infty,\infty)$**; Range:** $(-1,\infty)$**.** The first piece, $h(x)=x^2+2$, has the domain $(-\infty,0]$, and the second piece, $h(x)=3x-1$, has a domain $(0,\infty)$. The range of $h(x)=x^2+2$ is $[2,\infty)$, and the range of $h(x)=3x-1$ is $(-1,\infty)$.

3. **See the following graph.**

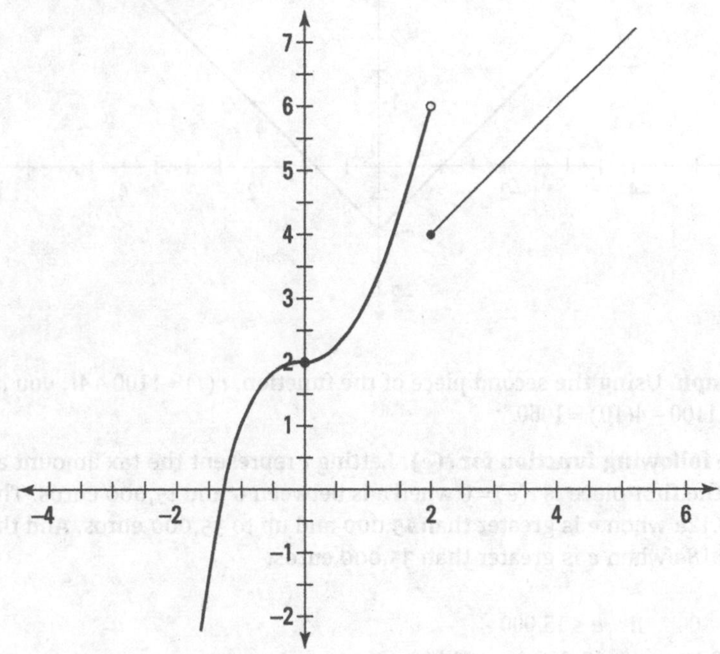

(4) **See the following graph.**

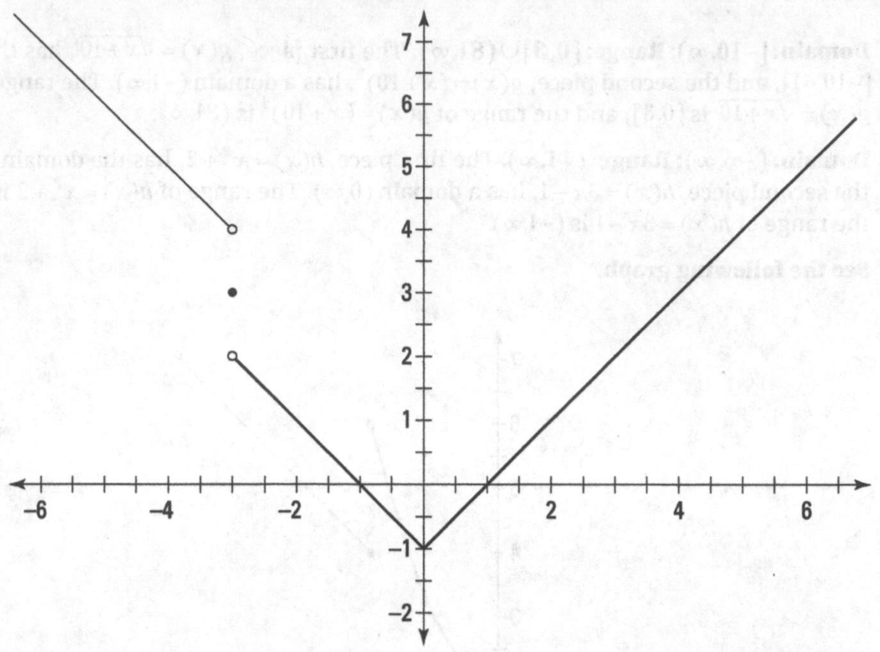

(5) **1,060 mph.** Using the second piece of the function, $v(t) = 1100 - 4t$, you have $v(10) = 1100 - 4(10) = 1060$.

(6) **See the following function for $t(e)$.** Letting t represent the tax amount and e the number of euros, the first piece is $t(e) = 0$ when e is between 0 and 15,000 euros. The second piece is $t(e) = 0.12e$ when e is greater than 15,000 and up to 35,000 euros. And the third piece is $t(e) = 0.18e$ when e is greater than 35,000 euros.

$$t(e) = \begin{cases} 0 & \text{if} \quad e \leq 15{,}000 \\ 0.12e & \text{if} \quad 15{,}000 < e \leq 35{,}000 \\ 0.18e & \text{if} \quad e > 35{,}000 \end{cases}$$

(7) **−2 and −1.** $[-1.9] = -2$, which is the largest integer smaller than −1.9, and $\lceil -1.9 \rceil = -1$, which is the smallest integer larger than −1.9.

(8) **2 and 3.** $\lfloor 2.203 \rfloor = 2$, which is the largest integer smaller than 2.203, and $\lceil 2.203 \rceil = 3$, which is the smallest integer larger than 2.203.

(9) **\$53.** Using the function, you get $R(8.1) = 8 + 5\lceil 8.1 \rceil = 8 + 5(9) = 8 + 45 = 53$. Using the ceiling function, you use the smallest integer larger than 8.1, which is 9.

(10) **\$200.** Using the ceiling function, the cost is $20\lceil 9.5 \rceil = 20(10) = 200$.

If you're ready to test your skills a bit more, take the following chapter quiz that incorporates all the chapter topics.

Whaddya Know? Chapter 7 Quiz

Quiz time! Complete each problem to test your knowledge on the various topics covered in this chapter. You can then find the solutions and explanations in the next section.

1 **Graph the function** $g(x)=\begin{cases} 2x+3 & \text{if } x<-2 \\ x^2 & \text{if } -2\le x<1 \\ |x|-1 & \text{if } x\ge 1 \end{cases}$.

2 **Find** $h(-3.4)$ **if** $h(x)=\begin{cases} [x] & \text{if } x<-2 \\ \lceil x\rceil & \text{if } x\ge -2 \end{cases}$.

3 $\lfloor -11.3\rfloor =$

4 **Find** $f(-3)$ **if** $f(x)=\begin{cases} x^2+6 & \text{if } x\le -3 \\ \sqrt{1-x} & \text{if } x>-3 \end{cases}$.

5 **Write a piece-wise function describing the following graph.**

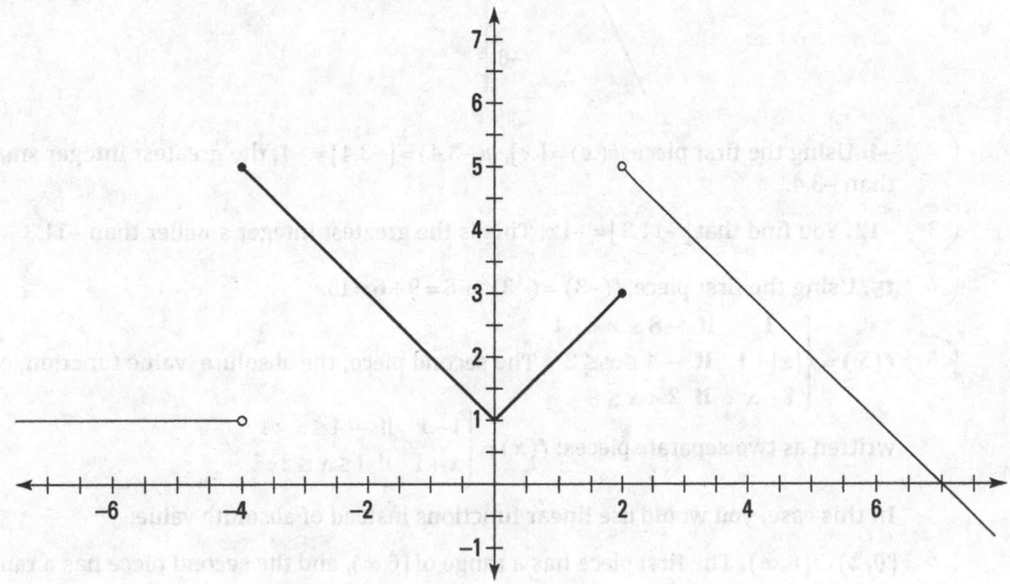

6 **Find the range of** $f(x)=\begin{cases} x^2+6 & \text{if } x\le -3 \\ \sqrt{1-x} & \text{if } x>-3 \end{cases}$.

7 $\lceil 17.4\rceil =$

8 **Find the domain of** $f(x)=\begin{cases} x^2+6 & \text{if } x\le -3 \\ \sqrt{1-x} & \text{if } x>-3 \end{cases}$.

Answers to Chapter 7 Quiz

1. **See the following figure.** The piece-wise function contains part of a line, part of a parabola, and part of an absolute value function.

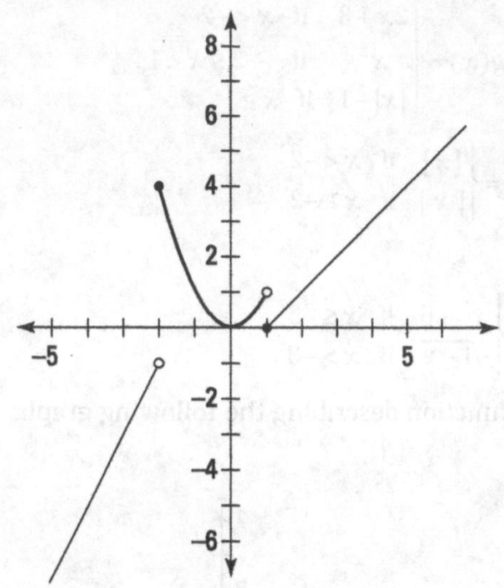

2. **–4.** Using the first piece, $h(x) = [x]$, $h(-3.4) = [-3.4] = -4$, the greatest integer smaller than –3.4.

3. **–12.** You find that $\lfloor -11.3 \rfloor = -12$. This is the greatest integer smaller than –11.3.

4. **15.** Using the first piece, $f(-3) = (-3)^2 + 6 = 9 + 6 = 15$.

5. $f(x) = \begin{cases} 1 & \text{if } -8 \le x < -4 \\ |x| + 1 & \text{if } -4 \le x \le 2 \\ 7 - x & \text{if } 2 < x \le 8 \end{cases}$. The second piece, the absolute-value function, could also be written as two separate pieces: $f(x) = \begin{cases} 1 - x & \text{if } -4 \le x < 1 \\ x + 1 & \text{if } 1 \le x \le 2 \end{cases}$.

   In this case, you would use linear functions instead of absolute value.

6. $[0,2) \cup [6,\infty)$. The first piece has a range of $[6,\infty)$, and the second piece has a range of $[0,2)$. So the range of the function is the union of these two ranges: $[0,2) \cup [6,\infty)$.

7. **18.** You find that $\lceil 17.4 \rceil = 18$. This is the smallest integer greater than 17.4.

8. $(-\infty,1]$. The domain of the first piece is $(-\infty,-3]$, and the domain of the second piece is $(-3,1]$. The union of these two domains is $(-\infty,1]$.

3

The Essentials of Trigonometry

IN THIS UNIT . . .

Reviewing angles and angle types, along with their properties.

Determining trig function values using degrees and radians.

Graphing basic trig functions.

Transforming trig function graphs.

IN THIS CHAPTER

» Discovering alternate trig function definitions

» Inserting triangles on the unit circle

» Calculating trig functions on the unit circle

Chapter 8

Circling In on Angles

In this chapter, you find right triangles drawn on the coordinate plane (x- and y-axes). Moving right triangles onto the coordinate plane introduces many more interesting concepts such as evaluating trig functions and solving trig equations. Also, you become acquainted with a very handy tool known as the unit circle.

The unit circle is extremely important in the real world and in mathematics; for instance, you're at its mercy whenever you fly in an airplane. Pilots use the unit circle, along with vectors, to fly airplanes in the correct direction and over the correct distance. Imagine the disaster that would result if a pilot tried to land a plane a bit to the left of the runway!

In this chapter, you work on building the unit circle as you review the basics of angles in radians and degrees as they're found in triangles. With that information, you can place the triangles onto the unit circle (which is also located in the coordinate plane) to solve the problems at the end of this chapter. (You find more on these ideas as you move into graphing trig functions in Chapter 10.)

Introducing Radians and Relating to Degrees

When you first studied geometry, you probably measured every angle in degrees, based on a portion of a 360° circle around a point. As it turns out, the number 360 was picked to represent the degrees in a circle only for convenience.

TECHNICAL STUFF

What's the convenience of the number 360, you ask? Well, you can divide a circle into many different, equal parts by using the number 360, because it's divisible by 2, 3, 4, 5, 6, 8, 9, 10, 12, 15, 18, 20, 24, 30, 36, 40, 45 . . . and these are just the numbers less than 50! Basically, the number 360 is pretty darn flexible for performing calculations.

The radian was introduced as an angle measure to make some of the computations easier and nicer. The word *radian* is based on the same root word as radius, which is the building block of a circle. An angle measurement of 360°, or a complete circle, is equal to 2π radians, which breaks down in the same way that degrees do.

In pre-calculus, you draw angles with their vertices at the origin of the coordinate plane $(0, 0)$, and you place one side on the positive x-axis (this side is called the *initial side* of the angle, and it is always in this location). The other side of the angle extends from the origin to anywhere on the coordinate plane (this side is the *terminal side*). An angle whose initial side lies on the positive x-axis is said to be in *standard position*.

REMEMBER

If you move from the initial side to the terminal side in a counterclockwise direction, the angle has a *positive measure*. If you move from the initial side to the terminal side in a clockwise direction, you say that this angle has a *negative measure*.

A positive/negative discussion of angles brings up another related and important point: co-terminal angles. *Co-terminal angles* are angles that have different measures, but their terminal sides lie in the same spot. These angles can be found by adding or subtracting 360° (or 2π radians) from an angle as many times as you want. Infinitely many co-terminal angles exist, which becomes quite handy in future chapters!

Q. Name four angles that are co-terminal with a 60° angle.

EXAMPLE

A. Co-terminal angles have the same initial side on the positive x-axis and the same terminal side. If you add 360° to 60°, you get an angle measuring 420°. Yes, you go round-in-circles, but that's fine! Add another 360° and you have 780°, and another is 1140°. For the last co-terminal angle, I'm going to subtract 360° from the 60° angle and get –300°. Refer to Figure 8-1 for a look at four of these angles.

YOUR TURN

1. Name three positive co-terminal angles of the angle 45°.

2. Name three negative co-terminal angles of the angle 90°.

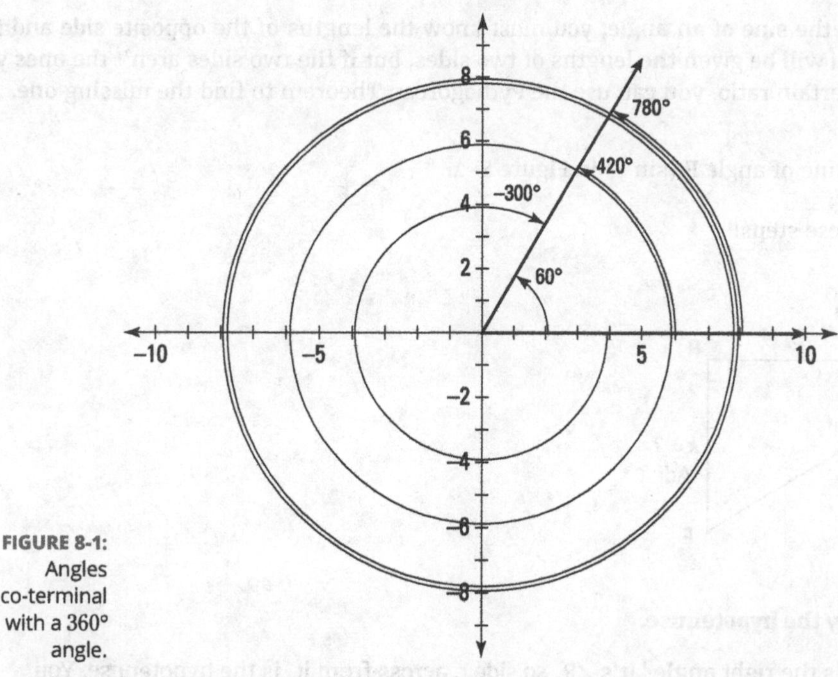

FIGURE 8-1: Angles co-terminal with a 360° angle.

Trig Ratios: Taking Right Triangles a Step Further

Reach back into your brain for a second and recall that a *ratio* is the comparison of two things. If a pre-calculus class has 20 guys and 14 gals, the ratio of guys to gals is $\frac{20}{14}$, and because that's a fraction, the ratio reduces to $\frac{10}{7}$. Ratios are important in many areas of life. For instance, if you have 20 people at a cookout with only 10 burgers, the ratio tells you that you have a problem!

Because trigonometric functions are so important in pre-calculus, you need to understand ratios. In this section, you see three very important ratios in right triangles — sine, cosine, and tangent — as well as three not-so-vital but still important ratios — cosecant, secant, and cotangent. These ratios are all *functions*, where an angle is the input, and a real number is the output. Each function looks at an angle of a right triangle, known or unknown, and then uses the definition of its specific ratio to help you find missing information in the triangle quickly and easily. To round out this section, I show you how to use inverse trig functions to solve for unknown angles in a right triangle.

Following the sine

In a right triangle, the *sine* of an acute angle named theta is defined as the ratio of the length of the opposite leg to the length of the hypotenuse. In symbols, you write $\sin\theta$. Here's what the ratio looks like: $\sin\theta = \frac{\text{opposite}}{\text{hypotenuse}}$.

In order to find the sine of an angle, you must know the lengths of the opposite side and the hypotenuse. You will be given the lengths of two sides, but if the two sides aren't the ones you need to find a certain ratio, you can use the Pythagorean Theorem to find the missing one.

EXAMPLE

Q. Find the sine of angle F (sin F) in Figure 8-2.

A. Follow these steps:

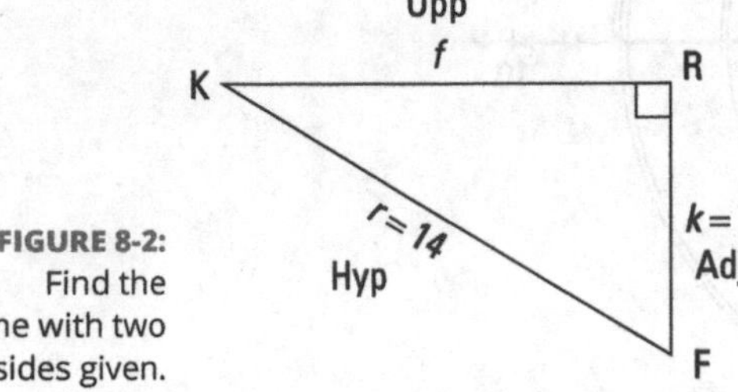

FIGURE 8-2: Find the sine with two sides given.

1. **Identify the hypotenuse.**

   Where's the right angle? It's ∠R, so side *r*, across from it, is the hypotenuse. You can label it "Hyp."

2. **Locate the opposite side.**

   Look at the angle in question, which is ∠F here. Which side is across from it? Side *f* is the opposite leg. You can label it "Opp."

3. **Label the adjacent side.**

   The only side that's left, side *k*, has to be the adjacent leg. You can label it "Adj."

4. **Locate the two sides that you use in the trig ratio.**

   Because you are finding the sine of ∠F, you need the opposite side and the hypotenuse. For this triangle, $(\text{leg})^2 + (\text{leg})^2 = (\text{hypotenuse})^2$ becomes $f^2 + k^2 = r^2$. Plug in what you know to get $f^2 + 7^2 = 14^2$. When you solve this for *f*, you get $f = 7\sqrt{3}$.

5. **Find the sine.**

   With the information from Step 4, you can find that $\sin F = \frac{\text{opposite}}{\text{hypotenuse}} = \frac{7\sqrt{3}}{14} = \frac{\sqrt{3}}{2}$.

Looking for a cosine

The *cosine* of an angle theta, or $\cos\theta$, is defined as the ratio of the length of the adjacent leg to the length of the hypotenuse, or $\cos\theta = \frac{\text{adjacent}}{\text{hypotenuse}}$. Finding a cosine is similar to finding a sine.

EXAMPLE

Q. A ladder leans against a building, creating an angle of 75° with the ground. The base of the ladder is 3 feet away from the building. How long is the ladder?

A. Did your heart just sink when you realized this is a . . . *word problem?* No problem! Just follow these steps to solve; here, you're looking for the length of the ladder:

1. **Draw a picture so you can see a familiar shape.**

   Figure 8-3 represents the ladder leaning against the building.

   The right angle is formed between the building and the ground, because otherwise the building would be crooked and fall down. Because you know where the right angle is, you know that the hypotenuse is the ladder itself. The given angle is down on the ground, which means the opposite leg is the distance on the building from where the ladder touches to the ground. The third side, the adjacent leg, is the distance the ladder rests from the building.

2. **Set up a trigonometry equation, using the information from the picture.**

   You know that the adjacent side is 3 feet, and you're looking for the length of the ladder, or the hypotenuse. Therefore, you have to use the cosine ratio, because it's the ratio of the adjacent leg to the hypotenuse. You have $\cos 75° = \frac{\text{adjacent}}{\text{hypotenuse}} = \frac{3}{x}$. The building has nothing to do with this problem right now, other than it's what's holding up the ladder.

   Why do you use 75° in the cosine function? Because you actually know how big the angle is; you don't have to use θ to represent an unknown angle.

3. **Solve for the unknown variable.**

   Multiply the unknown x to both sides of the equation to get $x\cos 75° = 3$. The $\cos 75°$ is just a number. When you plug it into your calculator, you get a decimal answer (make sure you set your calculator to degree mode before attempting to do this problem). Now divide both sides by $\cos 75°$ to isolate x; you get $x = \frac{3}{\cos 75°}$. Using the value of $\cos 75°$ from your calculator, you get $x = \frac{3}{\cos 75°} \approx \frac{3}{0.25882} \approx 11.5911$, which means the ladder is about 11.6 feet long.

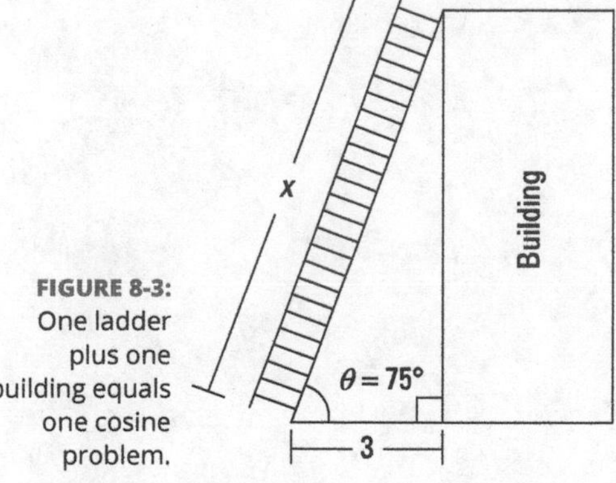

FIGURE 8-3: One ladder plus one building equals one cosine problem.

Going out on a tangent

The tangent of an angle theta, or $\tan\theta$, is the ratio of the opposite leg to the adjacent leg. Here's what it looks like in equation form: $\tan\theta = \frac{\text{opposite}}{\text{adjacent}}$. And here's an example of how you can use the tangent of an angle.

EXAMPLE

Q. **Imagine for a moment that you're an engineer. You're working with a 39-foot tower with a wire attached to the top of it. The wire needs to attach to the ground and make an angle of 80° with the ground to keep the tower from moving. How far is it from the base of the tower to where the wire should attach to the ground?**

A. Follow these steps:

1. **Draw a diagram that represents the given information.**

   Figure 8-4 shows the wire, the tower, and the known information.

2. **Set up a trigonometry equation, using the information from the picture.**

   For this problem, you must set up the trig equation that features tangent, because the opposite side is the length of the tower, the hypotenuse is the wire, and the adjacent side is what you need to find. You get $\tan 80° = \frac{\text{opposite}}{\text{adjacent}} = \frac{39}{x}$.

3. **Solve for the unknown.**

   Multiply both sides by the unknown x to get $x\tan 80° = 39$. Divide both sides by the $\tan 80°$ to get $x = \frac{39}{\tan 80°}$. Use your calculator and simplify to get $x \approx \frac{39}{5.6713} \approx 6.8768$. The wire attaches to the ground about 6.88 feet from the base of the tower to form the 80° angle.

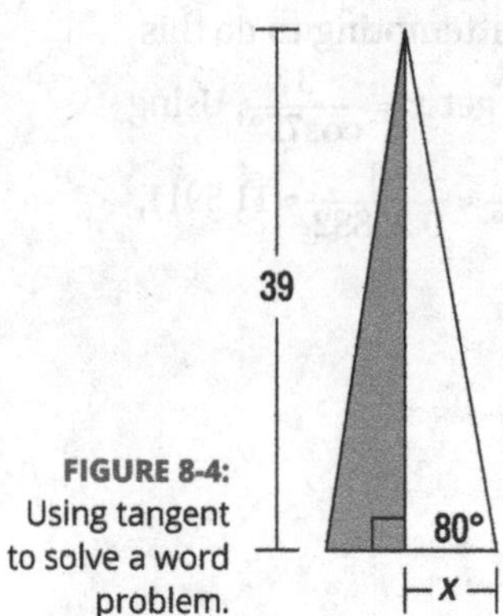

FIGURE 8-4: Using tangent to solve a word problem.

YOUR TURN

Use the right triangle ABC to solve the following.

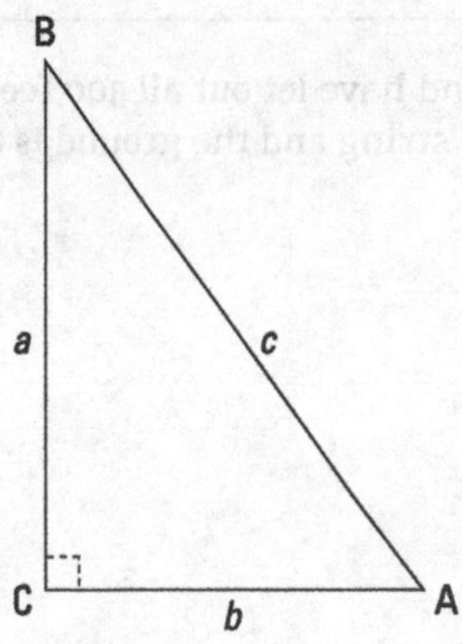

3 **In triangle ABC, if *b*, the side opposite angle B, measures 6 units and the hypotenuse measures 10 units, then what is $\tan B$?**

4 **In triangle ABC, if the side opposite angle A measures 24 units and *b*, the side adjacent to angle A, measures 7 units, then what is $\sin A$?**

5 **Instead of walking on the sidewalk from A to C (side *b*) and then from C to B (side *a*), you're going to cut across the empty lot from A to B (side *c*). Angle A measures 60° and it's 100 feet from A to C. How far is it from A to B?**

 6 You are flying a kite and have let out all 300 feet of string (the distance from A to B). If the angle between the string and the ground is 80°, then how high off the ground is your kite?

Discovering the flip side: Reciprocal trig functions

Three additional trig ratios — cosecant, secant, and cotangent — are called *reciprocal functions* because they're the reciprocals of sine, cosine, and tangent. These three functions open up three more ways in which you can solve equations in pre-calculus. The following list breaks down these functions and how you use them:

- ***Cosecant*, or $\csc\theta$, is the reciprocal of sine.** The reciprocal of a is $\frac{1}{a}$, so $\csc\theta = \frac{1}{\sin\theta}$. And, because $\sin\theta = \frac{\text{opposite}}{\text{hypotenuse}}$, $\csc\theta = \frac{\text{hypotenuse}}{\text{opposite}}$.
- ***Secant*, or $\sec\theta$, is the reciprocal of cosine.** It has a formula similar to cosecant: $\sec\theta = \frac{1}{\cos\theta} = \frac{\text{hypotenuse}}{\text{adjacent}}$.

WARNING

A common mistake is to think that secant is the reciprocal of sine and that cosecant is the reciprocal of cosine, but the previous bullets illustrate the truth.

- ***Cotangent*, or $\cot\theta$, is the reciprocal of tangent.** (How's that for obvious?) You should have the hang of this if you've looked at the previous bullets: $\cot\theta = \frac{1}{\tan\theta} = \frac{\text{adjacent}}{\text{opposite}}$.

  Two other important ratios involving the tangent and cotangent are $\tan\theta = \frac{\sin\theta}{\cos\theta}$ and $\cot\theta = \frac{\cos\theta}{\sin\theta}$.

REMEMBER

Secant, cosecant, and cotangent are all reciprocals, but you won't find a button for them on your calculator. You must use their reciprocals — sine, cosine, and tangent. Don't get confused and use the $\sin^{-1}$, $\cos^{-1}$, and $\tan^{-1}$ buttons, either. Those buttons are for inverse trig functions, which I describe in the following section. So, if you want the cosecant of 60°, using your calculator, first find the sine of 60° and then use the reciprocal button, x^{-1}.

Working in reverse: Inverse trig functions

Many functions have an inverse. An *inverse function* basically undoes a function. The trigonometry functions sine, cosine, and tangent all have inverses, and they're often called *arcsin*, *arccos*, and *arctan*.

In trig functions, the angle θ is the input, and the output is the number representing the ratio of the sides of a triangle. If you're given the ratio of the sides and need to find an angle, you use the inverse trig function.

- **Inverse sine (arcsin):** $\theta = \sin^{-1}\left(\frac{\text{opposite}}{\text{hypotenuse}}\right)$
- **Inverse cosine (arccos):** $\theta = \cos^{-1}\left(\frac{\text{adjacent}}{\text{hypotenuse}}\right)$
- **Inverse tangent (arctan):** $\theta = \tan^{-1}\left(\frac{\text{opposite}}{\text{adjacent}}\right)$

Defining ranges of inverse trig functions

Functions have specific domains and ranges. And this, of course, carries over to our inverse trig functions. When you input a value, you expect to get one angle measure. And this happens because each inverse trig function has a specific, defined range. Here are the functions and their ranges:

- $f(x) = \sin^{-1}(x), [-90°, 90°]$, which is quadrants I and IV.
- $f(x) = \cos^{-1}(x), [0°, 180°]$, which is quadrants I and II.
- $f(x) = \tan^{-1}(x), [-90°, 90°]$, which is quadrants I and IV.
- $f(x) = \cot^{-1}(x), [0°, 180°]$, which is quadrants I and II.
- $f(x) = \sec^{-1}(x), [0°, 90°) \cup (90°, 180°]$, which is quadrants I and II except for 90°.
- $f(x) = \csc^{-1}(x), [-90°, 0°) \cup (0°, 90°]$, which is quadrants I and IV except for 0°.

EXAMPLE

Q. Find the angle θ in degrees in a right triangle if you know that the $\tan\theta = 1.7$.

A. Follow these steps:

1. **Isolate the trig function on one side and move everything else to the other.**

   This step is done already. Tangent is on the left and the decimal 1.7 is on the right: $\tan\theta = 1.7$.

2. **Isolate the variable.**

   You're given the ratio for the trig function and have to find the angle. To work backward and figure out the angle, use some algebra. You have to undo the tangent function, which means using the inverse tangent function on both sides: $\tan^{-1}(\tan\theta) = \tan^{-1}(1.7)$. This equation simplifies to $\theta = \tan^{-1}(1.7)$. In words, this equation says that theta is the angle whose tangent is 1.7.

3. **Solve the simplified equation.**

   $\theta = \tan^{-1}(1.7)$ is solved with a calculator. When you do this, you get $\theta = 59.53445508$ or about 59.53°.

REMEMBER

Be sure to read the problem carefully so you know whether the angle you're looking for should be expressed in degrees or radians. Set your calculator to the correct mode.

YOUR TURN

 Find the angle (in degrees) whose sine is about 0.707.	 Find the angle (in degrees) whose cosine is –0.5.

Understanding How Trig Ratios Work on the Coordinate Plane

The unit circle that you find in this chapter lies on the coordinate plane — the same plane you've been graphing on since algebra. The *unit circle* is a very small circle centered at the origin (0,0). The radius of the unit circle is 1, which is why it's called the unit circle. To carry out the work of the rest of this chapter, all the angles specified will be drawn on the coordinate plane. The ratios from earlier can be found, again, in the unit circle.

To put angles on the coordinate plane, essentially all you do is look at the trig ratios in terms of *x* and *y* values rather than opposite, adjacent, and hypotenuse. Redefining these ratios to fit the coordinate plane (sometimes called the *point-in-the-plane* definition) makes visualizing the differences easier. Some of the angles, for instance, are larger than 180°, but using the new definitions allows you to create a right triangle by using a point and the *x*-axis. You then use the new ratios to find missing sides of right triangles and/or trig function values of angles.

When a point (x, y) exists on a coordinate plane, you can calculate all the trig functions for the angle corresponding to the point.

EXAMPLE

Q. Find the six trig functions corresponding to the point $(-4, -6)$ on the coordinate plane.

A. Follow these steps:

1. **Locate the point on the coordinate plane and connect it to the origin, using a straight line.**

   Plot the point $(-4,-6)$ and draw a line segment from the point to the origin. This line segment is your hypotenuse of a right triangle and its length is the radius *r* (see Figure 8-5).

2. **Draw a perpendicular line connecting the given point to the *x*-axis, creating a right triangle.**

   The lengths of the legs of the right triangle are −4 and −6. Don't let the negative signs scare you; the lengths of the sides are still 4 and 6. The negative signs just reveal the location of that point or directions on the coordinate plane.

3. **Find the length of the hypotenuse *r* by using the distance formula or the Pythagorean Theorem.**

   The distance you want to find is the length of *r* from Step 1. The distance formula between a point (*x*, *y*) and the origin (0,0) is $r = \sqrt{(x-0)^2 + (y-0)^2} = \sqrt{x^2 + y^2}$.

   REMEMBER

   This equation implies the principal or positive root only, so the hypotenuse for these point-in-the-plane triangles is always positive.

   For this example, you get $r = \sqrt{(-4)^2 + (-6)^2} = \sqrt{52}$, which simplifies to $2\sqrt{13}$. Check out what the triangle looks like in Figure 8-5.

4. **Determine the trig function values, using their alternate definitions.**

   With the labels from Figure 8-5, you get the following formulas:

   - $\sin\theta = \frac{y}{r}$ implies that $\csc\theta = \frac{r}{y}$.
   - $\cos\theta = \frac{x}{r}$ implies that $\sec\theta = \frac{r}{x}$.
   - $\tan\theta = \frac{y}{x}$ implies that $\cot\theta = \frac{x}{y}$.

   Substitute the numbers from Figure 8-5 to pinpoint the trig values:

   - $\sin\theta = \frac{-6}{2\sqrt{13}}$

   Simplified and rationalized:

   $\sin\theta = \frac{-6}{2\sqrt{13}} = -\frac{3}{\sqrt{13}} = -\frac{3\sqrt{13}}{13}$

   - $\cos\theta = \frac{-4}{2\sqrt{13}}$

   Simplified and rationalized:

   $\cos\theta = \frac{-4}{2\sqrt{13}} = -\frac{2}{\sqrt{13}} = -\frac{2\sqrt{13}}{13}$

   - $\tan\theta = \frac{-6}{-4} = \frac{3}{2}$
   - $\cot\theta = \frac{2}{3}$
   - $\sec\theta = -\frac{\sqrt{13}}{2}$
   - $\csc\theta = -\frac{\sqrt{13}}{3}$

TIP

Notice that the rules of trig functions and their reciprocals still apply. For example, if you know $\sin\theta$, you automatically know $\csc\theta$ because they're reciprocals.

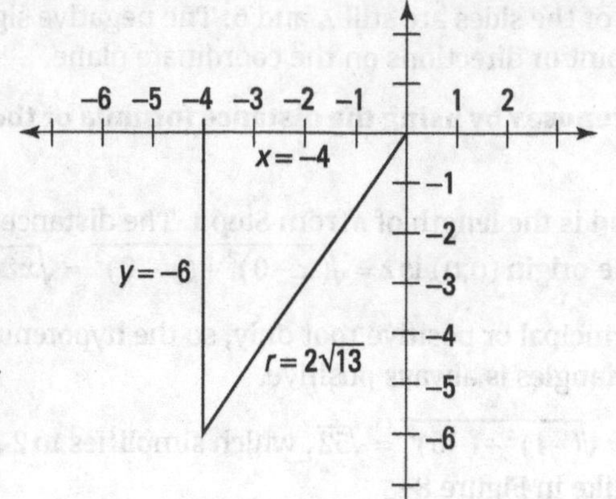

FIGURE 8-5: Finding the hypotenuse of a right triangle when given a point on the plane.

REMEMBER

When the point you're given is a point on one of the axes, you can still find all the trig function values. For instance, if the point is on the x-axis, the cosine and the radius have the same absolute value (because the cosine can be negative but the radius can't). If the point is on the positive x-axis, the cosine is 1 and the sine is 0; if the point is on the negative x-axis, the cosine is −1. Similarly, if the point is on the y-axis, the sine value and the radius are of the same absolute value; the sine will be either 1 or −1, and the cosine will always be 0.

YOUR TURN

9 Find the six trig functions corresponding to the point $(5, 12)$ on the coordinate plane.

10 Find the six trig functions corresponding to the point $(-1, 1)$ on the coordinate plane.

Digesting Special Triangle Ratios

You see two particular triangles over and over again in trigonometry; they are often called the *45er* and the old 30 – 60. In fact, you see them so often that it's often recommended that you just memorize their ratios. Sound difficult? Relax! You see how to compute the ratios in this section. (And, yes, you've probably seen these triangles before in geometry.)

The 45er: Isosceles right triangle

All 45° – 45° – 90° triangles have sides that are in a unique ratio. The two legs are the exact same length, and the hypotenuse is that length times $\sqrt{2}$. Figure 8-6 shows the ratio. (If you look at the 45° – 45° – 90° triangle in radians, you have $\frac{\pi}{4} - \frac{\pi}{4} - \frac{\pi}{2}$. Either way, it's still the same ratio.)

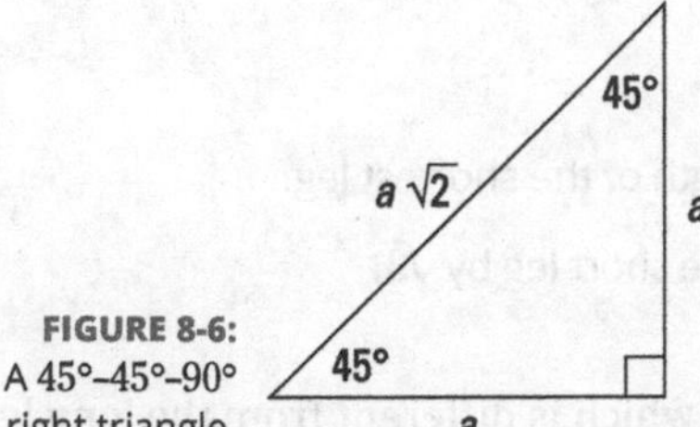

FIGURE 8-6: A 45°–45°–90° right triangle.

Why is this triangle important? Because any time you're given one side of a 45er triangle, you can figure out the other two sides relatively quickly. When you complete calculations with this type of triangle, it will fall into one of two categories.

- **Type 1: You're given one leg.**

  Because you know both legs are equal, you know the length of both the legs. You can find the hypotenuse by multiplying this length by $\sqrt{2}$.

- **Type 2: You're given the hypotenuse.**

  Divide the hypotenuse by $\sqrt{2}$ to find the legs (which are equal).

EXAMPLE

Q. The diagonal in a square is 16 centimeters long. How long is each side of the square?

A. Draw it out first. Figure 8-7 shows the square.

The diagonal of a square divides the angles into 45° pieces, so you have the hypotenuse of a 45er triangle. To find the legs, divide the hypotenuse by $\sqrt{2}$. When you do, you get $\frac{16}{\sqrt{2}}$.

Rationalizing the denominator, $\frac{16}{\sqrt{2}} \cdot \frac{\sqrt{2}}{\sqrt{2}} = \frac{16\sqrt{2}}{2} = 8\sqrt{2}$, which is the measure of each side of the square.

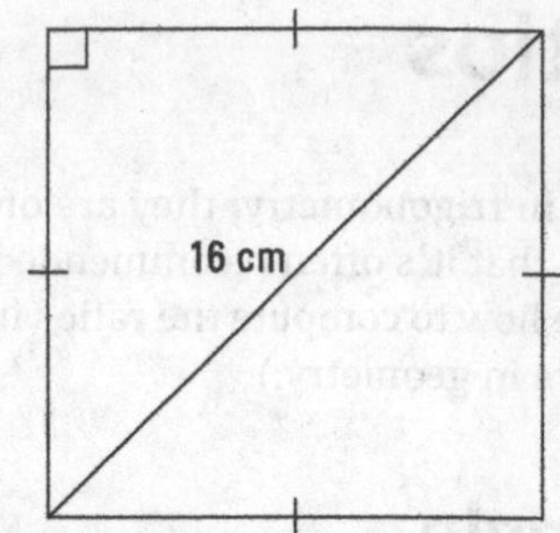

FIGURE 8-7: A square with a diagonal.

The 30–60–90 right triangle

All 30°–60°–90° triangles have sides with the same basic ratio. If you look at the 30°–60°–90° triangle in radians, it's read $\frac{\pi}{6}-\frac{\pi}{3}-\frac{\pi}{2}$.

- The shortest leg is across from the 30° angle.
- The length of the hypotenuse is always two times the length of the shortest leg.
- You can find the length of the longer leg by multiplying the short leg by $\sqrt{3}$.

Note: The hypotenuse is the longest side in a right triangle, which is different from the long leg. The long leg is the leg opposite the 60° angle.

Figure 8-8 illustrates the ratio of the sides for the 30–60–90 triangle.

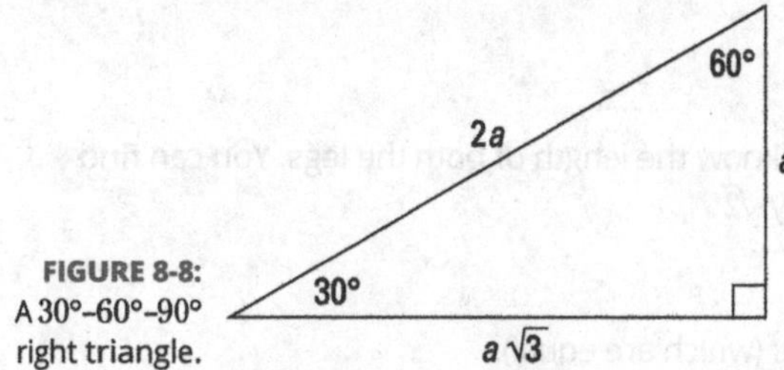

FIGURE 8-8: A 30°–60°–90° right triangle.

If you know one side of a 30–60–90 triangle, you can find the other two sides by using shortcuts. Here are the three situations you come across when doing these calculations.

- **Type 1: You know the short leg (the side across from the 30° angle).** Double its length to find the hypotenuse. You can multiply the short side by $\sqrt{3}$ to find the long leg.
- **Type 2: You know the hypotenuse.** Divide the hypotenuse by 2 to find the short side. Multiply this answer by $\sqrt{3}$ to find the long leg.
- **Type 3: You know the long leg (the side across from the 60° angle).** Divide this side by $\sqrt{3}$ to find the short side. Double that figure to find the hypotenuse.

EXAMPLE

Q. **In the triangle TRI in Figure 8-9, the hypotenuse is 14 inches long; how long are the other sides?**

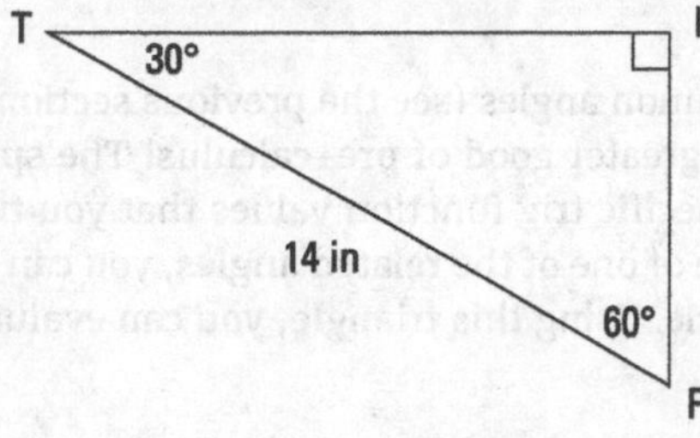

FIGURE 8-9: Finding the other sides of a 30-60-90 triangle when you know the hypotenuse.

A. **Because you have the hypotenuse TR = 14, you can divide by 2 to get the short side: RI = 7. Now, multiply this length by $\sqrt{3}$ to get the long side: IT = $7\sqrt{3}$.**

YOUR TURN

11 **A ladder is leaning against a building at a 60° angle and is 14 feet from the base of the building. How long is the ladder?**

12 **You've created a triangular garden that is a perfect isosceles right triangle. If one of the shorter sides is 100 feet long, how much fencing will you need to go completely around the garden?**

Triangles and the Unit Circle: Working Together for the Common Good

Rejoice in the fusion of right triangles, common angles (see the previous section), and the unit circle, because they come together for the greater good of pre-calculus! The special right triangles play an important role in finding specific trig function values that you find on the unit circle. Specifically, if you know the measure of one of the related angles, you can make a special right triangle that will fit onto the unit circle. Using this triangle, you can evaluate all kinds of trig functions without a calculator!

REMEMBER

All congruent angles (angles with the same measure) have the same values for the different trig functions. Some non-congruent angles also have identical values for certain trig functions; you can use a reference angle to find out the measures for these angles.

Brush up on the special right triangles before attempting to evaluate the functions in this section. Although many of the values look identical, looks can be deceiving. The numbers may be the same, but the signs and locations of these numbers change as you move around the unit circle.

Placing the major angles correctly

In this section, you see the unit-circle angles and the special right triangles put together to create a neat little package: the full unit circle. The special triangles in the unit circle create points on the coordinate plane. Regardless of how long the sides are that make up a particular angle in a triangle, the trig function values for that specific angle are always the same.

REMEMBER

The hypotenuse of a right triangle in a unit circle is always 1, and the calculations that involve the triangles are much easier to compute. Because of the unit circle, you can draw *any* angle with *any* measurement, and all right triangles with the same reference angle are the same size.

Starting in quadrant I: Calculate the points to plot

Look at an angle marked 30° in the unit circle (see Figure 8-10) and follow these steps to build a triangle out of it — similar to the steps from the section, "Understanding How Trig Ratios Work on the Coordinate Plane."

1. **Draw the terminal side of the angle from the origin to the circle.**

   The terminal side of a 30° angle should be in the first quadrant, and the size of the angle should be rather small. In fact, it should be one-third of the way between 0° and 90°. You can use a protractor to get an accurate slant.

2. **Draw a perpendicular line connecting the point where the terminal side intersects the circle to the x-axis, creating a right triangle.**

   The triangle's hypotenuse is the radius of the unit circle; one of its legs is on the x-axis; and the other leg is parallel to the y-axis. You can see what this 30°–60°–90° triangle looks like in Figure 8-10.

3. **Find the length of the hypotenuse.**

   The radius of the unit circle is always 1, which means the hypotenuse of the triangle is also 1.

4. **Find the lengths of the other sides.**

   To find the other two sides, you use the techniques discussed in the 30°–60°–90° triangle section. Find the short leg first by dividing the length of the hypotenuse by 2, which gives you $\frac{1}{2}$. To find the long leg, multiply $\frac{1}{2}$ by $\sqrt{3}$ to get $\frac{\sqrt{3}}{2}$.

5. **Identify the point on the unit circle.**

   The unit circle is on the coordinate plane, centered at the origin. So, each of the points on the unit circle has unique coordinates. You can now name the point corresponding to an angle of 30° on the circle:

   $$\left(\frac{\sqrt{3}}{2}, \frac{1}{2}\right)$$

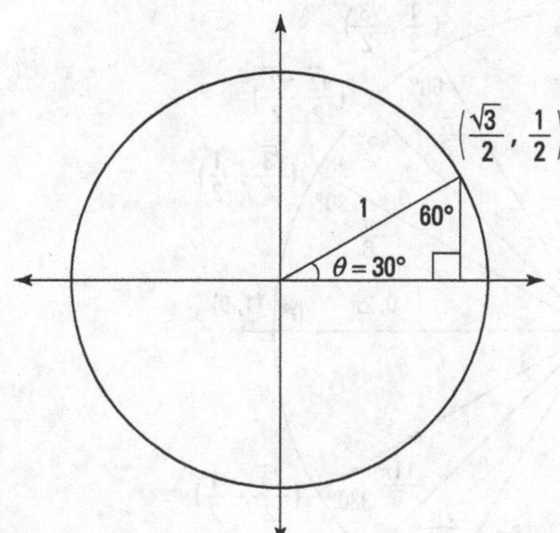

FIGURE 8-10: A 30°–60°–90° triangle drawn on the unit circle.

TIP

After going through the previous steps, you can easily find the points of other angles on the unit circle as well. For instance:

- Look at the point on the circle marked 45°. You can draw a triangle from it, using Steps 1 and 2. Its hypotenuse is still 1, the radius of the unit circle. To find the length of the legs of a 45°–45°–90° triangle, you divide the hypotenuse by $\sqrt{2}$. You then rationalize the denominator to get $\frac{\sqrt{2}}{2}$. You can now name this point on the circle $\left(\frac{\sqrt{2}}{2}, \frac{\sqrt{2}}{2}\right)$.
- Moving counterclockwise to the 60° angle, you can create a triangle with Steps 1 and 2. If you look closely, you'll realize that the 30° angle is at the top, so the short side is the side on the *x*-axis. That makes the point at 60° $\left(\frac{1}{2}, \frac{\sqrt{3}}{2}\right)$ due to the radius of 1 (divide 1 by 2 and multiply $\frac{1}{2}$ by $\sqrt{3}$).

Moving along to the other quadrants

Quadrants II to IV in the coordinate plane are just mirror images of the first quadrant (see the previous section). However, the signs of the coordinates are different because the points on the unit circle are on different locations of the plane:

- In quadrant I, both *x* and *y* values are positive.
- In quadrant II, *x* is negative and *y* is positive.
- In quadrant III, both *x* and *y* are negative.
- In quadrant IV, *x* is positive and *y* is negative.

TIP

The good news is that you never have to memorize the whole unit circle. You can simply apply the basics of what you know about right triangles and the unit circle! Figure 8-11 shows the whole pizza pie of the unit circle.

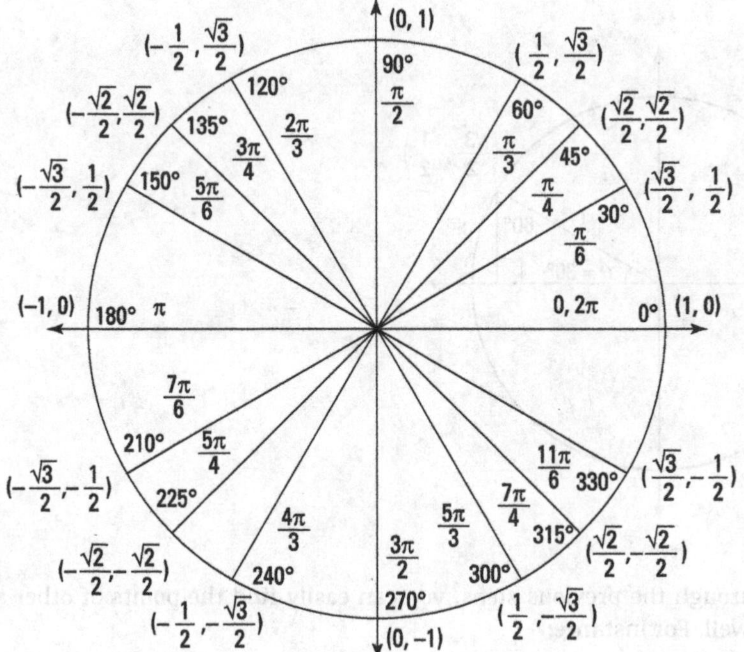

FIGURE 8-11: The whole unit circle.

Retrieving trig-function values on the unit circle

The previous sections on angles and the unit circle are preliminary. You need to be comfortable with the unit circle and the special triangles in it so that you can evaluate trig functions quickly and easily, which you do in this section. You don't want to waste precious moments constructing the entire unit circle just to evaluate a couple of angles. And the more comfortable you are with trig ratios and the unit circle as a whole, the less likely you are to make an error with a negative sign or get trig values mixed up.

Finding values for the six trig functions

Sometimes you need to evaluate the six trig functions — sine, cosine, tangent, cosecant, secant, and cotangent — for a single value on the unit circle. For each angle on the unit circle, three other related angles have similar trig function values. The only difference is that the signs of these values change, depending on which quadrant the angle is in. When the angle isn't related to one of the special angles on the unit circle, you'll probably have to use your calculator. Just take advantage of the special triangle relationships that are shown in this chapter.

REMEMBER

The point-in-the-plane definition of cosine in a right triangle is $\cos\theta = \frac{x}{r}$. Because the hypotenuse r is always 1 in the unit circle, the x value *is* the cosine value. And if you remember the alternate definition of sine — $\sin\theta = \frac{y}{r}$ — you'll realize that the y value is the sine value. Therefore, any point anywhere on the unit circle is always $(\cos\theta, \sin\theta)$. Talk about putting all the pieces together!

Determining the values of the tangent, cotangent, secant, and cosecant requires a little more effort than the sine and cosine. For many angles on the unit circle, evaluating these functions requires some careful work with fractions and square roots. Remember to always rationalize the denominator for any fraction in your final answer. Also, remember that any number divided by 0 is undefined. The tangent and secant functions, for instance, are undefined when the cosine value (x-coordinate) is 0. Similarly, the cotangent and cosecant values are undefined when the sine value (y-coordinate) is 0.

EXAMPLE

Q. Evaluate the six trigonometric functions of $\theta = 225°$ on the unit circle.

A. Follow these steps.

1. **Draw the picture.**

   When you're asked to find the trig function of an angle, you don't have to draw out a unit circle every time. Instead, use your smarts to figure out the picture. For this example, 225° is 45° more than 180°. Draw out a 45°–45°–90° triangle in the third quadrant only (see the earlier section, "Placing the major angles correctly").

2. **Fill in the lengths of the legs and the hypotenuse.**

   Use the rules of the 45er triangle. The coordinate of the point at 225° is $\left(-\frac{\sqrt{2}}{2}, -\frac{\sqrt{2}}{2}\right)$.

   Figure 8-12 shows the triangle, as well as all the information you need to evaluate the six trig functions.

REMEMBER

   Be careful! Use what you know about the positive and negative axes on the coordinate plane to help you. Because the triangle is in quadrant III, both the x and y values should be negative.

3. **Find the sine of the angle.**

   The sine of an angle is the y value, or the length of the vertical line that extends from the point on the unit circle to the x-axis. For 225°, the y value is $-\frac{\sqrt{2}}{2}$, so $\sin(225°) = -\frac{\sqrt{2}}{2}$.

4. **Find the cosine of the angle.**

   The cosine value is the *x* value, so it must be $-\frac{\sqrt{2}}{2}$.

5. **Find the tangent of the angle.**

   To find the tangent of an angle on the unit circle, use the tangent's alternate definition: $\tan\theta = \frac{y}{x}$. Another way of looking at it is that $\tan\theta = \frac{\sin\theta}{\cos\theta}$ because in the unit circle, the *y* value is the sine and the *x* value is the cosine. So, if you know the sine and cosine of any angle, you also know the tangent. (Thanks, unit circle!) The sine and the cosine of 225° are both $-\frac{\sqrt{2}}{2}$. Therefore, you can divide the sine by the cosine to get the tangent of 225°, which is 1.

6. **Find the cosecant of the angle.**

   The cosecant of any angle is the reciprocal of the sine. Using what you determined in Step 1, $\sin(225°) = -\frac{\sqrt{2}}{2}$. The reciprocal is $-\frac{2}{\sqrt{2}}$, which rationalizes to $-\sqrt{2}$. Hence, $\csc 225° = -\sqrt{2}$.

7. **Find the secant of the angle.**

   The secant of any angle is the reciprocal of the cosine. Because $\cos 225°$ is also $-\frac{\sqrt{2}}{2}$, found in Step 4, $\sec 225° = -\sqrt{2}$.

8. **Find the cotangent of the angle.**

   The cotangent of an angle is the reciprocal of the tangent. From Step 5, $\tan 225° = 1$. So, $\cot 225° = \frac{1}{1} = 1$. Easy as pie!

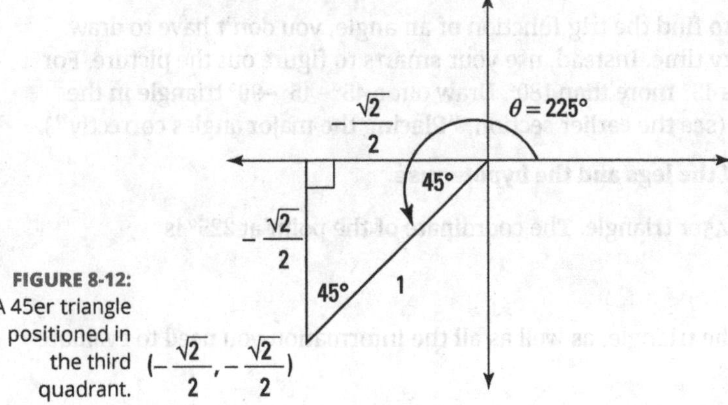

FIGURE 8-12: A 45er triangle positioned in the third quadrant.

TIP

The tangent is always the slope of the radius *r*. This easy calculation gives you a nice check for your work. Because the radius of the unit circle (the hypotenuse of the triangle) in the previous problem slants up, it has a positive slope, as does the tangent value.

The shortcut: Finding trig values of the 30°, 45°, and 60° families

Good news! There is a shortcut that can help you avoid some of the work of the previous section. You'll have to do less memorizing when you realize that certain special angles (and, therefore, their special triangles) on the unit circle always follow the same ratio of the sides. All you have to do is use the quadrants of the coordinate plane to figure out the signs. Solving trig-function problems on the unit circle will be so much easier after this section!

Perhaps you already figured out the shortcut by looking at Figure 8-11. If not, here are the families on the unit circle (for *any* family, the hypotenuse *r* is always 1).

- **The first family is the $\frac{\pi}{6}$ family (multiples of 30°).** Any angle with the denominator of 6 has these qualities:
  - The longer leg is the *x* leg, $\frac{\sqrt{3}}{2}$.
  - The shorter leg is the *y* leg, $\frac{1}{2}$.
- **The second family is the $\frac{\pi}{3}$ family (multiples of 60°).** Any angle with the denominator of 3 has these qualities:
  - The shorter leg is the *x* leg, $\frac{1}{2}$.
  - The longer leg is the *y* leg, $\frac{\sqrt{3}}{2}$.
- **The last family is the $\frac{\pi}{4}$ family (multiples of 45°).** Any angle with the denominator of 4 has this quality: The two legs are equal in length, $\frac{\sqrt{2}}{2}$.

YOUR TURN

Use Figure 8-11 to find the following:

13 $\sin 120°$

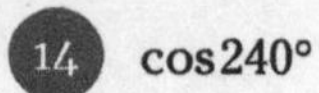

14 $\cos 240°$

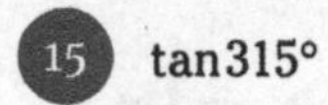

15 $\tan 315°$

16 $\csc 390°$

Practice Questions Answers and Explanations

1. **405°, 765°, 1125°.** These are the smallest three positive co-terminal angles of the angle 45°.

2. **-270°, -630°, -990°.** These are the largest negative co-terminal angles of the angle 90°.

3. $\frac{3}{4}$**.** First, find the measure of side a. Using the Pythagorean Theorem, $a^2+6^2=10^2 \rightarrow a^2=100-36=64 \rightarrow a=\pm 8$, Using the positive solution, you have $a=8$. The $\tan B$ is the opposite over the adjacent, so $\tan B=\frac{6}{8}=\frac{3}{4}$. See the following figure.

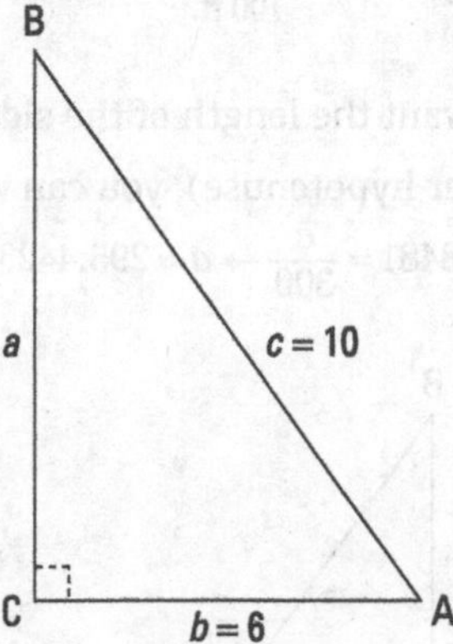

4. $\frac{24}{25}$**.** First, find the measure of side c. Using the Pythagorean Theorem, $24^2+7^2=c^2 \rightarrow c^2=576+49=625 \rightarrow c=\pm 25$. Using the positive solution, you have $c=25$. The $\sin A$ is the opposite over the hypotenuse, so $\sin A=\frac{24}{25}$. See the following figure.

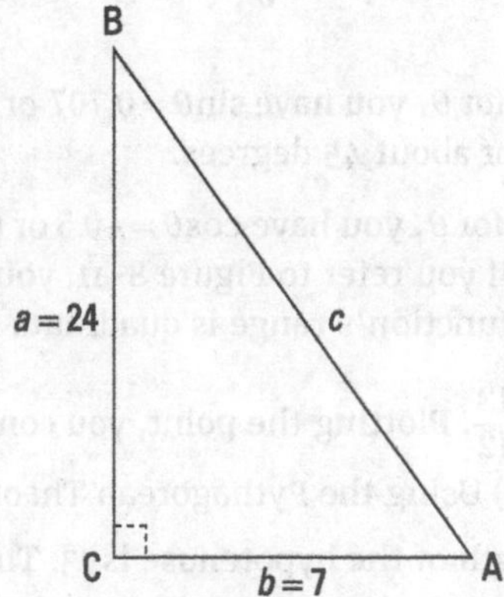

5. **200 feet.** You want the length of the hypotenuse. Using the cosine of angle A, since it is the ratio of the adjacent to the hypotenuse, you have $\cos 60°=\frac{100}{c}$. Your calculator tells you that $\cos 60°=0.5$, so now you have $0.5=\frac{100}{c} \rightarrow c=\frac{100}{0.5} \rightarrow c=200$. See the following figure.

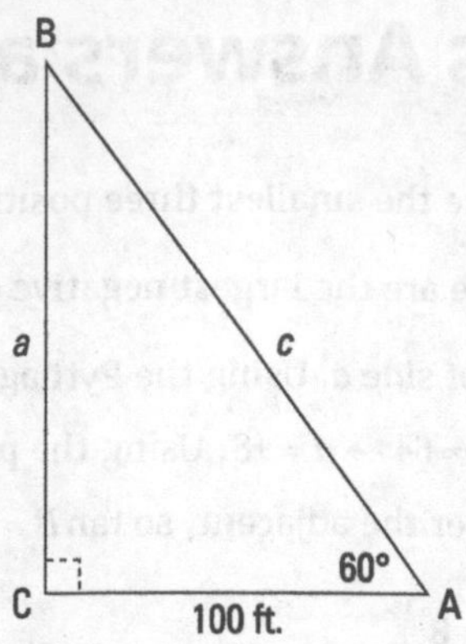

6. **295.4 feet.** This time, you want the length of the side of the triangle opposite the 80° angle. Using the sine (opposite over hypotenuse), you can write $\sin 80° = \frac{a}{300}$. Your calculator gives you $\sin 80° \approx 0.98481$. So, $0.98481 = \frac{a}{300} \to a \approx 295.44233$. See the following figure.

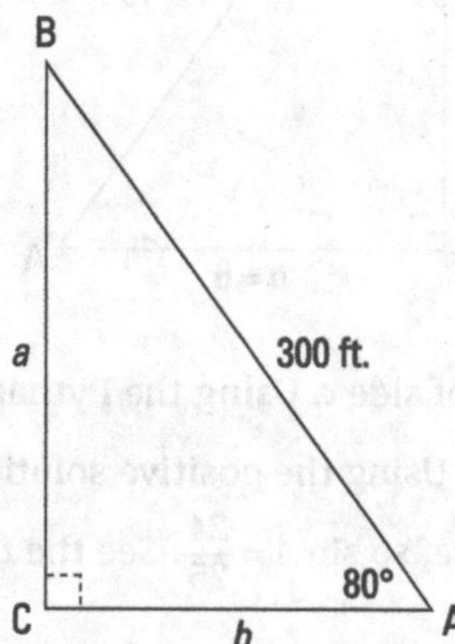

7. **45°.** Solving $\sin\theta = 0.707$ for θ, you have $\sin\theta = 0.707$ or $\theta = \sin^{-1} 0.707$. Your calculator tells you that $\theta = 44.99134834$ or about 45 degrees.

8. **120°.** Solving $\cos\theta = -0.5$ for θ, you have $\cos\theta = -0.5$ or $\theta = \cos^{-1}(-0.5)$. Your calculator tells you that $\theta = 120$ degrees. If you refer to Figure 8-11, you see that $\cos\theta = -0.5$ for both 120° and 240°. The inverse cosine function's range is quadrants I and II, so you choose the 120°.

9. $\frac{12}{13}, \frac{5}{13}, \frac{12}{5}, \frac{5}{12}, \frac{13}{5}, \frac{13}{12}$. Plotting the point, you construct a right triangle in quadrant I. (See the following figure.) Using the Pythagorean Theorem, $5^2 + 12^2 = c^2 \to c^2 = 25 + 144 = 169 \to c = \pm 13$. So, the length of the hypotenuse is 13. The six trig functions are $\sin = \frac{\text{opp}}{\text{hyp}} = \frac{12}{13}$, $\cos = \frac{\text{adj}}{\text{hyp}} = \frac{5}{13}$, $\tan = \frac{\text{opp}}{\text{adj}} = \frac{12}{5}$, $\cot = \frac{\text{adj}}{\text{opp}} = \frac{5}{12}$, $\sec = \frac{\text{hyp}}{\text{adj}} = \frac{13}{5}$, $\csc = \frac{\text{hyp}}{\text{opp}} = \frac{13}{12}$.

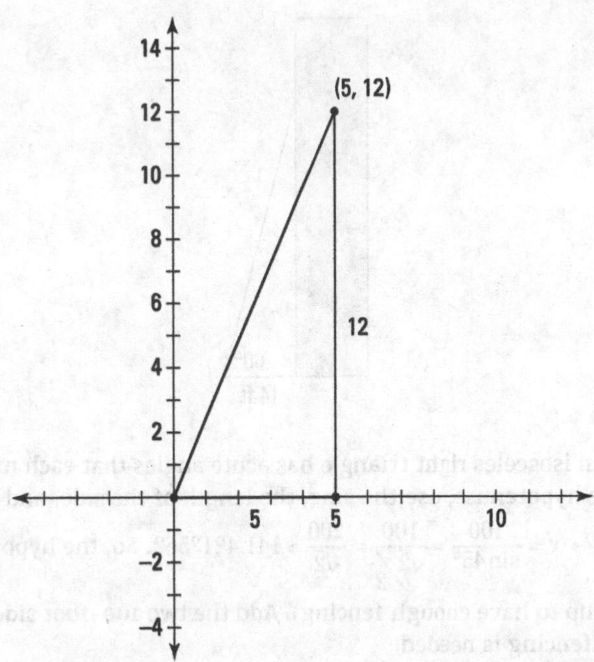

10 $\frac{\sqrt{2}}{2}, -\frac{\sqrt{2}}{2}, -1, -1, -\sqrt{2}, \sqrt{2}$. **Plotting the point, you construct a right triangle in quadrant II. (See the following figure.) Using the Pythagorean Theorem, $1^2 + 1^2 = c^2 \rightarrow c^2 = 1 + 1 = 2 \rightarrow c = \pm\sqrt{2}$. So, the length of the hypotenuse is $\sqrt{2}$. The six trig functions are**

$$\sin = \frac{\text{opp}}{\text{hyp}} = \frac{1}{\sqrt{2}} = \frac{\sqrt{2}}{2},\ \cos = \frac{\text{adj}}{\text{hyp}} = \frac{-1}{\sqrt{2}} = -\frac{\sqrt{2}}{2},\ \tan = \frac{\text{opp}}{\text{adj}} = \frac{1}{-1} = -1,\ \cot = \frac{\text{adj}}{\text{opp}} = \frac{-1}{1} = -1,$$

$$\sec = \frac{\text{hyp}}{\text{adj}} = \frac{\sqrt{2}}{-1} = -\sqrt{2},\ \csc = \frac{\text{hyp}}{\text{opp}} = \frac{\sqrt{2}}{1} = \sqrt{2}.$$

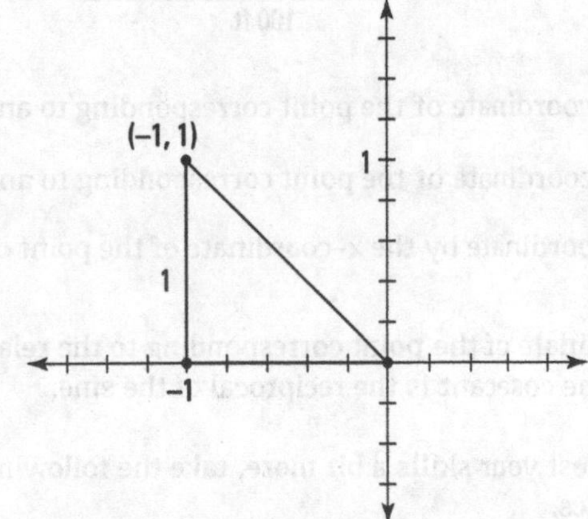

11 **28 feet.** The 14-foot length forms the adjacent side of the triangle, and you want the length of the hypotenuse. Look at the following figure. The adjacent side and hypotenuse form the cosine, so you write $\cos 60° = \frac{14}{x}$ and solve for the length of the ladder: $x = \frac{14}{\cos 60°} = \frac{14}{0.5} = 28$.

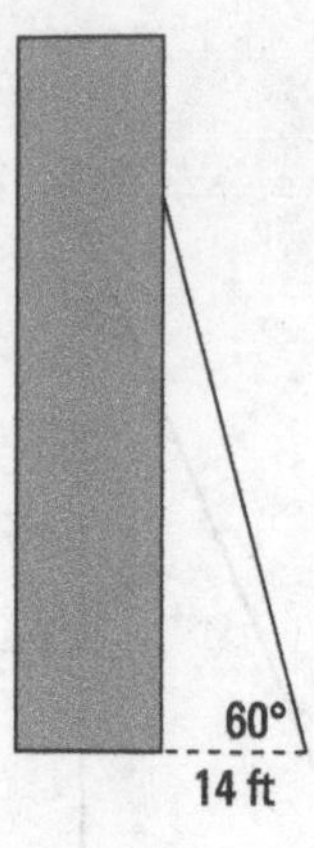

12. **341.5 feet.** An isosceles right triangle has acute angles that each measure 45°. To find the length of the hypotenuse, use the sine, the length of the side, and solve for that length, x: $\sin 45° = \frac{100}{x} \rightarrow x = \frac{100}{\sin 45°} = \frac{100}{\sqrt{2}/2} = \frac{200}{\sqrt{2}} \approx 141.4213562$. So, the hypotenuse is about 141.5 feet long (round up to have enough fencing). Add the two 100-foot sides to the hypotenuse, and 341.5 feet of fencing is needed.

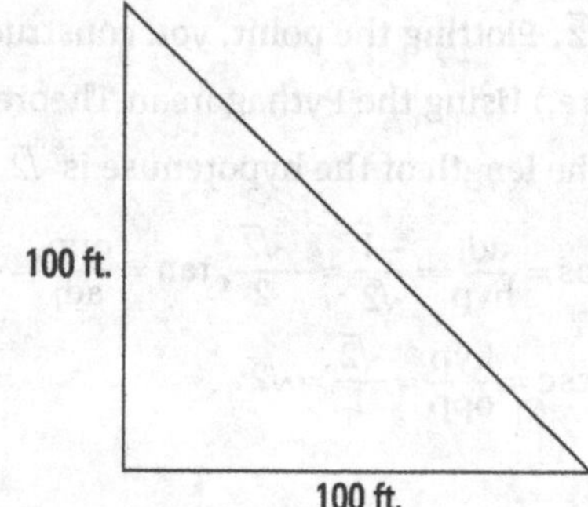

13. $\frac{\sqrt{3}}{2}$**.** This is the y-coordinate of the point corresponding to an angle of 120°.
14. $-\frac{1}{2}$**.** This is the x-coordinate of the point corresponding to an angle of 240°.
15. **−1.** Divide the y-coordinate by the x-coordinate of the point corresponding to an angle of 315°.
16. **2.** Use the x-coordinate of the point corresponding to the relative angle 30° and find its reciprocal, since the cosecant is the reciprocal of the sine.

If you're ready to test your skills a bit more, take the following chapter quiz that incorporates all the chapter topics.

Whaddya Know? Chapter 8 Quiz

Quiz time! Complete each problem to test your knowledge on the various topics covered in this chapter. You can then find the solutions and explanations in the next section.

1 **Find the area of a right triangle with one leg measuring $4\sqrt{3}$ and the hypotenuse measuring $8\sqrt{3}$.**

2 **An angle measuring 500° is co-terminal with what positive angle measuring less than 360°?**

3 **What is the length of the hypotenuse of a right triangle with legs measuring 8 inches and 9 inches?**

4 **Solve for x: $9\tan x = -\sqrt{3}$.**

5 **What is the perimeter of a right triangle that has an acute angle measuring 30° and a hypotenuse of 32 yards?**

6 **An angle measuring $-\frac{17\pi}{4}$ is co-terminal with what positive angle measuring less than 2π?**

7 **What is the measure of the smallest angle in a right triangle where one leg measures 15 inches and the hypotenuse measures 39 inches?**

Answers to Chapter 8 Quiz

1. **$24\sqrt{3}$.** Using the Pythagorean Theorem, $(4\sqrt{3})^2 + x^2 = (8\sqrt{3})^2 \rightarrow 16\cdot 3 + x^2 = 64\cdot 3 \rightarrow$ $x^2 = 192 - 48 = 144$. Solving for x in $x^2 = 144$, you use the positive solution: $x = 12$. This is the length of the other leg. The area is $A = \frac{1}{2}bh$, where b and h are the perpendicular sides, so $A = \frac{1}{2}bh = \frac{1}{2}(12)4\sqrt{3} = 24\sqrt{3}$.

2. **140°.** Subtracting: $500 - 360 = 140$.

3. **About 12 inches.** Using the Pythagorean Theorem, $8^2 + 9^2 = x^2 \rightarrow 64 + 81 = x^2 \rightarrow 145 = x^2$. Using only the positive solution, $x = \sqrt{145} \approx 12.04$.

4. **About −0.19 degrees.** Divide each side by 9. Then solve for x. $\tan x = \frac{-\sqrt{3}}{9} \rightarrow x = \tan^{-1}\left(\frac{-\sqrt{3}}{9}\right)$. Using your calculator, you find that $\tan x \approx -0.19245$ degrees. Then $x \approx \tan^{-1}(-0.19245) \approx$ -0.19013 degrees or about 359.8 degrees.

5. **About 75.7 yards.** In a 30–60–90 right triangle, the length of the hypotenuse is twice that of the shorter leg, so the shorter leg measures 16 yards. The longer leg is $\sqrt{3}$ times the length of the shorter leg, so, in this triangle, it measures $16\sqrt{3}$ yards. The perimeter is the sum of the lengths: $32 + 16 + 16\sqrt{3} = 48 + 16\sqrt{3}$ yards or about 75.7 yards.

6. **$\frac{7\pi}{4}$.** Adding 2π three times, you have $-\frac{17\pi}{4} + 6\pi = -\frac{17\pi}{4} + \frac{24\pi}{4} = \frac{7\pi}{4}$.

7. **About 22.62 degrees.** First, find the measure of the missing leg using the Pythagorean Theorem: $15^2 + x^2 = 39^2 \rightarrow 225 + x^2 = 1521 \rightarrow x^2 = 1296 \rightarrow x = 36$. This is a right triangle whose legs are 15 and 36 and hypotenuse is 39 (a nice Pythagorean triple). (See the following figure.) The smallest angle is opposite the smallest side. The sine of that angle is the measure of that opposite side divided by the measure of the hypotenuse. So, $\sin x = \frac{15}{39} = \frac{5}{13}$. Solving for x, $x = \sin^{-1}\left(\frac{5}{13}\right)$. Using your calculator, $x \approx 22.61986$ degrees.

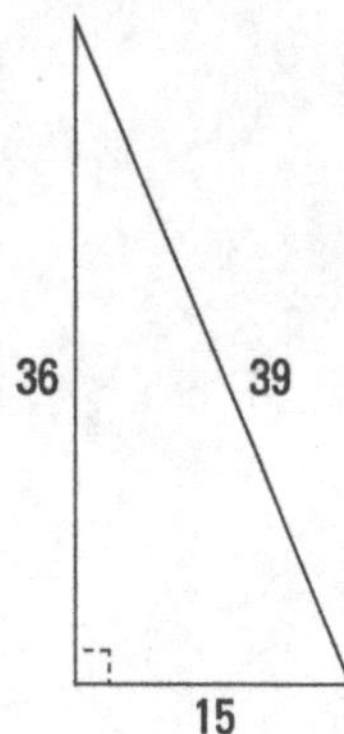

IN THIS CHAPTER

» Placing triangles on the unit circle

» Solving practical problems using angle values

» Measuring the length of arcs

Chapter 9 Homing In on the Friendliest Angles

In this chapter, the trig functions described earlier in the book are taken a bit farther in their properties and uses. The most commonly found angles in examples and applications are 30°, 45°, 60°, and 90°, and their multiples. Why? Because their function values are so cooperative! Other angles occur in real life, and you can use a calculator to find the values needed, but it's just easier to present the nicer, friendlier angles here!

Building on the Unit Circle

The unit circle is a vital part of the study of trigonometry. The angles are measured counterclockwise if the angle measure is positive, and clockwise if it is negative. You can spin around several times to create a really large or really small angle. But the nice thing is that the function values you find on the unit circle are from the coordinates of the points. The points don't change; you just have to match the angle with the correct coordinates.

Familiarizing yourself with the most common angles

In pre-calculus, you often want function values of the most common angles, so dedicating to memory exactly what some of these values are isn't a bad idea. The values you'll see in the following table are the exact values, not an approximate decimal value that you often see on calculators. These three angles help you find the trig function values for those special (or more common) angles on the unit circle.

In Table 9-1, you see the function values for sine, cosine, and tangent.

Table 9-1 Function Values for Basic Angles

	$0° = 0$ rad	$30° = \frac{\pi}{6}$ rad	$45° = \frac{\pi}{4}$ rad	$60° = \frac{\pi}{3}$ rad	$90° = \frac{\pi}{2}$ rad
sin	0	$\frac{1}{2}$	$\frac{\sqrt{2}}{2}$	$\frac{\sqrt{3}}{2}$	1
cos	1	$\frac{\sqrt{3}}{2}$	$\frac{\sqrt{2}}{2}$	$\frac{1}{2}$	0
tan	0	$\frac{\sqrt{3}}{3}$	1	$\sqrt{3}$	undefined

Notice the nice pattern in the function values. For the sine, think of the values as being of the form $\frac{\sqrt{n}}{2}$, where n goes from 0 to 4 moving from left to right: $\frac{\sqrt{0}}{2} = 0, \frac{\sqrt{1}}{2} = \frac{1}{2}, \frac{\sqrt{2}}{2}, \frac{\sqrt{3}}{2}, \frac{\sqrt{4}}{2} = 1$. And the cosine is just the backward version. You really don't even need the row for the tangent, because each of the values is the result of dividing the sine by the cosine!

Q. Find $\sec 60°$

EXAMPLE **A.** From the table (which you have already memorized), you find $\cos 60° = \frac{1}{2}$. Since $\sec x = \frac{1}{\cos x}$, you just find the reciprocal of $\frac{1}{2}$ and have $\sec 60° = 2$.

Q. Find $\cot 90°$

EXAMPLE **A.** From the table, you find that $\tan 90°$ is undefined. You know that $\cot x = \frac{1}{\tan x}$ and that the tangent came from dividing the sine by the cosine. Just flip that division, and you find that $\cot 90° = 0$.

YOUR TURN

1 Find $\csc 45°$

2 Find $\cot 60°$

Finding the reference angle to solve for functions of angles

A *reference angle* of angle θ is the smallest acute angle that can be drawn from the terminal side of θ up or down to the x-axis. The functions of the reference angle will have the same absolute value as the acute angle it's related to. You can use these function values and determine the sign based on the quadrant. In the following list, θ is a particular angle and θ' is its acute reference angle (see Figure 9-1).

- **Quadrant I:** $\theta = \theta'$ because the reference angle and the solution angle are the same.
- **Quadrant II:** $\theta = \pi - \theta' = 180° - \theta'$ because θ falls short of π by however much the reference angle is.
- **Quadrant III:** $\theta = \pi + \theta' = 180° + \theta'$ because the angle is greater than π.
- **Quadrant IV:** $\theta = 2\pi - \theta' = 360° - \theta'$ because θ falls short of a full circle by however much the reference angle is.

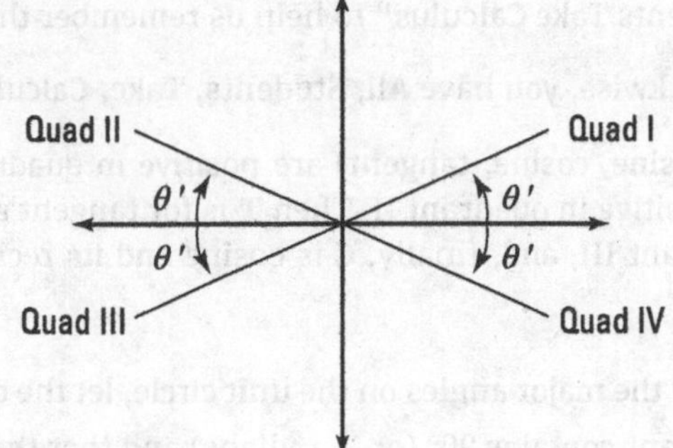

FIGURE 9-1: Finding the solution angle, given the reference angle.

EXAMPLE

Q. Find the reference angles for angles measuring 300°, 210°, and $\frac{5\pi}{4}$.

A. $300° = 360° - 60°$, so its reference angle is 60°.

$210° = 180° + 30°$, so its reference angle is 30°.

$\frac{5\pi}{4} = \pi + \frac{\pi}{4}$, so its reference angle is $\frac{\pi}{4}$.

YOUR TURN

3 Find the reference angle related to 240°.

4. Find the reference angle related to $\frac{11\pi}{6}$.

Relating reference angles to their parents

Using the exact values of the sine, cosine, and tangent of our favorite angles, you can determine the exact values of the functions of any angle that is a multiple of one of them. All you need is the reference angle, the value from the table, and the location of the angle on the plane. My trig teacher told us that "All Students Take Calculus" to help us remember the positions of the letters: $\frac{S \mid A}{T \mid C}$. Reading counterclockwise, you have **All**, **Students**, **Take**, **Calculus**. This refers to the fact that **All** three functions (sine, cosine, tangent) are positive in quadrant I. The **S** says that sine and its reciprocal are positive in quadrant II. Then **T** is for tangent and its reciprocal, which are both positive, in quadrant III, and, finally, **C** is cosine and its reciprocal is positive in quadrant IV.

TIP

When determining the locale of all the major angles on the unit circle, let the quadrants be your guide. Remember that each quadrant contains 90° (or $\frac{\pi}{2}$ radians) and that the measures of the angles increase as you move counterclockwise around the vertex. With that information and a little bit of math, you can figure out the location of the angle you need.

EXAMPLE

Q. Find the exact value of $\sin 240°$.

A. The angle 240° is in quadrant III. Its reference angle is 60°. From the table, $\sin 60° = \frac{\sqrt{3}}{2}$. And, since the sine is negative in quadrant III, $\sin 240° = -\frac{\sqrt{3}}{2}$.

EXAMPLE

Q. Find the exact value of $\sec 300°$.

A. The angle 300° is in quadrant IV. Its reference angle is 60°. The secant is the reciprocal of cosine, so looking at the table, you find that $\cos 60° = \frac{1}{2}$. And the cosine is positive in quadrant IV. Now you just need the reciprocal of the value of the cosine. Since $\cos 60° = \frac{1}{2}$, you have $\sec 60° = 2$.

YOUR TURN

5 Find the exact value of $\cos 210°$.

6 Find the exact value of $\tan 120°$.

7 Find the exact value of $\sin \frac{3\pi}{4}$.

8 Find the exact value of $\cot 330°$.

Drawing uncommon angles

Many times in your journey through trig (actually, all the time), drawing a figure will help you solve a given problem. Refer to any example problem in this chapter for an illustration of how to use a picture to help. Trig always starts with the basics of drawing angles, so that by the time you get to the problems, the actual drawing stuff is second nature.

TIP

When drawing angles on the coordinate plane, you will zoom in on their *reference angles.* A *reference angle* is always an acute angle. It's the smallest angle you can make between the terminal side of the angle you're working on and the *x*-axis. You move up or down from the terminal side to the axis.

To draw angles on the coordinate plane, begin by sketching their terminal sides in the correct places. Then, by drawing a vertical line up or down to the *x*-axis, you can make right triangles that fit in the unit circle, with smaller angles that are more familiar to you.

What do you do if you're asked to draw an angle that has a measure greater than 360°? Or a negative measure? How about both? Your head must be spinning! No worries; this section gives you the steps.

EXAMPLE

Q. Draw a –570° angle.

A. Here's what you do:

1. **Find a co-terminal angle by adding 360°.**

   Adding 360° to –570° gives you –210°.

2. **If the angle is still negative, keep adding 360° until you get a positive angle in standard position.**

   Adding 360° to –210° gives you 150°. This is your co-terminal angle.

3. **Find a reference angle (an angle between 0 and 90 degrees).**

   The reference angle for 150° is 30°. This tells you which type of triangle you want to refer to (see Figure 9-2). The angle 30° is 150° less than 180° and is just above the negative *x*-axis.

4. **Draw the angle you created in Step 2.**

   You need to draw the terminal side of a –570° angle, so be careful which way your arrow points and how many times you travel around the unit circle before you stop at the terminal side.

   This angle starts at 0 on the *x*-axis and moves in a clockwise direction, because you're finding a negative angle. Figure 9-2 shows what the finished angle looks like.

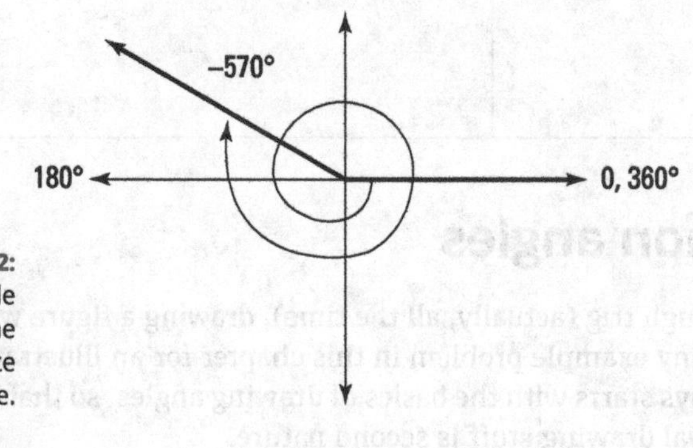

FIGURE 9-2: A –570° angle on the coordinate plane.

YOUR TURN

9 Draw the angle measuring 790° on the coordinate plane.

10 Draw the angle measuring –390° on the coordinate plane.

Solving Equations Using Inverse Functions

You can incorporate reference angles (see the previous section) into some other pre-calculus techniques to solve trig equations. One such technique is factoring. You've been factoring since algebra (and in Chapter 5), so this process shouldn't be anything new. When confronted with an equation that's equal to 0 and a trig function that's being squared, or when you have two different trig functions that are being multiplied together, you should try to use factoring to get your solution first. After factoring, you can use the zero-product property (see Chapter 1) to set each factor equal to 0 and then solve them separately.

EXAMPLE

Q. Solve for x: $2\sin^2 x + \sin x - 1 = 0$. Determine all solutions when $0 \le x \le 2\pi$.

A. Follow these steps:

1. **Let a variable equal the trig ratio and rewrite the equation to simplify.**

   Let $u = \sin x$ and rewrite the equation as $2u^2 + u - 1 = 0$.

2. **Check to make sure that the equation factors.**

   Remember to always check for greatest common factor first. Refer to the factoring information in Chapter 5.

3. **Factor the quadratic.**

   The equation $2u^2 + u - 1 = 0$ factors to $(u+1)(2u-1) = 0$.

4. **Switch the variables back to trig functions.**

   Rewriting your factored trig equation gives you $(\sin x + 1)(2\sin x - 1) = 0$.

5. **Use the zero-product property to solve.**

   If $\sin x = -1$, then solving for x, you have $x = \sin^{-1}(-1)$, and this occurs when $x = \frac{3\pi}{2}$; if $\sin x = \frac{1}{2}$, then you write $x = \sin^{-1}\left(\frac{1}{2}\right)$. The sine is equal to $\frac{1}{2}$ at both $x = \frac{\pi}{6}$ and $x = \frac{5\pi}{6}$. So, if $0 \le x \le 2\pi$, then $x = \frac{\pi}{6}, \frac{5\pi}{6}, \frac{3\pi}{2}$.

EXAMPLE **Q.** Solve for x: $4\sin^2 x - 3 = 0$. Determine all possible values of x for $0 \le x \le 360°$.

A. Follow these steps:

1. **Isolate the trig expression.**

   For $4\sin^2 x - 3 = 0$, add 3 to each side and divide by 4 on both sides to get $\sin^2 x = \frac{3}{4}$.

2. **Take the square root of both sides.**

   Don't forget to take the positive and negative square roots, which gives you $\sin x = \pm\frac{\sqrt{3}}{2}$.

3. **Write the related inverse function.**

   When you have $\sin x = \pm\frac{\sqrt{3}}{2}$, you can say that $x = \sin^{-1}\left(\frac{\sqrt{3}}{2}\right)$ or $x = \sin^{-1}\left(-\frac{\sqrt{3}}{2}\right)$.

4. **Find the solutions.**

   Working with $x = \sin^{-1}\left(\frac{\sqrt{3}}{2}\right)$, this is true when $x = 60°$ and $x = 120°$ Now, working with the negative solution, $x = \sin^{-1}\left(-\frac{\sqrt{3}}{2}\right)$. This is true when $x = 240°$ and $x = 300°$.

   So $x = 60°, 120°, 240°, 300°$.

Q. Solve for x: $\tan^2 x + \frac{2\sqrt{3}}{3}\tan x - 1 = 0$. Determine all possible values of x in terms of radian measures.

A. Follow these steps:

1. **Factor the quadratic.**

   Let $u = \tan x$ and rewrite the equation as $u^2 + \frac{2\sqrt{3}}{3}u - 1 = 0$. Since you're working with the tangent, take advantage of the two tangent values $\sqrt{3}$ and $\frac{\sqrt{3}}{3}$. The quadratic factors into $(u + \sqrt{3})\left(u - \frac{\sqrt{3}}{3}\right) = 0$.

2. **Switch the variables back to trig functions.**

   $(\tan x + \sqrt{3})\left(\tan x - \frac{\sqrt{3}}{3}\right) = 0$

3. **Use the zero-product property to solve.**

   When you have $\tan x + \sqrt{3} = 0$, then $\tan x = -\sqrt{3}$, and you can say that $x = \tan^{-1}(-\sqrt{3})$. Between 0 and 2π this is true when $x = \frac{2\pi}{3}$ and $x = \frac{5\pi}{3}$. And when $\tan x - \frac{\sqrt{3}}{3} = 0$, you have $x = \tan^{-1}\left(\frac{\sqrt{3}}{3}\right)$. Between 0 and 2π this is true when $x = \frac{\pi}{6}$ and $x = \frac{7\pi}{6}$.

4. **Write expressions indicating all the solutions.**

   When $x = \frac{2\pi}{3}$ and $x = \frac{5\pi}{3}$, then all the angles with the same values are written: $x = \frac{(3k+2)\pi}{3}$ where k is an integer. And when $x = \frac{\pi}{6}$ and $x = \frac{7\pi}{6}$, then all the angles with the same values are written: $x = \frac{(6k+1)\pi}{6}$ where k is an integer.

YOUR TURN

11 **Solve $2\sin x = -\sqrt{3}$. Determine all possible values of x in terms of degree measures.**

Solve $-11\tan x = 0$. Determine all solutions when $0 \le x \le 360°$.

13 **Solve $\cos^2 x - \frac{1}{2} = 0$. Determine all solutions when $0 \le x \le 2\pi$.**

Solve $\sin^2 x = \sin x$. Determine all possible values of x in terms of radian measures.

TIP

You can use your knowledge of trig functions to make an educated guess about how many solutions an equation can have. If the equation has sine or cosine values that are greater than 1 or less than −1, for instance, then the equation has no solutions.

When you see a trig equation that asks you to solve for an unknown variable, you move backward from what you're given to arrive at a solution that makes sense. This solution should be in the form of an angle measurement, and the location of the angle should be in the correct quadrant. Knowledge of the unit circle comes in handy here because you'll be thinking of angles that fulfill the requirements of the given equation.

Measuring Arcs: When the Circle Is Put in Motion

Knowing how to calculate the circumference of a circle and, in turn, the length of an *arc* — a portion of the circumference — is important because you can use that information to analyze the motion of an object moving in a circle.

An arc can come from a *central angle*, which is an angle whose vertex is located at the center of the circle. You can measure an arc in two different ways.

- **As an angle:** The measure of an arc in degrees or radians is the same as the measure of the central angle that intercepts it.
- **As a length:** The length of an arc is directly proportional to the circumference of the circle and is dependent on both the central angle and the radius of the circle.

If you think back to geometry, you may remember that the formula for the circumference of a circle is $C = 2\pi r$, with r standing for the radius. Also recall that a circle has 360° or 2π radians. So if you need to find the length of an arc, you need to figure out what part of the whole circumference (or what fraction) you're looking at.

TECHNICAL STUFF

You use the following formula to calculate the arc length; θ represents the measure of the angle in degrees, and s represents arc length, as shown in Figure 9-3: $s = \frac{\theta}{360} \cdot 2\pi r$.

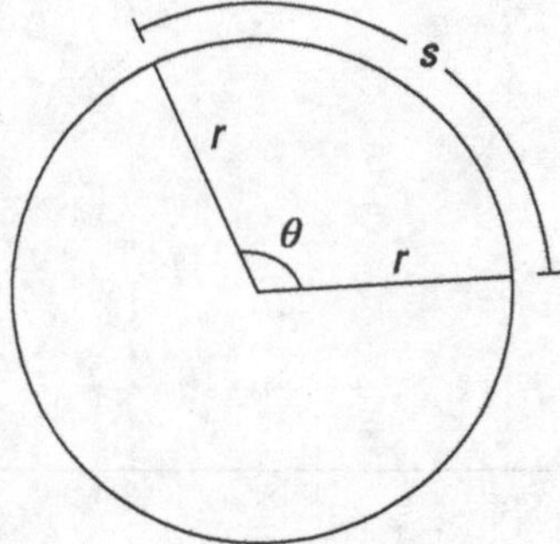

FIGURE 9-3: The variables involved in computing arc length.

If the given angle is in radians, the 2π cancels and its arc length is $s = \frac{\theta}{\cancel{2\pi}} \cdot \cancel{2\pi} r = \theta r$.

EXAMPLE

Q. Find the length of an arc with an angle measurement of 40° if the circle has a radius of 10.

A. Use the following steps:

1. **Assign variable names to the values in the problem.**

   The angle measurement here is 40°, which is θ. The radius is 10, which is r.

2. **Plug the known values into the formula.**

   This step gives you $s = \frac{40}{360} \cdot 2\pi(10)$.

3. **Simplify to solve the formula.**

   You first get $s = \frac{1}{9} \cdot 20\pi$ or $s = \frac{20\pi}{9}$.

   Figure 9-4 shows what this arc looks like.

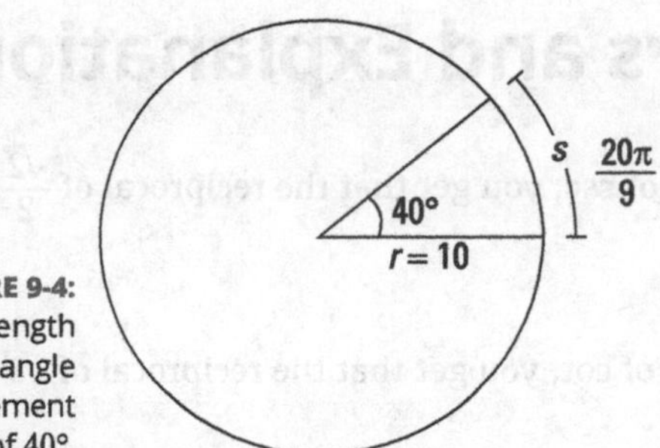

FIGURE 9-4: The arc length for an angle measurement of 40°.

EXAMPLE

Q. **Find the measure of the central angle of a circle, in radians, with an arc length of 28π and a radius of 16.**

A. This time, you must solve for θ (the formula is $s = \theta r$ when dealing with radians):

1. **Plug in what you know to the radian formula.**

   This gives you $28\pi = \theta \cdot 16$.

2. **Divide both sides by 16.**

   Your formula looks like this: $\frac{28\pi}{16} = \frac{\theta \cdot 16}{16}$.

3. **Reduce the fraction.**

   You get $\frac{{}^{7}\cancel{28}\pi}{{}_{4}\cancel{16}} = \frac{\theta \cdot \cancel{16}}{\cancel{16}}$.

   You're left with the solution $\theta = \frac{7\pi}{4}$.

YOUR TURN

15 Find the length of an arc with an angle measurement of $\frac{5\pi}{6}$ radians if the circle has a radius of 5.

16 Find the measure of the central angle of a circle, in degrees, with an arc length of $\frac{\pi}{2}$ and a radius of 9.

Practice Questions Answers and Explanations

(1) $\sqrt{2}$. Using $\sin 45° = \frac{\sqrt{2}}{2}$, which is the reciprocal of csc, you get that the reciprocal of $\frac{\sqrt{2}}{2}$ is $\frac{2}{\sqrt{2}} = \frac{2}{\sqrt{2}} \cdot \frac{\sqrt{2}}{\sqrt{2}} = \frac{2\sqrt{2}}{2} = \sqrt{2}$.

(2) $\frac{\sqrt{3}}{3}$. Using $\tan 60° = \sqrt{3}$, which is the reciprocal of cot, you get that the reciprocal of $\sqrt{3}$ is $\frac{1}{\sqrt{3}} = \frac{1}{\sqrt{3}} \cdot \frac{\sqrt{3}}{\sqrt{3}} = \frac{\sqrt{3}}{3}$.

(3) **60°.** $180 + \boxed{60} = 240$.

(4) $\frac{\pi}{6}$. $2\pi - \boxed{\frac{\pi}{6}} = \frac{11\pi}{6}$.

(5) $-\frac{\sqrt{3}}{2}$. The reference angle is 30°, and $\cos 30° = \frac{\sqrt{3}}{2}$. The cosine is negative in quadrant III.

(6) $-\sqrt{3}$. The reference angle is 60°, and $\tan 60° = \sqrt{3}$. The tangent is negative in quadrant II.

(7) $\frac{\sqrt{2}}{2}$. The reference angle is $\frac{\pi}{4}$, and $\sin\frac{\pi}{4} = \frac{\sqrt{2}}{2}$. The sine is positive in quadrant II.

(8) $-\sqrt{3}$. The reference angle is 30°. You find the value of the tangent function and use the reciprocal: $\tan 30° = \frac{\sqrt{3}}{3}$, and its reciprocal is $\sqrt{3}$. Since the tangent (and cotangent) are negative in quadrant IV, $\cot 330° = -\sqrt{3}$.

(9) **See the following figure.** Making 2 complete counterclockwise rotations brings you to 720°, which is back to the positive x-axis. Then 70° more creates an angle in quadrant I.

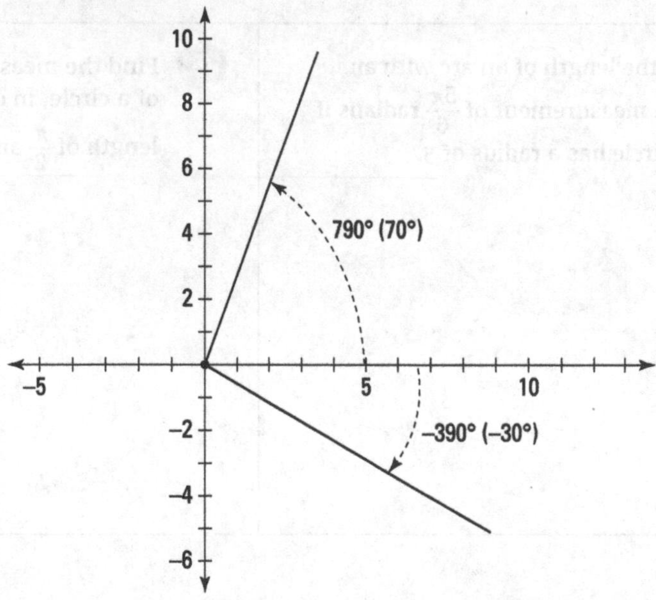

(10) **See the above figure.** Making one complete clockwise rotation brings you back to the positive x-axis. Then rotating 30° more creates an angle in quadrant IV.

11 **$x = 240° + 360k°$ or $x = 300° + 360k°$.** Divide each side by 2 to get $\sin x = -\frac{\sqrt{3}}{2}$. Then solve for x in $x = \sin^{-1}\left(-\frac{\sqrt{3}}{2}\right)$. The two angles between 0 and 360° whose sign is $-\frac{\sqrt{3}}{2}$ are 240° and 300°. All the multiples of 240° are: $x = 240° + 360k°$ where k is an integer. And all the multiples of 300° are: $x = 300° + 360k°$, where k is an integer.

12 **0° or 180°.** Divide each side by −11. Then solve $x = \tan^{-1}(0)$. This happens when $x = 0°$ or $180°$.

13 $\frac{\pi}{4}, \frac{3\pi}{4}$. Add $\frac{1}{2}$ to each side and then find the square root: $\cos^2 x = \frac{1}{2} \rightarrow \cos x = \pm\sqrt{\frac{1}{2}} = \pm\frac{\sqrt{2}}{2}$, then $x = \cos^{-1}\left(\pm\frac{\sqrt{2}}{2}\right) = \frac{\pi}{4}, \frac{3\pi}{4}$.

14 $x = k\pi$ **or** $x = \frac{(4k+1)\pi}{2}$. Subtract $\sin x$ from each side and then factor: $\sin^2 x - \sin x = \sin x(\sin x - 1) = 0$. Setting the two factors equal to 0, you have $\sin x = 0$ and $\sin x = 1$. The solutions are $x = \sin^{-1}(0) = 0, \pi$ and $x = \sin^{-1}(1) = \frac{\pi}{2}$. The multiples of 0 and π are written $x = k\pi$, where k is an integer. And the multiples of $\frac{\pi}{2}$ are written $x = \frac{(4k+1)\pi}{2}$, where k is an integer.

15 $\frac{25}{6}\pi$. Using the formula $s = \theta r$, you have $s = \frac{5\pi}{6}(5) = \frac{25}{6}\pi$.

16 **10°.** Using the formula $s = \theta r$, you have $\frac{\pi}{2} = \theta \cdot 9$. Solving for θ, $\frac{\pi}{18} = \theta$. Change the angle measure in radians to degrees. Since $2\pi = 360°$, $\pi = 180° \rightarrow \frac{\pi}{18} = \frac{180°}{18} = 10°$.

If you're ready to test your skills a bit more, take the following chapter quiz that incorporates all the chapter topics.

Whaddya Know? Chapter 9 Quiz

Quiz time! Complete each problem to test your knowledge on the various topics covered in this chapter. You can then find the solutions and explanations in the next section.

1. **What is the arc length created by a 135° angle in a circle with a radius of 10 feet?**
2. $\sec 60° =$
3. **Find the reference angle related to $\frac{4\pi}{3}$.**
4. **Find the exact value of $\cos\frac{5\pi}{6}$.**
5. **Find the reference angle related to tan(−600°).**
6. **Solve for x: $4\cos^4 x = \cos^2 x$, when $0 \le x \le 2\pi$.**
7. **Find the exact value of $\csc 90°$.**
8. **Find the exact value of $\sec\frac{5\pi}{3}$.**
9. **Solve for all values of x where x is in degrees: $3\tan x = -3$.**
10. **What is the arc length created by a $\frac{3\pi}{4}$ radian angle in a circle with a radius of 6?**
11. **Solve for x: $2\sin^2 x - \sin x = 1$, when $0 \le x \le 360°$.**
12. **Find the reference angle related to 300°.**
13. **Find the exact value of $\sin 225°$.**

Answers to Chapter 9 Quiz

1. $\approx$ **23.56 ft.** Using the formula $s = \frac{\theta}{360} \cdot 2\pi r$, you have $s = \frac{135}{360} \cdot 2\pi \cdot 10 = \frac{135}{18} \cdot \pi = \frac{15\pi}{2} \approx 23.56$.

2. **2.** Since the secant is the reciprocal of cosine, use $\cos 60° = \frac{1}{2}$ and find its reciprocal value: 2.

3. $\frac{\pi}{3}$. This is because $\pi + \boxed{\frac{\pi}{3}} = \frac{4\pi}{3}$.

4. $-\frac{\sqrt{3}}{2}$. The reference angle for $\frac{5\pi}{6}$ is $\frac{\pi}{6}$. Then, using the table, $\cos\frac{\pi}{6} = \frac{\sqrt{3}}{2}$. Since $\frac{5\pi}{6}$ is in quadrant II, the cosine is negative.

5. **60°.** Add 360 degrees twice to get $-600 + 360 + 360 = 120$. The reference angle for 120° is 60°, since $120° = 180° - 60°$.

6. $\frac{\pi}{2}, \frac{3\pi}{2}, \frac{\pi}{3}, \frac{2\pi}{3}, \frac{4\pi}{3}, \frac{5\pi}{3}$. First, subtract $\cos^2 x$ from each side, and then factor it from each term: $4\cos^4 x - \cos^2 x = 0 \rightarrow \cos^2 x(4\cos^2 x - 1) = 0$. Set the two factors equal to 0 and solve for x. When $\cos^2 x = 0$, you have $\cos x = 0$. Solving for x, $x = \cos^{-1} 0 = \frac{\pi}{2}, \frac{3\pi}{2}$. Next, when $4\cos^2 x - 1 = 0 \rightarrow 4\cos^2 x = 1 \rightarrow \cos^2 x = \frac{1}{4} \rightarrow \cos x = \pm\sqrt{\frac{1}{4}} = \pm\frac{1}{2}$. Solving for x: first, $x = \cos^{-1}\left(\frac{1}{2}\right) = \frac{\pi}{3}, \frac{5\pi}{3}$, and lastly, $x = \cos^{-1}\left(-\frac{1}{2}\right) = \frac{2\pi}{3}, \frac{4\pi}{3}$.

7. **1.** Since cosecant is the reciprocal of sine, first find the sine of 90°: $\sin 90° = 1$. The reciprocal of 1 is 1.

8. **2.** First, find $\cos\frac{5\pi}{3}$. The reference angle is $\frac{\pi}{3}$, and $\cos\frac{\pi}{3} = \frac{1}{2}$. An angle of $\frac{5\pi}{3}$ is in quadrant IV, so the cosine is positive, giving you $\cos\frac{5\pi}{3} = \frac{1}{2}$. The reciprocal then is 2.

9. $x = 135° + 180k°$. First, divide each side by 3, which gives you $\tan x = -1$. Solving for x, $x = \tan^{-1}(-1)$. This is true when $x = 135°$ or 315°. For all the possible angles, write $x = 135° + 180k°$ where k is an integer.

10. $\frac{9\pi}{2}$. Using the formula $s = \theta r$, you have $s = \frac{3\pi}{4} \cdot 6 = \frac{9\pi}{2}$.

11. **90°, 330°.** First, subtract 1 from each side of the equation, and then factor the quadratic: $2\sin^2 x - \sin x = 1 \rightarrow 2\sin^2 x - \sin x - 1 = 0 \rightarrow (2\sin x + 1)(\sin x - 1) = 0$. Set each factor equal to 0 and solve for x. When $2\sin x + 1 = 0 \rightarrow 2\sin x = -1 \rightarrow \sin x = -\frac{1}{2}$. Solving for x: $x = \sin^{-1}\left(-\frac{1}{2}\right) = 330°$. And when $\sin x - 1 = 0 \rightarrow \sin x = 1$. Solving for x: $x = \sin^{-1}(1) = 90°$.

12. **60°.** This is because $360° - \boxed{60°} = 300°$.

13. $-\frac{\sqrt{2}}{2}$. The reference angle for 225° is 45°, and $\sin 45° = \frac{\sqrt{2}}{2}$. The sine is negative in quadrant III, so $\sin 225° = -\frac{\sqrt{2}}{2}$.

IN THIS CHAPTER

» Plotting the sine and cosine parent graphs

» Picturing tangent and cotangent

» Charting secant and cosecant

Chapter 10
Picturing Basic Trig Functions and Reciprocal Functions

"Graph the trig function . . ." This command sends shivers down the spines of many otherwise brave mathematics students. But there's really nothing to fear, because graphing functions can be easy. Graphing functions is simply a matter of inserting the x value (from the domain) in place of the function's variable and solving the equation to get the y value (in the range). You continue with that calculation until you have enough points to plot. When do you know you have enough? When your graph has a clear line, ray, curve, or what-have-you.

You've dealt with functions before in math, but up until now, the input of a function was typically x. In trig functions, however, the input of the function is usually θ, which is basically just another variable to use but indicates an angle measure. This chapter shows you how to graph trig functions by using different values for θ.

REMEMBER

In Chapter 8, you use two ways of measuring angles: degrees and radians. Lucky for you, you now get to focus solely on radians when graphing trig functions. More often than not, mathematicians graph in radians when working with trig functions, and this presentation will continue that tradition — until someone comes up with a better way, of course.

Drafting the Sine and Cosine Parent Graphs

The trig functions, especially sine and cosine, displayed their usefulness in Chapter 9. After putting them under the microscope, you're now ready to begin graphing them. After you discover the basic shape of the sine and cosine graphs, you can begin to graph more complicated versions, using the same transformations you discover in Chapter 4 when performing them on the algebraic functions.

Knowing how to graph trig functions allows you to measure the movement of objects that move back and forth or up and down in a regular interval, such as pendulums. Sine and cosine as functions are perfect ways of expressing this type of movement, because their graphs are repetitive and they oscillate (like a wave). The following sections illustrate these facts.

Sketching sine

Sine graphs move in waves. The waves crest and fall over and over again forever, because you can keep plugging in values for θ forever. In this section you see how to construct the parent graph for the sine function, $f(\theta) = \sin\theta$ (for more on parent graphs, refer to Chapters 3 and 4).

TIP

Because all the values of the sine function come from the unit circle, you should be pretty comfy with the unit circle before proceeding with this work. If you're not, refer to Chapter 8 for a brush-up.

You can graph any trig function in four or five steps. Here are the steps to construct the graph of the parent function $f(\theta) = \sin\theta$:

1. **Find the values for domain and range.**

REMEMBER

   No matter what you put into the sine function, you get a number as output, because the angle measures of θ can rotate around the unit circle in either direction an infinite number of times. Therefore, the domain of sine is all real numbers, or $(-\infty, \infty)$.

   On the unit circle, the *y* values are your sine values — what you get after plugging the value of θ into the sine function. Because the radius of the unit circle is 1, the *y* values can't be more than 1 or less than −1 — your range for the sine function. So in the *x*-direction, the wave (or *sinusoid,* in math language) goes on forever, and in the *y*-direction, the sinusoid oscillates only between −1 and 1, including these values. In interval notation, you write this as $[-1, 1]$.

2. **Calculate the graph's *x*-intercepts.**

   When you graphed lines in algebra, the *x*-intercepts occurred when $y = 0$. In this case, the sine function gives you the *y* value. Find out where the graph crosses the *x*-axis by finding unit circle values where sine is 0. In one full rotation, the graph crosses the *x*-axis three times: once at 0, once at π, and once at 2π. You now know that three of the coordinate points are $(0,0)$, $(\pi,0)$, and $(2\pi,0)$. These are the *x*-intercepts.

3. **Calculate the graph's maximum and minimum points.**

   To complete this step, use your knowledge of the range from Step 1. You know that the highest value of y is 1. Where does this happen? At $\frac{\pi}{2}$. You now have another coordinate point at $\left(\frac{\pi}{2},1\right)$. You also can see that the lowest value of y is -1, when x is $\frac{3\pi}{2}$. Hence, you have another coordinate point: $\left(\frac{3\pi}{2},-1\right)$.

4. **Sketch the graph of the function.**

   Using the five key points as a guide, connect the points with a smooth, round curve. Figure 10-1 shows the parent graph of sine approximately.

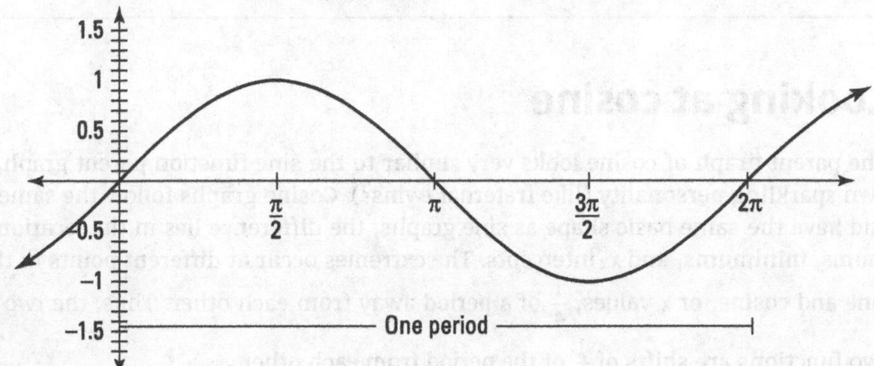

FIGURE 10-1: The parent graph of sine, $f(\theta)=\sin\theta$.

REMEMBER

The parent graph of the sine function has a couple of important characteristics worth noting:

- **It repeats itself every 2π radians.** This repetition occurs because 2π radians is one trip around the unit circle — called the *period* of the sine graph — and after that, you start to go around again. Usually, you're asked to draw the graph to show one period of the function, because in this period you capture all possible values for sine before it starts repeating over and over again. The graph of sine is called *periodic* because of this repeating pattern.
- **It's symmetrical about the origin (thus, in math speak, it's an *odd function*).** The sine function has 180°-point symmetry about the origin. If you look at it upside down, the graph looks exactly the same. The official math definition of an *odd function,* though, is $f(-x)=-f(x)$ for every value of x in the domain. In other words, if you put in an opposite input, you get an opposite output. For example, $\sin\frac{\pi}{6}=\frac{1}{2}$, but if you look at $\sin\left(-\frac{\pi}{6}\right)$, you get $-\frac{1}{2}$.

YOUR TURN

Sketch the graph of $f(\theta)=\sin\theta$ from -2π to 2π. Name all the x-intercepts.

2 In your sketch of $f(\theta) = \sin\theta$, draw the line $y = \frac{1}{2}$. Name all the points where the sine curve intersects with the line.

Looking at cosine

The parent graph of cosine looks very similar to the sine function parent graph, but it has its own sparkling personality (like fraternal twins?). Cosine graphs follow the same basic pattern and have the same basic shape as sine graphs; the difference lies in the location of the maximums, minimums, and x-intercepts. The extremes occur at different points in the domains of sine and cosine, or x values, $\frac{1}{4}$ of a period away from each other. Thus, the two graphs of the two functions are shifts of $\frac{1}{4}$ of the period from each other.

Just as with the sine graph, you use the five key points of graphing trig functions to get the parent graph of the cosine function. If necessary, you can refer to the unit circle for the cosine values to start with (see Chapter 8). As you work more with these functions, your dependence on the unit circle should decrease until eventually you don't need it at all. Here are the steps:

1. **Find the values for domain and range.**

   As with sine graphs (see the previous section), the domain of cosine is all real numbers, and its range is $-1 \le y \le 1$, or $[-1,1]$.

2. **Calculate the graph's x-intercepts.**

   Referring to the unit circle, find where the graph crosses the x-axis by finding unit circle values of 0. It crosses the x-axis twice — once at $\frac{\pi}{2}$ and then at $\frac{3\pi}{2}$. Those crossings give you two coordinate points: $\left(\frac{\pi}{2},0\right)$ and $\left(\frac{3\pi}{2},0\right)$.

3. **Calculate the graph's maximum and minimum points.**

   Using your knowledge of the range for cosine from Step 1, you know the highest value that y can be is 1, which happens twice for cosine within one period — once at 0 and once at 2π (see Figure 10-2), giving you two maximums: $(0,1)$ and $(2\pi,1)$. The minimum value that y can be is -1, which occurs at π. You now have another coordinate pair at $(\pi,-1)$.

4. **Sketch the graph of the function.**

   Figure 10-2 shows the full parent graph of cosine with the five key points plotted.

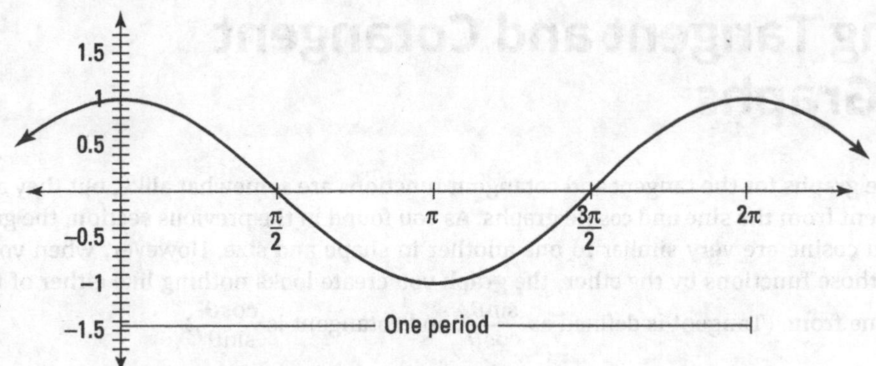

FIGURE 10-2: The parent graph of cosine, $f(\theta) = \cos\theta$.

REMEMBER

The cosine parent graph has a couple of characteristics worth noting:

- **It repeats every 2π radians.** This repetition means it's a periodic function, so its waves rise and fall in the graph (see the previous section for the full explanation).
- **It's symmetrical about the *y*-axis (in mathematical dialect, it's an *even function*).** Unlike the sine function, which has 180° symmetry, cosine has *y*-axis symmetry. In other words, you can fold the graph in half at the *y*-axis and it matches exactly. The formal definition of an even function is $f(x) = f(-x)$; if you plug in the opposite input, you'll get the same output.

  For example, $\cos\left(\frac{\pi}{6}\right) = \frac{\sqrt{3}}{2}$ and $\cos\left(-\frac{\pi}{6}\right) = \frac{\sqrt{3}}{2}$.

  Even though the input sign changed, the output sign stayed the same. This always happens with the cosine of any θ value and its opposite.

YOUR TURN

 **3** **Sketch the graph of $f(\theta) = \cos\theta$ from -2π to 2π. Name all the x-intercepts.**

 **4** **In your sketch of $f(\theta) = \cos\theta$, draw the line $y = -1$. Name all those minimum points where the cosine curve intersects with the line.**

Graphing Tangent and Cotangent Parent Graphs

The graphs for the tangent and cotangent functions are somewhat alike, but they are quite different from the sine and cosine graphs. As you found in the previous section, the graphs of sine and cosine are very similar to one another in shape and size. However, when you divide one of those functions by the other, the graph you create looks nothing like either of the graphs it came from. (Tangent is defined as $\frac{\sin\theta}{\cos\theta}$, and cotangent is $\frac{\cos\theta}{\sin\theta}$.)

The hardest part of graphing tangent and cotangent comes from the fact that they both have asymptotes in their graphs, which happens often with rational functions, and they continue on to positive and negative infinity (see Chapter 3).

REMEMBER

The tangent graph has an asymptote wherever the cosine is 0, and the cotangent graph has an asymptote wherever the sine is 0. Keeping these asymptotes separate from one another helps you draw your graphs.

The tangent and cotangent functions have parent graphs just like any other function. Using the graphs of these functions, you can make the same types of transformations that apply to the parent graphs of any function. The following sections plot the parent graphs of tangent and cotangent.

Tackling tangent

TIP

The easiest way to remember how to graph the tangent function is to remember that $\tan\theta = \frac{\sin\theta}{\cos\theta}$ (see Chapter 8 to review).

Because $\cos\theta = 0$ for various values of θ, some interesting things happen to tangent's graph. When the denominator of a fraction is 0, the fraction is *undefined.* Therefore, the graph of tangent jumps over an asymptote, which is where the function is undefined, at each of these places.

Table 10-1 presents θ, $\sin\theta$, $\cos\theta$, and $\tan\theta$. It shows the roots (or zeros), the asymptotes (where the function is undefined), and the behavior of the graph in between certain key points on the unit circle.

TABLE 10-1 Finding Out Where $\tan\theta$ Is Undefined

θ	0	$0<\theta<\frac{\pi}{2}$	$\frac{\pi}{2}$	$\frac{\pi}{2}<\theta<\pi$	π	$\pi<\theta<\frac{\sqrt{3}\pi}{2}$	$\frac{\sqrt{3}\pi}{2}$	$\frac{\sqrt{3}\pi}{2}<\theta<2\pi$	2π
$\sin\theta$	0	positive	1	positive	0	negative	–1	negative	0
$\cos\theta$	1	positive	0	negative	–1	negative	0	positive	1
$\tan\theta$	0	positive	undefined	negative	0	positive	undefined	negative	0

To plot the parent graph of a tangent function, you start out by finding the vertical asymptotes. Those asymptotes give you some structure from which you can fill in the missing points.

1. **Find the vertical asymptotes so you can find the domain.**

   In order to find the domain of the tangent function, you have to locate the vertical asymptotes. The first asymptote occurs when $\theta = \frac{\pi}{2}$, and it repeats every π radians (see the unit circle in Chapter 8). (***Note:*** The period of the tangent graph is π radians, which is different from that of sine and cosine.) Between 0 and 2π, the tangent function has asymptotes when $\theta = \frac{\pi}{2}$ and $\theta = \frac{3\pi}{2}$.

TIP

   The easiest way to write how often the asymptotes reoccur is by describing when the tangent is undefined: $\theta \neq \frac{\pi}{2} + n\pi$ where n is an integer. You write "$+n\pi$" because the period of tangent is π radians, so if an asymptote is at $\frac{\pi}{2}$ and you add or subtract π, you automatically find the next asymptote.

2. **Determine values for the range.**

   Recall that the tangent function can be defined as $\frac{\sin\theta}{\cos\theta}$.

   Both of these values can be decimals. The closer you get to the values where $\cos\theta = 0$, the smaller the number on the bottom of the fraction gets and the larger the value of the overall fraction gets — in either the positive or negative direction.

REMEMBER

   The range of tangent has no restrictions; you aren't stuck between 1 and −1, like with sine and cosine. In fact, the ratios are any and all numbers. The range is $(-\infty, \infty)$.

3. **Calculate the graph's *x*-intercepts.**

   Tangent's parent graph has roots (it crosses the *x*-axis) at 0, π, and 2π.

   You can find these values by setting $\frac{\sin\theta}{\cos\theta}$ equal to 0 and then solving. The *x*-intercepts for the parent graph of tangent are located wherever the sine value is 0.

4. **Figure out what's happening to the graph between the intercepts and the asymptotes.**

   The graph in the first quadrant of tangent is positive and moves upward toward the asymptote at $\frac{\pi}{2}$, because all sine and cosine values are positive in the first quadrant. The graph in quadrant II is negative because sine is positive and cosine is negative. In quadrant III it's positive because both sine and cosine are negative, and in quadrant IV it's negative because sine is negative and cosine is positive.

   Note: A tangent graph has no maximum or minimum points.

Figure 10-3 shows what the parent graph of tangent looks like when you put it all together.

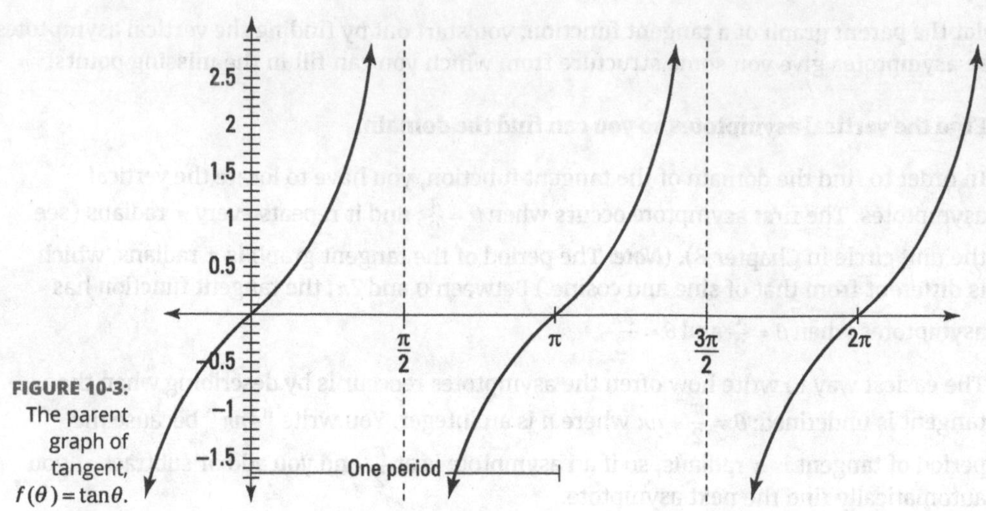

FIGURE 10-3: The parent graph of tangent, $f(\theta) = \tan\theta$.

Clarifying cotangent

The parent graphs of sine and cosine are very similar because the values are exactly the same; they just occur for different values of θ. Similarly, the parent graphs of tangent and cotangent are comparable because they both have asymptotes and x-intercepts. The only differences you can see between tangent and cotangent are the values of θ where the asymptotes and x-intercepts occur. You can find the parent graph of the cotangent function, $\cot\theta = \frac{\cos\theta}{\sin\theta}$, by using the same techniques you use to find the tangent parent graph (see the previous section).

TIP

Table 10-2 shows θ, $\cos\theta$, $\sin\theta$, and $\cot\theta$ so that you can see both the x-intercepts and the asymptotes in comparison. These points help you find the general shape of your graph so that you have a nice place to start.

TABLE 10-2 Spotting Where cot θ Is Undefined

θ	0	$0 < \theta < \frac{\pi}{2}$	$\frac{\pi}{2}$	$\frac{\pi}{2} < \theta < \pi$	π
$\cos\theta$	1	positive	0	negative	-1
$\sin\theta$	0	positive	1	positive	0
$\cot\theta$	undefined	positive	0	negative	undefined
θ		$\pi < \theta < \frac{\sqrt{3}\pi}{2}$	$\frac{\sqrt{3}\pi}{2}$	$\frac{\sqrt{3}\pi}{2} < \theta < 2\pi$	2π
$\cos\theta$		negative	0	positive	1
$\sin\theta$		negative	-1	negative	0
$\cot\theta$		positive	0	negative	undefined

To sketch the full parent graph of cotangent, follow these steps:

1. **Find the vertical asymptotes so you can find the domain.**

   Because cotangent is the quotient of cosine divided by sine, and $\sin\theta$ is sometimes 0, the graph of the cotangent function has asymptotes, just like with tangent. However, these asymptotes occur whenever $\sin\theta = 0$. The asymptotes of $\cot\theta$ are at 0, π, and 2π.

   The cotangent parent graph repeats every π units. Its domain is based on its vertical asymptotes: the first one comes at 0 and then repeats every π radians. The domain, in other words, is everything, except that $\theta \neq n\pi$, where n is an integer.

2. **Find the values for the range.**

   Similar to the tangent function, you can define cotangent as $\frac{\cos\theta}{\sin\theta}$.

   Both of these values can be decimals. The range of cotangent also has no restrictions; the ratios are any and all numbers: $(-\infty, \infty)$. The closer you get to the values where $\sin\theta = 0$, the smaller the number on the bottom of the fraction is and the larger the value of the overall fraction — in either the positive or negative direction.

3. **Determine the x-intercepts.**

   The roots (or zeros) of cotangent between 0 and 2π occur wherever the cosine value is 0: at $\frac{\pi}{2}$ and $\frac{3\pi}{2}$.

4. **Evaluate what happens to the graph between the x-intercepts and the asymptotes.**

   The positive and negative values in the four quadrants stay the same as in tangent, but the asymptotes change the graph. You can see the full parent graph for cotangent in Figure 10-4.

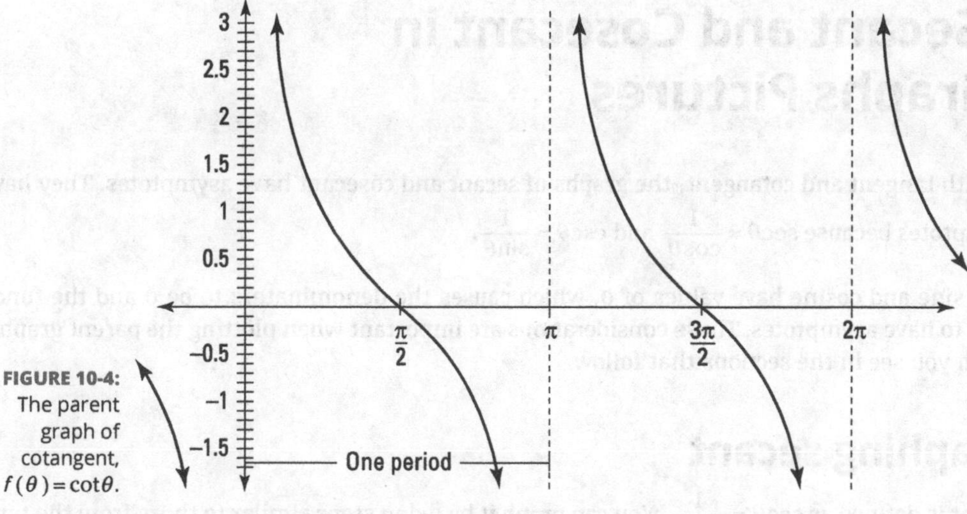

FIGURE 10-4: The parent graph of cotangent, $f(\theta) = \cot\theta$.

YOUR TURN

5 **Sketch the graph of $f(\theta) = \tan\theta$ from -2π to 2π. Name all the x-intercepts.**

6 **Sketch the graph of $f(\theta) = \cot\theta$ from -2π to 2π. Then draw the line $y = 1$ and name all the points where the cotangent curve intersects with the line.**

Putting Secant and Cosecant in Parent Graphs Pictures

As with tangent and cotangent, the graphs of secant and cosecant have asymptotes. They have asymptotes because $\sec\theta = \frac{1}{\cos\theta}$ and $\csc\theta = \frac{1}{\sin\theta}$.

Both sine and cosine have values of 0, which causes the denominators to be 0 and the functions to have asymptotes. These considerations are important when plotting the parent graphs, which you see in the sections that follow.

Graphing secant

Secant is defined as $\sec\theta = \frac{1}{\cos\theta}$. You can graph it by using steps similar to those from the tangent and cotangent sections.

REMEMBER

The cosine graph crosses the x-axis on the interval $[0, 2\pi]$ at two places, so the secant graph has two asymptotes, which divide the period interval into three smaller sections. The parent secant graph doesn't have any x-intercepts.

Follow these steps to create the parent graph of secant:

1. **Find the asymptotes of the secant graph.**

   Because secant is the reciprocal of cosine (see Chapter 8), any place on the cosine graph where the value is 0 creates an asymptote on the secant graph. (And any point with 0 in the denominator is undefined.) Finding these points first helps you define the rest of the graph. The graph of cosine has values of 0 at both $\frac{\pi}{2}$ and $\frac{3\pi}{2}$. So the graph of secant has asymptotes at those same values. Figure 10-5 shows the graph of $\cos\theta$ and the asymptotes.

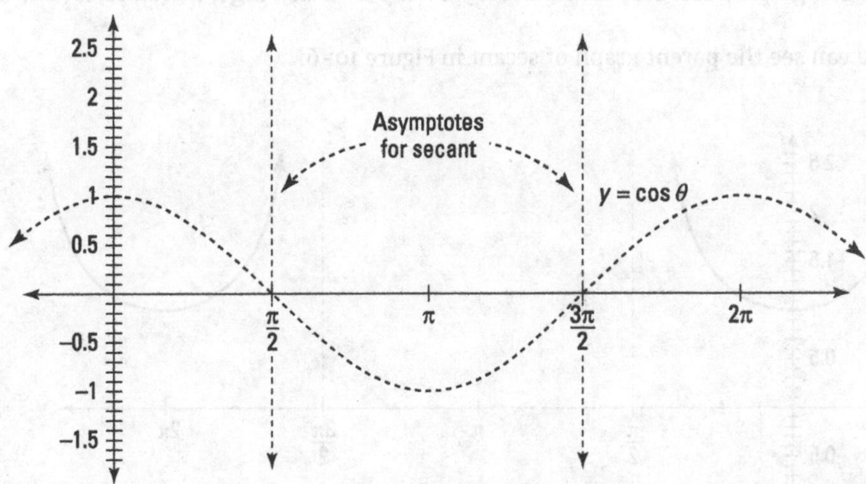

FIGURE 10-5: The graph of cosine reveals the asymptotes of secant.

2. **Calculate what happens to the graph at the first interval between the asymptotes.**

   The period of the parent cosine graph starts at 0 and ends at 2π. You need to figure out what the graph does in between the following points:

   - Zero and the first asymptote at $\frac{\pi}{2}$
   - The two asymptotes in the middle
   - The second asymptote and the end of the graph at 2π

   Start on the interval $\left(0, \frac{\pi}{2}\right)$. The graph of cosine goes from 1, into fractional values, and all the way down to 0. Secant takes the reciprocal of all these values and ends on this first interval at the asymptote. The graph goes higher and higher rather than lower, because as the fractions in the cosine function get smaller, their reciprocals in the secant function get bigger.

3. **Repeat Step 2 for the second interval $\left(\frac{\pi}{2}, \frac{3\pi}{2}\right)$.**

   If you refer to the cosine graph, you see that halfway between $\frac{\pi}{2}$ and $\frac{3\pi}{2}$, there's a low point of −1. Its reciprocal will be the high point of the graph of the secant in this interval. So secant's graph falls as the cosine gets closer to 0.

4. **Repeat Step 2 for the last interval $\left(\frac{3\pi}{2}, 2\pi\right)$.**

   This interval is a mirror image of what happens in the first interval.

5. **Find the domain and range of the graph.**

   Its asymptotes are at $\frac{\pi}{2}$ and repeat every π, so the domain of secant, where n is an integer, is all θ except that $\theta \neq \frac{\pi}{2} + n\pi$.

   The graph exists only for numbers $y \geq 1$ or $y \leq -1$. Its range, therefore, is $(-\infty, -1] \cup [1, \infty)$.

You can see the parent graph of secant in Figure 10-6.

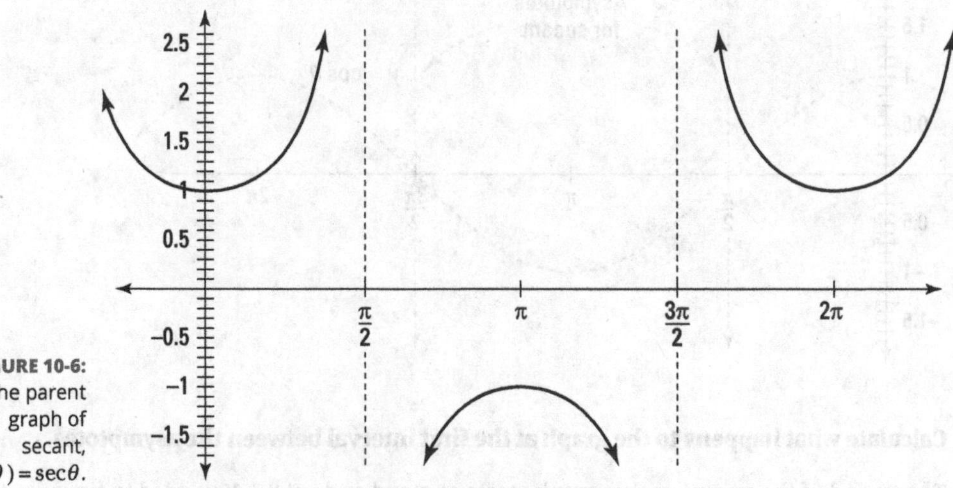

FIGURE 10-6: The parent graph of secant, $f(\theta) = \sec\theta$.

Checking out cosecant

Cosecant is almost exactly the same as secant because it's the reciprocal of sine (as opposed to cosine). Anywhere sine has a value of 0, you see an asymptote in the cosecant graph. Because the sine graph crosses the x-axis three times on the interval $[0, 2\pi]$, you have three asymptotes and two sub-intervals to graph.

REMEMBER

The parent graph of the cosecant function has no x-intercepts, so don't bother looking for them.

The following list explains how to graph cosecant:

1. **Find the asymptotes of the graph.**

   Because cosecant is the reciprocal of sine, any place on sine's graph where the value is 0 creates an asymptote on cosecant's graph. The parent graph of sine has values of 0 at 0, π, and 2π. So cosecant has three asymptotes.

2. **Calculate what happens to the graph at the first interval between 0 and π.**

   The period of the parent sine graph starts at 0 and ends at 2π. You can figure out what the graph does in between the first asymptote at 0 and the second asymptote at π.

   The graph of sine goes from 0 to 1 and then back down again. Cosecant takes the reciprocal of these values, which causes the graph to get bigger where the values of sine are smaller.

3. **Repeat for the second interval $(\pi, 2\pi)$.**

   If you refer to the sine graph, you see that it goes from 0 down to −1 and then back up again. Because cosecant is the reciprocal, its graph gets bigger in the negative direction as it approaches the asymptotes.

4. **Find the domain and range of the graph.**

   Cosecant's asymptotes start at 0 and repeat every π. Its domain is $\theta \neq n\pi$, where n is an integer. The graph also exists for numbers $y \geq 1$ or $y \leq -1$. Its range, therefore, is $(-\infty, -1] \cup [1, \infty)$.

You can see the full graph of $f(\theta) = \csc\theta$ along with its reciprocal $\sin\theta$ in Figure 10-7.

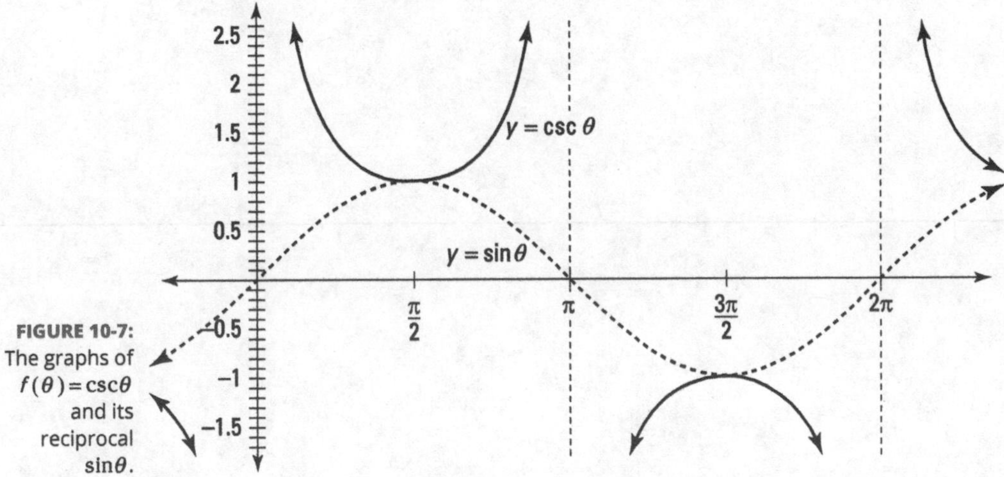

FIGURE 10-7: The graphs of $f(\theta) = \csc\theta$ and its reciprocal $\sin\theta$.

YOUR TURN

7 **Sketch the graph of $f(\theta) = \sec\theta$ from -2π to 2π. Find the distance between two consecutive minimum points (points above the x-axis). See Chapter 1 for the distance formula.**

8 **Sketch the graph of $f(\theta) = \csc\theta$ from -2π to 2π. Find the distance between a minimum point and a maximum point closest to it. See Chapter 1 for the distance formula.**

Practice Questions Answers and Explanations

1. **See the following sketch; the x-intercepts are $(-2\pi,0)$, $(-\pi,0)$, $(0,0)$, $(\pi,0)$, $(2\pi,0)$.** The intercepts are easily recognized on the graph of the function.

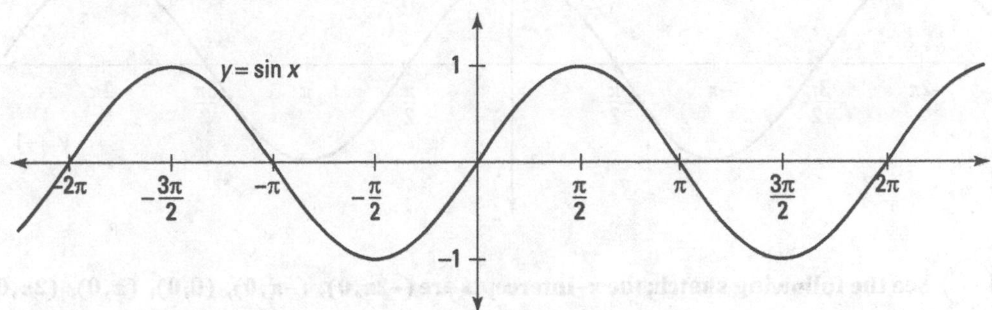

2. **See the following sketch; the x-intercepts are $\left(-\frac{11\pi}{6},\frac{1}{2}\right)$, $\left(-\frac{7\pi}{6},\frac{1}{2}\right)$, $\left(\frac{\pi}{6},\frac{1}{2}\right)$, $\left(\frac{5\pi}{6},\frac{1}{2}\right)$.** The points of intersection aren't easily recognizable from the graph. Refer to Chapter 9 for a chart of basic function values for help in identifying the points.

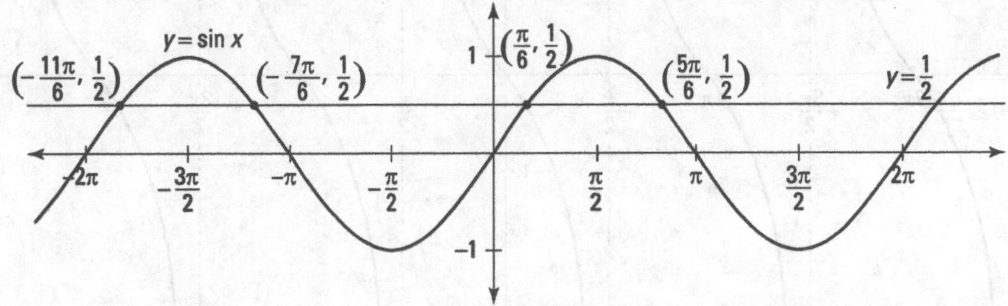

3. **See the following sketch; the x-intercepts are $\left(-\frac{3\pi}{2},0\right)$, $\left(-\frac{\pi}{2},0\right)$, $\left(\frac{\pi}{2},0\right)$, $\left(\frac{3\pi}{2},0\right)$.**

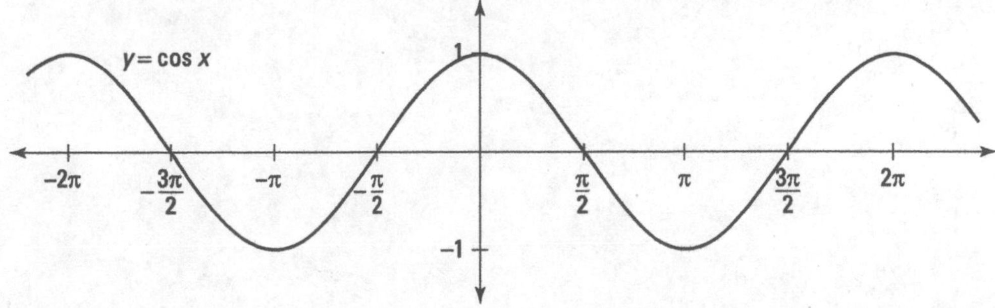

4. **See the following sketch; the minimum points are $(-\pi, 0)$, $(\pi, 0)$.** There are only two minimum values between -2π and 2π.

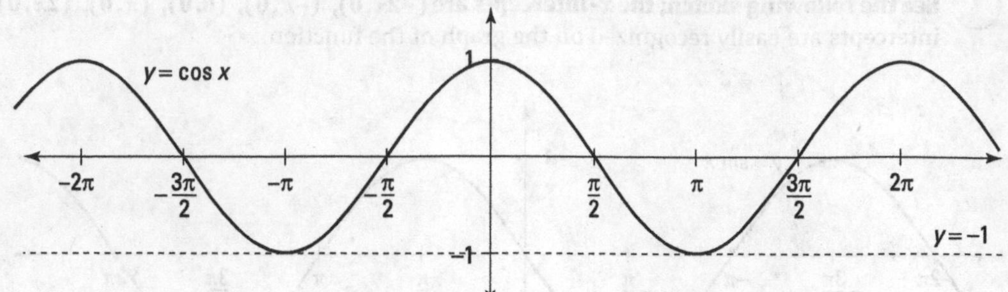

5. **See the following sketch; the *x*-intercepts are $(-2\pi, 0)$, $(-\pi, 0)$, $(0, 0)$, $(\pi, 0)$, $(2\pi, 0)$.**

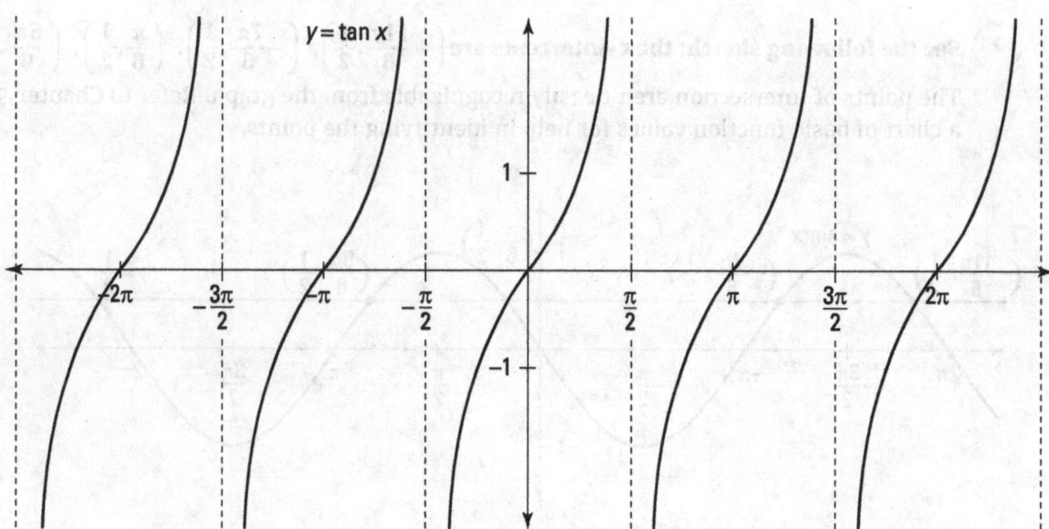

6. **See the following sketch; the points are $\left(-\frac{7\pi}{4},1\right)$, $\left(-\frac{3\pi}{4},1\right)$, $\left(\frac{\pi}{4},1\right)$, $\left(\frac{5\pi}{4},1\right)$.**

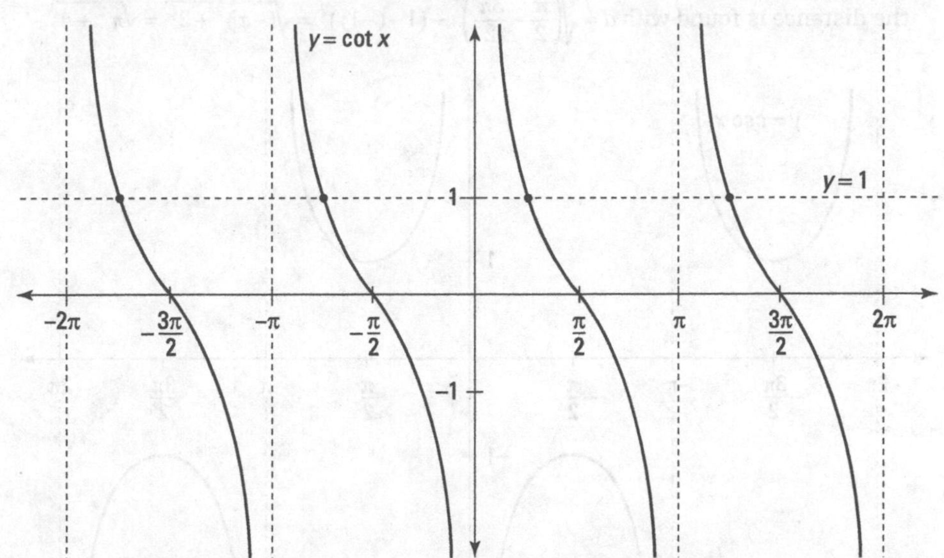

7. **See the following sketch; the distance is 2π.** Choosing the minimum points at (0,1) and $(2\pi,1)$, the distance is found with $d=\sqrt{(0-2\pi)^2+(1-1)^2}=\sqrt{4\pi^2+0}=2\pi$.

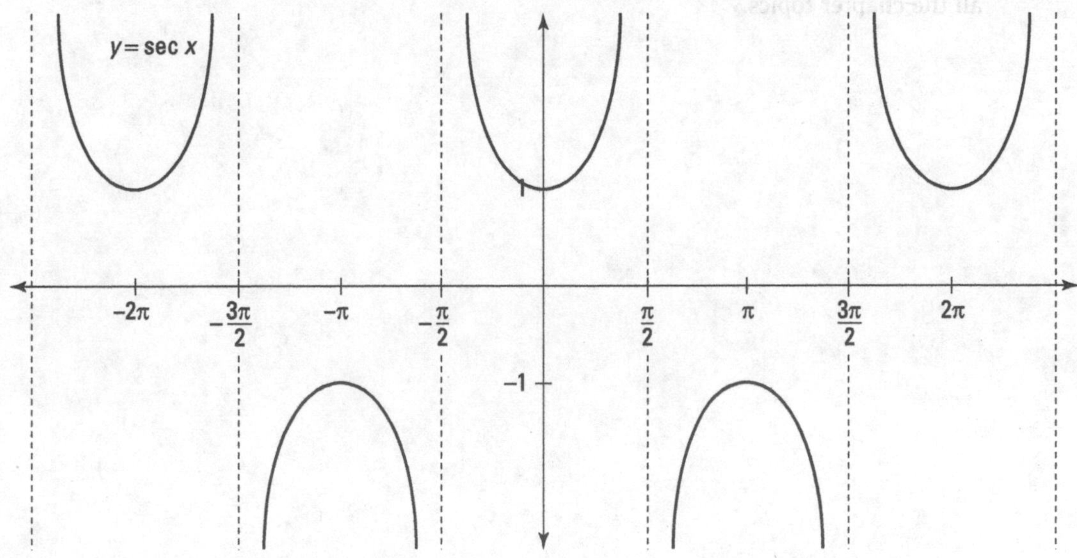

8. **See the following sketch; the distance is $\sqrt{\pi^2+4}$.** Choosing the points at $\left(\frac{\pi}{2},1\right)$ and $\left(\frac{3\pi}{2},-1\right)$, the distance is found with $d=\sqrt{\left(\frac{\pi}{2}-\frac{3\pi}{2}\right)^2+(1-(-1))^2}=\sqrt{(-\pi)^2+2^2}=\sqrt{\pi^2+4}$.

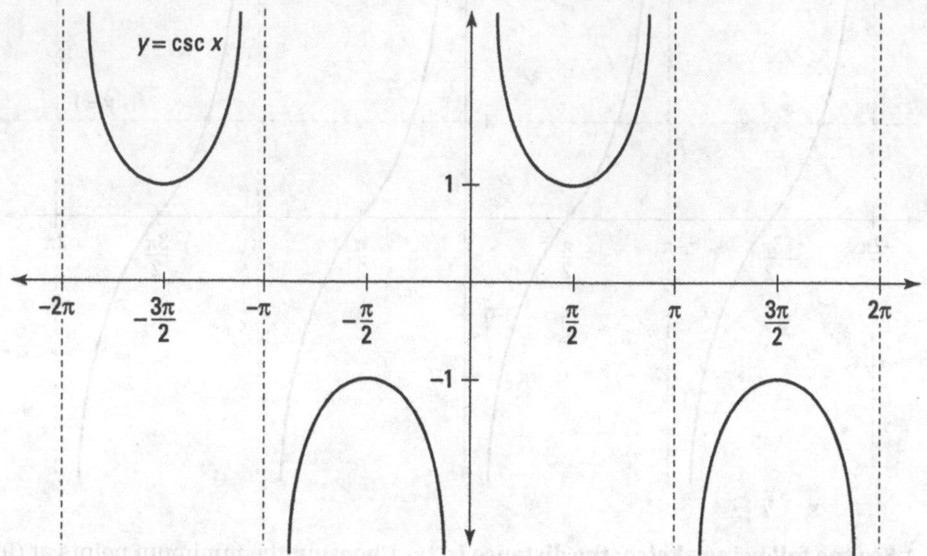

If you're ready to test your skills a bit more, take the following chapter quiz that incorporates all the chapter topics.

Whaddya Know? Chapter 10 Quiz

Quiz time! Complete each problem to test your knowledge on the various topics covered in this chapter. You can then find the solutions and explanations in the next section.

1. **Which of the six parent trig functions are even?**

2. **Which of the six parent trig functions have no y-intercept?**

3. **Write a rule for all the x-intercepts of the tangent function.**

4. **Which of the six parent trig function graphs contain the point (0,1)?**

5. **Sketch the graphs of $y = \sin x$ and $y = \cos x$ on the same graph from -2π to 2π. Where do they intersect?**

6. **Which of the six parent trig functions have no x-intercept?**

7. **Which of the six parent trig functions are odd?**

8. **Write a rule for all the x-intercepts of the cotangent function.**

9. **Which of the six parent trig function graphs contain the point (0,−1)?**

Answers to Chapter 10 Quiz

1. **Cosine, secant.** An even function is symmetric about the y-axis. The cosine and secant are the only two even functions.

2. **Secant, cosecant.** The range of secant and cosecant is $(-\infty,-1]\cup[1,\infty)$. The other four functions have ranges from $(-\infty,\infty)$.

3. $\boldsymbol{x = n\pi,\ n \in \{\text{integers}\}}$. Some of the x-intercepts are $(-2\pi,0)$, $(-\pi,0)$, $(0,0)$, $(\pi,0)$, $(2\pi,0)$. Letting n represent any integer, a rule for the x-intercepts would be $x = n\pi$.

4. **Cosine, secant.** The graphs of these two functions are the only ones containing the point $(0,1)$.

5. **See the following figure; they intersect at** $\left(-\frac{7\pi}{4},\frac{\sqrt{2}}{2}\right)$, $\left(-\frac{3\pi}{4},-\frac{\sqrt{2}}{2}\right)$, $\left(\frac{\pi}{4},\frac{\sqrt{2}}{2}\right)$, $\left(\frac{5\pi}{4},-\frac{\sqrt{2}}{2}\right)$.
   The exact intersection values are difficult to determine from the graph. Refer to the unit circle figure in Chapter 8 for help identifying the points of intersection.

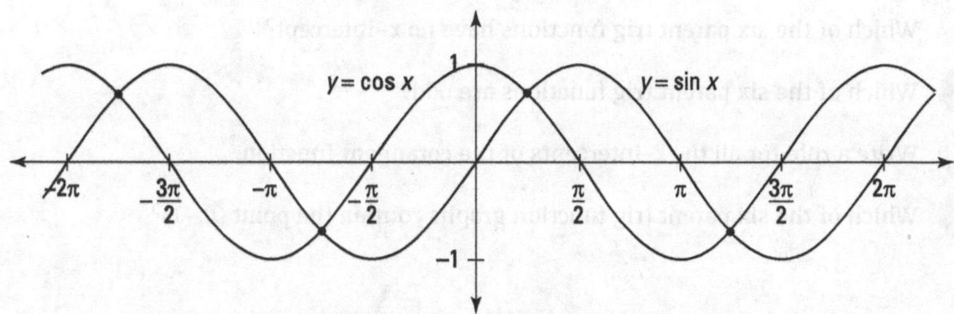

6. **Secant, cosecant.** The ranges of secant and cosecant do not include 0.

7. **Sine, tangent, cosecant, cotangent.** An odd function is symmetric about the origin.

8. $\boldsymbol{x = \frac{(2n+1)\pi}{2},\ n \in \{\text{integers}\}}$. Some of the x-intercepts are $\left(-\frac{3\pi}{2},0\right)$, $\left(-\frac{\pi}{2},0\right)$, $\left(\frac{\pi}{2},0\right)$, $\left(\frac{3\pi}{2},0\right)$.
   Letting n represent any integer, $2n+1$ represents odd integers. A rule for the x-intercepts would be $x = \frac{(2n+1)\pi}{2}$.

9. **None.** None of the parent graphs contain the point $(0,-1)$.

IN THIS CHAPTER

» Plotting and transforming parent functions by stretching and flattening

» Picturing and changing parent functions by shifting about

» Altering parent functions with reflections

Chapter 11 Graphing and Transforming Trig Functions

In Chapter 10, you are introduced to the six *parent* trig functions, how they behave, where they exist, what their features are, and how they look when graphed.

This chapter shows you how to start with the *parent graphs* — the foundation on which all the other graphs are built. From there, you can stretch a trig function graph, move it around on the coordinate plane, or flip and shrink it. All of this is covered in this chapter.

Transforming Trig Graphs

The basic parent graphs open the door to many more advanced and interesting graphs, which ultimately have more real-world applications. Usually, the graphs of functions that model real-world situations are stretched, shrunk, or even shifted to an entirely different location on the coordinate plane. The good news is that the *transformation* of trig functions follows the same set of general guidelines as the transformations you see in Chapter 4.

The rules for graphing transformed trig functions are actually pretty simple. When asked to graph a more-complicated trig function, you can take the parent graph (which you know from the previous sections) and alter it in some way to find the more complex graph. Basically, you can change each parent graph of a trig function in four ways:

- **Stretching vertically.** When dealing with the graph for sine and cosine functions, a vertical transformation changes the graph's height, also known as the *amplitude*, or its steepness.
- **Flattening horizontally.** This transformation makes it move faster or slower, which affects its horizontal length.
- **Shifting (translating) up, down, left, or right.** The parent function doesn't change shape; it just moves around the graph.
- **Reflecting across the *x*- or *y*-axis or some other line.** The shape of the graph doesn't change; it just flips over a line.

The following sections cover how to transform the parent trig graphs. However, before you move on to transforming these graphs, make sure you're comfortable with the parent graphs from Chapter 10.

A general format for the transformation of a trig function is $g(\theta) = a \cdot f[p(\theta - h)] + k$, where g is the transformed function, f is the parent trig function, a is the vertical transformation, p is the horizontal transformation, h is the horizontal shift, and k is the vertical shift.

Messing with sine and cosine graphs

The sine and cosine graphs look similar to a spring. If you pull the ends of this spring, all the points are farther apart from one another; in other words, the spring is stretched. If you push the ends of the spring together, all the points are closer together; in other words, the spring is shrunk. So the graphs of sine and cosine look and act a lot like a spring, but this spring can be changed both horizontally *and* vertically; aside from pulling the ends or pushing them together, you can make the spring taller or shorter. Now that's some spring!

In this section, you see how to alter the parent graphs for sine and cosine using both vertical and horizontal stretches and shrinks. You also see how to move the graph around the coordinate plane using translations (which can be both vertical and horizontal).

Changing the amplitude

Multiplying a sine or cosine by a constant changes the graph of the parent function; specifically, you change the amplitude of the graph. When measuring the height of a graph of sine or cosine, you measure the distance between the maximum crest and the minimum wave. Smack dab in the middle of that measurement is a horizontal line called the *sinusoidal axis*. *Amplitude* is the measure of the distance from the sinusoidal axis to the maximum or the minimum. Figure 11-1 illustrates this point further.

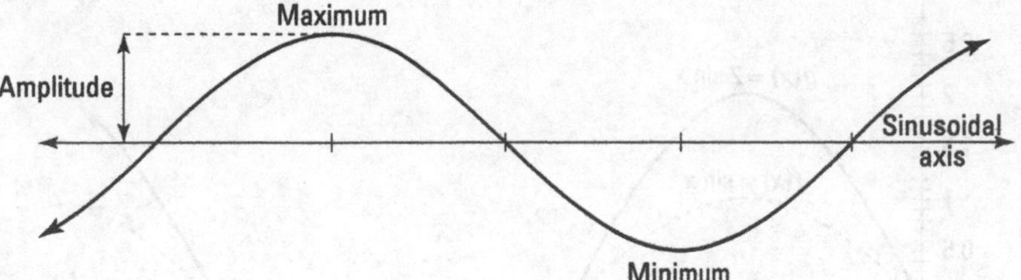

FIGURE 11-1: The sinusoidal axis and amplitude of a trig function graph.

By multiplying a trig function by certain values, you can make the graph taller or shorter:

- **Positive values of the multiplier a that are greater than 1 make the height of the graph greater.** Basically, $2\sin\theta$ makes the graph taller than $\sin\theta$; $5\sin\theta$ makes it even taller, and so on. For example, if $g(\theta) = 2\sin\theta$, you multiply the height of the original sine graph by 2 at every point. Every place on the graph, therefore, is twice as tall as the original. (Except, of course, when $\sin\theta = 0$.)
- **Fractional values between 0 and 1 make the graph shorter.** You can say that $\frac{1}{2}\sin\theta$ is shorter than $\sin\theta$, and $\frac{1}{5}\sin\theta$ is even shorter. For example, if $h(\theta) = \frac{1}{5}\sin\theta$, you multiply the parent graph's height by $\frac{1}{5}$ at each point, making it that much shorter.

The change of amplitude affects the range of the function as well, because the maximum and minimum values of the graph change. Before you multiply a sine or cosine function by 2, for instance, its graph oscillates between −1 and 1; afterward, it moves between −2 and 2.

WARNING

Sometimes you multiply a trigonometric function by a negative number. That negative number doesn't make the amplitude negative, however! Amplitude is a measure of distance, and distance can't be negative. You can't walk −5 feet, for instance, no matter how hard you try. Even if you walk backward, you still walk 5 feet. Similarly, if $k(\theta) = -5\sin\theta$, its amplitude is still 5. The negative sign just flips the graph upside down. This process is discussed later in this chapter.

EXAMPLE

Q. Graph the functions $g(\theta) = 2\sin\theta$ and $k(\theta) = -\frac{1}{5}\sin\theta$.

A. Figure 11-2 illustrates what the graphs of sine look like after the transformations. Figure 11-2a is the graph of $f(\theta) = \sin\theta$ and $g(\theta) = 2\sin\theta$; Figure 11-2b compares $f(\theta) = \sin\theta$ with $k(\theta) = -\frac{1}{5}\sin\theta$.

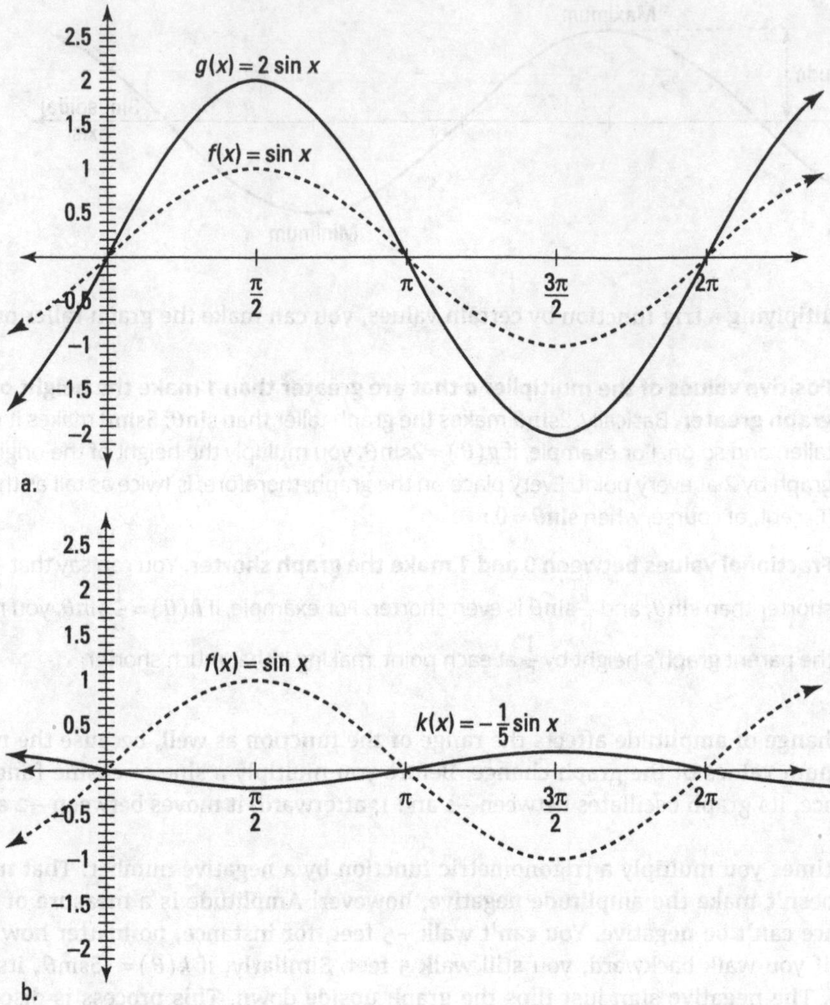

FIGURE 11-2: The graphs of transformations of sine.

YOUR TURN

1 Graph the function $h(\theta) = 3\cos\theta$.

2 Graph the function $m(\theta) = \frac{1}{3}\cos\theta$.

Altering the period

The *period* of the parent graphs of sine and cosine is 2π, which is once around the unit circle (for more about sketching sine, see Chapter 10). Sometimes in trig, the variable θ, not the function, gets multiplied by a constant. This action affects the period of the trig function graph. For example, $g(x) = \cos 2x$ makes the graph repeat itself twice in the same amount of time; in other words, the graph moves twice as fast. Think of it like fast-forwarding a video. Figures 11-3a and 11-3b show the cosine function with various period changes.

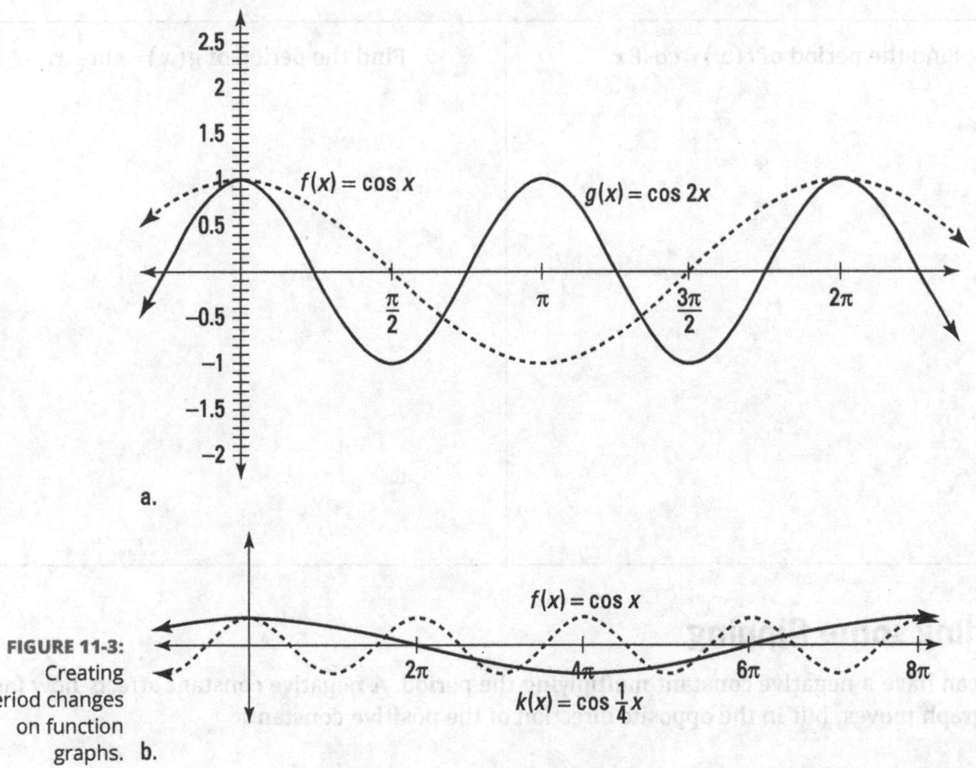

FIGURE 11-3: Creating period changes on function graphs.

To find the period of $g(x) = \cos 2x$, think of the general equation for the transformation of a trig function, $g(x) = a \cdot f[p(x-h)] + k$. The 2, which replaces the p in this case, is what creates the horizontal transformation. You divide the period of the function in question by p. In this case, dividing 2π by 2, you get that the period equals π, so the graph finishes its trip at π. Each point along the x-axis also moves at twice the speed.

You can make the graph of a trig function move faster or slower with different constants:

- **Positive values of period greater than 1 make the graph repeat itself more and more frequently.**
- **Fraction values between 0 and 1 make the graph repeat itself less frequently.**

Q. Find the period of $k(x) = \cos\frac{1}{4}x$.

EXAMPLE

A. You can find its period by dividing 2π by $\frac{1}{4}$. Solving for period gets you 8π. Before, the graph finished at 2π; now it waits to finish at 8π, which slows it down by $\frac{1}{4}$. Figure 11-3b shows you how the original function $f(x) = \cos x$ and the new function $k(x) = \cos\frac{1}{4}x$ compare.

YOUR TURN

 Find the period of $f(x) = \cos 8x$.

 Find the period of $g(x) = \sin\frac{1}{3}x$.

Adding some flipping

You can have a negative constant multiplying the period. A negative constant affects how fast the graph moves, but in the opposite direction of the positive constant.

Q. Determine the changes occurring in $q(x) = \sin(-3x)$.

EXAMPLE

A. Starting with the change in period, look at $p(x) = \sin 3x$. The period of $p(x)$ is $\frac{2\pi}{3}$. And then the negative sign in $q(x) = \sin(-3x)$ acts the same as a reflection over the vertical axis. (You find more on reflections in Chapter 4.) The graph of $p(x)$ has a period of $\frac{2\pi}{3}$ and is reflected over the y-axis. Figure 11-4 illustrates this transformation clearly.

WARNING

Don't confuse amplitude and period when graphing trig functions. For example, $f(x) = 2\sin x$ and $g(x) = \sin 2x$ affect the graph differently: $f(x) = 2\sin x$ makes it taller, and $g(x) = \sin 2x$ makes it move faster.

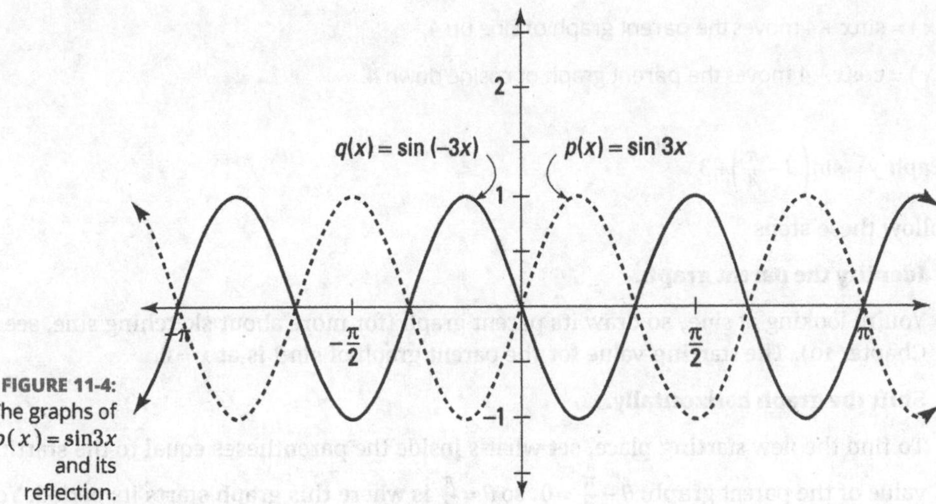

FIGURE 11-4: The graphs of $p(x) = \sin 3x$ and its reflection.

YOUR TURN

 Graph $f(x) = \cos 4x$

 Graph $g(x) = -\sin\frac{1}{2}x$

Shifting the waves on the coordinate plane

The movement of a parent graph around the coordinate plane is another type of transformation known as a translation or a *shift*. For this type of transformation, every point on the parent graph is moved somewhere else on the coordinate plane. A translation doesn't affect the overall shape of the graph; it only changes its location on the plane. In this section, you see how to take the parent graphs of sine and cosine and shift them both horizontally and vertically.

REMEMBER

Did you pick up on the rules for shifting a function horizontally and vertically from Chapter 4? If not, go back and check them out, because they're important for sine and cosine graphs as well.

You see the horizontal and vertical shifts indicated with $\sin(x-h)+k$, or $\cos(x-h)+k$. The variable h represents the horizontal shift of the graph, and k represents the vertical shift of the graph. The sign makes a difference in the direction of the movement. For example,

- $f(x) = \sin(x-3)$ moves the parent graph of sine to the right by 3.
- $g(x) = \cos(x+2)$ moves the parent graph of cosine to the left by 2.

- $k(x) = \sin x + 4$ moves the parent graph of sine up 4.
- $p(x) = \cos x - 4$ moves the parent graph of cosine down 4.

EXAMPLE

Q. Graph $y = \sin\left(\theta - \frac{\pi}{4}\right) + 3$

A. Follow these steps:

1. **Identify the parent graph.**

   You're looking at sine, so draw its parent graph (for more about sketching sine, see Chapter 10). The starting value for the parent graph of $\sin\theta$ is at $x = 0$.

2. **Shift the graph horizontally.**

   To find the new starting place, set what's inside the parentheses equal to the starting value of the parent graph: $\theta - \frac{\pi}{4} = 0$, so $\theta = \frac{\pi}{4}$ is where this graph starts its period. You move every point on the parent graph to the right by $\frac{\pi}{4}$. Figure 11-5 shows what you have so far.

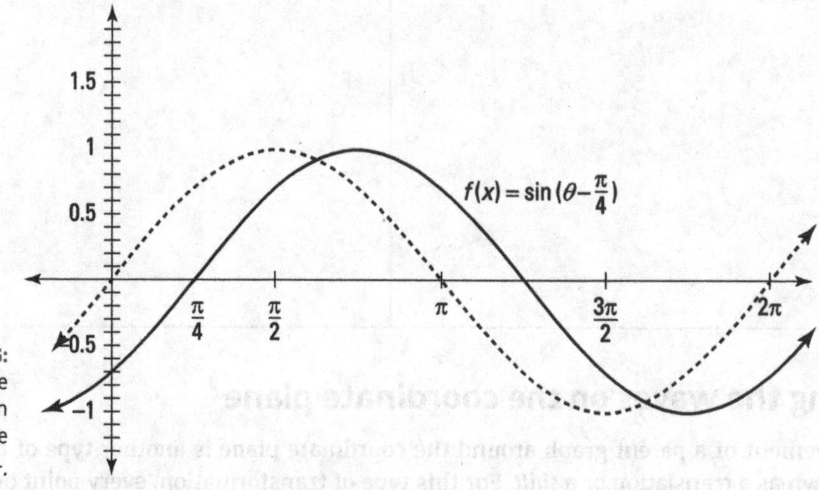

FIGURE 11-5: Shifting the parent graph of sine to the right by $\frac{\pi}{4}$.

3. **Move the graph vertically.**

   The sinusoidal axis of the graph moves up 3 units in this function, so shift all the points of the parent graph this direction now. You can see this shift in Figure 11-6. You're actually seeing both shifts — the one to the right plus the one upward.

4. **State the domain and range of the transformed graph, if asked.**

   The domain and range of a function may be affected by a transformation. When this happens, you may be asked to state the new domain and range. Usually, you can visualize the range of the function easily by looking at the graph. Two factors that change the range are a vertical transformation (stretch or shrink) and a vertical translation.

   Keep in mind that the range of the parent sine graph is $[-1, 1]$. Shifting the parent graph up 3 units makes the range of $y = \sin\left(\theta - \frac{\pi}{4}\right) + 3$ shift up 3 units also. Therefore, the new range is $[2, 4]$. The domain of this function isn't affected; it's still $(-\infty, \infty)$.

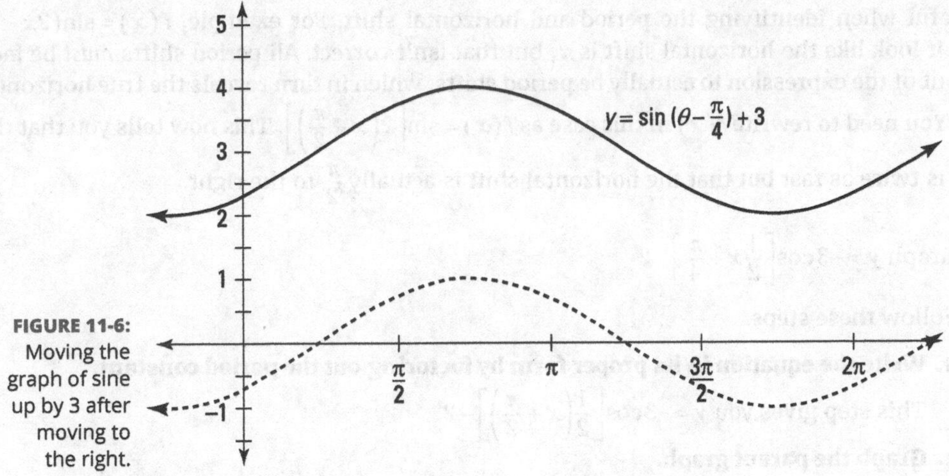

FIGURE 11-6: Moving the graph of sine up by 3 after moving to the right.

YOUR TURN

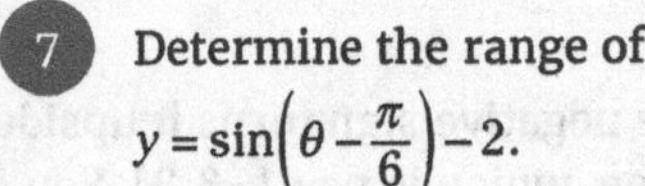

7 Determine the range of $y = \sin\left(\theta - \frac{\pi}{6}\right) - 2$.

8 Determine the range of $y = \cos(\theta) + 7$.

Combining transformations in one fell swoop

When you're asked to graph a trig function with multiple transformations, it's suggested that you do them in this order:

1. **Change the amplitude.**
2. **Change the period.**
3. **Shift the graph horizontally.**
4. **Shift the graph vertically.**

TECHNICAL STUFF

The equations that combine all the transformations into one are as follows: $g(x) = a \cdot f[p(x-h)] + k$.

And, specifically for the sine and cosine, you have $g(x) = a \cdot \sin[p(x-h)] + k$ and $g(x) = a \cdot \cos[p(x-h)] + k$.

The absolute value of the variable *a* is the amplitude. To find the new period for sine or cosine, you take 2π and divide by *p* to find the period. The variable *h* is the horizontal shift, and *k* is the vertical shift.

REMEMBER

Be careful when identifying the period and horizontal shift. For example, $f(x)=\sin(2x-\pi)$ makes it look like the horizontal shift is π, but that isn't correct. All period shifts *must* be factored out of the expression to actually be period shifts, which in turn reveals the true horizontal shifts. You need to rewrite $f(x)$ in this case as $f(x)=\sin\left[2\left(x-\frac{\pi}{2}\right)\right]$. This now tells you that the period is twice as fast but that the horizontal shift is actually $\frac{\pi}{2}$ to the right.

EXAMPLE

Q. Graph $y=-3\cos\left[\frac{1}{2}x+\frac{\pi}{4}\right]-2$

A. Follow these steps.

1. **Write the equation in its proper form by factoring out the period constant.**

   This step gives you $y=-3\cos\left[\frac{1}{2}\left(x+\frac{\pi}{2}\right)\right]-2$.

2. **Graph the parent graph.**

   Graph the original cosine function as you know it.

3. **Change the amplitude.**

   This graph has an amplitude of 3, and the negative sign turns it upside down. The amplitude change affects the graph's range, which is now [–3, 3]. You can see the amplitude change and flip over the x-axis in Figure 11-7.

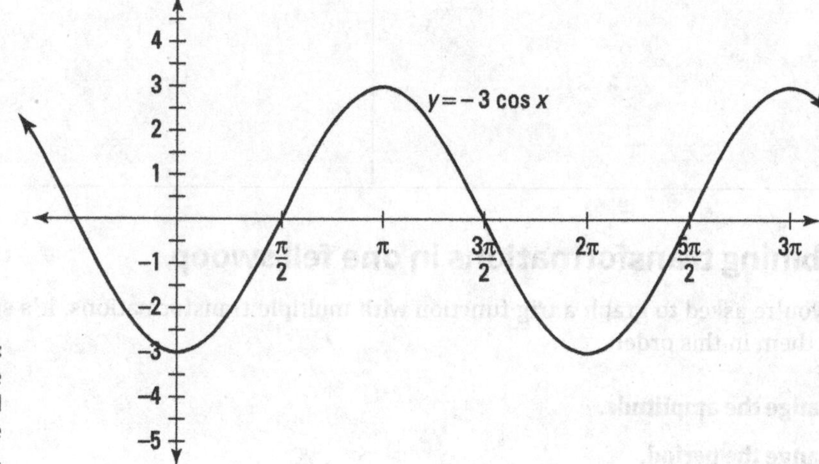

FIGURE 11-7: Changing the amplitude and flipping the cosine.

4. **Alter the period.**

   The constant $\frac{1}{2}$ affects the period. Dividing 2π by $\frac{1}{2}$ gives you the period of 4π. The graph moves half as fast and finishes at 4π. Figure 11-8 shows the results of the amplitude change, flip, and period change.

5. **Shift the graph horizontally.**

   When you factored out the period constant in Step 1, you discovered that the horizontal shift is to the left $\frac{\pi}{2}$.

6. **Shift the graph vertically.**

   Because of the –2 you see in Step 1, this graph moves down two positions.

7. **State the new domain and range.**

The functions of sine and cosine are defined for all values of θ. The domain for the cosine function is all real numbers, or $(-\infty,\infty)$. The range of the graph of $y=-3\cos\left[\frac{1}{2}\left(x+\frac{\pi}{2}\right)\right]-2$ has been stretched because of the amplitude change and shifted down. To find the range of a function that has been shifted vertically, you add or subtract the vertical shift (-2) from the altered range based on the amplitude. For this problem, the range of the transformed cosine function is $[-3-2, 3-2]$, or $[-5, 1]$.

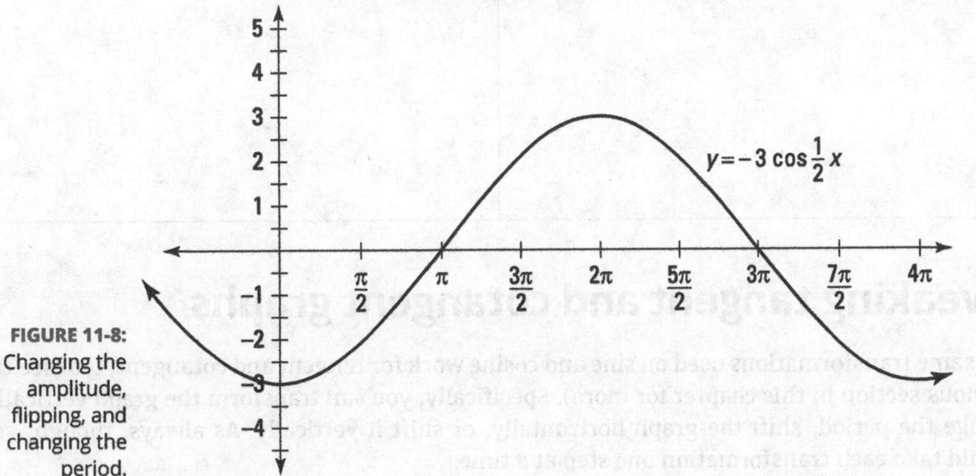

FIGURE 11-8: Changing the amplitude, flipping, and changing the period.

Now, to complete the entire graph and its transformations, take the altered cosine graph from Figure 11-8 and move its "starting point" to the left by $\frac{\pi}{2}$ and down 2 units. The completely transformed graph goes through the point $\left(\frac{\pi}{2},-2\right)$, which is left and down from $(\pi,0)$, and through $\left(\frac{5\pi}{2},-2\right)$, which is left and down from $(3\pi,0)$. It takes just a few anchor points to get you sailing on your way. The completed graph is shown in Figure 11-9.

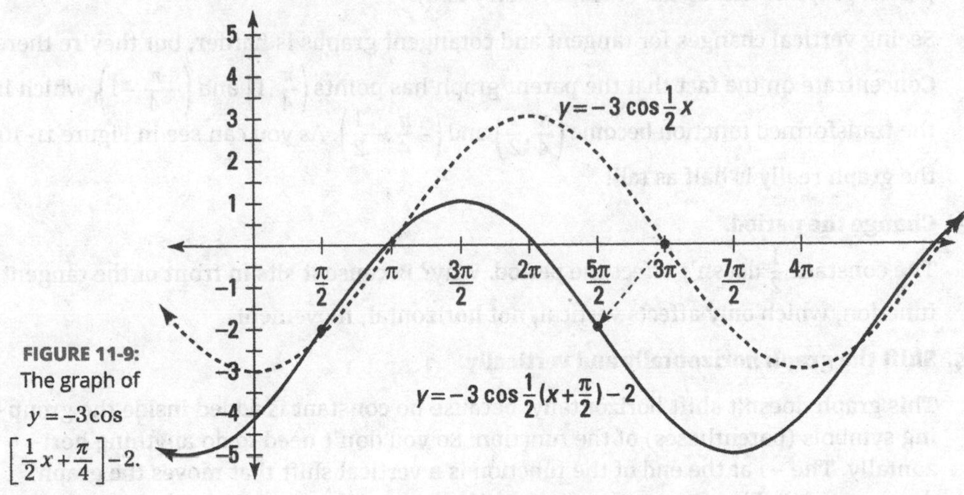

FIGURE 11-9: The graph of $y=-3\cos\left[\frac{1}{2}x+\frac{\pi}{4}\right]-2$.

YOUR TURN

9 Graph $y=2\sin\left[3\left(x-\frac{\pi}{2}\right)\right]+1$

10 Graph $y=-3\cos\left[\frac{1}{4}(x-\pi)\right]+1$

Tweaking tangent and cotangent graphs

The same transformations used on sine and cosine work for tangent and cotangent, too (see the previous section in this chapter for more). Specifically, you can transform the graph vertically, change the period, shift the graph horizontally, or shift it vertically. As always, though, you should take each transformation one step at a time.

EXAMPLE

Q. Graph $f(\theta)=\frac{1}{2}\tan\theta-1$

A. Follow these steps:

1. **Sketch the parent graph for tangent.**
2. **Shrink or stretch the parent graph.**

   The vertical shrink is $\frac{1}{2}$ for every point on this function, so each point on the tangent parent graph is half as tall (except when y is 0).

   Seeing vertical changes for tangent and cotangent graphs is harder, but they're there. Concentrate on the fact that the parent graph has points $\left(\frac{\pi}{4},1\right)$ and $\left(-\frac{\pi}{4},-1\right)$, which in the transformed function become $\left(\frac{\pi}{4},\frac{1}{2}\right)$ and $\left(-\frac{\pi}{4},-\frac{1}{2}\right)$. As you can see in Figure 11-10, the graph really is half as tall!

3. **Change the period.**

   The constant $\frac{1}{2}$ doesn't affect the period. Why? Because it sits in front of the tangent function, which only affects vertical, not horizontal, movement.

4. **Shift the graph horizontally and vertically.**

   This graph doesn't shift horizontally, because no constant is added inside the grouping symbols (parentheses) of the function. So you don't need to do anything horizontally. The −1 at the end of the function is a vertical shift that moves the graph down one unit. Figure 11-11 shows the shift.

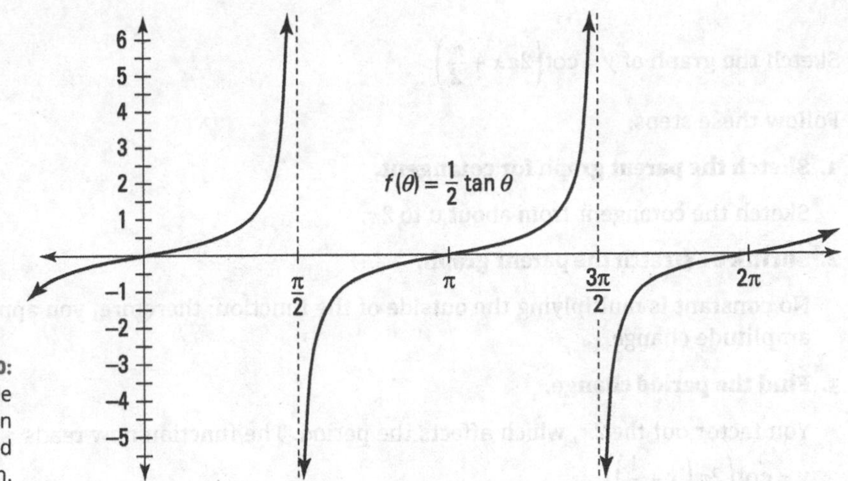

FIGURE 11-10: The amplitude is half as tall in the changed graph.

5. State the transformed function's domain and range, if asked.

Because the range of the tangent function is all real numbers, transforming its graph doesn't affect the range, only the domain. The domain of the tangent function isn't all real numbers because of the asymptotes. The domain of the example function hasn't been affected by the transformations, however. The domain consists of all θ except $\theta \neq \frac{\pi}{2} + n\pi$, where n is an integer.

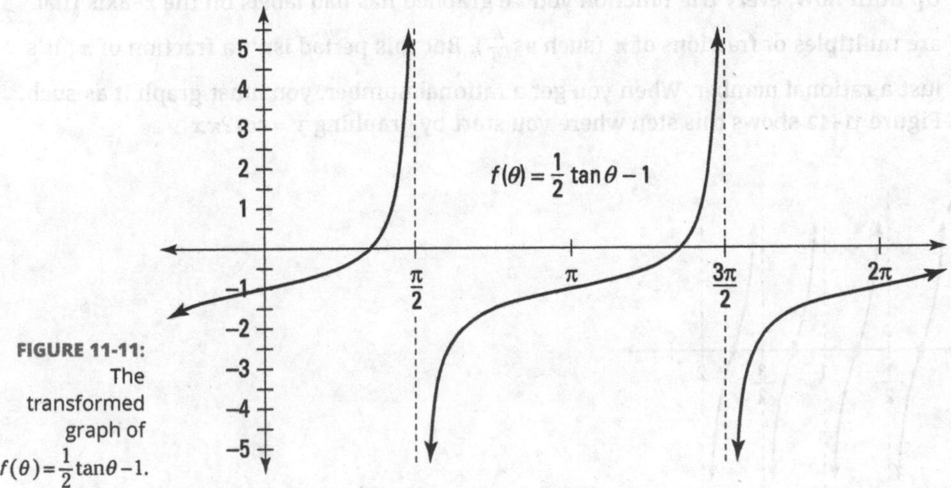

FIGURE 11-11: The transformed graph of $f(\theta) = \frac{1}{2}\tan\theta - 1$.

Now that you've graphed the basics, you can graph a function that has a period change, with a multiple of π in the formula. In the following example, you see the variable x being multiplied by π. You know this graph has a period change because you see a π multiplying by the variable. This constant changes the period of the function, which in turn changes the distance between the asymptotes. In order for the graph to show this change correctly, you must factor this constant out of the parentheses. Take the transformation one step at a time.

EXAMPLE

Q. Sketch the graph of $y=\cot\left(2\pi x+\frac{\pi}{2}\right)$.

A. Follow these steps:

1. **Sketch the parent graph for cotangent.**

   Sketch the cotangent from about 0 to 2π.

2. **Shrink or stretch the parent graph.**

   No constant is multiplying the outside of the function; therefore, you apply no amplitude change.

3. **Find the period change.**

   You factor out the 2π, which affects the period. The function now reads $y=\cot\left(2\pi\left(x+\frac{1}{4}\right)\right)$.

   The period of the parent function for cotangent is π. Therefore, instead of dividing 2π by the period constant to find the period change (like you did for the sine and cosine graphs), you must divide π by the period constant. This step gives you the period for the transformed cotangent function.

   Dividing π by 2π, you get a period of $\frac{1}{2}$ for the transformed function. The graph of this function begins to repeat starting at $\frac{1}{2}$, which is different from $\frac{\pi}{2}$, so be careful when you're labeling your graph.

REMEMBER

Up until now, every trig function you've graphed has had labels on the x-axis that are multiples or fractions of π (such as $\frac{\pi}{2}$). But this period isn't a fraction of π; it's just a rational number. When you get a rational number, you must graph it as such. Figure 11-12 shows this step where you start by graphing $y=\cot 2\pi x$.

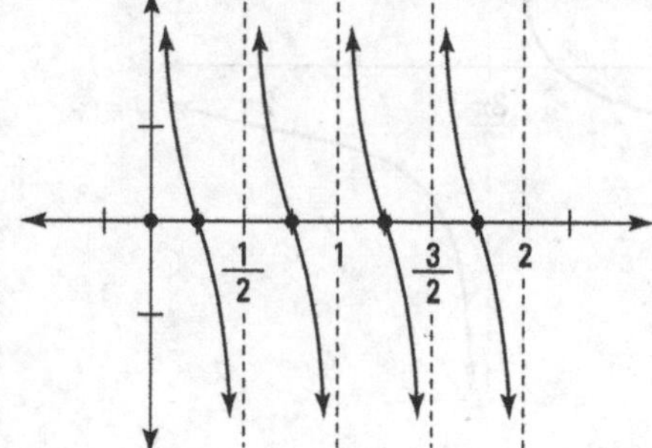

FIGURE 11-12: The graph of $y=\cot 2\pi x$ shows a period of $\frac{1}{2}$.

4. **Determine the horizontal and vertical shifts.**

   Because you've already factored the period constant, you can see that the horizontal shift is to the left by $\frac{1}{4}$. Figure 11-13 shows this transformation on the graph, where the vertical asymptotes have moved by $\frac{1}{4}$ unit.

   No constant is being added to or subtracted from this function on the outside, so the graph doesn't experience a vertical shift.

5. **State the transformed function's domain and range, if asked.**

   The horizontal shift affects the domain of this graph. To find the first asymptote, set $2\pi x+\frac{\pi}{2}=0$ (setting the period shift equal to the original first asymptote). You find that $x=-\frac{1}{4}$ is your new asymptote. The graph repeats every $\frac{1}{2}$ radian because of its period. So the domain is all x except $x\neq\frac{n}{2}-\frac{1}{4}$, where n is an integer. The graph's range isn't affected: $(-\infty,\infty)$.

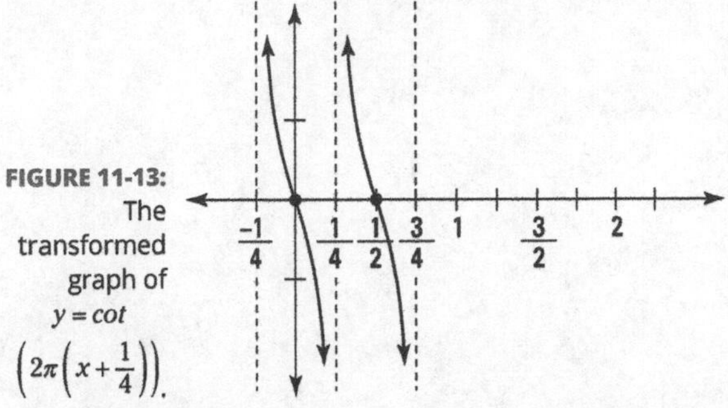

FIGURE 11-13: The transformed graph of $y=\cot\left(2\pi\left(x+\frac{1}{4}\right)\right)$.

YOUR TURN

11 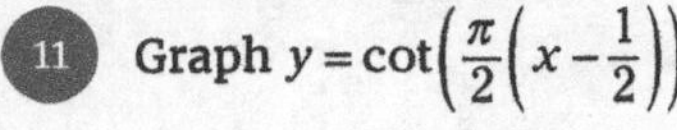 Graph $y=\cot\left(\frac{\pi}{2}\left(x-\frac{1}{2}\right)\right)$

12 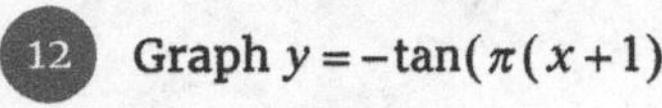 Graph $y=-\tan(\pi(x+1))$

Transforming the graphs of secant and cosecant

REMEMBER

To graph transformed secant and cosecant graphs, your best bet is to graph their reciprocal functions and transform them first. The reciprocal functions, sine and cosine, are easier to graph because they don't have as many complex parts (no asymptotes, basically). If you can graph the reciprocals first, you can deal with the more complicated pieces of the secant/cosecant graphs last.

EXAMPLE

Q. Graph $f(\theta)=\frac{1}{4}\sec\theta-1$

A. Use the following steps:

1. **Graph the transformed reciprocal function.**

   Look at the reciprocal function for secant, which is cosine. Pretend just for a bit that you're graphing $f(\theta)=\frac{1}{4}\cos\theta-1$. Follow all the rules for graphing the cosine graph in order to end up with a graph that looks like the one in Figure 11-14.

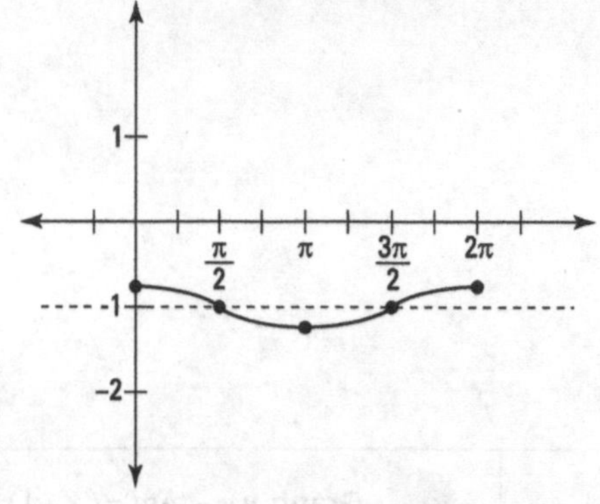

FIGURE 11-14: Graphing the cosine function first.

2. **Sketch the asymptotes of the transformed reciprocal function.**

   Wherever the transformed graph of $\cos\theta$ crosses its sinusoidal axis, you have an asymptote in $\sec\theta$. You see that $\cos\theta=0$ when $\theta=\frac{\pi}{2}$ and $\theta=\frac{3\pi}{2}$.

3. **Find out what the graph looks like between each asymptote.**

   Now that you've identified the asymptotes, you simply figure out what happens on the intervals between them. The finished graph ends up looking like the one in Figure 11-15.

4. **State the domain and range of the transformed function.**

   Because the new transformed function may have different asymptotes than the parent function for secant and it may be shifted up or down, you may be required to state the new domain and range.

   This example, $f(\theta)=\frac{1}{4}\sec\theta-1$ has asymptotes at $\theta=\frac{\pi}{2}$ and $\theta=\frac{3\pi}{2}$, and so on, repeating every π radians. Therefore, the domain is restricted to not include these values and is written $\theta\neq\frac{\pi}{2}+n\pi$, where n is an integer. In addition, the range of this function changes because the transformed function is shorter than the parent function and has been shifted down 2. The range has two separate intervals, $\left(-\infty,-\frac{5}{4}\right]$ and $\left[-\frac{3}{4},\infty\right)$.

You can graph a transformation of the cosecant graph by using the same steps you use when graphing the secant function, only this time you use the sine function to guide you.

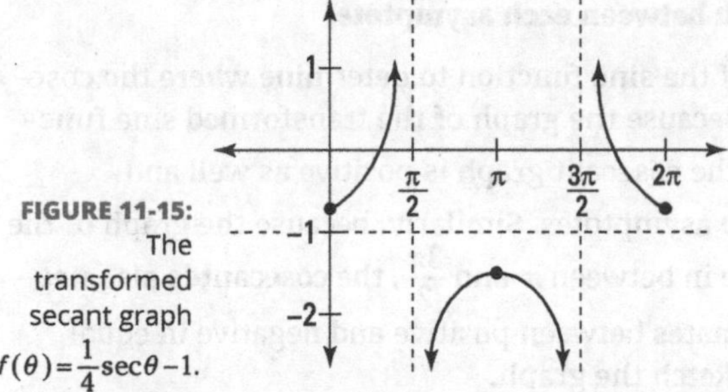

FIGURE 11-15: The transformed secant graph $f(\theta) = \frac{1}{4}\sec\theta - 1$.

REMEMBER

The shape of the transformed cosecant graph should be very similar to the secant graph, except the asymptotes are in different places. For this reason, be sure you're graphing with the help of the sine graph (to transform the cosecant graph) and the cosine function (to guide you for the secant graph).

EXAMPLE

Q. Graph the transformed cosecant $g(\theta) = \csc(2\theta - \pi) + 1$.

A. Use the following steps:

1. Graph the transformed reciprocal function.

Look first at the reciprocal function $g(\theta) = \sin(2\theta - \pi) + 1$. The rules to transforming a sine function tell you to first factor out the 2 and get $g(\theta) = \sin\left(2\left(\theta - \frac{\pi}{2}\right)\right) + 1$. It has a horizontal shrink of 2, a horizontal shift of $\frac{\pi}{2}$ to the right, and a vertical shift of up 1. Figure 11-16 shows the transformed sine graph.

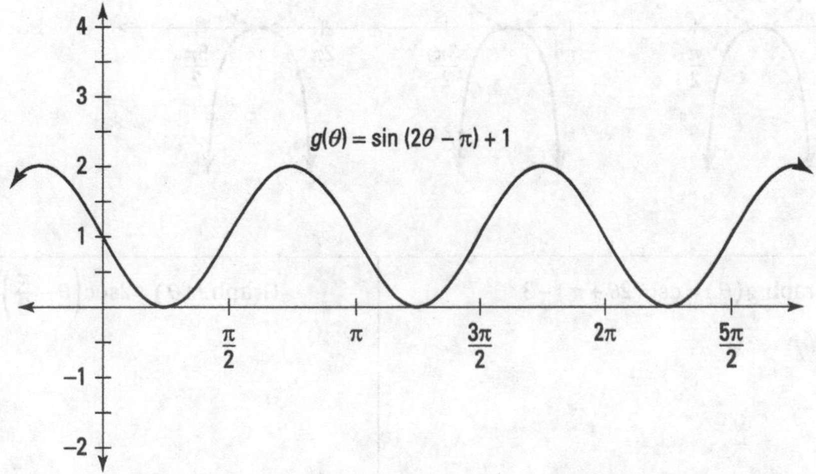

FIGURE 11-16: A transformed sine graph.

2. Sketch the asymptotes of the reciprocal function.

The sinusoidal axis that runs through the middle of the sine function is the line $y = 1$. Therefore, an asymptote of the cosecant graph exists everywhere the transformed sine function crosses this line. The asymptotes of the cosecant graph are at $\frac{\pi}{2}$ and π, and repeat every $\frac{\pi}{2}$ radians.

3. **Figure out what happens to the graph between each asymptote.**

   You can use the transformed graph of the sine function to determine where the cosecant graph is positive and negative. Because the graph of the transformed sine function is positive in between $\frac{\pi}{2}$ and π, the cosecant graph is positive as well and extends up when getting closer to the asymptotes. Similarly, because the graph of the transformed sine function is negative in between π and $\frac{3\pi}{2}$, the cosecant is also negative in this interval. The graph alternates between positive and negative in equal intervals for as long as you want to sketch the graph.

   Figure 11-17 shows the transformed cosecant graph.

4. **State the new domain and range.**

   The domain of the transformed cosecant function is all values of θ except for the values that are asymptotes. From the graph, you can see that the domain is all values of θ, where $\theta \neq \frac{\pi}{2} + \frac{n\pi}{2}$ (where n is an integer). The range of the transformed cosecant function is also split up into two intervals: $(-\infty, 0] \cup [2, \infty)$.

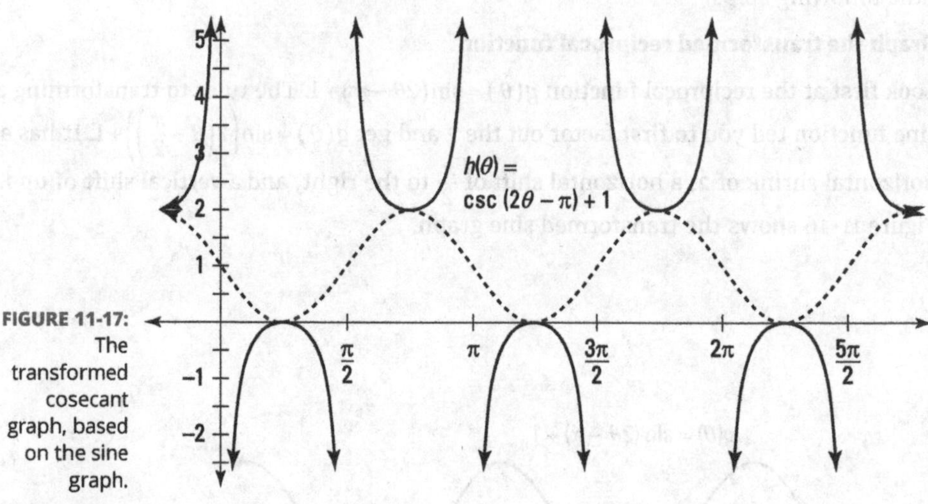

FIGURE 11-17: The transformed cosecant graph, based on the sine graph.

YOUR TURN

13 Graph $g(\theta) = \csc(2\theta + \pi) - 3$

14 Graph $f(\theta) = 2\sec\left(\theta - \frac{\pi}{2}\right)$

Practice Questions Answers and Explanations

1. **See the following graph.** The period is the same as in the parent graph, $f(x) = \cos x$, and the x-intercepts are the same. The graph steepens by a multiple of 3, with each y-coordinate 3 times that in the parent graph.

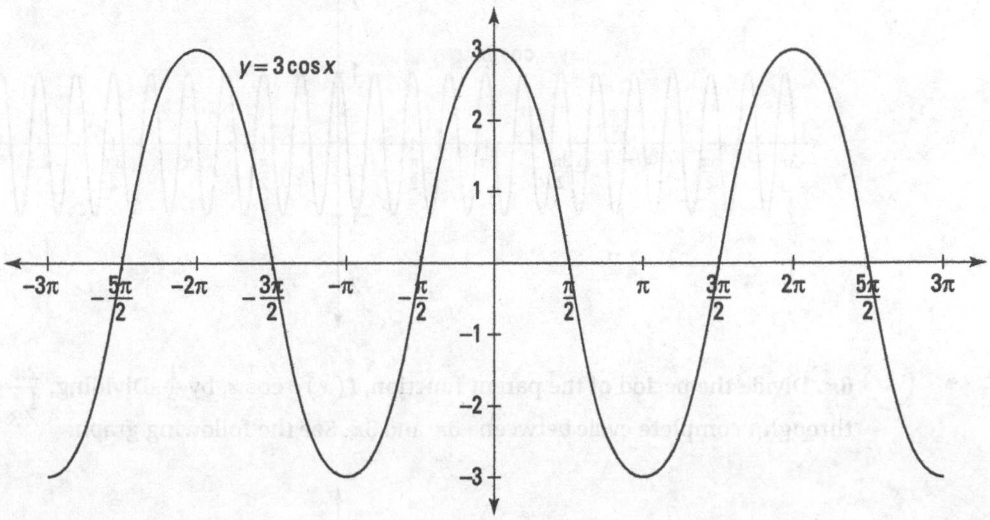

2. **See the following graph.** The period is the same as in the parent graph, $f(x) = \cos x$, and the x-intercepts are the same. The graph flattens by a multiple of $\frac{1}{3}$, with each y-coordinate $\frac{1}{3}$ of that in the parent graph.

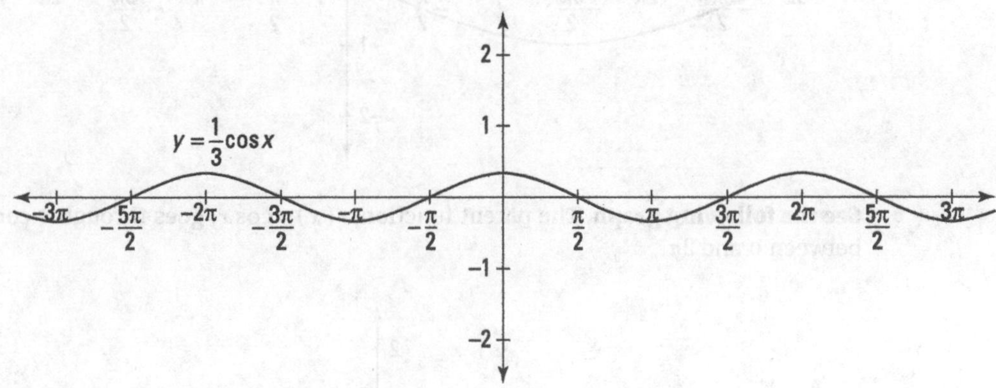

3. $\frac{\pi}{4}$. Divide the period of the parent function, $f(x)=\cos x$, by 8. Dividing, $\frac{2\pi}{8}=\frac{\pi}{4}$. So the function has a period of just $\frac{\pi}{4}$. It goes through 8 complete cycles between 0 and 2π. See the following graph.

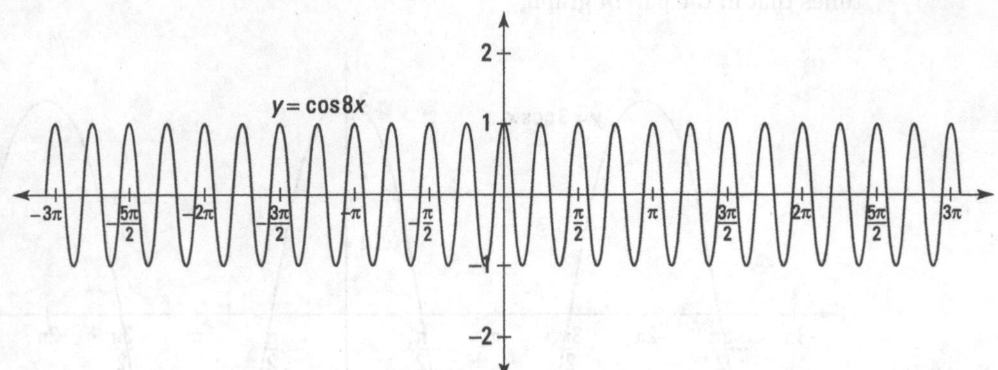

4. 6π. Divide the period of the parent function, $f(x)=\cos x$, by $\frac{1}{3}$. Dividing, $\frac{2\pi}{1/3}=6\pi$. It goes through 1 complete cycle between -3π and 3π. See the following graph.

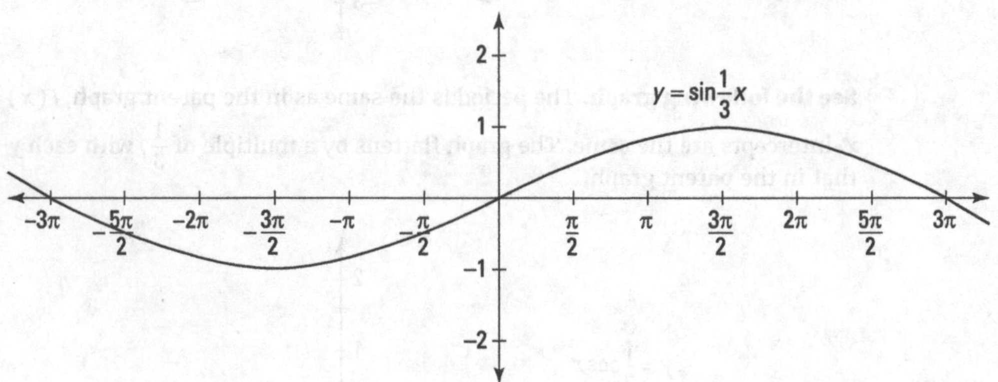

5. **See the following graph.** The parent function, $f(x)=\cos x$, goes through 4 complete cycles between 0 and 2π.

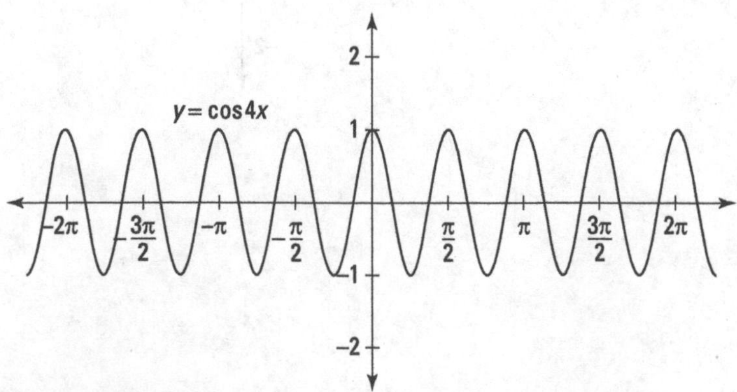

6. **See the following graph.** The parent function, $f(x) = \sin x$, goes through only half its 2π cycle between 0 and 2π. Its period is 4π. And the negation of the sine function creates a graph that is a reflection over the x-axis.

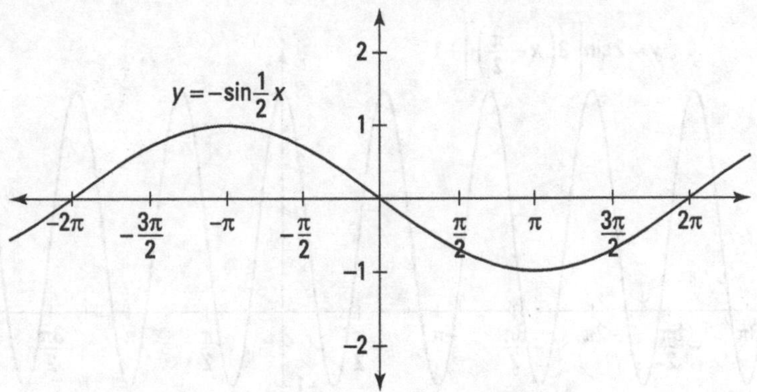

7. **[−3,−1].** The parent graph drops by 2 units, so, instead of a range of [−1,1], the range is now [−3,−1]. See the following graph.

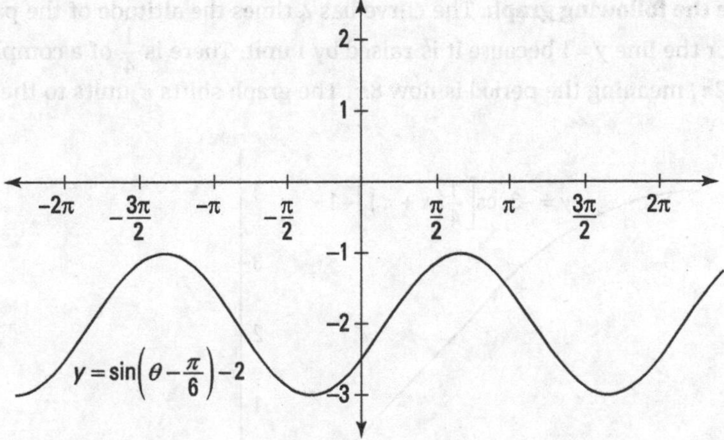

8. **[6,8].** The parent graph rises by 7 units, so, instead of a range of [−1,1], the range is now [6,8]. See the following graph.

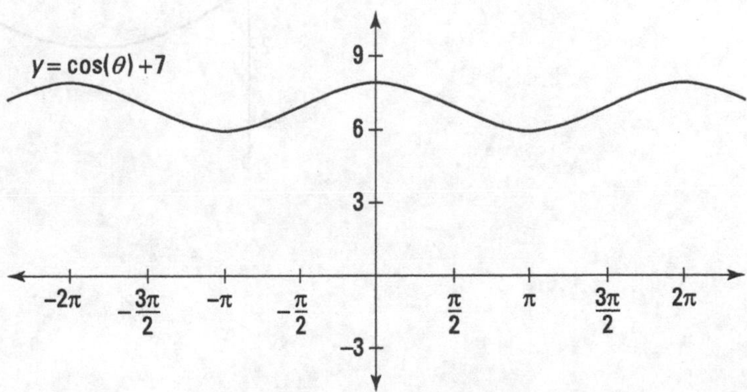

(9) **See the following graph.** The curve has 2 times the altitude of the parent graph. There are 3 complete cycles in each period of 2π, so the period is $\frac{2\pi}{3}$. The graph shifts $\frac{\pi}{2}$ units to the right, and it is raised by 1 unit. The range is 2 above and below the line $y=1$ or $[-1,3]$.

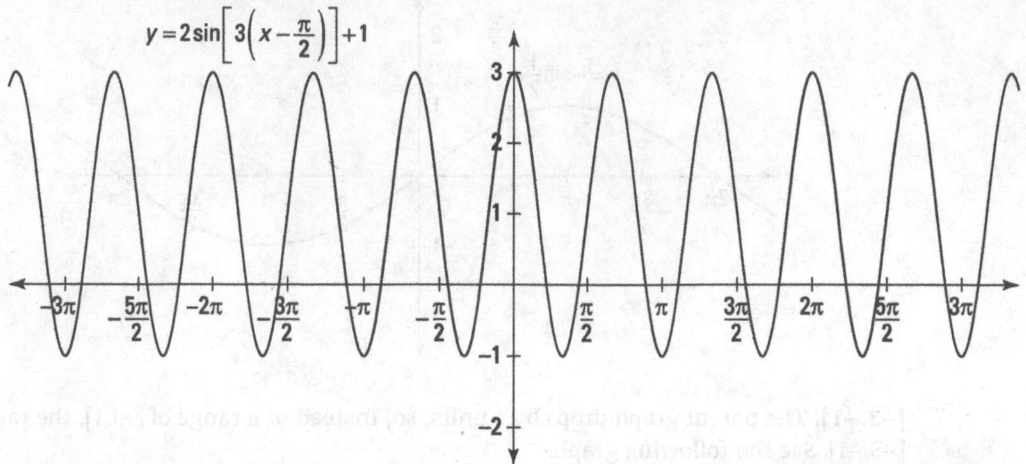

(10) **See the following graph.** The curve has 4 times the altitude of the parent graph and is flipped over the line $y=1$ because it is raised by 1 unit. There is $\frac{1}{4}$ of a complete cycle in each period of 2π, meaning the period is now 8π. The graph shifts π units to the right.

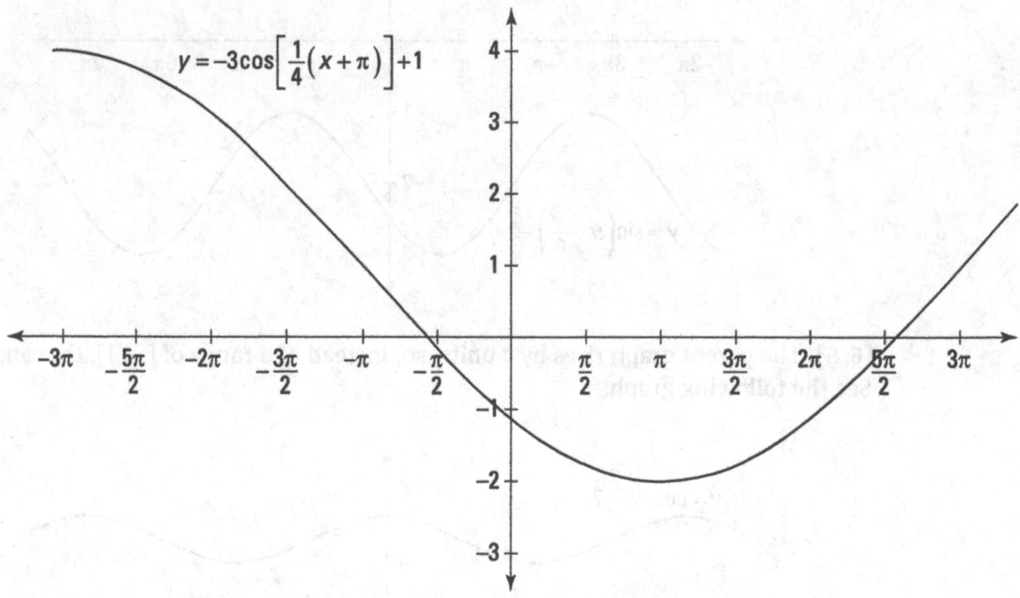

11 **See the following graph.** There is a $\frac{\pi}{2}$ complete cycle in each period of π, meaning the period is now $\frac{1}{2}$. The graph shifts $\frac{1}{2}$ unit to the right.

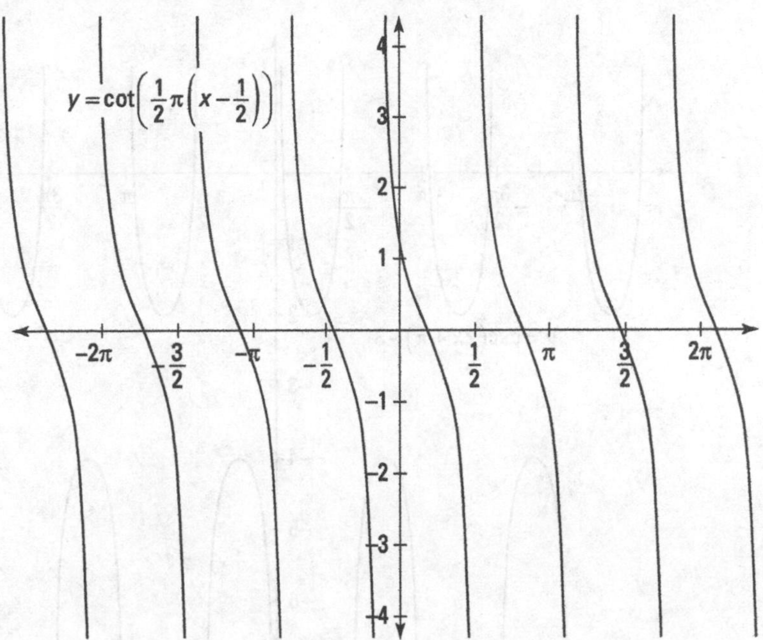

12 **See the following graph.** The curve is flipped over the x-axis. The period is now 1, the result of dividing the parent graph's period of π by π. The graph shifts 1 unit to the left.

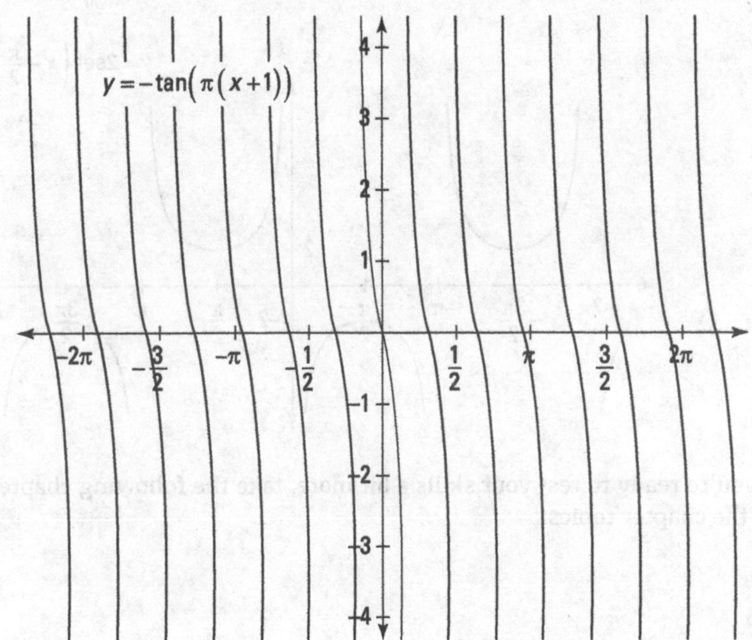

13 **See the following graph.** First, factor the 2 from the two terms in the parentheses to get the function equation $g(\theta) = \csc\left(2\left(\theta + \frac{\pi}{2}\right)\right) - 3$. The period is π. The graph shifts $\frac{\pi}{2}$ units to the left and drops 3 units.

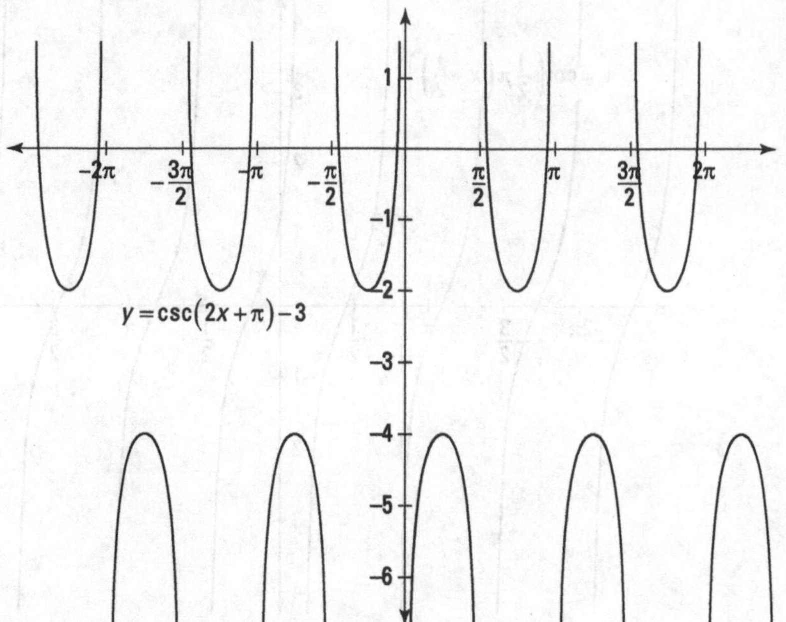

14 **See the following graph.** The curve is 2 times as steep as the parent graph. The graph shifts $\frac{\pi}{2}$ units to the right.

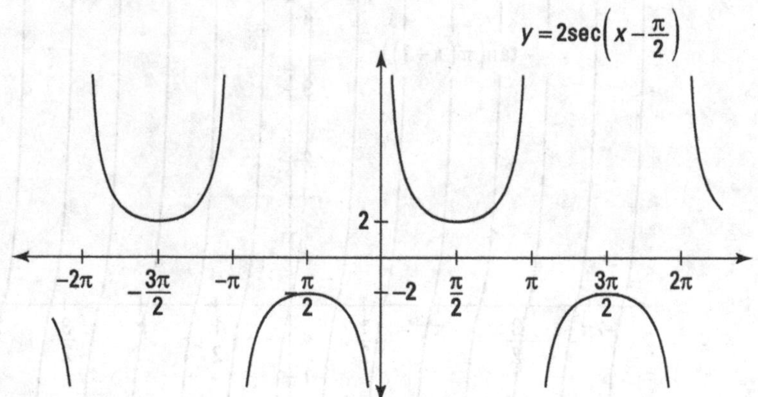

If you're ready to test your skills a bit more, take the following chapter quiz that incorporates all the chapter topics.

Whaddya Know? Chapter 11 Quiz

Quiz time! Complete each problem to test your knowledge on the various topics covered in this chapter. You can then find the solutions and explanations in the next section.

1. Determine the range of the function $f(x) = 3\sin x - 2$.
2. Graph the function $f(x) = 2\sin x$.
3. Graph the function $f(x) = 2\sin x + 1$.
4. Graph the function $f(x) = 2\sin\left(x - \frac{\pi}{2}\right) + 1$.
5. Graph the function $f(x) = 2\sin 3\left(x - \frac{\pi}{2}\right) + 1$.
6. Graph the function $f(x) = -2\sin 3\left(x - \frac{\pi}{2}\right) + 1$.
7. Graph the function $g(x) = \frac{1}{3}\tan x$.
8. Graph the function $g(x) = -2\cot x$.
9. Graph the function $g(x) = 2\sec x - 1$.
10. Graph the function $g(x) = \frac{1}{4}\csc\left(x + \frac{\pi}{3}\right)$.
11. Determine the range of the function $f(x) = -\frac{1}{2}\cos\left(x - \frac{\pi}{2}\right)$.

Answers to Chapter 11 Quiz

1. **[−5,1]. The multiplier of 3 increases the altitude, and the −2 drops the curve by 2, so the resulting range is [−5,1].**

2. **See the following graph.**

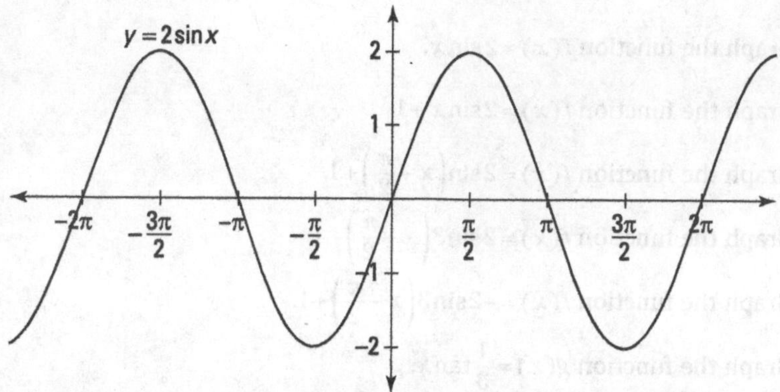

The amplitude is increased to 2, making the graph steeper.

3. **See the following graph.**

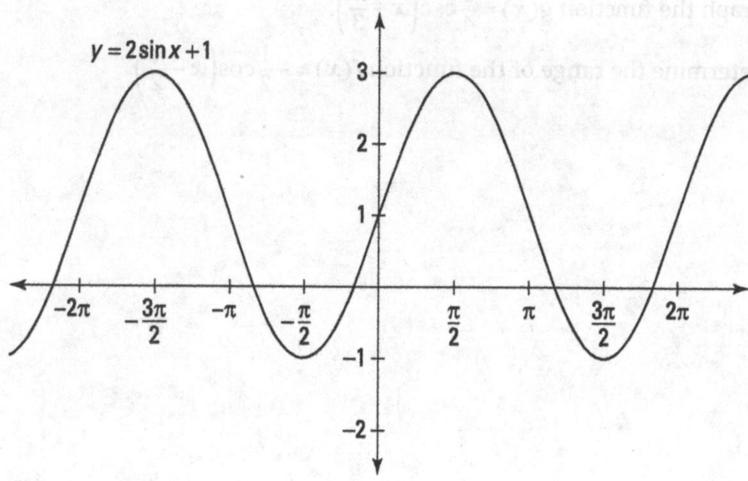

The amplitude is increased to 2, making the graph steeper. Adding 1 raises the entire graph by 1 unit, changing the range to [−1,3].

4. **See the following graph.**

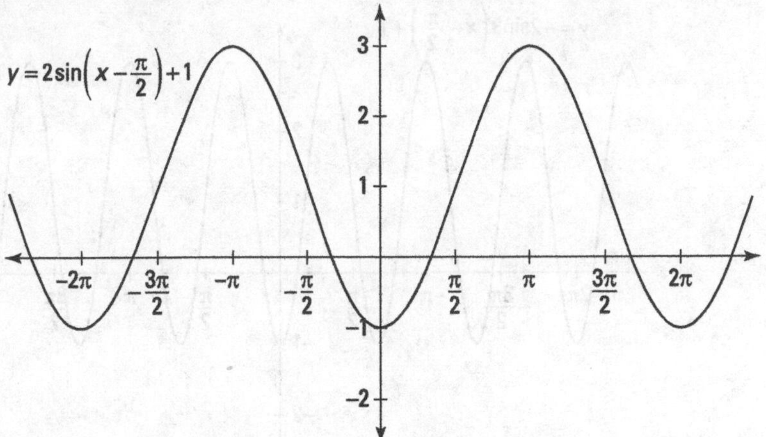

The amplitude is increased to 2, making the graph steeper. Subtracting the $\frac{\pi}{2}$ slides the graph to the right. Adding 1 raises the entire graph by 1 unit, changing the range to $[-1,3]$.

5. **See the following graph.**

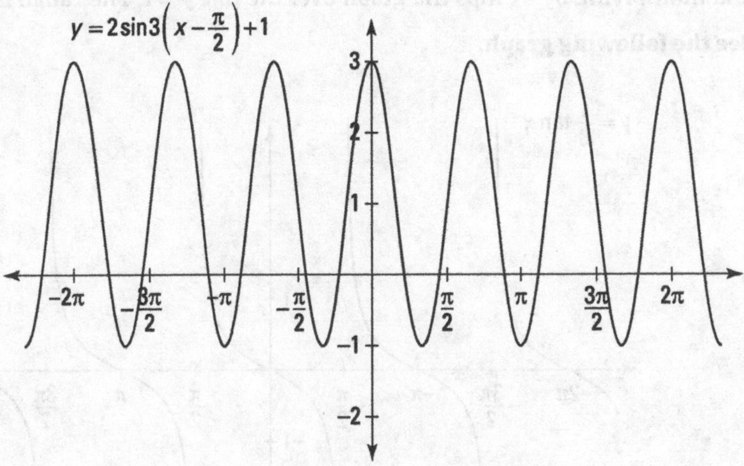

The 3 multiplier creates 3 instances of the usual sine curve in a 2π interval — or you can say that the period is now $\frac{2\pi}{3}$. The amplitude is increased to 2, making the graph steeper. Subtracting the $\frac{\pi}{2}$ slides the graph to the right. Adding 1 raises the entire graph by 1 unit, changing the range to $[-1,3]$.

6. **See the following graph.**

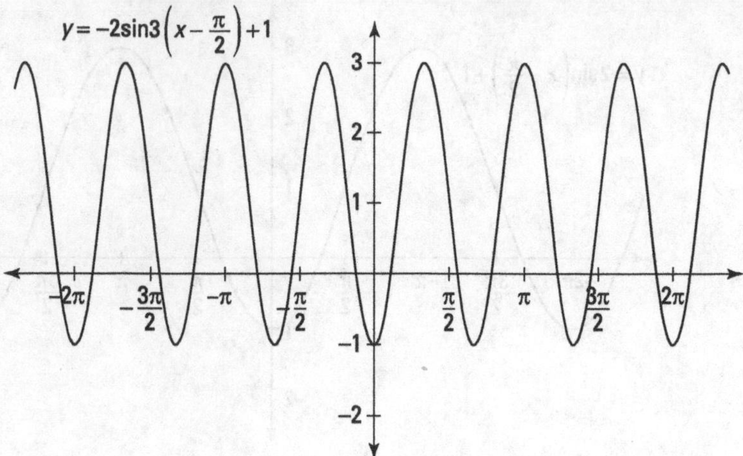

The 3 multiplier creates 3 instances of the usual sine curve in a 2π interval — or you can say that the period is now $\frac{2\pi}{3}$. The amplitude is increased to 2, making the graph steeper. Subtracting the $\frac{\pi}{2}$ slides the graph to the right. Adding 1 raises the entire graph by 1 unit, and multiplying by −1 flips the graph over the line $y = 1$. The range is still $[-1,3]$.

7. **See the following graph.**

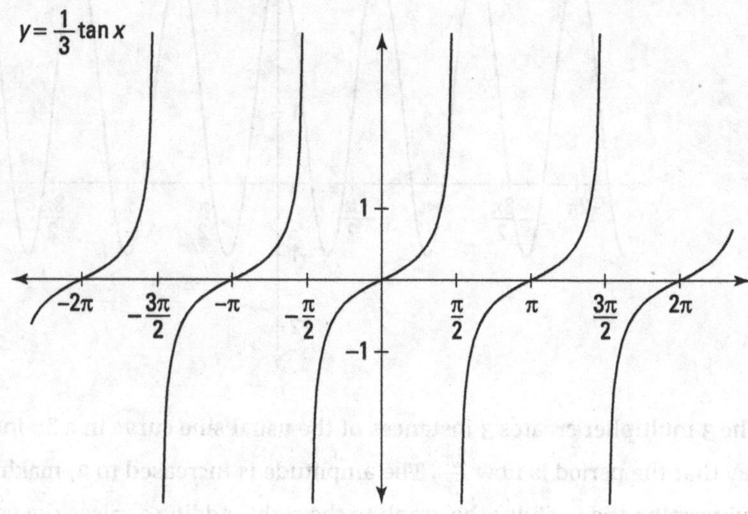

The multiplier of $\frac{1}{3}$ flattens the graph of the parent function.

8. **See the following graph.**

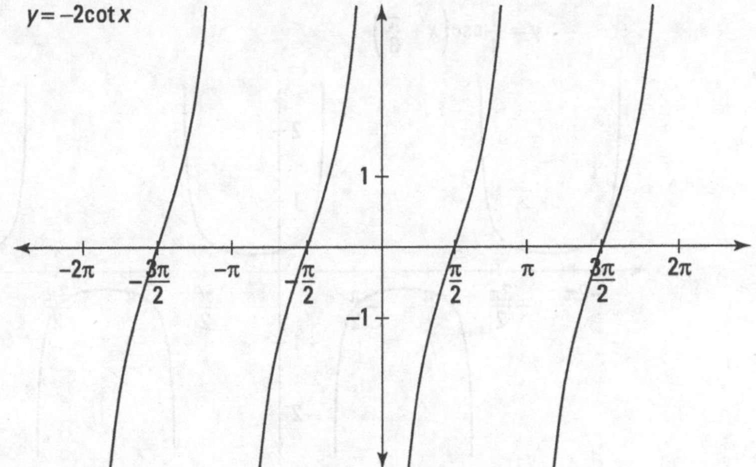

The multiplier of −2 steepens the parent graph and flips it over the *x*-axis.

9. **See the following graph.**

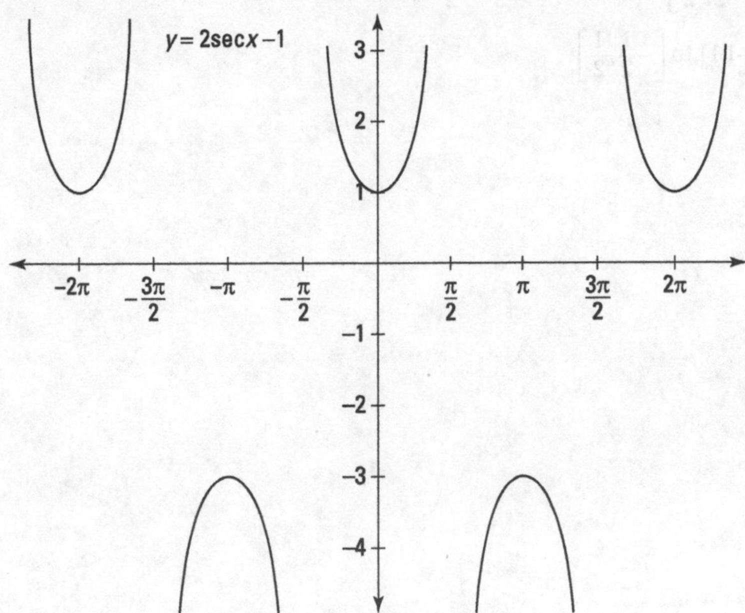

The multiplier of 2 multiplies every *y*-coordinate by 2, which steepens the parent graph, and the −1 drops the graph by 1.

(10) **See the following graph.**

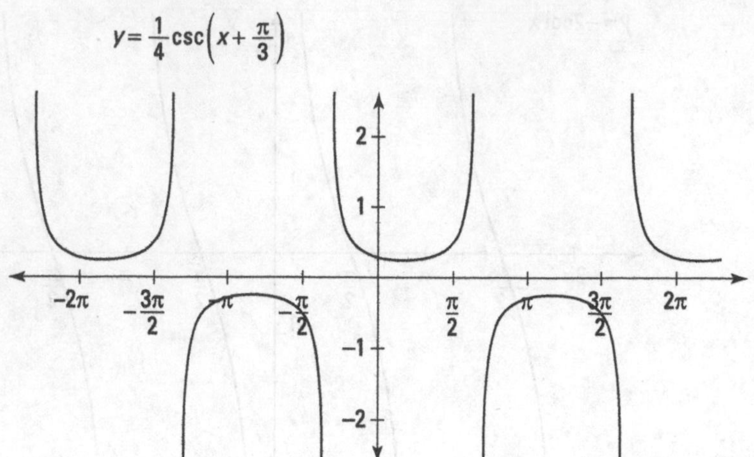

The $\frac{1}{4}$ multiplier reduces the value of every y-coordinate, and the $\frac{\pi}{3}$ slides the graph that much to the left.

(11) $\left[-\frac{1}{2},\frac{1}{2}\right]$. **The multiplier of $\frac{1}{2}$ reduces the range of the graph of the parent function**

$[-1,1]$ to $\left[-\frac{1}{2},\frac{1}{2}\right]$.

4

Identities and Special Triangles

IN THIS UNIT . . .

Working with basic trig identities.

Moving on to more advanced identities.

Solving oblique triangles.

IN THIS CHAPTER

» Reviewing the basics of solving trig equations

» Simplifying and proving expressions with fundamental trig identities

» Dealing with trig proofs

Chapter 12 Identifying with Trig Identities: The Basics

In this chapter and the next, you work on simplifying trig expressions. After performing all the possible algebraic simplification, you can often use basic trig identities to further simplify expressions and prove more complicated identities. This chapter covers basic identities, which are statements that are always true and that you use throughout an entire equation to help you simplify the problem, one expression at a time, prior to solving it. And you're in luck, because trig has many basic identities to choose from.

The biggest challenge you'll find when simplifying trig expressions by using trig identities, though, is knowing when to stop. Enter *proofs*, which give you an end goal so that you know when you've hit a stopping point. You use proofs when you need to show that two expressions are equal even though they look completely different. (If you thought you were done with proofs when you moved past geometry, think again!)

TIP

Here is some really important advice: Know (learn, memorize) these basic identities forward, backward, and upside down, because you'll use them frequently in higher mathematics. They'll make your life easier when things get complicated. If you commit the basic identities to memory but can't remember the more complex identities presented in Chapter 13 (which is highly likely), you can simply use a combination of the basic identities to derive a new identity that suits your situation. This chapter shows you the way.

In this chapter, I focus on two main ideas that are central to your studies: simplifying expressions and proving identities. These concepts share one common theme: They both involve the trig functions you're introduced to back in Chapter 10.

Keeping the End in Mind: A Quick Primer on Identities

The techniques used to get to the end of each type of problem in this chapter are very similar. However, the final results of the two main ideas — simplifying expressions and proving complicated identities — are different, and that's what I discuss in this section.

- **Simplifying:** Simplifying algebraic expressions with real numbers and variables is nothing new to you; you base your simplification processes on properties of algebra and real numbers that you *know* are always true. In this chapter, you find some new rules to play by so that you can work with trig problems of all different sorts. Think of the basic trig identities as tools in your toolbox to help you build your mathematically unique house. One tool on its own doesn't do you any good. However, when you put the identities together, all the things you can do with them may surprise you! (For instance, you may change a trig expression with many different functions into one simple function, or perhaps even change it to a simple number such as 0 or 1.)
- **Proving:** Trig proofs have equations, and your job is to make one side look exactly like the other. Usually, you make the more-complicated side look like the less-complicated side. Sometimes, though, if you can't get one side to look like the other, you can "cheat" and work on the other side for a while (or both at the same time). In the end, though, you need to show that one side transforms into the other.

Lining Up the Means to the End: Basic Trig Identities

If you're reading this book straight through, you'll likely recognize some of the identities in this chapter, because they were already covered in earlier chapters. You see trig functions in Chapter 10 as the ratios between the sides of right triangles. By definition, the reciprocal trig functions form identities, because they're true for all angle values. The full discussion of identities is found in this chapter because the identities weren't really necessary to do the mathematical calculations in the earlier chapters. Now, however, you're ready to expand your horizons and work with some more-complicated (yet still basic) identities.

In the following sections, you'll see the most basic (and most useful) identities. With this information, you can manipulate complicated trig expressions into expressions that are much simpler and more user-friendly. This simplification process takes a bit of practice. However, after you master simplifying trig expressions, proving complex identities and solving complicated equations will be a breeze.

REMEMBER

If each step you take to simplify, prove, or solve a problem with trig is based on an identity (and executed correctly), you're pretty much guaranteed to get the right answer; the particular route you take to get there doesn't matter. However, fundamental math skills still apply; you can't just throw out math rules willy-nilly. Following are some important, fundamental rules that people often forget when working with identities:

- Dividing a fraction by another fraction is the same as multiplying by its reciprocal.
- To add or subtract two fractions, you must find the common denominator.
- Factoring out the greatest common factor and writing trinomials in a factored format leads to many possibilities.

Reciprocal and ratio identities

When you're asked to simplify an expression involving cosecant, secant, or cotangent, you usually change the expression to functions that involve sine, cosine, or tangent, respectively. You do this step so that you can cancel factors and simplify the problem. When you change functions in this manner, you're using the *reciprocal identities* or the *ratio identities.* (Technically, the identities are statements about trig functions that just happen to be considered identities as well because they help you simplify expressions.)

TECHNICAL STUFF

Here are the reciprocal identities:

- $\csc\theta = \frac{1}{\sin\theta}$
- $\sec\theta = \frac{1}{\cos\theta}$
- $\cot\theta = \frac{1}{\tan\theta}$

Here are the ratio identities:

- $\tan\theta = \frac{\sin\theta}{\cos\theta}$
- $\cot\theta = \frac{\cos\theta}{\sin\theta}$

REMEMBER

You can write every trig ratio as a combination of sines and/or cosines, so changing all the functions in an equation to sines and cosines is the simplifying strategy that works most often. Also, dealing with sines and cosines is usually easier when you're looking for a common denominator for fractions. From there, you can use what you know about fractions to simplify as much as you can. You'll often first do some combining of like terms, maybe two cosecant or cotangent terms that can be added together. And there are other types of identities that will jump out at you to use before the big switch. But keep in mind that changing everything to the three basic functions makes things quite clear.

Simplifying an expression with reciprocal identities

Look for opportunities to use reciprocal identities whenever the problem you're given contains secant, cosecant, or cotangent. You can write all these functions in terms of sine and cosine, and sines and cosines are always the best place to start.

EXAMPLE **Q.** Simplify this expression: $\frac{\cos\theta\csc\theta}{\cot\theta}$.

A. Follow these steps:

1. **Change all the functions into versions of the sine and cosine functions using reciprocal and ratio identities.**

   Because this problem involves a cosecant and a cotangent, you use the reciprocal identity $\csc\theta = \frac{1}{\sin\theta}$ and the ratio identity $\cot\theta = \frac{\cos\theta}{\sin\theta}$.

   The substitution gives you $\frac{\cos\theta\left(\frac{1}{\sin\theta}\right)}{\frac{\cos\theta}{\sin\theta}} = \frac{\frac{\cos\theta}{\sin\theta}}{\frac{\cos\theta}{\sin\theta}}$.

2. **Change the division of fractions to multiplication by writing as numerator times the reciprocal of the denominator.**

   $$\frac{\frac{\cos\theta}{\sin\theta}}{\frac{\cos\theta}{\sin\theta}} = \frac{\cos\theta}{\sin\theta}\cdot\frac{\sin\theta}{\cos\theta}$$

3. **Multiply.**

   $$\frac{\cancel{\cos\theta}}{\sin\theta}\cdot\frac{\sin\theta}{\cancel{\cos\theta}} = \frac{1}{\cancel{\sin\theta}}\cdot\frac{\cancel{\sin\theta}}{1} = 1$$

   The sines and cosines cancel, and you end up getting 1 as your answer. Can't get much simpler than that!

YOUR TURN

1 Simplify the expression: $\sec\theta\cot\theta + \csc\theta$.

2 Simplify the expression: $\sin\theta(\cot\theta - \csc\theta)$.

Working backward: Using reciprocal identities to prove equalities

You are sometimes asked to prove complicated identities, because the process of proving those identities helps you to wrap your brain around the conceptual side of math. Oftentimes you'll be asked to prove identities that involve the secant, cosecant, or cotangent functions. Whenever you see these functions in a proof, the reciprocal and ratio identities often are the best places to start. Without the reciprocal identities, you can go in circles all day without getting anywhere.

EXAMPLE

Q. Prove that $\tan\theta\csc\theta = \sec\theta$.

A. You work with the left side of the equation only. (You usually start with the more "complicated" side.) Follow these simple steps:

1. **Convert both functions on the left side to sines and cosines.**

   The equation now looks like this: $\frac{\sin\theta}{\cos\theta}\cdot\frac{1}{\sin\theta} = \sec\theta$.

2. **Simplify.**

   $\frac{\cancel{\sin\theta}}{\cos\theta}\cdot\frac{1}{\cancel{\sin\theta}} = \sec\theta$ or $\frac{1}{\cos\theta}\cdot\frac{1}{1} = \frac{1}{\cos\theta} = \sec\theta$

   This gives you $\frac{1}{\cos\theta} = \sec\theta$.

3. **Use the reciprocal identity to change the term on the left.**

   Because $\sec\theta = \frac{1}{\cos\theta}$, this equation becomes $\sec\theta = \sec\theta$. Bingo!

YOUR TURN

 Prove that $\cot\theta\cdot\sec\theta = \csc\theta$.	 Prove that $\sec\theta\cdot\sin\theta\cdot\cot\theta = 1$.

Pythagorean identities

The *Pythagorean identities* are among the most useful identities because they simplify second-degree expressions so nicely. When you see a trig function that's squared ($\sin^2$, $\cos^2$, and so on), keep these identities in mind. They're built from previous knowledge of right triangles and the alternate trig function values (which are covered in Chapter 10). Recall that when a right triangle sits in the unit circle, the x-leg is $\cos\theta$, the y-leg is $\sin\theta$, and the hypotenuse is 1. Because you know that $\text{leg}^2 + \text{leg}^2 = \text{hypotenuse}^2$, thanks to the Pythagorean Theorem, you also know that $\sin^2\theta + \cos^2\theta = 1$. In this section, you see where these important identities come from and then how to use them.

TECHNICAL STUFF

The three Pythagorean identities are as follows:

- $\sin^2\theta + \cos^2\theta = 1$
- $1 + \cot^2\theta = \csc^2\theta$
- $\tan^2\theta + 1 = \sec^2\theta$

TIP

To limit the amount of memorizing you have to do, you can use the first Pythagorean identity to derive the other two:

If you divide every term of $\sin^2\theta + \cos^2\theta = 1$ by $\sin^2\theta$, you get $\frac{\sin^2\theta}{\sin^2\theta} + \frac{\cos^2\theta}{\sin^2\theta} = \frac{1}{\sin^2\theta}$, which simplifies to the second Pythagorean identity:

$$1 + \cot^2\theta = \csc^2\theta \text{ (because of the ratio identity, } \frac{\sin^2\theta}{\sin^2\theta} = 1 \text{ and } \frac{\cos^2\theta}{\sin^2\theta} = \cot^2\theta)$$

$$\frac{1}{\sin^2\theta} = \csc^2\theta \text{ (because of the reciprocal identity)}$$

When you divide every term of $\sin^2\theta + \cos^2\theta = 1$ by $\cos^2\theta$, you get $\frac{\sin^2\theta}{\cos^2\theta} + \frac{\cos^2\theta}{\cos^2\theta} = \frac{1}{\cos^2\theta}$, which simplifies to the third Pythagorean identity:

$$\tan^2\theta + 1 = \sec^2\theta \text{ (because } \frac{\sin^2\theta}{\cos^2\theta} = \tan^2\theta \text{ and } \frac{\cos^2\theta}{\cos^2\theta} = 1 \text{ and } \frac{1}{\cos^2\theta} = \sec^2\theta)$$

Putting the Pythagorean identities into action

You normally use Pythagorean identities if you know one function and are looking for another. For example, if you know the sine ratio, you can use the first Pythagorean identity from the previous section to find the cosine ratio. In fact, you can find whatever you're asked to find if all you have is the value of one trig function and the understanding of what quadrant the angle θ is in.

EXAMPLE

Q. Given that $\sin\theta = \frac{24}{25}$ and $\frac{\pi}{2} < \theta < \pi$, find $\cos\theta$.

A. Use the following steps:

1. **Plug what you know into the appropriate Pythagorean identity.**

   Because you're using sine and cosine, you use the first identity: $\sin^2\theta + \cos^2\theta = 1$. Plug in the values you know to get $\left(\frac{24}{25}\right)^2 + \cos^2\theta = 1$.

2. **Isolate the trig function with the variable on one side.**

   First, square the sine value to get $\frac{576}{625}$, giving you $\frac{576}{625} + \cos^2\theta = 1$. Subtract $\frac{576}{625}$ from both sides (***Hint:*** You need to find a common denominator): $\cos^2\theta = 1 - \frac{576}{625} = \frac{49}{625}$.

3. **Find the square root of the sides you need to solve.**

   You now have $\cos\theta = \pm\frac{7}{25}$. But, because of the constraint $\frac{\pi}{2} < \theta < \pi$ that you're given in the problem, there's only one solution.

4. **Draw a picture of the unit circle so you can visualize the angle.**

   Because $\frac{\pi}{2} < \theta < \pi$, the angle lies in quadrant II, so the cosine of θ must be negative. You have your answer: $\cos\theta = -\frac{7}{25}$, as shown in Figure 12-1.

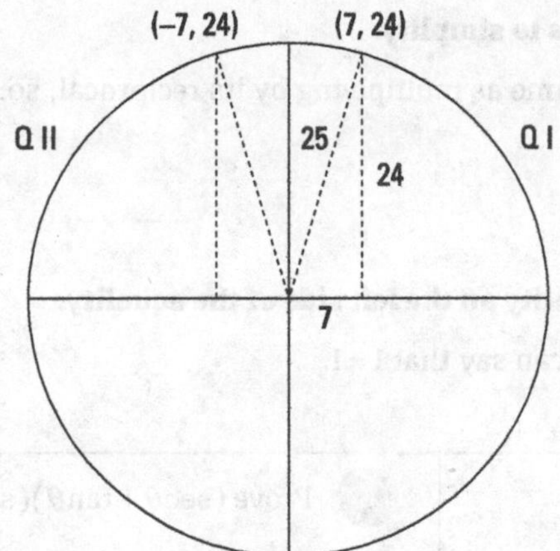

FIGURE 12-1: The angle in quadrant II.

YOUR TURN

Given that $\cos\theta = -\frac{5}{13}$ and $\frac{3\pi}{2} < \theta < 2\pi$, find $\sin\theta$.

Given that $\sin\theta = -\frac{11}{61}$ and $\pi < \theta < \frac{3\pi}{2}$, find $\tan\theta$.

Using the Pythagorean identities to prove an equality

The Pythagorean identities pop up frequently in trig proofs. Pay attention and look for trig functions being squared. Try changing them to a Pythagorean identity and see whether anything interesting happens. This section shows you how one proof can involve a Pythagorean identity. And after you change sines and cosines, the proof simplifies and makes your job that much easier.

EXAMPLE

Q. Prove $\frac{\sin x}{\csc x} + \frac{\cos x}{\sec x} = 1$

A. Follow these steps:

1. **Convert all the functions in the equality to sines and cosines.**

$$\frac{\sin x}{\frac{1}{\sin x}} + \frac{\cos x}{\frac{1}{\cos x}} = 1$$

2. **Use the properties of fractions to simplify.**

Dividing by a fraction is the same as multiplying by its reciprocal, so:

$$\sin x \cdot \frac{\sin x}{1} + \cos x \cdot \frac{\cos x}{1} = 1$$
$$\sin^2 x + \cos^2 x = 1$$

3. **Identify the Pythagorean identity on the left side of the equality.**

Because $\sin^2\theta + \cos^2\theta = 1$, you can say that $1 = 1$.

YOUR TURN

 Prove $\csc\theta\cos^2\theta + \sin\theta = \csc\theta$

 Prove $(\sec\theta + \tan\theta)(\sec\theta - \tan\theta) = 1$

 Prove $\cos^2\theta - \sin^2\theta = 1 - 2\sin^2\theta$

 Prove $\sin^2\theta\left(\tan^2\theta + 1\right) = \tan^2\theta$

Even/odd identities

TECHNICAL STUFF

Because sine, cosine, and tangent are basic to trig functions, their functions can be defined as even or odd as well (see Chapter 3). Sine and tangent are both odd functions, and cosine is an even function. In other words,

- $\sin(-\theta) = -\sin\theta$
- $\cos(-\theta) = \cos\theta$
- $\tan(-\theta) = -\tan\theta$

These identities will appear in problems that ask you to simplify an expression, prove an identity, or solve an equation. The big red flag this time? You want the variables representing the angles to be positive, so they'll all be the same. When $\tan(-x)$, for example, appears somewhere in an expression, it should usually be changed to $-\tan x$.

Simplifying expressions with even/odd identities

Mostly, you use even/odd identities for graphing purposes, but you may see them in simplifying problems as well. (You can find the graphs of the trigonometric equations in Chapter 10 if you need a refresher.) You use an even/odd identity to simplify any expression where $-x$ (or whatever variable you see) is inside the trig function.

EXAMPLE

Q. Simplify $[1+\sin(-x)][1-\sin(-x)]$

A. Use the following steps:

1. **Get rid of all the $-x$ values inside the trig functions.**

   You see two $\sin(-x)$ functions, so you replace them both with $-\sin x$ to get $[1+(-\sin x)][1-(-\sin x)]$.

2. **Simplify the new expression.**

   First, rewrite the two negative signs within the brackets to get $[1-\sin x][1+\sin x]$, and then FOIL these two binomials to get $1-\sin^2 x$.

3. **Look for any combination of terms that could give you a Pythagorean identity.**

   Whenever you see a function squared, you should think of the Pythagorean identities. Looking back at the section, "Pythagorean identities," you see that $1-\sin^2 x$ is the same as $\cos^2 x$. Now the expression is fully simplified as $\cos^2 x$:

   $$[1+(-\sin x)][1-(-\sin x)] = \cos^2 x$$

Proving an equality with even/odd identities

When asked to prove an identity, if you see a negative variable inside a trig function, you automatically use an even/odd identity. First, replace all trig functions with negative angles. Then simplify the trig expression to make one side look like the other side. Here's just one example of how this works.

EXAMPLE

Q. Prove this identity: $\frac{\cos(-x)-\sin(-x)}{\sin x}-\frac{\cos(-x)+\sin(-x)}{\cos x}=\sec x\csc x$.

A. Use the following steps:

1. **Replace all negative angles and their trig functions using the even/odd identity that applies.**

$$\frac{\cos x-(-\sin x)}{\sin x}-\frac{\cos x-\sin x}{\cos x}=\sec x\csc x$$

$$\frac{\cos x+\sin x}{\sin x}-\frac{\cos x-\sin x}{\cos x}=\sec x\csc x$$

2. **Simplify the new expression.**

The left side is considered to be the more complicated side, so that's where you start working. In order to subtract the fractions, you first must find a common denominator. However, before doing that, look at the two fractions. These fractions can be split up into the sum of two fractions. When you do this step first, certain terms simplify and make your job much easier when the time comes to work with the fractions. Breaking up the fractions, you get $\frac{\cos x}{\sin x}+\frac{\sin x}{\sin x}-\left(\frac{\cos x}{\cos x}-\frac{\sin x}{\cos x}\right)=\sec x\csc x$, which quickly simplifies to $\frac{\cos x}{\sin x}+1-1+\frac{\sin x}{\cos x}=\sec x\csc x\rightarrow\frac{\cos x}{\sin x}+\frac{\sin x}{\cos x}=\sec x\csc x$.

Now you need a common denominator for the left side. The common denominator is $\sin x\cos x$.

Multiplying the first term by $\frac{\cos x}{\cos x}$ and the second term by $\frac{\sin x}{\sin x}$, you have

$$\frac{\cos x}{\sin x}\cdot\frac{\cos x}{\cos x}+\frac{\sin x}{\cos x}\cdot\frac{\sin x}{\sin x}=\sec x\csc x$$

$$\frac{\cos^2 x}{\sin x\cos x}+\frac{\sin^2 x}{\sin x\cos x}=\sec x\csc x.$$

You can now add the fractions and get $\frac{\cos^2 x+\sin^2 x}{\sin x\cos x}=\sec x\csc x$.

Here is a Pythagorean identity in its finest form! Note that $\sin^2\theta+\cos^2\theta=1$ is the most frequently used of the Pythagorean identities. This equation then simplifies to

$$\frac{1}{\sin x\cos x}=\sec x\csc x.$$

Using the reciprocal identities, you get

$$\frac{1}{\sin x}\cdot\frac{1}{\cos x}=\sec x\csc x$$

$$\csc x\cdot\sec x=\sec x\csc x.$$

Multiplication is commutative, so the identity $\sec x\csc x=\sec x\csc x$ is proven.

YOUR TURN

11 Simplify $\frac{\cot^2(-x)-1}{-\cot^2 x}$

12 Prove $\frac{\tan(-x)}{\sec(-x)}\cdot\sin(-x)=\sin^2 x$

Co-function identities

If you take the graph of sine and shift it to the left or right, it looks exactly like the cosine graph (see Chapter 11). The same is true for secant and cosecant. Tangent and cotangent require a shift and a flip to look exactly alike. And that's the basic premise of *co-function identities*: they say that the sine and cosine functions have the same values, but those values are shifted slightly on the coordinate plane when you look at one function compared to the other. You have experience with all six of the trig functions, as well as their relationships to one another. The only difference is that in this section, they are introduced formally as *identities*.

TECHNICAL STUFF

The following list of co-function identities illustrates this point:

- $\sin\theta=\cos\left(\frac{\pi}{2}-\theta\right)$ $\quad\cos\theta=\sin\left(\frac{\pi}{2}-\theta\right)$
- $\tan\theta=\cot\left(\frac{\pi}{2}-\theta\right)$ $\quad\cot\theta=\tan\left(\frac{\pi}{2}-\theta\right)$
- $\sec\theta=\csc\left(\frac{\pi}{2}-\theta\right)$ $\quad\csc\theta=\sec\left(\frac{\pi}{2}-\theta\right)$

Putting co-function identities to the test

The co-function identities are great to use whenever you see $\frac{\pi}{2}$ inside the grouping parentheses. You may see functions in trig expressions such as $\sin\left(\frac{\pi}{2}-x\right)$. If the quantity inside the trig function looks like $\left(\frac{\pi}{2}-x\right)$ or $(90°-\theta)$, you'll know to use the co-function identities.

EXAMPLE

Q. Simplify $\frac{\cos x}{\cos\left(\frac{\pi}{2}-x\right)}$.

A. $\cot x$. Follow these steps:

1. **Look for co-function identities and substitute.**

   You first realize that $\cos\left(\frac{\pi}{2}-x\right)=\sin x$ because of the co-function identity. That means you can substitute $\sin x$ in for $\cos\left(\frac{\pi}{2}-x\right)$ to get $\frac{\cos x}{\sin x}$.

2. Look for other substitutions you can make.

Because of the ratio identity for cotangent, $\frac{\cos x}{\sin x} = \cot x$.

So $\frac{\cos x}{\cos\left(\frac{\pi}{2} - x\right)}$ simplifies to cot *x*.

Proving an equality by employing the co-function identities

Co-function identities also pop up in trig proofs. The following identity uses both co-function and even/odd identities.

EXAMPLE

Q. Prove this equality: $\frac{\csc\left(\frac{\pi}{2} - x\right)}{\tan(-x)} = -\csc x$.

A. Follow these steps:

1. Replace any trig functions that contain $\frac{\pi}{2}$ with the appropriate co-function identity.

Replacing $\csc\left(\frac{\pi}{2} - x\right)$ with sec *x*, you get $\frac{\sec x}{\tan(-x)} = -\csc x$.

2. Simplify the new expression.

You have many trig identities at your disposal, and you may use any of them at any given time. Now is the perfect time to use an even/odd identity for tangent: $\frac{\sec x}{-\tan x} = -\csc x$.

Then use the reciprocal identity for secant and the ratio identity for tangent to get

$$\frac{\frac{1}{\cos x}}{-\frac{\sin x}{\cos x}} = -\csc x.$$

Now multiply the numerator by the reciprocal of the denominator:

$$\frac{1}{\cos x}\left(-\frac{\cos x}{\sin x}\right) = -\csc x$$

$$\frac{1}{\cancel{\cos x}}\left(-\frac{\cancel{\cos x}}{\sin x}\right) = -\csc x$$

$$-\frac{1}{\sin x} = -\csc x$$

Use the reciprocal identity for csc *x* on the left side to get $-\csc x = -\csc x$.

YOUR TURN

13 Simplify $\sin\left(\frac{\pi}{2} - x\right) + \tan\left(\frac{\pi}{2} - x\right) \cdot \cos\left(\frac{\pi}{2} - x\right)$

14 Prove $\dfrac{\cos\left(\dfrac{\pi}{2}-x\right)}{\sin\left(\dfrac{\pi}{2}-x\right)}\cdot\cot\left(\dfrac{\pi}{2}-x\right)=\tan^2 x$

Periodicity identities

Periodicity identities illustrate how shifting the graph of a trig function by one period to the left or right results in the same function. (You see periods and periodicity identities when graphing trig functions in Chapter 10.) The functions of sine, cosine, secant, and cosecant repeat every 2π radians; tangent and cotangent, on the other hand, repeat every π radians.

The following identities show how the different trig functions repeat:

- $\sin(\theta+2\pi)=\sin\theta$ $\quad\cos(\theta+2\pi)=\cos\theta$
- $\tan(\theta+\pi)=\tan\theta$ $\quad\cot(\theta+\pi)=\cot\theta$
- $\sec(\theta+2\pi)=\sec\theta$ $\quad\csc(\theta+2\pi)=\csc\theta$

Seeing how the periodicity identities work to simplify equations

Similar to the co-function identities, you use the periodicity identities when you find the expressions $(x+2\pi)$ or $(x-2\pi)$ inside a trig function. Because adding (or subtracting) 2π radians from an angle gives you a new angle in the same position, you can use that idea to form an identity. For tangent and cotangent only, adding or subtracting π radians from the angle gives you the same result, because the period of the tangent and cotangent function is π.

Q. Simplify $\sin(2\pi+x)+\cos(2\pi+x)\cot(\pi+x)$

A. $\csc x$. Follow these steps:

1. **Replace all trig functions with the appropriate periodicity identity.**

   You get $\sin x+\cos x\cot x$.

2. **Simplify the new expression using the ratio identity for cot x.**

   $$\sin x+\cos x\cdot\frac{\cos x}{\sin x}=\sin x+\frac{\cos^2 x}{\sin x}$$

To find a common denominator before adding the fractions, multiply the first term by $\frac{\sin x}{\sin x}$. You get: $\sin x \cdot \frac{\sin x}{\sin x} + \frac{\cos^2 x}{\sin x} = \frac{\sin^2 x}{\sin x} + \frac{\cos^2 x}{\sin x}$.

Add them together and simplify, using the Pythagorean identity:

$\frac{\sin^2 x}{\sin x} + \frac{\cos^2 x}{\sin x} = \frac{\sin^2 x + \cos^2 x}{\sin x} = \frac{1}{\sin x}$.

Finally, using the reciprocal identity, $\frac{1}{\sin x} = \csc x$.

Proving an equality with the periodicity identities

Using the periodicity identities also comes in handy when you need to prove an equality that includes the expression $(x + 2\pi)$ or the addition (or subtraction) of the period.

EXAMPLE

Q. Prove $[\sec(2\pi + x) - \tan(\pi + x)][\csc(2\pi + x) + 1] = \cot x$

A. Follow these steps:

1. **Replace all trig functions with the appropriate periodicity identity.**

   You're left with $[\sec x - \tan x][\csc x + 1] = \cot x$.

2. **Simplify the new expression.**

   For this example, the best place to start is to FOIL (see Chapter 4); you multiply the two factors together to get four separate terms: $\sec x \csc x + \sec x - \tan x \csc x - \tan x = \cot x$.

   Now convert all terms on the left to sines and cosines to get $\frac{1}{\cos x} \cdot \frac{1}{\sin x} + \frac{1}{\cos x} - \frac{\sin x}{\cos x} \cdot \frac{1}{\sin x} - \frac{\sin x}{\cos x} = \cot x$.

   Reduce the third term, simplify, then find a common denominator and add the fractions:

   $$\frac{1}{\cos x} \cdot \frac{1}{\sin x} + \frac{1}{\cos x} - \frac{\cancel{\sin x}}{\cos x} \cdot \frac{1}{\cancel{\sin x}} - \frac{\sin x}{\cos x} = \cot x$$

   $$\frac{1}{\sin x \cos x} + \cancel{\frac{1}{\cos x}} - \cancel{\frac{1}{\cos x}} - \frac{\sin x}{\cos x} = \cot x$$

   $$\frac{1}{\sin x \cos x} - \frac{\sin^2 x}{\sin x \cos x} = \cot x$$

   $$\frac{1 - \sin^2 x}{\sin x \cos x} = \cot x$$

3. **Apply any other applicable identities.**

   The Pythagorean identity $\sin^2\theta + \cos^2\theta = 1$ can also be written as $\cos^2\theta = 1 - \sin^2\theta$, so replace $1 - \sin^2 x$ with $\cos^2 x$. Cancel one of the cosines in the numerator (because it's squared) with the cosine in the denominator to get

   $$\frac{\cos^{\cancel{2}} x}{\sin x \cancel{\cos x}} = \cot x$$

   $$\frac{\cos x}{\sin x} = \cot x$$

   Finally, using the ratio identity for cot x, this equation simplifies to $\cot x = \cot x$.

YOUR TURN

15 Simplify $1-\sin(2\pi+x)\cdot\cot(\pi+x)\cdot\cos(2\pi+x)$

16 Prove $\dfrac{\sec(x+2\pi)}{\csc(x+2\pi)}=\tan x$

Tackling Difficult Trig Proofs: Some Techniques to Know

Historically speaking, proofs in geometry and trigonometry are not anyone's favorite part when studying those subjects. But, with a little help, they don't seem quite so formidable — and even become fun! So far in this chapter, you've seen proofs that require only a few basic steps to complete. Now you'll see how to tackle the more complicated proofs. The techniques here are based on ideas you've dealt with before in your math journey. Okay, so a few trig functions are thrown into the discussion, but why should that scare you?

TIP

One tip will always help you when you're faced with complicated trig proofs that require multiple identities: *Always* check your work and review all the identities you know to make sure that you haven't forgotten to simplify something.

REMEMBER

The goal in proofs is to make one side of the given equation look like the other side through a series of steps, all of which are based on identities, properties, and definitions. All decisions you make must be based on the rules. Here's an overview of the techniques you'll see in this section.

- **Fractions in proofs:** These types of proofs allow you to use every rule you've ever learned regarding fractions and their operations. Also, you have many identities involving fractions that can help.
- **Factoring:** Degrees higher than 1 on a trig function often are great indicators that you may need to do some factoring. And if you see several sets of parentheses, you may get to do some multiplying. Always check, though, that the power doesn't involve an expression that's connected to a Pythagorean identity.

- **Square roots:** When roots show up in a proof, sooner or later you may have to square both sides to get things moving.
- **Working on both sides at once:** Sometimes you may get stuck while working on one side of a proof. At that point, it's okay to work on the other side to see how to get the two sides to match.

Dealing with demanding denominators

Fractions are a way of life. They're here to stay. And they're especially prevalent in trigonometry. When you are dealing with trig proofs, fractions inevitably pop up. So just take a deep breath. Even if you don't mind fractions, you should still read this section because it shows you specifically how to work with fractions in trig proofs. Three main types of proofs you'll work with have fractions:

- Proofs where you end up creating fractions
- Proofs that start off with fractions
- Proofs that require multiplying by a conjugate to deal with a fraction

Each one of these types in this section comes with an example proof so you can see what to do.

Creating fractions when working with reciprocal identities

Converting all the functions to sines and cosines can make a trig proof easier. When terms are being multiplied, this conversion usually allows you to cancel and simplify to your heart's content so that one side of the equation ends up looking just like the other, which is the goal. But when the terms are adding or subtracting, you may create fractions where none were before. This is especially true when dealing with secant and cosecant, because you create fractions when you convert them (respectively) to $\frac{1}{\cos x}$ and $\frac{1}{\sin x}$. The same is true for tangent when you change it to $\frac{\sin x}{\cos x}$, and cotangent when it becomes $\frac{\cos x}{\sin x}$.

TIP

When choosing which side of the identity to work on, you usually opt for the side that has more terms. It's easier to add two terms together to create one than to try to figure out how to break one term into two.

EXAMPLE

Q. Prove $\sec^2 t + \csc^2 t = \sec^2 t \csc^2 t$

A. Follow these steps:

1. **Convert the trig functions on the left to sines and cosines.**

   On the left side, you now have $\frac{1}{\cos^2 t} + \frac{1}{\sin^2 t} = \sec^2 t \csc^2 t$.

2. **Find the common denominator of the two fractions and rewrite them.**

   This multiplication gives you $\frac{\sin^2 t}{\sin^2 t \cos^2 t} + \frac{\cos^2 t}{\sin^2 t \cos^2 t} = \sec^2 t \csc^2 t$.

3. **Add the two fractions.**

$$\frac{\sin^2 t + \cos^2 t}{\sin^2 t\cos^2 t} = \sec^2 t\csc^2 t$$

4. **Simplify the expression with a Pythagorean identity in the numerator.**

$$\frac{1}{\sin^2 t\cos^2 t} = \sec^2 t\csc^2 t$$

5. **Use reciprocal identities to rewrite without a fraction.**

$$\frac{1}{\sin^2 t}\cdot\frac{1}{\cos^2 t} = \sec^2 t\csc^2 t \rightarrow \csc^2 t\cdot\sec^2 t = \sec^2 t\csc^2 t$$

Multiplication is commutative, so you have $\sec^2 t\csc^2 t = \sec^2 t\csc^2 t$.

Starting off with fractions

When the expression you're given begins with fractions, most of the time you have to add (or subtract) them to get things to simplify. Here's one example of a proof where doing just that gets the ball rolling.

EXAMPLE

Q. Simplify $\frac{\cos t}{1+\sin t} + \frac{\sin t}{\cos t}$.

A. **sec *t*. You have to find the LCD to add the two fractions in order to simplify this expression. Follow these steps:**

1. **Find the common denominator and add.**

The least common denominator is $(1+\sin t)\cos t$, the product of the two denominators:

$$\frac{\cos t}{1+\sin t}\cdot\frac{\cos t}{\cos t} + \frac{\sin t}{\cos t}\cdot\frac{1+\sin t}{1+\sin t} = \frac{\cos^2 t}{(1+\sin t)\cos t} + \frac{\sin t + \sin^2 t}{(1+\sin t)\cos t}.$$

Adding the two fractions together, you get $\frac{\cos^2 t + \sin t + \sin^2 t}{(1+\sin t)\cos t}$.

2. **Look for any trig identities and substitute.**

You can rewrite the numerator as $\frac{\sin t + \sin^2 t + \cos^2 t}{(1+\sin t)\cos t}$.

Applying the Pythagorean identity, $\sin^2\theta + \cos^2\theta = 1$, to the second two terms in the numerator, you get $\frac{\sin t + 1}{(1+\sin t)\cos t}$.

3. **Cancel or reduce the fraction.**

After the top and the bottom are completely factored (see Chapter 5), you can cancel terms: $\frac{\cancel{1+\sin t}}{(\cancel{1+\sin t})\cos t} = \frac{1}{\cos t}$.

4. **Apply the reciprocal identity.**

$$\frac{1}{\cos t} = \sec t$$

So the original expression $\frac{\cos t}{1+\sin t} + \frac{\sin t}{\cos t}$ simplifies to sec *t*.

Multiplying by a conjugate

When one side of a proof is a fraction with a binomial in its denominator, always consider multiplying by the conjugate before you do anything else. Most of the time, this technique allows you to simplify, because it creates the difference of two squares in the denominator.

EXAMPLE

Q. Rewrite this expression without a fraction: $\frac{\sin x}{\sec x - 1}$.

A. $\cot x + \cot x \cos x$. Follow these steps:

1. **Multiply by the conjugate of the denominator.**

   The conjugate of $a+b$ is $a-b$, and vice versa. So you multiply both the numerator and denominator by $\sec x + 1$. This step gives you $\frac{\sin x}{\sec x - 1} \cdot \frac{\sec x + 1}{\sec x + 1}$.

2. **FOIL the conjugates.**

   $$\frac{\sin x(\sec x + 1)}{\sec^2 x - 1}$$

   If you've been following along all through this chapter, the bottom should look awfully familiar. One of those Pythagorean identities? Yep!

3. **Change any identities to their simpler form.**

   Using the Pythagorean identity $\tan^2\theta + 1 = \sec^2\theta$ and substituting on the bottom, you get $\frac{\sin x(\sec x + 1)}{\tan^2 x + 1 - 1} = \frac{\sin x(\sec x + 1)}{\tan^2 x}$.

4. **Change every trig function to sines and cosines.**

   Here, it gets more complex: $\frac{\sin x\left(\frac{1}{\cos x} + 1\right)}{\frac{\sin^2 x}{\cos^2 x}}$.

5. **Multiply the numerator times the reciprocal of the denominator.**

   $$\sin x\left(\frac{1}{\cos x} + 1\right) \cdot \frac{\cos^2 x}{\sin^2 x} = \frac{\sin x}{1}\left(\frac{1}{\cos x} + 1\right) \cdot \frac{\cos^2 x}{\sin^2 x}$$

6. **Cancel what you can from the expression.**

   The sine on the top cancels one of the sines on the bottom, leaving you with the following equation: $\frac{\cancel{\sin x}}{1}\left(\frac{1}{\cos x} + 1\right) \cdot \frac{\cos^2 x}{\sin^{\cancel{2}} x} = \left(\frac{1}{\cos x} + 1\right) \cdot \frac{\cos^2 x}{\sin x}$.

7. **Distribute and watch what happens!**

   Through cancellations, you go from $\frac{1}{\cancel{\cos x}} \cdot \frac{\cos^{\cancel{2}} x}{\sin x} + 1 \cdot \frac{\cos^2 x}{\sin x} = \frac{\cos x}{\sin x} + \frac{\cos^2 x}{\sin x}$ to $\frac{\cos x}{\sin x} + \frac{\cos x}{\sin x} \cdot \cos x$.

   Using a ratio identity, this expression finally simplifies to $\cot x + \cot x \cos x$. And if you're asked to take it even a step further, you can factor to get $\cot x(1 + \cos x)$.

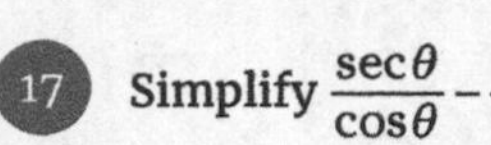

YOUR TURN

17 Simplify $\dfrac{\sec\theta}{\cos\theta} - \dfrac{\tan\theta}{\cot\theta}$

18 Prove $\dfrac{\sin x}{1+\cos x} + \dfrac{1+\cos x}{\sin x} = \dfrac{2}{\sin x}$

19 Simplify $\dfrac{1+\cos t}{1-\cos t} + \dfrac{1-\cos t}{1+\cos t}$

20 Prove $1 - \dfrac{\cos^2 x}{1+\sin x} = \sin x$

Going solo on each side

Sometimes doing work on both sides of a proof, one side at a time, leads to a quicker solution, because in order to prove a very complicated identity, you may need to complicate the expression even further before it can begin to simplify. However, you should take this action only in dire circumstances after every other technique has failed.

The main idea here is that you work on the more complicated side first, stop when you just can't go any further, and then switch to working on the other side. By switching back and forth, your goal is to make the two sides of the proof meet in the middle somewhere. They should end up with the same expression on each side.

EXAMPLE

Q. Prove this identity: $\frac{1+\cot x}{\cot x}=\tan x+\csc^2 x-\cot^2 x$.

A. It's hard to say which side is more complicated, but, for this example, consider the fraction to be your first line of attack.

1. **Break up the fraction by writing each term in the numerator over the term in the denominator, separately.**

   You now have $\frac{1}{\cot x}+\frac{\cot x}{\cot x}=\tan x+\csc^2 x-\cot^2 x$.

2. **Use a reciprocal identity and reduce the fraction to simplify.**

   You now have $\tan x+1=\tan x+\csc^2 x-\cot^2 x$.

   You've come to the end of the road on the left side. The expression is now so simplified that it would be hard to expand it again to look like the right side, so you should turn to the right side and simplify it.

WARNING

You may be tempted to subtract tan x from each side, but don't. You can work on one side or the other, but not both at the same time.

3. **Look for any applicable trig identities on the right side.**

   You use the Pythagorean identity $\csc^2\theta=1+\cot^2\theta$, replacing the $\csc^2 x$ term.

   You now have $\tan x+1=\tan x+\left(1+\cot^2 x\right)-\cot^2 x$. This simplifies to $\tan x+1=\tan x+1$.

4. **Rewrite the proof starting on one side and ending up like the other side.**

   This time, working on the right side, steps are used from the previous work to change the right side to what's on the left side.

$$\begin{aligned}\frac{1+\cot x}{\cot x}&=\tan x+\csc^2 x-\cot^2 x\\&=\tan x+\left(1+\cot^2 x\right)-\cot^2 x\\&=\tan x+1\\&=\frac{1}{\cot x}+\frac{\cot x}{\cot x}\\&=\frac{1+\cot x}{\cot x}\end{aligned}$$

YOUR TURN

21 Prove the identity by working on both sides of the equation: $\cot^2 x-\cos^2 x=\cos^2 x\cot^2 x$.

22 Prove the identity by working on both sides of the equation: $\frac{\sec x-\csc x}{\sec x+\csc x}=\frac{\tan x-1}{\tan x+1}$.

Practice Questions Answers and Explanations

1. **$2\csc\theta$. First, apply the reciprocal and ratio identities: $\sec\theta\cot\theta + \csc\theta = \frac{1}{\cos\theta}\cdot\frac{\cos\theta}{\sin\theta} + \frac{1}{\sin\theta}$.**

   Reduce the first term and add the two terms together: $\frac{1}{\cancel{\cos\theta}}\cdot\frac{\cancel{\cos\theta}}{\sin\theta} + \frac{1}{\sin\theta} = \frac{1}{\sin\theta} + \frac{1}{\sin\theta} = \frac{2}{\sin\theta}$.

   The result $\frac{1}{\sin\theta}$ can be considered as simplified, but writing the answer without a fraction, you apply the reciprocal identity and get $\csc\theta$.

2. **$\cos\theta - 1$. Apply the reciprocal and ratio identities to the two terms in the parentheses, and then distribute the sine over the terms:**

   $$\sin\theta(\cot\theta - \csc\theta) = \sin\theta\left(\frac{\cos\theta}{\sin\theta} - \frac{1}{\sin\theta}\right) = \cancel{\sin\theta}\cdot\frac{\cos\theta}{\cancel{\sin\theta}} - \cancel{\sin\theta}\cdot\frac{1}{\cancel{\sin\theta}} = \cos\theta - 1$$

3. **$\csc\theta = \csc\theta$. First, apply the reciprocal and ratio identities to the factors on the left: $\cot\theta\cdot\sec\theta = \csc\theta \rightarrow \frac{\cos\theta}{\sin\theta}\cdot\frac{1}{\cos\theta} = \csc\theta$. Reduce and multiply; then apply the reciprocal identity: $\frac{\cancel{\cos\theta}}{\sin\theta}\cdot\frac{1}{\cancel{\cos\theta}} = \csc\theta \rightarrow \frac{1}{\sin\theta} = \csc\theta \rightarrow \csc\theta = \csc\theta$.**

4. **$1 = 1$. First, apply the reciprocal and ratio identities to the factors on the left:**

   $\sec\theta\cdot\sin\theta\cdot\cot\theta = 1 \rightarrow \frac{1}{\cos\theta}\cdot\sin\theta\cdot\frac{\cos\theta}{\sin\theta} = 1$. Reduce the fractions and multiply:

   $\frac{1}{\cancel{\cos\theta}}\cdot\cancel{\sin\theta}\cdot\frac{\cancel{\cos\theta}}{\cancel{\sin\theta}} = 1 \rightarrow 1 = 1$.

5. **$-\frac{12}{13}$. Using the Pythagorean identity, replace $\cos\theta$ with the given value and solve for $\sin\theta$.**

   $$\sin^2\theta + \cos^2\theta = 1 \rightarrow \sin^2\theta + \left(-\frac{5}{13}\right)^2 = 1 \rightarrow \sin^2\theta + \frac{25}{169} = 1 \rightarrow \sin^2\theta = 1 - \frac{25}{169}$$

   $$\rightarrow \sin^2\theta = \frac{144}{169} \rightarrow \sin\theta = \pm\sqrt{\frac{144}{169}} = \pm\frac{12}{13}$$

   Since θ is in quadrant IV, where the sine is negative, you have $\sin\theta = -\frac{12}{13}$.

6. **$\frac{11}{60}$. Using the Pythagorean identity, replace $\sin\theta$ with the given value and solve for $\cos\theta$.**

   $$\sin^2\theta + \cos^2\theta = 1 \rightarrow \left(-\frac{11}{61}\right)^2 + \cos^2\theta = 1 \rightarrow \frac{121}{3721} + \cos^2\theta = 1 \rightarrow \cos^2\theta = 1 - \frac{121}{3721}$$

   $$\rightarrow \cos^2\theta = \frac{3600}{3721} \rightarrow \cos\theta = \pm\sqrt{\frac{3600}{3721}} = \pm\frac{60}{61}.$$

   Since θ is in quadrant III, where the cosine is negative, you have $\cos\theta = -\frac{60}{61}$. Using the ratio identity, solve for the tangent: $\tan\theta = \frac{\sin\theta}{\cos\theta} = \frac{-\frac{11}{61}}{-\frac{60}{61}} = -\frac{11}{\cancel{61}}\left(-\frac{\cancel{61}}{60}\right) = \frac{11}{60}$.

7. **$\csc\theta = \csc\theta$. First, apply the reciprocal identity to the first term:**

   $$\csc\theta\cos^2\theta + \sin\theta = \csc\theta \rightarrow \frac{1}{\sin\theta}\cdot\cos^2\theta + \sin\theta = \csc\theta \rightarrow \frac{\cos^2\theta}{\sin\theta} + \sin\theta = \csc\theta$$

Find a common denominator for the two terms by multiplying the second term by $\frac{\sin\theta}{\sin\theta}$:

$$\frac{\cos^2\theta}{\sin\theta}+\sin\theta=\csc\theta \to \frac{\cos^2\theta}{\sin\theta}+\sin\theta\cdot\frac{\sin\theta}{\sin\theta}=\csc\theta \to \frac{\cos^2\theta}{\sin\theta}+\frac{\sin^2\theta}{\sin\theta}=\csc\theta$$

Add the two terms together and apply the Pythagorean identity:

$$\to \frac{\cos^2\theta+\sin^2\theta}{\sin\theta}=\csc\theta \to \frac{1}{\sin\theta}=\csc\theta$$

Applying the reciprocal identity, you have $\csc\theta=\csc\theta$.

8 **$1=1$. Multiply the two binomials together, since they're the sum and difference of the same two terms: $(\sec\theta+\tan\theta)(\sec\theta-\tan\theta)=1 \to \sec^2\theta-\tan^2\theta=1$. Now apply the Pythagorean identity $\tan^2\theta+1=\sec^2\theta$ to get $\sec^2\theta-\tan^2\theta=1 \to \tan^2\theta+1-\tan^2\theta=1 \to 1=1$.**

9 **$1-2\sin^2\theta=1-2\sin^2\theta$. Apply the Pythagorean identity to the left side by first solving for $\cos^2\theta$: $\sin^2\theta+\cos^2\theta=1 \to \cos^2\theta=1-\sin^2\theta$. Now substitute the result into the original equation to get $\cos^2\theta-\sin^2\theta=1-2\sin^2\theta \to 1-\sin^2\theta-\sin^2\theta=1-2\sin^2\theta \to 1-2\sin^2\theta=1-2\sin^2\theta$.**

10 **$\tan^2\theta=\tan^2\theta$. Replace the expression in the parentheses using the Pythagorean identity. Then rewrite that value using the reciprocal identity: $\sin^2\theta(\tan^2\theta+1)=\tan^2\theta \to \sin^2\theta(\sec^2\theta)=\tan^2\theta \to \sin^2\theta\left(\frac{1}{\cos^2\theta}\right)=\tan^2\theta$.**

Multiply and then apply the ratio identity: $\sin^2\theta\left(\frac{1}{\cos^2\theta}\right)=\tan^2\theta \to \frac{\sin^2\theta}{\cos^2\theta}=\tan^2\theta \to \tan^2\theta=\tan^2\theta$.

11 **$\sec^2 x$. Apply the even/odd identity, and then write the expression as two fractions: $\frac{\cot^2(-x)-1}{-\cot^2 x}=\frac{-\cot^2 x-1}{-\cot^2 x}=\frac{-\cot^2 x}{-\cot^2 x}+\frac{-1}{-\cot^2 x}=1+\frac{1}{\cot^2 x}$. Rewrite the second term using the reciprocal identity. Then use the Pythagorean identity to finish: $1+\frac{1}{\cot^2 x}=1+\tan^2 x=\sec^2 x$.**

Another approach, rather than writing the two fractions, is to factor out -1 in the numerator and then apply the Pythagorean identity $\cot^2 x+1=\csc^2 x$. You then have $\frac{-\cot^2 x-1}{-\cot^2 x}=\frac{-(\cot^2 x+1)}{-\cot^2 x}=\frac{\csc^2 x}{\cot^2 x}$. Now apply the reciprocal and ratio identities, perform the division, and apply the reciprocal identity: $\frac{\csc^2 x}{\cot^2 x}=\frac{\frac{1}{\sin^2 x}}{\frac{\cos^2 x}{\sin^2 x}}=\frac{1}{\cancel{\sin^2 x}}\cdot\frac{\cancel{\sin^2 x}}{\cos^2 x}=\frac{1}{\cos^2 x}=\sec^2 x$.

12 **$\sin^2\theta=\sin^2\theta$. Apply the even/odd identities. Then apply the ratio identity and reciprocal identity, and then multiply:**

$$\frac{\tan(-x)}{\sec(-x)}\cdot\sin(-x)=\cos^2 x \to \frac{-\tan x}{\sec x}\cdot(-\sin x)=\cos^2 x \to \frac{-\frac{\sin x}{\cos x}}{\frac{1}{\cos x}}\cdot(-\sin x)=\cos^2 x$$

$$\to -\frac{\sin x}{\cos x}\cdot\frac{\cos x}{1}\cdot\frac{-\sin x}{1}=\cos^2 x \to -\frac{\sin x}{\cancel{\cos x}}\cdot\frac{\cancel{\cos x}}{1}\cdot\frac{-\sin x}{1}=\cos^2 x \to \sin^2 x=\sin^2 x$$

13 **$2\cos x$. First, apply the co-function identities: $\sin\left(\frac{\pi}{2}-x\right)+\tan\left(\frac{\pi}{2}-x\right)\cdot\cos\left(\frac{\pi}{2}-x\right) \to \cos x+\cot x\cdot\sin x$. Next, apply the ratio identity and multiply the two factors in the second term: $\cos x+\cot x\cdot\sin x=\cos x+\frac{\cos x}{\cancel{\sin x}}\cdot\cancel{\sin x}=\cos x+\cos x=2\cos x$.**

14 **$\tan^2\theta = \tan^2\theta$. Apply the co-function identities, and then apply the ratio identity:**

$$\frac{\cos\left(\frac{\pi}{2}-x\right)}{\sin\left(\frac{\pi}{2}-x\right)}\cdot\cot\left(\frac{\pi}{2}-x\right)=\tan^2 x \to \frac{\sin x}{\cos x}\cdot\tan x=\tan^2 x \to \tan x\cdot\tan x=\tan^2 x \to \tan^2 x=\tan^2 x$$

15 **$\sin^2 x$. First, apply the periodicity identities: $1-\sin(2\pi+x)\cdot\cot(\pi+x)\cdot\cos(2\pi+x)=$ $1-\sin x\cdot\cot x\cdot\cos x$. Use the ratio identity and multiply the factors in the second term: $1-\sin x\cdot\cot x\cdot\cos x=1-\cancel{\sin x}\cdot\frac{\cos x}{\cancel{\sin x}}\cdot\cos x=1-\cos^2 x$. Finally, use the Pythagorean identity: $\sin^2 x+\cos^2 x=1\to\sin^2 x=1-\cos^2 x$.**

16 **$\tan x=\tan x$. Apply the two periodicity identities, then reciprocal identities. Multiply the numerator of the fraction times the reciprocal of the denominator. And, finally, apply the ratio identity.**

$$\frac{\sec(x+2\pi)}{\csc(x+2\pi)}=\frac{\sec x}{\csc x}=\frac{\frac{1}{\cos x}}{\frac{1}{\sin x}}=\frac{1}{\cos x}\cdot\frac{\sin x}{\cos x}=\frac{\sin x}{\tan x}=\tan x$$

17 **1. Use the reciprocal identities in each denominator: $\frac{\sec(x+2\pi)}{\csc(x+2\pi)}=\frac{\sec x}{\csc x}=\frac{\frac{1}{\cos x}}{\frac{1}{\sin x}}=$ $\frac{1}{\cos x}\cdot\frac{\sin x}{1}=\frac{\sin x}{\cos x}=\tan x$. Now perform the two divisions to get $\frac{\frac{\sec\theta}{1}}{\frac{1}{\sec\theta}}-\frac{\frac{\tan\theta}{1}}{\frac{1}{\tan\theta}}=$. Replace the $\sec\theta\cdot\frac{\sec\theta}{1}-\tan\theta\cdot\frac{\tan\theta}{1}=\sec^2\theta-\tan^2\theta$ first term using the Pythagorean identity: $\sec^2\theta-\tan^2\theta=\tan^2\theta+1-\tan^2\theta=1$.**

18 **$\frac{2}{\sin x}=\frac{2}{\sin x}$. Add the two fractions together after finding a common denominator:**

$$\frac{\sin x}{1+\cos x}+\frac{1+\cos x}{\sin x}=\frac{2}{\sin x}\to\frac{\sin x}{1+\cos x}\cdot\frac{\sin x}{\sin x}+\frac{1+\cos x}{\sin x}\cdot\frac{1+\cos x}{1+\cos x}=\frac{2}{\sin x}$$

$$\to\frac{\sin^2 x}{\sin x(1+\cos x)}+\frac{(1+\cos x)^2}{\sin x(1+\cos x)}=\frac{2}{\sin x}\to\frac{\sin^2 x+(1+\cos x)^2}{\sin x(1+\cos x)}=\frac{2}{\sin x}$$

Square the binomial, and then apply the Pythagorean identity:

$$\to\frac{\sin^2 x+1+2\cos x+\cos^2 x}{\sin x(1+\cos x)}=\frac{2}{\sin x}\to\frac{\sin^2 x+\cos^2 x+1+2\cos x}{\sin x(1+\cos x)}=\frac{2}{\sin x}$$

$$\to\frac{1+1+2\cos x}{\sin x(1+\cos x)}=\frac{2}{\sin x}\to\frac{2+2\cos x}{\sin x(1+\cos x)}=\frac{2}{\sin x}$$

Factor the numerator, and then reduce the fraction:

$$\to\frac{2(1+\cos x)}{\sin x(1+\cos x)}=\frac{2}{\sin x}\to\frac{2\cancel{(1+\cos x)}}{\sin x\cancel{(1+\cos x)}}=\frac{2}{\sin x}\to\frac{2}{\sin x}=\frac{2}{\sin x}$$

19 **$\frac{2(2-\sin^2 t)}{\sin^2 t}$ or $4\csc^2 t-2$. Multiply each fraction by the conjugate of its denominator:**

$$\frac{1+\cos t}{1-\cos t}+\frac{1-\cos t}{1+\cos t}=\frac{1+\cos t}{1-\cos t}\cdot\frac{1+\cos t}{1+\cos t}+\frac{1-\cos t}{1+\cos t}\cdot\frac{1-\cos t}{1-\cos t}=\frac{(1+\cos t)^2}{1-\cos^2 t}+\frac{(1-\cos t)^2}{1-\cos^2 t}.$$

Now square both numerators and add the two fractions together: $\frac{(1+\cos t)^2}{1-\cos^2 t}+\frac{(1-\cos t)^2}{1-\cos^2 t}=\frac{1+2\cos t+\cos^2 t+1-2\cos t+\cos^2 t}{1-\cos^2 t}=\frac{2+2\cos^2 t}{1-\cos^2 t}$. Factor the numerator and use the Pythagorean identity on the denominator: $\frac{2+2\cos^2 t}{1-\cos^2 t}=\frac{2(1+\cos^2 t)}{\sin^2 t}$. Now use the Pythagorean identity on the numerator: $\frac{2(1+\cos^2 t)}{\sin^2 t}=\frac{2(1+(1-\sin^2 t))}{\sin^2 t}=\frac{2(2-\sin^2 t)}{\sin^2 t}$.

You can stop here, or you can create two terms by breaking this into two fractions: $\frac{2(2-\sin^2 t)}{\sin^2 t}=\frac{4-2\sin^2 t}{\sin^2 t}=\frac{4}{\sin^2 t}-\frac{2\sin^2 t}{\sin^2 t}=4\csc^2 t-2$. Sometimes the simplifying is just carried out until you have the format you want.

(20) $\sin x=\sin x$. Write the left side as a single fraction: $1-\frac{\cos^2 x}{1+\sin x}=\sin x\rightarrow\frac{1+\sin x}{1+\sin x}-\frac{\cos^2 x}{1+\sin x}=\sin x\rightarrow\frac{1+\sin x-\cos^2 x}{1+\sin x}=\sin x$. Now apply the Pythagorean identity and simplify the numerator: $\rightarrow\frac{1+\sin x-(1-\sin^2 x)}{1+\sin x}=\sin x\rightarrow\frac{\sin x+\sin^2 x}{1+\sin x}=\sin x$.

Factor the numerator and reduce the fraction: $\rightarrow\frac{\sin x(\cancel{1+\sin x})}{\cancel{1+\sin x}}=\sin x\rightarrow\sin x=\sin x$.

(21) $\frac{\cos^4 x}{\sin^2 x}=\frac{\cos^4 x}{\sin^2 x}$. First, apply the ratio identity: $\cot^2 x-\cos^2 x=\cos^2 x\cdot\cot^2 x\rightarrow\frac{\cos^2 x}{\sin^2 x}-\cos^2 x=\cos^2 x\cdot\frac{\cos^2 x}{\sin^2 x}$. On the left, write the two terms in one fraction, and on the right, do the multiplication: $\rightarrow\frac{\cos^2 x}{\sin^2 x}-\frac{\sin^2 x\cos^2 x}{\sin^2 x}=\frac{\cos^4 x}{\sin^2 x}\rightarrow\frac{\cos^2 x-\sin^2 x\cos^2 x}{\sin^2 x}=\frac{\cos^4 x}{\sin^2 x}$. Now factor the numerator and apply the Pythagorean identity: $\rightarrow\frac{\cos^2 x(1-\sin^2 x)}{\sin^2 x}=\frac{\cos^4 x}{\sin^2 x}\rightarrow\frac{\cos^2 x(\cos^2 x)}{\sin^2 x}=\frac{\cos^4 x}{\sin^2 x}\rightarrow\frac{\cos^4 x}{\sin^2 x}=\frac{\cos^4 x}{\sin^2 x}$.

(22) $\frac{\sin x-\cos x}{\sin x+\cos x}=\frac{\sin x-\cos x}{\sin x+\cos x}$. On the left side of the equation, apply the reciprocal identities; on the right side, apply the ratio identity: $\frac{\sec x-\csc x}{\sec x+\csc x}=\frac{\tan x-1}{\tan x+1}\rightarrow\frac{\frac{1}{\cos x}-\frac{1}{\sin x}}{\frac{1}{\cos x}+\frac{1}{\sin x}}=\frac{\frac{\sin x}{\cos x}-1}{\frac{\sin x}{\cos x}+1}$.

In each numerator and denominator, write the two terms as one fraction after finding the common denominator:

$$\rightarrow\frac{\frac{1}{\cos x}\cdot\frac{\sin x}{\sin x}-\frac{1}{\sin x}\cdot\frac{\cos x}{\cos x}}{\frac{1}{\cos x}\cdot\frac{\sin x}{\sin x}+\frac{1}{\sin x}\cdot\frac{\cos x}{\cos x}}=\frac{\frac{\sin x}{\cos x}-1\cdot\frac{\cos x}{\cos x}}{\frac{\sin x}{\cos x}+1\cdot\frac{\cos x}{\cos x}}\rightarrow\frac{\frac{\sin x-\cos x}{\sin x\cos x}}{\frac{\sin x+\cos x}{\sin x\cos x}}=\frac{\frac{\sin x-\cos x}{\cos x}}{\frac{\sin x+\cos x}{\cos x}}$$

Now perform the division on both the right and left by multiplying the numerator times the reciprocal of the denominator:

$$\rightarrow\frac{\sin x-\cos x}{\cancel{\sin x\cos x}}\cdot\frac{\cancel{\sin x\cos x}}{\sin x+\cos x}=\frac{\sin x-\cos x}{\cancel{\cos x}}\cdot\frac{\cancel{\cos x}}{\sin x+\cos x}\rightarrow\frac{\sin x-\cos x}{\sin x+\cos x}=\frac{\sin x-\cos x}{\sin x+\cos x}$$

If you're ready to test your skills a bit more, take the following chapter quiz that incorporates all the chapter topics.

Whaddya Know? Chapter 12 Quiz

Quiz time! Complete each problem to test your knowledge on the various topics covered in this chapter. You can then find the solutions and explanations in the next section.

1. Prove $\sec x - \tan x \sin x = \dfrac{1}{\sec x}$
2. Simplify $\dfrac{\sec x \sin x}{\tan x + \cot x}$
3. Prove $\cos^2\theta - \sin^2\theta = 1 - 2\sin^2\theta$
4. Prove $\dfrac{\sec^2 t}{\sec^2 t - 1} = \csc^2 t$
5. Prove the identity by working on both sides of the equation: $\dfrac{\cot^2 z}{\csc z} \cdot \sec^2 z = \tan z \cos z \csc^2 z$.
6. Prove $\dfrac{\sec x + \csc x}{\tan x + \cot x} = \sin x + \cos x$
7. Prove $\dfrac{1}{1 - \sin x} - \dfrac{1}{1 + \sin x} = 2\sec x \tan x$
8. Simplify $\dfrac{\sec\theta - \cos\theta}{\sec\theta}$
9. Prove $\dfrac{\sec x}{\cos x} - \dfrac{\tan x}{\cot x} = 1$
10. If $\cos x = -\dfrac{40}{41}$ and $\dfrac{\pi}{2} < x < \pi$, then what is $\sin x$?
11. Prove $\dfrac{1 + \cos x}{\sin x} = \csc x + \cot x$
12. Working on both sides of the equation, prove: $\csc x + \sin(-x) = \cos(-x)\cot x$.

Answers to Chapter 12 Quiz

1. $\frac{1}{\sec x} = \frac{1}{\sec x}$. **First, apply the reciprocal identity and ratio identity to the terms on the left:** $\sec x - \tan x \sin x = \frac{1}{\sec x} \rightarrow \frac{1}{\cos x} - \frac{\sin x}{\cos x} \cdot \sin x = \frac{1}{\sec x}$. **Simplify and subtract to create one fraction:** $\rightarrow \frac{1}{\cos x} - \frac{\sin^2 x}{\cos x} = \frac{1}{\sec x} \rightarrow \frac{1 - \sin^2 x}{\cos x} = \frac{1}{\sec x}$. **Now apply the Pythagorean identity and reduce the fraction:** $\rightarrow \frac{\cos^2 x}{\cos x} = \frac{1}{\sec x} \rightarrow \cos x = \frac{1}{\sec x}$. **Apply the reciprocal identity on the left:** $\rightarrow \frac{1}{\sec x} = \frac{1}{\sec x}$.

2. $\sin^2 x$. **Apply the reciprocal identity and ratio identities:** $\frac{\sec x \sin x}{\tan x + \cot x} = \frac{\frac{1}{\cos x} \cdot \sin x}{\frac{\sin x}{\cos x} + \frac{\cos x}{\sin x}}$.

   Multiply the terms in the numerator and add the terms in the denominator after finding a common denominator:

   $$\frac{\frac{1}{\cos x} \cdot \sin x}{\frac{\sin x}{\cos x} + \frac{\cos x}{\sin x}} = \frac{\frac{\sin x}{\cos x}}{\frac{\sin x}{\cos x} \cdot \frac{\sin x}{\sin x} + \frac{\cos x}{\sin x} \cdot \frac{\cos x}{\cos x}} = \frac{\frac{\sin x}{\cos x}}{\frac{\sin^2 x}{\sin x \cos x} + \frac{\cos^2 x}{\sin x \cos x}} = \frac{\frac{\sin x}{\cos x}}{\frac{\sin^2 x + \cos^2 x}{\sin x \cos x}}$$

   Apply the Pythagorean identity, and then multiply the numerator by the reciprocal of the denominator: $\frac{\frac{\sin x}{\cos x}}{\frac{\sin^2 x + \cos^2 x}{\sin x \cos x}} = \frac{\frac{\sin x}{\cos x}}{\frac{1}{\sin x \cos x}} = \frac{\sin x}{\cancel{\cos x}} \cdot \frac{\sin x \cancel{\cos x}}{1} = \sin^2 x$.

3. $1 - 2\sin^2 \theta = 1 - 2\sin^2 \theta$. **Apply the Pythagorean identity and simplify:**

   $$\cos^2\theta - \sin^2\theta = 1 - 2\sin^2\theta \rightarrow (1 - \sin^2\theta) - \sin^2\theta = 1 - 2\sin^2\theta \rightarrow 1 - 2\sin^2\theta = 1 - 2\sin^2\theta$$

4. $\csc^2 t = \csc^2 t$. **Apply the Pythagorean identity, and then apply the reciprocal and ratio identities. Multiply the numerator by the reciprocal of the denominator:**

   $$\frac{\sec^2 t}{\sec^2 t - 1} = \csc^2 t \rightarrow \frac{\sec^2 t}{\tan^2 t} = \csc^2 t \rightarrow \frac{\frac{1}{\cos^2 t}}{\frac{\sin^2 t}{\cos^2 t}} = \csc^2 t \rightarrow \frac{1}{\cancel{\cos^2 t}} \cdot \frac{\cancel{\cos^2 t}}{\sin^2 t} = \csc^2 t \rightarrow \frac{1}{\sin^2 t} = \csc^2 t$$

   Now apply the reciprocal identity, and you have $\rightarrow \csc^2 t = \csc^2 t$.

5. $\frac{1}{\sin z} = \frac{1}{\sin z}$. **First, apply the ratio and reciprocal identities to each factor:**

   $\frac{\cot^2 z}{\csc z} \cdot \sec^2 z = \tan z \cos z \csc^2 z \rightarrow \frac{\frac{\cos^2 z}{\sin^2 z}}{\frac{1}{\sin z}} \cdot \frac{1}{\cos^2 z} = \frac{\sin z}{\cos z} \cos z \frac{1}{\sin^2 z}$. **Now write each side as a product:**

   $$\rightarrow \frac{\cos^2 z}{\sin^2 z} \cdot \frac{\sin z}{1} \cdot \frac{1}{\cos^2 z} = \frac{\sin z}{\cos z} \cdot \frac{\cos z}{1} \cdot \frac{1}{\sin^2 z} \rightarrow \frac{\cancel{\cos^2 z}}{\sin^2 z} \cdot \frac{\sin z}{1} \cdot \frac{1}{\cancel{\cos^2 z}} = \frac{\sin z}{\cancel{\cos z}} \cdot \frac{\cancel{\cos z}}{1} \cdot \frac{1}{\sin^2 z}$$

   Finish reducing: $\rightarrow \frac{1}{\sin^2 z} \cdot \frac{\sin z}{1} = \frac{\sin z}{1} \cdot \frac{1}{\sin^2 z} \rightarrow \frac{1}{\sin^{\cancel{2}} z} \cdot \frac{\cancel{\sin z}}{1} = \frac{\cancel{\sin z}}{1} \cdot \frac{1}{\sin^{\cancel{2}} z} \rightarrow \frac{1}{\sin z} = \frac{1}{\sin z}$.

6 **$\sin x + \cos x = \sin x + \cos x$.** Apply the reciprocal and ratio identities: $\frac{\sec x + \csc x}{\tan x + \cot x} =$

$\sin x + \cos x \rightarrow \frac{\frac{1}{\cos x} + \frac{1}{\sin x}}{\frac{\sin x}{\cos x} + \frac{\cos x}{\sin x}} = \sin x + \cos x$. Add the fractions in both the numerator and denominator after first finding the common denominators. Then apply the Pythagorean identity:

$$\rightarrow \frac{\frac{1}{\cos x}\cdot\frac{\sin x}{\sin x} + \frac{1}{\sin x}\cdot\frac{\cos x}{\cos x}}{\frac{\sin x}{\cos x}\cdot\frac{\sin x}{\sin x} + \frac{\cos x}{\sin x}\cdot\frac{\cos x}{\cos x}} = \sin x + \cos x \rightarrow \frac{\frac{\sin x + \cos x}{\sin x \cos x}}{\frac{\sin^2 x + \cos^2 x}{\sin x \cos x}} = \sin x + \cos x \rightarrow$$

$$\frac{\frac{\sin x + \cos x}{\sin x \cos x}}{\frac{1}{\sin x \cos x}} = \sin x + \cos x$$

Multiply the numerator by the reciprocal of the denominator and simplify:

$$\rightarrow \frac{\sin x + \cos x}{\cancel{\sin x \cos x}} \cdot \frac{\cancel{\sin x \cos x}}{1} = \sin x + \cos x \rightarrow \sin x + \cos x = \sin x + \cos x$$

7 **$2\sec x \tan x = 2\sec x \tan x$.** Add the two fractions together after finding the common denominator:

$$\frac{1}{1-\sin x} - \frac{1}{1+\sin x} = 2\sec x \tan x \rightarrow \frac{1}{1-\sin x}\cdot\frac{1+\sin x}{1+\sin x} - \frac{1}{1+\sin x}\cdot\frac{1-\sin x}{1-\sin x} = 2\sec x \tan x$$

$$\rightarrow \frac{1+\sin x}{(1-\sin x)(1+\sin x)} - \frac{1-\sin x}{(1-\sin x)(1+\sin x)} = 2\sec x \tan x$$

Next, multiply the denominators and apply the Pythagorean identity:

$$\rightarrow \frac{1+\sin x}{1-\sin^2 x} - \frac{1-\sin x}{1-\sin^2 x} = 2\sec x \tan x \rightarrow \frac{1+\sin x}{\cos^2 x} - \frac{1-\sin x}{\cos^2 x} = 2\sec x \tan x$$

Subtract and break the result into the product of three fractions:

$$\rightarrow \frac{1+\sin x-(1-\sin x)}{\cos^2 x} = 2\sec x \tan x \rightarrow \frac{2\sin x}{\cos^2 x} = 2\sec x \tan x \rightarrow \frac{2}{1}\cdot\frac{1}{\cos x}\cdot\frac{\sin x}{\cos x} = 2\sec x \tan x$$

And, finally, apply the reciprocal and ratio identities: $\rightarrow 2\sec x \tan x = 2\sec x \tan x$.

8 **$\sin^2\theta$.** Break up the fraction into two fractions. Then apply the reciprocal identity, multiply the numerator by the reciprocal of the denominator, and, finally, apply the Pythagorean identity:

$$\frac{\sec\theta - \cos\theta}{\sec\theta} = \frac{\sec\theta}{\sec\theta} - \frac{\cos\theta}{\sec\theta} = 1 - \frac{\cos\theta}{\frac{1}{\cos\theta}} = 1 - \frac{\cos\theta}{1}\cdot\frac{\cos\theta}{1} = 1 - \cos^2\theta = \sin^2\theta$$

9 **1=1.** Apply the reciprocal and ratio identities. Then multiply the numerators times the reciprocals of the denominators and simplify:

$$\frac{\sec x}{\cos x}-\frac{\tan x}{\cot x}=1\to\frac{\frac{1}{\cos x}}{\cos x}-\frac{\frac{\sin x}{\cos x}}{\frac{\cos x}{\sin x}}=1\to\frac{1}{\cos x}\cdot\frac{1}{\cos x}-\frac{\sin x}{\cos x}\cdot\frac{\sin x}{\cos x}=1\to\frac{1}{\cos^2 x}-\frac{\sin^2 x}{\cos^2 x}=1$$

Write the subtraction problem in one fraction. Then apply the Pythagorean identity:

$$\to\frac{1-\sin^2 x}{\cos^2 x}=1\to\frac{\cos^2 x}{\cos^2 x}=1\to 1=1$$

10 $\frac{9}{41}$. Using the Pythagorean Theorem, $\sin^2 x+\cos^2 x=1$, you have $\sin^2 x+\left(-\frac{40}{41}\right)^2=1\to\sin^2 x=1-\frac{1600}{1681}=\frac{81}{1681}$. Finding the square root, $\sin x=\pm\sqrt{\frac{81}{1681}}=\pm\frac{9}{41}$. Since the angle, x, is in quadrant II, the sine is positive, so $\sin x=\frac{9}{41}$.

11 $\csc x+\cot x=\csc x+\cot x$. Write the fraction as the sum of two fractions. Then apply the reciprocal and ratio identities:

$$\frac{1+\cos x}{\sin x}=\csc x+\cot x\to\frac{1}{\sin x}+\frac{\cos x}{\sin x}=\csc x+\cot x\to\csc x+\cot x=\csc x+\cot x$$

12 $\frac{\cos^2 x}{\sin x}=\frac{\cos^2 x}{\sin x}$. Working on both sides of the equation, first apply the opposite angle identities: $\csc x+\sin(-x)=\cos(-x)\cot x\to\csc x-\sin x=\cos x\cot x$. Next, apply the reciprocal and ratio identities: $\to\frac{1}{\sin x}-\sin x=\cos x\cdot\frac{\cos x}{\sin x}\to\frac{1}{\sin x}-\frac{\sin^2 x}{\sin x}=\frac{\cos^2 x}{\sin x}\to\frac{1-\sin^2 x}{\sin x}=\frac{\cos^2 x}{\sin x}$.

Finally, apply the Pythagorean identity: $\to\frac{\cos^2 x}{\sin x}=\frac{\cos^2 x}{\sin x}$.

IN THIS CHAPTER

» **Applying the sum and difference formulas of trig functions**

» **Utilizing double-angle formulas**

» **Cutting angles in two with half-angle formulas**

» **Changing from products to sums and back**

» **Tossing aside exponents with power-reducing formulas**

Chapter 13
Advancing with Advanced Identities

Prior to the invention of calculators (not as long ago as you may imagine), people had only one way to calculate the exact trig values for angles not shown on the unit circle: using advanced identities. Even now, most calculators give you only an approximation of the trig value, not the exact one. Exact values are important to trig calculations and to their applications. Engineers designing a bridge, for example, don't want an almost correct value — and neither should you, for that matter.

This chapter is the meat and potatoes of pre-calculus identities: It contains the bulk of the formulas that you need to know for calculus. It also builds on the basic identities discussed in Chapter 12. Advanced identities provide you with opportunities to determine values that you couldn't calculate before — like finding the exact value of the sine of 15°, or figuring out the sine or cosine of the sum of angles without actually knowing the function value of the angles. This information is truly helpful when you get to calculus, which takes these calculations to another level (a level at which you integrate and differentiate by using these identities).

Finding Trig Functions of Sums and Differences

Long ago, some fantastic mathematicians found identities that hold true when adding and subtracting angle measures from special triangles ($30°-60°-90°$ right triangles and $45°-45°-90°$ right triangles; see Chapter 9). The focus is to find a way to rewrite an angle as a sum or difference of two "convenient" angles. Those mathematicians were curious; they could find the trig values for the special triangles but wanted to know how to deal with other angles that aren't part of the special triangles on the unit circle. They could solve problems with multiples of 30° and 45°, but they knew nothing about all the other angles that couldn't be formed that way!

Constructing these angles was simple; however, evaluating trig functions for them proved to be a bit more difficult. So they put their collective minds together and discovered the sum and difference identities discussed in this section. Their only problem was that they still couldn't find the trig values of many other angles using the sum $(a+b)$ and difference $(a-b)$ formulas.

This section takes the information covered in earlier chapters, such as calculating trig values of special angles, to the next level. You are advanced to identities that allow you to find trig values of angles that are multiples of 15°.

Note: You'll never be asked to find the sine of 87°, for example, using trig identities. If the angle can't be written as the sum or difference of special angles, then you resort to using a calculator. If you can break down the given angle into the sum or difference of two angles whose functions are known, you have it made in the shade. (If you can't express the angle as the sum or difference of special angles, you have to find some other way to solve the problem — such as doubling angles, halving angles, and finding products of angles. You'll find more on this later in this chapter.)

REMEMBER

For the most part, when you're presented with advanced identity problems, you'll usually be asked to work with angles in radians rather than degrees. This section starts with calculations in degrees because they're easier to manipulate (whole numbers rather than fractions). Then you switch to radians and see how to make the formulas work with them, too.

Searching out the sine of $(a \pm b)$

Using the special right triangles (see Chapter 9), which have points on the unit circle that are easy to identify, you can find the sines of 30° and 45° angles (among others). However, no point on the unit circle allows you to find trig values at angle measures that aren't special or the multiple of a special angle (such as the sine of 15°) directly. The sine for such an angle isn't given as a point on the circle; it just isn't one of the nicely labeled points. Don't despair, because this is where advanced identities help you out.

TECHNICAL STUFF

Notice that $45°-30°=15°$, and $45°+30°=75°$. For the angles you can rewrite as the sum or difference of special angles, here are the sum and difference formulas for sine:

- $\sin(a+b)=\sin a\cos b+\cos a\sin b$
- $\sin(a-b)=\sin a\cos b-\cos a\sin b$

WARNING

You can't rewrite the $\sin(a+b)$ as $\sin a+\sin b$. You can't distribute the sine values inside the parentheses, because sine isn't a multiplication operation; therefore, the distributive property doesn't apply (like it does to real numbers). Sine is a function, not a number or variable.

WARNING

You have more than one way to combine unit circle angles to get a requested angle. You can also write $\sin 75°$ as $\sin(135°-60°)$ or $\sin(225°-150°)$. After you find a way to rewrite an angle as a sum or difference, roll with it. Use the one that works for you!

Calculating in degrees

Measuring angles in degrees for the sum and difference formulas is easier than measuring in radians, because adding and subtracting degrees is much easier than adding and subtracting radians. Adding and subtracting angles in radians requires finding a common denominator. Moreover, evaluating trig functions requires you to work backward from a common denominator to split the angle into two fractions with different denominators. You'll see the computations in both degrees and radians, so you'll be ready for whatever comes your way.

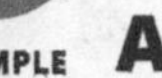

EXAMPLE

Q. Find the sine of 75°.

A. $\frac{\sqrt{2}+\sqrt{6}}{4}$. Use these steps:

1. **Rewrite the angle, using the special angles from right triangles (see Chapter 9).**

   One way to rewrite 75° is $30°+45°$.

2. **Choose the appropriate sum or difference formula.**

   The computation in Step 1 uses addition, so you want to use the sum formula for sine: $\sin(a+b)=\sin a\cos b+\cos a\sin b$.

3. **Plug the information you know into the formula.**

   You know that $\sin 75°=\sin(30°+45°)$; therefore, $a=30°$ and $b=45°$. The formula gives you $\sin(30°+45°)=\sin 30°\cos 45°+\cos 30°\sin 45°$.

4. **Use the unit circle (see Chapter 9) to look up the sine and cosine values you need.**

   You now have $\sin 30°\cos 45°+\cos 30°\sin 45°=\frac{1}{2}\cdot\frac{\sqrt{2}}{2}+\frac{\sqrt{3}}{2}\cdot\frac{\sqrt{2}}{2}$.

5. **Multiply and simplify to find the final answer.**

   You end up with $\frac{1}{2}\cdot\frac{\sqrt{2}}{2}+\frac{\sqrt{3}}{2}\cdot\frac{\sqrt{2}}{2}=\frac{\sqrt{2}}{4}+\frac{\sqrt{6}}{4}=\frac{\sqrt{2}+\sqrt{6}}{4}$. This represents an exact number, not a decimal approximation.

Calculating in radians

You can put the concept of sum and difference formulas to work using radians. Here you're asked to find the trig value of a specific angle that isn't readily marked on the unit circle (but still is a multiple of 15° or $\frac{\pi}{12}$ radians). Prior to choosing the appropriate formula (Step 2 from the previous section), you simply break the angle into either the sum or the difference of two angles from the unit circle. Refer to the unit circle and notice the angles in radians in Figure 13-1. You see that all the denominators are different, which makes adding and subtracting them a challenge. You must find a common denominator to perform the operations. For the angles shown on the unit circle, the common denominator is 12, as you can see in Figure 13-1.

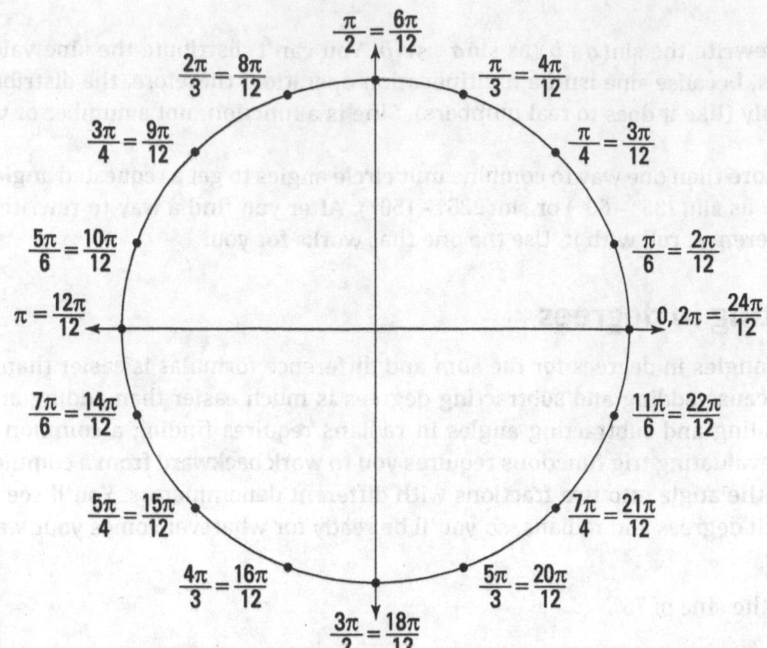

FIGURE 13-1: The unit circle showing angles in radians with common denominators.

REMEMBER

Figure 13-1 comes in handy only for sum and difference formulas, because finding a common denominator is something you do only when you're adding or subtracting fractions.

EXAMPLE

Q. Find the exact value of $\sin\frac{\pi}{12}$.

A. $\frac{\sqrt{6}-\sqrt{4}}{4}$. Follow these steps:

1. **Rewrite the angle in question, using the special angles in radians with common denominators.**

   From Figure 13-1, you want a way to add or subtract two angles so that, in the end, you get $\frac{\pi}{12}$. In this case, you can rewrite $\frac{\pi}{12}$ as $\frac{\pi}{4}-\frac{\pi}{6}=\frac{3\pi}{12}-\frac{2\pi}{12}$.

2. **Choose the appropriate sum/difference formula.**

   Because the operation is subtraction, you need to use the difference formula.

3. **Plug the information you know into the chosen formula.**

   You know the following equality: $\sin\frac{\pi}{12}=\sin\left(\frac{3\pi}{12}-\frac{2\pi}{12}\right)=\sin\left(\frac{\pi}{4}-\frac{\pi}{6}\right)$.

   Substitute as follows into the difference formula: $\sin\left(\frac{\pi}{4}-\frac{\pi}{6}\right)=\sin\frac{\pi}{4}\cos\frac{\pi}{6}-\cos\frac{\pi}{4}\sin\frac{\pi}{6}$.

4. **Use the unit circle to look up the sine and cosine values that you need.**

   You now have $\sin\left(\frac{\pi}{4}-\frac{\pi}{6}\right)=\frac{\sqrt{2}}{2}\cdot\frac{\sqrt{3}}{2}-\frac{\sqrt{2}}{2}\cdot\frac{1}{2}$.

5. **Multiply and simplify to get your final answer.**

   You end up with the following answer: $\sin\frac{\pi}{12}=\frac{\sqrt{6}}{4}-\frac{\sqrt{2}}{4}=\frac{\sqrt{6}-\sqrt{4}}{4}$.

YOUR TURN

 Find the exact value of $\sin\frac{7\pi}{12}$ using the sum formula.

 Find the exact value of $\sin 165°$ using the difference formula.

Applying the sum and difference formulas to proofs

The goal when dealing with trig proofs in this chapter is the same as the goal when dealing with them in Chapter 12: You need to make one side of a given equation look like the other. This section contains info on how to deal with sum and difference formulas in a proof.

EXAMPLE

 Prove $\sin(x+y)+\sin(x-y)=2\sin x\cos y$

 Follow these steps:

1. **Look for identities in the equation.**

   In this example, you can see the opening terms of the sum and difference identities, $\sin(a+b)=\sin a\cos b+\cos a\sin b$ and $\sin(a-b)=\sin a\cos b-\cos a\sin b$.

2. **Substitute for the identities.**

   $\sin x\cos y+\cos x\sin y+(\sin x\cos y-\cos x\sin y)=2\sin x\cos y$

3. **Simplify to get the proof.**

   Two terms cancel: $\sin x\cos y+\cancel{\cos x\sin y}+\sin x\cos y-\cancel{\cos x\sin y}=2\sin x\cos y$.

   And this simplifies to $\sin x\cos y+\sin x\cos y=2\sin x\cos y$.

   Combine the like terms to finish your proof:

   $2\sin x\cos y=2\sin x\cos y$

YOUR TURN

 Prove $\sin(-a-b)=-\sin(a+b)$

 Prove $\sin\left(a+\frac{\pi}{2}\right)=\cos a$

Calculating the cosine of $(a \pm b)$

After you familiarize yourself with the sum and difference formulas for sine, you can easily apply your newfound knowledge to calculate the sums and differences of cosines, because the formulas look very similar to each other. When working with sums and differences for sines and cosines, you're simply plugging in given values for variables. Just make sure you use the correct formula based on the information you're given in the question.

TECHNICAL STUFF

Here are the formulas for the sum and difference of cosines:

- $\cos(a+b) = \cos a \cos b - \sin a \sin b$
- $\cos(a-b) = \cos a \cos b + \sin a \sin b$

Applying the formulas to find the cosine of the sum or difference of two angles

The sum and difference formulas for cosine (and sine) can do more than calculate a trig value for an angle not marked on the unit circle (at least for angles that are multiples of 15°). They can also be used to find the cosine of the sum or difference based on information given about the two angles. For such problems, you'll be given two angles (call them A and B), the sine or cosine of A and B, and the quadrant(s) in which the two angles are located. You aren't given the angle measure, just information about the angle.

EXAMPLE

Q. Find the exact value of $\cos(A+B)$, given that $\cos(A) = -\frac{3}{5}$, with A in quadrant II of the coordinate plane, and $\sin(B) = -\frac{7}{25}$, with B in quadrant III.

A. $\cos(A+B) = \frac{4}{5}$. Use the following steps:

1. **Choose the appropriate formula and substitute the information you know to determine the missing information.**

   If $\cos(A+B) = \cos A \cos B - \sin A \sin B$, then substitutions result in this equation: $\cos(A+B) = \left(-\frac{3}{5}\right)\cos B - \sin A\left(-\frac{7}{25}\right)$.

   To proceed any further, you need to find cos B and sin A.

2. **Draw pictures representing right triangles in the quadrant(s).**

   You need to draw one triangle for angle A in quadrant II and one for angle B in quadrant III. Using the definition of sine as $\frac{\text{opp}}{\text{hyp}}$ and cosine as $\frac{\text{adj}}{\text{hyp}}$, Figure 13-2 shows these triangles using the values from the cosine and sine given. Notice that the value of a leg is missing in each triangle.

3. **To find the missing values, use the Pythagorean Theorem (once for each triangle; see Chapter 8).**

   The missing leg in Figure 13-2a is 4, and the missing leg in Figure 13-2b is −24.

4. **Determine the missing trig ratios to use in the sum/difference formula.**

   You use the definition of cosine and the fact that the angle is in quadrant III to find that $\cos(B) = -\frac{24}{25}$, and the definition of sine and that the angle is in quadrant II to find that $\sin(A) = \frac{4}{5}$.

5. **Substitute the missing trig ratios into the sum/difference formula and simplify.**

   You now can write the equation: $\cos(A+B) = \left(-\frac{3}{5}\right)\left(-\frac{24}{25}\right) - \left(\frac{4}{5}\right)\left(-\frac{7}{25}\right)$.

   Follow the order of operations to get $\cos(A+B) = \frac{72}{125} + \frac{28}{125} = \frac{100}{125} = \frac{4}{5}$.

   This equation simplifies to $\cos(A+B) = \frac{4}{5}$.

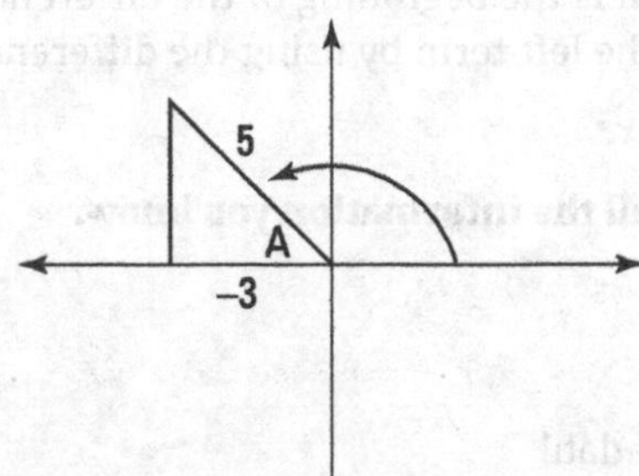

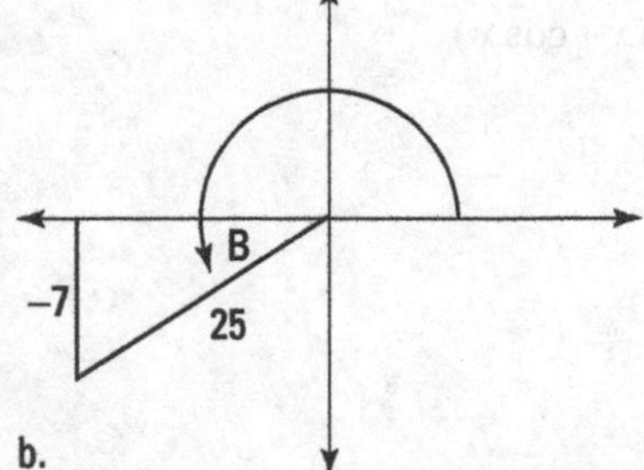

FIGURE 13-2: Drawing pictures helps you visualize the missing pieces of info.

YOUR TURN

5 Find the exact value of $\cos(A+B)$, given that $\sin(A) = -\frac{5}{13}$, with A in quadrant IV of the coordinate plane, and $\cos(B) = -\frac{12}{13}$, with B in quadrant II.

6 Find the exact value of $\cos(A-B)$, given that $\sin(A) = -\frac{24}{25}$, with A in quadrant III of the coordinate plane, and $\cos(B) = \frac{7}{25}$, with B in quadrant IV.

Applying the cosine sum and difference formulas to proofs

You can prove the co-function identities from Chapter 12 by using the sum and difference formulas for cosine.

Q. Prove $\cos\left(\frac{\pi}{2}-x\right)=\sin x$

EXAMPLE **A.** Follow these steps:

1. **Outline the given information.**

   You start with $\cos\left(\frac{\pi}{2}-x\right)=\sin x$.

2. **Look for sum and/or difference identities for cosine.**

   In this case, the left side of the equation is the beginning of the difference formula for cosine. Therefore, you can rewrite the left term by using the difference formula for cosines: $\cos\frac{\pi}{2}\cos x+\sin\frac{\pi}{2}\sin x=\sin x$.

3. **Refer to the unit circle and substitute all the information you know.**

   $$0\cdot\cos x+1\cdot\sin x=\sin x$$
   $$0+\sin x=\sin x$$

   Your equation now says $\sin x=\sin x$. Ta-dah!

YOUR TURN

7 Prove that $\sin\left(x-\frac{\pi}{2}\right)-\cos\left(x-\frac{\pi}{2}\right)=-(\sin x+\cos x)$

8 Prove that $\left[\cos\left(x+\frac{\pi}{4}\right)\right]^2=\frac{1}{2}(1-\sin 2x)$

Taming the tangent of $(a \pm b)$

As with sine and cosine (see the previous sections of this chapter), you can rely on formulas to find the tangent of a sum or a difference of angles. The main difference is that you can't read tangents directly from the coordinates of points on the unit circle, as you can with sine and cosine, because each point represents $(\cos\theta, \sin\theta)$, as explained in Chapter 8.

All hope isn't lost, however, because tangent is defined as $\frac{\sin\theta}{\cos\theta}$; the sine of the angle is the y-coordinate and the cosine is the x-coordinate. You can express the tangent in terms of x and y on the unit circle as $\frac{y}{x}$.

TECHNICAL STUFF

Here are the formulas you need to find the tangent of a sum or difference of angles:

- $\tan(a+b) = \frac{\tan a + \tan b}{1 - \tan a \tan b}$
- $\tan(a-b) = \frac{\tan a - \tan b}{1 + \tan a \tan b}$

Two alternative forms, using sine and cosine, are

- $\tan(a+b) = \frac{\sin(a+b)}{\cos(a+b)}$
- $\tan(a-b) = \frac{\sin(a-b)}{\cos(a-b)}$

Applying the formulas to solve a common problem

The sum and difference formulas for tangent work in similar ways to the sine and cosine formulas. You can use the formulas to solve a variety of problems. In this section, you see how to find the tangent of an angle that isn't marked on the unit circle. You can do so as long as the angle can be written as the sum or difference of special angles.

EXAMPLE

Q. Find the exact value of $\tan 105°$.

A. $-(2+\sqrt{3})$ **or** $-2-\sqrt{3}$. Follow these steps:

1. **Rewrite the given angle, using the information from special right-triangle angles (see Chapter 8).**

   Refer to the unit circle in Chapter 8, noting that it's built from the special right triangles, to find a combination of angles that add or subtract to get $\tan 105°$. You can choose from $240° - 135°$, $330° - 225°$, and so on. In this example, a great choice is $60° + 45°$. So $\tan(105°) = \tan(60° + 45°)$.

   Because the angle is rewritten with addition, you need to use the sum formula for tangent.

2. **Plug the information you know into the appropriate formula.**

   $$\tan(60° + 45°) = \frac{\tan 60° + \tan 45°}{1 - \tan 60° \tan 45°}$$

3. **Use the unit circle to look up the sine and cosine values that you need.**

   To find $\tan 0°$, you locate 60° on the unit circle and use the sine and cosine values of its corresponding coordinates to calculate the tangent:

   $$\tan 60° = \frac{\sin 60°}{\cos 60°} = \frac{\frac{\sqrt{3}}{2}}{\frac{1}{2}} = \frac{\sqrt{3}}{\cancel{2}} \cdot \frac{\cancel{2}}{1} = \sqrt{3}.$$

   Follow the same process for $\tan 45°$: $\tan 45° = \frac{\sin 45°}{\cos 45°} = \frac{\frac{\sqrt{2}}{2}}{\frac{\sqrt{2}}{2}} = \frac{\cancel{\sqrt{2}}}{\cancel{2}} \cdot \frac{\cancel{2}}{\cancel{\sqrt{2}}} = 1.$

4. **Substitute the trig values from Step 3 into the formula.**

   This step gives you $\tan(60° + 45°) = \frac{\tan 60° + \tan 45°}{1 - \tan 60° \tan 45°} = \frac{\sqrt{3}+1}{1-\sqrt{3}\cdot 1} = \frac{\sqrt{3}+1}{1-\sqrt{3}}.$

5. **Rationalize the denominator.**

   You shouldn't leave the square root on the bottom of the fraction. Because the denominator is a binomial (the sum or difference of two terms), you can multiply by its conjugate. The conjugate of $a+b$ is $a-b$, and vice versa. So the conjugate of $1-\sqrt{3}$ is $1+\sqrt{3}$. Rationalizing, you get:

   $$\frac{\sqrt{3}+1}{1-\sqrt{3}} \cdot \frac{1+\sqrt{3}}{1+\sqrt{3}} = \frac{1+\sqrt{3}}{1-\sqrt{3}} \cdot \frac{1+\sqrt{3}}{1+\sqrt{3}}$$

   $$= \frac{1+2\sqrt{3}+3}{1-3} = \frac{4+2\sqrt{3}}{-2}$$

6. **Simplify the rationalized fraction by reducing.**

   $$\frac{4+2\sqrt{3}}{-2} = -(2+\sqrt{3}) \text{ or } -2-\sqrt{3}$$

Applying the sum and difference formulas to proofs

The sum and difference formulas for tangent are very useful if you want to prove a few of the basic identities from Chapter 9. For example, you can prove the co-function identities and periodicity identities by using the difference and sum formulas. If you see a sum or a difference inside a tangent function, you can try the appropriate formula to simplify things.

Q. Prove this identity: $\tan\left(\frac{\pi}{4} + x\right) = \frac{1+\tan x}{1-\tan x}.$

EXAMPLE **A.** $\frac{1+\tan x}{1-\tan x} = \frac{1+\tan x}{1-\tan x}$. Use the following steps:

1. **Look for identities for which you can substitute.**

   On the left side of the proof you can use the sum identity for tangent: $\tan(a+b) = \frac{\tan a + \tan b}{1 - \tan a \tan b}.$

   Working on the left side and substituting in values gives you the following equation:

   $$\frac{\tan\frac{\pi}{4} + \tan x}{1 - \tan\frac{\pi}{4}\tan x} = \frac{1+\tan x}{1-\tan x}.$$

2. **Use any applicable unit circle values to simplify the proof.**

From the unit circle (see Chapter 8), you see that $\tan\frac{\pi}{4}=1$, so you can plug in that value to get $\frac{1+\tan x}{1-1\cdot\tan x}=\frac{1+\tan x}{1-\tan x}$.

The identity is proved: $\frac{1+\tan x}{1-\tan x}=\frac{1+\tan x}{1-\tan x}$.

YOUR TURN

9. Find the exact value of $\tan 15°$.

10. Prove that $\tan(x+\pi)=\tan x$.

Doubling an Angle and Finding Its Trig Value

You use a *double-angle formula* to find the trig value of twice an angle. Sometimes you know the original angle; sometimes you don't. Working with double-angle formulas comes in handy when you need to solve trig equations or when you're given the sine, cosine, tangent, or other trig function of an angle and need to find the exact trig value of twice that angle without knowing the measure of the original angle. Isn't this your happy day?

Note: If you know the original angle in question, finding the sine, cosine, or tangent of twice that angle is easy; you can look it up on the unit circle or use your calculator to find the answer. However, if you don't have the measure of the original angle and you must find the exact value of twice that angle, the process isn't that simple. Read on!

Finding the sine of a doubled angle

To fully understand and be able to stow away the double-angle formula for sine, you should first understand where it comes from. (The double-angle formulas for sine, cosine, and tangent are extremely different from one another, although they can all be derived by using the sum formula.)

1. **Rewrite $\sin 2\theta$ as the sum of two angles.**

   $\sin 2\theta = \sin(\theta + \theta)$

2. **Use the sum formula for sine (see the section, "Searching out the sine of $(a \pm b)$") to expand the sine of the sum.**

   $\sin(\theta + \theta) = \sin\theta\cos\theta + \cos\theta\sin\theta$

3. **Simplify both sides.**

   $\sin 2\theta = 2\sin\theta\cos\theta$

TECHNICAL STUFF

This identity is called the *double-angle formula* for sine. If you're given an equation with more than one trig function and asked to solve for the angle, your best bet is to express the equation in terms of one trig function only. You often can achieve this by using the double-angle formula.

EXAMPLE

Q. Solve $4\sin 2x\cos 2x = 1$ for the value of *x*.

A. $x = \frac{\pi}{24}, \frac{13\pi}{24}, \frac{25\pi}{24}, \frac{37\pi}{24}$, and $x = \frac{5\pi}{24}, \frac{17\pi}{24}, \frac{29\pi}{24}, \frac{41\pi}{24}$. Notice that the equation isn't set equal to 0, so you can't factor it. Even if you subtract 1 from both sides to get 0 on the right, it can't be factored. Follow these steps:

1. **State the problem.**

   Solve for *x* in $4\sin 2x\cos 2x = 1$.

2. **Rewrite the equation to match a factor with a possible identity.**

   Factor out 2 on the left, and you have $2(2\sin 2x\cos 2x) = 1$.

   Inside the parentheses you have the results of applying the double-angle formula on $\sin 4x$. Notice that the two functions have angles that are half 4*x*, or 2*x*.

3. **Apply the correct formula.**

   The double-angle formula for sine allows you to replace what's in the parentheses and gives you $2(\sin 4x) = 1$.

4. **Simplify the equation and isolate the trig function.**

   Rewrite as $2\sin 4x = 1$, and then divide each side by 2 to get $\sin 4x = \frac{1}{2}$.

5. **Find all the solutions for the trig equation.**

   When is the sine of an angle equal to $\frac{1}{2}$? Referring to the unit circle, you find two points where the sine is $\frac{1}{2}$: when the angle measures $\frac{\pi}{6}$ and when it's $\frac{5\pi}{6}$. So $\sin 4x = \frac{1}{2}$ when 4*x* is $\frac{\pi}{6}$ or $\frac{5\pi}{6}$. Because of the 4 multiplier on the variable, you want the four corresponding angles — the original and the three rotations — for each angle. Add 2π to each angle three times.

For $4x = \frac{\pi}{6}$, you also have $4x = \frac{\pi}{6} + 2\pi = \frac{13\pi}{6}$, $4x = \frac{\pi}{6} + 4\pi = \frac{25\pi}{6}$, and $4x = \frac{\pi}{6} + 6\pi = \frac{37\pi}{6}$.

For $4x = \frac{5\pi}{6}$, you also have $4x = \frac{5\pi}{6} + 2\pi = \frac{17\pi}{6}$, $4x = \frac{5\pi}{6} + 4\pi = \frac{29\pi}{6}$, and $4x = \frac{5\pi}{6} + 6\pi = \frac{41\pi}{6}$.

Solve each equation by dividing by 4, and you have:

$$x = \frac{\pi}{24},\ x = \frac{13\pi}{24},\ x = \frac{25\pi}{24}, \text{ and } x = \frac{37\pi}{24}$$

$$x = \frac{5\pi}{24},\ x = \frac{17\pi}{24},\ x = \frac{29\pi}{24}, \text{ and } x = \frac{41\pi}{24}$$

And those eight solutions are just within one turn on the unit circle. You can add 2π to any of the solutions to get even more.

Calculating cosines for two

You can use three different formulas to find the value for cos $2x$ — the double-angle of cosine — so your job is to choose which one fits your specific problem the best. The double-angle formula for cosine comes from the sum formula, just like the double-angle formula for sine. If you can't remember the double-angle formula but you can remember the sum formula, just simplify $\cos(2x)$, which is the same as $\cos(x+x)$. Using the sum formula to create the double-angle formula for cosine yields $\cos 2x = \cos^2 x - \sin^2 x$. And you have two additional ways to express this by using Pythagorean identities (see Chapter 12):

- You can replace $\sin^2 x$ with $(1 - \cos^2 x)$ and simplify.
- You can replace $\cos^2 x$ with $(1 - \sin^2 x)$ and simplify.

TECHNICAL STUFF

Following are the formulas for the double-angle of cosine:

- $\cos 2\theta = \cos^2\theta - \sin^2\theta$
- $\cos 2\theta = 2\cos^2\theta - 1$
- $\cos 2\theta = 1 - 2\sin^2\theta$

Looking at what you're given and what you're asked to find usually will lead you toward the best formula for the situation. And hey, if you don't pick the right one at first, you have two more to try!

EXAMPLE

Q. If $\sec x = -\frac{15}{8}$, find the exact value of $\cos 2x$, given that x is in quadrant II of the coordinate plane.

A. $-\frac{97}{225}$. Follow these steps to solve:

1. **Use the reciprocal identity (see Chapter 12) to change secant to cosine.**

   Because secant doesn't appear in any of the identities involving a double angle, you have to complete this step first. Therefore, if $\sec x = -\frac{15}{8}$, then the reciprocal of the secant is equal to the cosine: $\cos x = -\frac{8}{15}$.

2. **Choose the appropriate double-angle formula.**

   Because you now know the cosine value, you choose the double-angle formula that just involves cosine: $\cos 2\theta = 2\cos^2\theta - 1$.

3. **Substitute the information you know into the formula.**

   You can plug cosine into the equation: $\cos 2x = 2\left(-\frac{8}{15}\right)^2 - 1$.

4. **Simplify the formula to solve.** $\cos 2x = 2\left(-\frac{8}{15}\right)^2 - 1 = 2\left(\frac{64}{225}\right) - 1 = \frac{128}{225} - 1 = \frac{128}{225} - \frac{225}{225} = -\frac{97}{225}$

YOUR TURN

11 If $\sin x = -\frac{5}{13}$, find the exact value of $\cos 2x$.

12 If $\csc x = -\frac{5}{4}$, find the exact value of $\sin 2x$, given that x is in quadrant III of the coordinate plane.

Squaring your cares away

As much as you may love radicals (the square root type, of course), when a square root appears inside a trig proof, you usually have to square both sides of the equation at some point to get where you need to go. Even though you usually want to work on just one side or the other to prove an identity, the radical on one side makes it hard to create the same expression on both sides.

EXAMPLE

Q. Prove that $2\sin^2 x - 1 = \sqrt{1-\sin^2 2x}$.

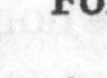

A. For this problem, you do need to try squaring both sides. Use the following steps:

1. **Square both sides.**

REMEMBER

Be careful when using this process of squaring both sides of the equation. In this case, you're fine, since the square of the sine will always be 1 or less, and you won't ever have a negative number under the radical.

You have $\left(2\sin^2 x - 1\right)^2 = \left(\sqrt{1-\sin^2 2x}\right)^2$.

On the left, you get a perfect square trinomial, and on the right, the square and root cancel one another. The result of the squaring equation gives you $4\sin^4 x - 4\sin^2 x + 1 = 1 - \sin^2 2x$.

2. **Look for identities.**

You can see a double angle on the right side. The notation means to square the sine of 2x: $\sin^2 2x = (\sin 2x)^2$.

So you rewrite the identity as $4\sin^4 x - 4\sin^2 x + 1 = 1 - (\sin 2x)^2$.

You then substitute in the double-angle identity and square:

$$4\sin^4 x - 4\sin^2 x + 1 = 1 - (2\sin x\cos x)^2$$
$$4\sin^4 x - 4\sin^2 x + 1 = 1 - 4\sin^2 x\cos^2 x$$

3. **Change the cosine term to a sine.**

Change the $\cos^2 x$ to $1-\sin^2 x$ by using the Pythagorean identity. You get $4\sin^4 x - 4\sin^2 x + 1 = 1 - 4\sin^2 x\left(1-\sin^2 x\right)$.

4. **Distribute and simplify.**

You end up with $4\sin^4 x - 4\sin x + 1 = 1 - 4\sin^2 x + 4\sin^4 x$.

Using the commutative and associative properties of equality (from Chapter 1), you get

$$4\sin^4 x - 4\sin x + 1 = 4\sin^4 x - 4\sin x + 1$$

and you have proven the identity.

YOUR TURN

13 Prove that $\sin x = \sqrt{\frac{1-\cos 2x}{2}}$.

14 Prove that $2\tan x = \sqrt{\frac{4\sin^2 x}{1-\sin^2 x}}$.

Having twice the fun with tangents

Unlike the formulas for cosine (see the section, "Calculating cosines for two"), tangent has just one double-angle formula. That's a relief, after dealing with cosine. The double-angle formula for tangent is used less often than the double-angle formulas for sine or cosine; however, you shouldn't overlook it just because it isn't as popular as its cooler counterparts!

The double-angle formula for tangent is derived by simplifying $\tan(x+x)$ with the sum formula. However, the simplification process is much more complicated here because it involves fractions.

TECHNICAL STUFF

The double-angle identity for tangent is:

$$\tan 2\theta = \frac{2\tan\theta}{1-\tan^2\theta}$$

REMEMBER

When solving equations for tangent, remember that the period for the tangent function is π. This detail is important — especially when you have to deal with more than one angle in an equation — because you usually need to find all the solutions on the interval $[0,2\pi)$. Double-angle equations have twice as many solutions in that interval as single-angle equations do.

EXAMPLE

Q. Find the solutions for $2\tan 2x + 2 = 0$ on the interval $[0,2\pi)$.

A. $x=\frac{3\pi}{8}$, $x=\frac{11\pi}{8}$, $x=\frac{7\pi}{8}$, $x=\frac{15\pi}{8}$. Follow these steps:

1. **Isolate the trig function.**

   Subtract 2 from both sides to get $2\tan 2x = -2$. Next, divide both sides of the equation by 2: $\tan 2x = -1$.

2. **Solve for the double-angle, 2x; when does $\tan 2x = -1$?**

   On the unit circle, the tangent is negative in the second and fourth quadrants. Moreover, the tangent is -1 at $\frac{3\pi}{4}$ and $\frac{7\pi}{4}$.

3. **List two rotations by adding 2π to each angle.**

   (The number of rotations is dictated by the coefficient on x.)

   Adding 2π, you get the four possibilities: $2x=\frac{3\pi}{4}$, $2x=\frac{11\pi}{4}$, $2x=\frac{7\pi}{4}$, and $2x=\frac{15\pi}{4}$.

4. **Find all the solutions on the required interval.**

   Divide each angle measure by 2 to solve for x.

   You get $x=\frac{3\pi}{8}$, $x=\frac{11\pi}{8}$, $x=\frac{7\pi}{8}$, and $x=\frac{15\pi}{8}$.

   Now you've found all the solutions.

YOUR TURN

 15 Prove that $\tan 2x = \dfrac{2\sin x\cos x}{1-2\sin^2 x}$.

 16 Find the solutions for $\tan^2 2x - 3 = 0$ on the interval $[0, 2\pi)$.

Taking Trig Functions of Common Angles Divided in Two

Some time ago, some mathematicians found ways to calculate half of an angle with an identity. As you can see from using the sum and difference identities earlier in this chapter, you now have the option to use *half-angle identities* to evaluate a trig function of an angle that isn't on the unit circle by using one that is. For example, 15°, which isn't on the unit circle, is half of 30°, which is on the unit circle. Cutting special angles on the unit circle in half gives you a variety of new angles that can't be achieved by using the sum and difference formulas or the double-angle formulas. Although the half-angle formulas won't give you all the angles of the unit circle, they certainly get you closer than you were before.

REMEMBER

The trick is knowing which type of identity serves your purpose best. Half-angle formulas are the better option when you need to find the trig values for any angle that can be expressed as half of another angle on the unit circle. For example, to evaluate a trig function of $\frac{\pi}{8}$, you can use the half-angle formula of $\frac{\pi}{4}$. Because no combination of sums or differences of special angles gets you $\frac{\pi}{8}$, you know to use a half-angle formula.

You also can find the values of trig functions for angles like $\frac{\pi}{16}$ or $\frac{\pi}{12}$, each of which is exactly half of the angles on the unit circle. Of course, these angles aren't the only types that the identities work for. You can continue to halve the trig-function value of half of any angle on the unit circle for the rest of your life (if you have nothing better to do). For example, 15° is half of 30°, and 7.5° is half of 15°.

REMEMBER

The half-angle formulas for sine, cosine, and tangent are as follows:

- $\sin\frac{\theta}{2} = \pm\sqrt{\frac{1-\cos\theta}{2}}$
- $\cos\frac{\theta}{2} = \pm\sqrt{\frac{1+\cos\theta}{2}}$
- $\tan\frac{\theta}{2} = \frac{1-\cos\theta}{\sin\theta} = \frac{\sin\theta}{1+\cos\theta}$

TECHNICAL STUFF

In the half-angle formula for sine and cosine, notice that ± appears in front of each radical (square root). Whether your answer is positive or negative depends on which quadrant the new angle (the half angle) is in. The half-angle formula for tangent doesn't have a ± sign in front, so this doesn't apply to tangent.

EXAMPLE

Q. Find $\sin 165°$

A. Follow these steps:

1. **Rewrite the trig function and the angle as half of a unit circle value.**

   First, realize that 165° is half of 330°, so you can rewrite the sine function as $\sin\left(\frac{330°}{2}\right)$.

2. **Determine the sign of the trig function.**

   Because 165° is in quadrant II of the coordinate plane, its sine value should be positive.

3. **Substitute the angle value into the identity.**

   The angle value 330° plugs in for x in the positive half-angle formula for sine. This gives you $\sin\frac{330°}{2} = \sqrt{\frac{1-\cos 330°}{2}}$.

4. **Replace cos x with its actual value.**

   Use the unit circle to find $\cos 330°$. Substituting that value into the equation gives you $\sin\frac{330°}{2} = \sqrt{\frac{1-\frac{\sqrt{3}}{2}}{2}}$.

5. **Simplify the half-angle formula to solve.**

   This approach has three steps:

   a. Rewrite the numerator under the radical as one term by subtracting:

   $$\sin\frac{330°}{2} = \sqrt{\frac{\frac{2}{2}-\frac{\sqrt{3}}{2}}{2}} = \sqrt{\frac{\frac{2-\sqrt{3}}{2}}{2}}.$$

   b. Multiply the numerator times the reciprocal of the denominator:

   $$= \sqrt{\frac{2-\sqrt{3}}{2}\cdot\frac{1}{2}} = \sqrt{\frac{2-\sqrt{3}}{4}}.$$

   c. Finally, simplify: $= \frac{\sqrt{2-\sqrt{3}}}{\sqrt{4}} = \frac{\sqrt{2-\sqrt{3}}}{2}$.

YOUR TURN

17 Find the exact value of $\cos\frac{\pi}{8}$.	**18** Find the exact value of $\tan\frac{\pi}{12}$.

A Glimpse of Calculus: Traveling from Products to Sums and Back

You've now reached the time-travel portion of the chapter, because all the information from here on comes into play mainly in calculus. In calculus, you'll get to integrate functions, which is much easier to do when you're dealing with sums of trig functions rather than products. The information in this section helps you prepare for the switch. Here you see how to express products as sums and how to transport from sums to products.

Expressing products as sums (or differences)

The process of the integration of two factors being multiplied together can be difficult, especially when you must deal with a mixture of trig functions. If you can break up a product into the sum of two different terms, each with its own trig function, doing the math becomes much easier. But you don't have to worry about making any big decisions right now. In pre-calculus, problems of this type usually say, "Express the product as a sum or difference." For the time being, you'll make the conversion from a product, and that will be the end of the problem.

You have three product-to-sum formulas to consider. The following list breaks down these formulas.

- Sine · Cosine: $\sin a\cos b = \frac{1}{2}[\sin(a+b)+\sin(a-b)]$
- Cosine · Cosine: $\cos a\cos b = \frac{1}{2}[\cos(a+b)+\cos(a-b)]$
- Sine · Sine: $\sin a\sin b = \frac{1}{2}[\cos(a-b)-\cos(a+b)]$

EXAMPLE

Q. Write the following as sums:

a) $6\cos q\sin 2q$

b) $\cos 6\theta\cos 3\theta$

c) $\sin 5x\sin 4x$

A. Using the formulas:

a) Rewrite this expression as $6\sin 2q\cos q$ (thanks to the commutative property) and then plug what you know into the formula to get

$6\sin 2q\cos q = 6\cdot\frac{1}{2}[\sin(2q+q)+\sin(2q-q)] = 3[\sin(3q)+\sin(q)]$.

b) You get $\cos 6\theta\cos 3\theta = \frac{1}{2}[\cos(6\theta+3\theta)+\cos(6\theta-3\theta)] = \frac{1}{2}[\cos(9\theta)+\cos(3\theta)]$.

c) You get $\sin 5x\sin 4x = \frac{1}{2}[\cos(5x-4x)-\cos(5x+4x)] = \frac{1}{2}[\cos(x)-\cos(9x)]$.

Transporting from sums (or differences) to products

On the flip side of the previous section, you need to familiarize yourself with a set of formulas that change sums to products. Sum-to-product formulas are useful to help you find the sum of two trig values that aren't on the unit circle. Of course, these formulas work only if the sum or difference of the two angles ends up being an angle from the special triangles from Chapter 8.

Here are the sum/difference-to-product identities:

» $\sin a+\sin b = 2\sin\left(\frac{a+b}{2}\right)\cos\left(\frac{a-b}{2}\right)$

» $\sin a-\sin b = 2\cos\left(\frac{a+b}{2}\right)\sin\left(\frac{a-b}{2}\right)$

» $\cos a+\cos b = 2\cos\left(\frac{a+b}{2}\right)\cos\left(\frac{a-b}{2}\right)$

» $\cos a-\cos b = -2\sin\left(\frac{a+b}{2}\right)\sin\left(\frac{a-b}{2}\right)$

For example, say you're asked to find the sum of the sines of two angles without a calculator — and the angles aren't any of your standard angles, but they're usable. Because you're asked to find the sum of two trig functions whose angles aren't special angles, you can change this to a product by using the sum-to-product formulas. This really only works, of course, if there are convenient angles whose sum and difference are what you're looking for.

EXAMPLE

Q. Find $\sin 105° + \sin 15°$ using a sum-to-product formula.

A. $\frac{\sqrt{6}}{2}$. Follow these steps:

1. **Change the sum to a product.**

   Because you're asked to find the sum of two sine functions, use this equation:

   $\sin a+\sin b = 2\sin\left(\frac{a+b}{2}\right)\cos\left(\frac{a-b}{2}\right)$.

Substituting in the angle measures, you get $\sin 105° + \sin 15° = 2\sin\left(\frac{105° + 15°}{2}\right)$ $\cos\left(\frac{105° - 15°}{2}\right)$.

2. **Simplify the result.**

   Combining like terms and dividing, you get $\sin 105° + \sin 15° = 2\sin 60° \cos 45°$. Those angles are unit circle values, so continue to the next step.

3. **Use the unit circle to simplify further.**

   $2\sin 60° \cos 45° = 2 \cdot \frac{\sqrt{3}}{2} \cdot \frac{\sqrt{2}}{2} = \frac{\sqrt{6}}{2}$

 19 Find the exact value of $\sin 75° - \sin 15°$.

 20 Find the exact value of $\cos 15° + \cos 165°$.

Eliminating Exponents with Power-Reducing Formulas

Power-reducing formulas allow you to get rid of exponents on trig functions so you can solve for an angle's measure. This ability will come in very handy when working on some higher-up math problems.

The formulas given here allow you to change from second-degree to first-degree expressions. In some cases, when the function is raised to the fourth power or higher, you may have to apply the power-reducing formulas more than once to eliminate all the exponents. You can use the following three power-reducing formulas to accomplish the elimination task:

- $\sin^2\theta = \frac{1 - \cos 2\theta}{2}$
- $\cos^2\theta = \frac{1 + \cos 2\theta}{2}$
- $\tan^2\theta = \frac{1 - \cos 2\theta}{1 + \cos 2\theta}$

EXAMPLE

Q. Express $\sin^4 x$ without exponents.

A. $\frac{1}{8}(3-4\cos 2x+\cos 4x)$. Follow these steps:

1. **Apply the power-reducing formula to the trig function.**

   First, realize that $\sin^4 x=\left(\sin^2 x\right)^2$. Because the problem requires the reduction of $\sin^4 x$, you apply the power-reducing formula twice. The first application gives you the following: $\left(\sin^2 x\right)^2=\left(\frac{1-\cos 2x}{2}\right)^2=\frac{1-\cos 2x}{2}\cdot\frac{1-\cos 2x}{2}$.

2. **Multiply the two fractions.**

   $$\left(\sin^2 x\right)^2=\frac{1-2\cos 2x+\cos^2 2x}{4}=\frac{1}{4}\left(1-2\cos 2x+\cos^2 2x\right)$$

3. **Apply the power-reducing formula again.**

   The term $\cos^2 2x$ means you must apply the power-reducing formula for cosine.

   Because writing a power-reducing formula inside a power-reducing formula is very confusing, find out what $\cos^2 2x$ is by itself first and then plug it back in:

   $$\cos^2 2x=\frac{1+\cos 2(2x)}{2}=\frac{1+\cos 4x}{2}$$

   $$\left(\sin^2 x\right)^2=\frac{1-2\cos 2x+\cos^2 2x}{4}=\frac{1}{4}\left(1-2\cos 2x+\frac{1+\cos 4x}{2}\right)$$

4. **Simplify to get your result.**

   Factor out $\frac{1}{2}$ from everything inside the brackets so that you don't have fractions both outside and inside the brackets. This step gives you $\left(\sin^2 x\right)^2=\frac{1}{8}(2-4\cos 2x+1+\cos 4x)$.

   Combine like terms to get $\left(\sin^2 x\right)^2=\frac{1}{8}(3-4\cos 2x+\cos 4x)$.

YOUR TURN

21 Express $1+2\cos^2\theta$ without exponents.

22 Express $\sin^2 x-\tan^2 x$ without exponents.

Practice Questions Answers and Explanations

1. $\frac{\sqrt{2}+\sqrt{6}}{4}$**. First, determine the angles you want to use:** $\sin\frac{7\pi}{12}=\sin\left(\frac{3\pi}{12}+\frac{4\pi}{12}\right)=\sin\left(\frac{\pi}{4}+\frac{\pi}{3}\right)$**.**
Now apply the formula:

$$\sin\left(\frac{\pi}{4}+\frac{\pi}{3}\right)=\sin\frac{\pi}{4}\cos\frac{\pi}{3}+\cos\frac{\pi}{4}\sin\frac{\pi}{3}=\left(\frac{\sqrt{2}}{2}\right)\left(\frac{1}{2}\right)+\left(\frac{\sqrt{2}}{2}\right)\left(\frac{\sqrt{3}}{2}\right)=\frac{\sqrt{2}}{4}+\frac{\sqrt{6}}{4}=\frac{\sqrt{2}+\sqrt{6}}{4}$$

2. $\frac{\sqrt{6}-\sqrt{2}}{4}$**. First, determine the angles you want to use:** $\sin 165°=\sin(210°-45°)$**. Now apply the formula:**

$$\sin(210°-45°)=\sin 210°\cos 45°-\cos 210°\sin 45°=\left(-\frac{1}{2}\right)\left(\frac{\sqrt{2}}{2}\right)-\left(-\frac{\sqrt{3}}{2}\right)\left(\frac{\sqrt{2}}{2}\right)=-\frac{\sqrt{2}}{4}+\frac{\sqrt{6}}{4}=\frac{\sqrt{6}-\sqrt{2}}{4}$$

3. $-\sin(a+b)=-\sin(a+b)$**. First, rewrite** $\sin(-a-b)$ **as** $\sin((-a)+(-b))$**. Now apply the formula for the sum, and you have** $\sin((-a)+(-b))=\sin(-a)\cos(-b)+\cos(-a)\sin(-b)$**. Apply the even/odd identities, and** $\sin(-a)\cos(-b)+\cos(-a)\sin(-b)=-\sin a\cos b+\cos a(-\sin b)=-(\sin a\cos b+\cos a\sin b)$**. Now replace the sum in the parentheses with** $\sin(a+b)$ **and the right side becomes** $-\sin(a+b)$**.**

4. $\cos a=\cos a$**. Apply the formula for the sum, and** $\sin\left(a+\frac{\pi}{2}\right)=\sin a\cos\frac{\pi}{2}+\cos a\sin\frac{\pi}{2}$**.**
Find the values of the two functions and insert them in the equation:
$\sin a\cos\frac{\pi}{2}+\cos a\sin\frac{\pi}{2}=\sin a(0)+\cos a(1)=\cos a$**.**

5. $-\frac{119}{169}$**. Completing the formula,** $\cos(A+B)=\cos A\cos B-\sin A\sin B$**. You need to find** $\cos A$ **and** $\sin B$**. Using the Pythagorean identity and the value for** $\sin A$**:**

$$\sin^2 A+\cos^2 A=1\rightarrow\left(-\frac{5}{13}\right)^2+\cos^2 A=1\rightarrow\cos^2 A=1-\frac{25}{169}=\frac{144}{169}$$

Finding the square root of both sides, $\cos A=\pm\frac{12}{13}$**. The value of cosine when A is in quadrant IV is** $\cos A=+\frac{12}{13}$**.**

Now for sin B: $\sin^2 B+\cos^2 B=1\rightarrow\sin^2 B+\left(-\frac{12}{13}\right)^2=1\rightarrow\sin^2 B=1-\frac{144}{169}=\frac{25}{169}$**. Finding the square root of both sides,** $\sin B=\pm\frac{5}{13}$**. And, since B is in quadrant II, the sine is positive:** $\sin B=+\frac{5}{13}$**. Returning to the formula for the cosine of a sum:**

$$\cos(A+B)=\frac{12}{13}\left(-\frac{12}{13}\right)-\left(-\frac{5}{13}\right)\left(\frac{5}{13}\right)=-\frac{144}{169}+\frac{25}{169}=-\frac{119}{169}$$

6. $\frac{527}{625}$**. Completing the formula,** $\cos(A-B)=\cos A\cos B+\sin A\sin B$**. You need to find** $\cos A$ **and** $\sin B$**. Using the Pythagorean identity and the value for** $\sin A$**:** $\sin^2 A+\cos^2 A=1\rightarrow\left(-\frac{24}{25}\right)^2+\cos^2 A=1\rightarrow\cos^2 A=1-\frac{576}{625}=\frac{49}{625}$

Finding the square root of both sides, $\cos A = \pm\frac{7}{25}$. The value of cosine when A is in quadrant III is $\cos A = -\frac{7}{25}$.

Now for sin B: $\sin^2 B + \cos^2 B = 1 \rightarrow \sin^2 B + \left(\frac{7}{25}\right)^2 = 1 \rightarrow \sin^2 B = 1 - \frac{49}{625} = \frac{576}{625}$. Finding the square root of both sides, $\sin B = \pm\frac{24}{25}$. And, since B is in quadrant VI, the sine is negative: $\sin B = -\frac{24}{25}$. Returning to the formula for the cosine of a difference:

$$\cos(A - B) = \left(-\frac{7}{25}\right)\left(\frac{7}{25}\right) + \left(-\frac{24}{25}\right)\left(-\frac{24}{25}\right) = -\frac{49}{625} + \frac{576}{625} = \frac{527}{625}$$

(7) **$-(\sin x + \cos x) = -(\sin x + \cos x)$.** Apply the difference formulas to the terms on the left:

$$\sin\left(x - \frac{\pi}{2}\right) - \cos\left(x - \frac{\pi}{2}\right) = \sin x \cos\frac{\pi}{2} - \cos x \sin\frac{\pi}{2} - \left(\cos x \cos\frac{\pi}{2} + \sin x \sin\frac{\pi}{2}\right)$$

Next, put in the function values, distribute, and simplify:

$$= \sin x(0) - \cos x(1) - (\cos x(0) + \sin x(1)) = -\cos x - \sin x = -(\cos x + \sin x) = -(\sin x + \cos x)$$

(8) **$\frac{1}{2}[1 - \sin 2x] = \frac{1}{2}[1 - \sin 2x]$.** Apply the sum formula on the left. Then put in the function values, simplify, and insert the double angle formula:

$$\left[\cos\left(x + \frac{\pi}{4}\right)\right]^2 = \left[\cos x \cos\frac{\pi}{4} - \sin x \sin\frac{\pi}{4}\right]^2 = \left[\cos x\left(\frac{\sqrt{2}}{2}\right) - \sin x\left(\frac{\sqrt{2}}{2}\right)\right]^2$$

$$= \left(\frac{\sqrt{2}}{2}\right)^2 [\cos x - \sin x]^2 = \frac{2}{4}\left[\cos^2 x - 2\sin x \cos x + \sin^2 x\right] = \frac{1}{2}\left[\sin^2 x + \cos^2 x - 2\sin x \cos x\right]$$

$$= \frac{1}{2}[1 - 2\sin x \cos x] = \frac{1}{2}[1 - \sin 2x].$$

(9) **$2 - \sqrt{3}$.** First, determine the angles you want to use: $\tan 15° = \tan(45° - 30°)$. Now apply the formula: $\tan(45° - 30°) = \frac{\tan 45° - \tan 30°}{1 + (\tan 45°)(\tan 30°)} = \frac{1 - \frac{\sqrt{3}}{3}}{1 + (1)\left(\frac{\sqrt{3}}{3}\right)} = \frac{\frac{3 - \sqrt{3}}{3}}{\frac{3 + \sqrt{3}}{3}} = \frac{3 - \sqrt{3}}{\cancel{3}} \cdot \frac{\cancel{3}}{3 + \sqrt{3}} = \frac{3 - \sqrt{3}}{3 + \sqrt{3}}$.

Rationalizing the denominator: $\frac{3 - \sqrt{3}}{3 + \sqrt{3}} \cdot \frac{3 - \sqrt{3}}{3 - \sqrt{3}} = \frac{9 - 6\sqrt{3} + 3}{9 - 3} = \frac{12 - 6\sqrt{3}}{6} = 2 - \sqrt{3}$.

(10) **$\tan x = \tan x$.** This is actually one of the periodicity identities! Apply the sum formula and determine the function values: $\tan(x + \pi) = \frac{\tan x - \tan\pi}{1 + (\tan x)(\tan\pi)} = \frac{\tan x - 0}{1 + (\tan x)(0)} = \frac{\tan x}{1} = \tan x$.

(11) **$\frac{119}{169}$.** Choose the double-angle formula that contains only the sine function. Then insert the value for the sine: $\cos 2x = 1 - 2\sin^2 x = 1 - 2\left(-\frac{5}{13}\right)^2 = 1 - 2\left(\frac{25}{169}\right) = 1 - \frac{50}{169} = \frac{119}{169}$.

12 $\frac{24}{25}$. **First, determine the value of the sine, so it can be inserted in the formula. If $\csc x = -\frac{5}{4}$, then its reciprocal is $\sin x = -\frac{4}{5}$. Since $\sin 2x = 2\sin x \cos x$, you also need the value of the cosine. Using the Pythagorean identity, $\sin^2 x + \cos^2 x = 1 \rightarrow \left(-\frac{4}{5}\right)^2 + \cos^2 x = 1 \rightarrow \cos^2 x = 1 - \frac{16}{25} = \frac{9}{25}$. Solving for cosine, $\cos x = \pm\sqrt{\frac{9}{25}} = \pm\frac{3}{5}$. Since x is in quadrant III of the coordinate plane, you choose the negative value for cosine: $\cos x = -\frac{3}{5}$. Now complete the formula: $\sin 2x = 2\sin x \cos x = 2\left(-\frac{4}{5}\right)\left(-\frac{3}{5}\right) = \frac{24}{25}$.**

13 **$\sin x = \sin x$. Apply the double-angle formula involving the sine:**

$$\sin x = \sqrt{\frac{1-\cos 2x}{2}} = \sqrt{\frac{1-(1-2\sin^2 x)}{2}} = \sqrt{\frac{1-1+2\sin^2 x}{2}} = \sqrt{\frac{2\sin^2 x}{2}} = \sqrt{\sin^2 x} = \sin x$$

14 **$4\tan^2 x = 4\tan^2 x$. Square both sides. Then replace the denominator on the right using a Pythagorean identity. Finally, use the reciprocal identity.**

$$2\tan x = \sqrt{\frac{4\sin^2 x}{1-\sin^2 x}} \rightarrow (2\tan x)^2 = \left(\sqrt{\frac{4\sin^2 x}{1-\sin^2 x}}\right)^2 \rightarrow 4\tan^2 x = \frac{4\sin^2 x}{1-\sin^2 x}$$

$$\rightarrow 4\tan^2 x = \frac{4\sin^2 x}{\cos^2 x} \rightarrow 4\tan^2 x = 4\tan^2 x$$

15 **$\tan 2x = \tan 2x$. The numerator is the formula for $\sin 2x$, and the denominator is one of the formulas for $\cos 2x$. Do the replacements and then apply the ratio identity:**

$$\tan 2x = \frac{2\sin x \cos x}{1-2\sin^2 x} = \frac{\sin 2x}{\cos 2x} = \tan 2x.$$

16 **$\frac{\pi}{6}, \frac{\pi}{3}, \frac{2\pi}{3}, \frac{5\pi}{6}, \frac{7\pi}{6}, \frac{4\pi}{3}, \frac{5\pi}{3}, \frac{11\pi}{6}$. Add 3 to each side of the equation. Then find the square root of both sides: $\tan^2 2x - 3 = 0 \rightarrow \tan^2 2x = 3 \rightarrow \tan 2x = \pm\sqrt{3}$. The angles for which the tangent is ±3 are $\frac{\pi}{3}, \frac{2\pi}{3}, \frac{4\pi}{3}, \frac{5\pi}{3}, \frac{7\pi}{3}, \frac{8\pi}{3}, \frac{10\pi}{3}, \frac{11\pi}{3}$, so $2x = \frac{\pi}{3}, \frac{2\pi}{3}, \frac{4\pi}{3}, \frac{5\pi}{3}, \frac{7\pi}{3}, \frac{8\pi}{3}, \frac{10\pi}{3}, \frac{11\pi}{3}$. Dividing through by 2, $x = \frac{\pi}{6}, \frac{\pi}{3}, \frac{2\pi}{3}, \frac{5\pi}{6}, \frac{7\pi}{6}, \frac{4\pi}{3}, \frac{5\pi}{3}, \frac{11\pi}{6}$.**

17 **$\frac{\sqrt{2+\sqrt{2}}}{2}$. Use the half-angle formula $\cos\frac{\theta}{2} = \pm\sqrt{\frac{1+\cos\theta}{2}}$, where $\theta = \frac{\pi}{4}$:**

$$\cos\frac{\pi}{8} = \cos\frac{\pi/4}{2} = \pm\sqrt{\frac{1+\cos\frac{\pi}{4}}{2}} = \pm\sqrt{\frac{1+\frac{\sqrt{2}}{2}}{2}} = \pm\sqrt{\frac{2+\sqrt{2}}{4}} = \pm\frac{\sqrt{2+\sqrt{2}}}{2}$$

Since $\frac{\pi}{8}$ is in quadrant I, you choose the positive answer.

(18) $2-\sqrt{3}$. Use the half-angle formula $\tan\frac{\theta}{2}=\frac{1-\cos\theta}{\sin\theta}$, where $\theta=\frac{\pi}{6}$:

$$\tan\frac{\pi/6}{2}=\frac{1-\cos\frac{\pi}{6}}{\sin\frac{\pi}{6}}=\frac{1-\frac{\sqrt{3}}{2}}{\frac{1}{2}}=\frac{\frac{2-\sqrt{3}}{2}}{\frac{1}{2}}=\frac{2-\sqrt{3}}{\not{2}}\cdot\frac{\not{2}}{1}=2-\sqrt{3}$$

(19) $\frac{\sqrt{2}}{2}$. Use the difference of sines formula:

$$\sin 75°-\sin 15°=2\cos\left(\frac{75°+15°}{2}\right)\sin\left(\frac{75°-15°}{2}\right)=2\cos\left(\frac{90°}{2}\right)\sin\left(\frac{60°}{2}\right)$$
$$=2\cos(45°)\sin(30°)=\not{2}\left(\frac{\sqrt{2}}{2}\right)\left(\frac{1}{\not{2}}\right)=\frac{\sqrt{2}}{2}$$

(20) **0.** Use the sum of cosines formula:

$$\cos 15°+\cos 165°=2\cos\left(\frac{15°+165°}{2}\right)\cos\left(\frac{15°-165°}{2}\right)=2\cos\left(\frac{180°}{2}\right)\cos\left(\frac{-150°}{2}\right)$$
$$=2\cos(90°)\cos(-75°)=2(0)\cos(-75°)=0$$

(21) $2+\cos 2\theta$. Starting with the double-angle formula for cosine, $\cos 2\theta=2\cos^2\theta-1$, solve for $2\cos^2\theta$ and you get that $2\cos^2\theta=\cos 2\theta+1$. Substituting, you have $1+2\cos^2\theta=1+\cos 2\theta+1=2+\cos 2\theta$. You also could have used the power-reducing formula to get the same result and save some time.

(22) $(\sin x-\tan x)(\sin x+\tan x)$. Factoring the difference of squares is the quickest, easiest method. But, for a different version, you can use the power-reducing formulas and get a different answer — one with double angles. Starting with the two power-reducing formulas, $\sin^2 x-\tan^2 x=\frac{1-\cos 2x}{2}-\frac{1-\cos 2x}{1+\cos 2x}$. You could stop here. If you try to add the fractions together, you need a common denominator, which will end up with an exponent again. You would have

$$\frac{1-\cos 2x}{2}\cdot\frac{1+\cos 2x}{1+\cos 2x}-\frac{1-\cos 2x}{1+\cos 2x}\cdot\frac{2}{2}=\frac{1-\cos^2 2x}{2(1+\cos 2x)}-\frac{2-2\cos 2x}{2(1+\cos 2x)}=\frac{-1+2\cos 2x-\cos^2 2x}{2(1+\cos 2x)}$$

which can be worked on further. Go with the simpler choice.

If you're ready to test your skills a bit more, take the following chapter quiz that incorporates all the chapter topics.

Whaddya Know? Chapter 13 Quiz

Quiz time! Complete each problem to test your knowledge on the various topics covered in this chapter. You can then find the solutions and explanations in the next section.

1 **Determine the exact value of $\cos(-15°)$.**

2 **Determine the exact value of $\sin\frac{7\pi}{12}+\sin\frac{\pi}{12}$.**

3 **Write $\cos^4\theta-\sin^4\theta$ without any exponents.**

4 **Determine the exact value of $\sin\frac{5\pi}{8}$.**

5 **Determine the value of $\tan 2x$ if $\cot x=\sqrt{3}$.**

6 **Solve $\sin x+\cos x=\sqrt{2}$ when $0<x<360°$.**

7 **Determine the exact value of $\sin(195°)$.**

8 **Prove that $\frac{\sin 2x}{\cos 2x+1}=\tan x$.**

9 **Prove that $\frac{\sin(x+\pi)}{\cos(x+\pi)}=\tan x$.**

10 **Solve for x in $\cos\left(\frac{x}{3}+\frac{\pi}{4}\right)=\frac{\sqrt{2}}{2}$ for $0<x<2\pi$.**

11 **Solve $\sin x\cos x=-\frac{\sqrt{3}}{2}$ for all x where $0<x<180°$.**

12 **Find the value of $\sin\frac{11\pi}{6}\cos\frac{4\pi}{3}$.**

Answers to Chapter 13 Quiz

1. $\frac{\sqrt{2}+\sqrt{6}}{4}$**. Using the cosine of the difference of angles with 45° and 60°:**

$$\cos(-15°)=\cos(45°-60°)=\cos 45°\cos 60°+\sin 45°\sin 60°=\frac{\sqrt{2}}{2}\cdot\frac{1}{2}+\frac{\sqrt{2}}{2}\cdot\frac{\sqrt{3}}{2}=\frac{\sqrt{2}+\sqrt{6}}{4}$$

2. $\frac{\sqrt{6}}{2}$**. Using the formula for the sum of sines:**

$$\sin\frac{7\pi}{12}+\sin\frac{\pi}{12}=2\sin\left(\frac{\frac{7\pi}{12}+\frac{\pi}{12}}{2}\right)\cos\left(\frac{\frac{7\pi}{12}-\frac{\pi}{12}}{2}\right)=2\sin\left(\frac{\frac{8\pi}{12}}{2}\right)\cos\left(\frac{\frac{6\pi}{12}}{2}\right)$$

$$=2\sin\left(\frac{\pi}{3}\right)\cos\left(\frac{\pi}{4}\right)=\cancel{2}\left(\frac{\sqrt{3}}{\cancel{2}}\right)\left(\frac{\sqrt{2}}{2}\right)=\frac{\sqrt{6}}{2}$$

3. $\cos 2\theta$**. This can be accomplished using the power-reducing identities, but an even easier way is to first factor the difference of squares, apply the Pythagorean identity, and then recognize that the result is the double-angle identity:**

$$\cos^4\theta-\sin^4\theta=(\cos^2\theta-\sin^2\theta)(\cos^2\theta+\sin^2\theta)=(\cos^2\theta-\sin^2\theta)\cdot 1=\cos 2\theta$$

An alternative would be to factor the difference of squares and create a product of two binomials. Here's what that would look like:

$$\cos^4\theta-\sin^4\theta=(\cos^2\theta-\sin^2\theta)(\cos^2\theta+\sin^2\theta)=(\cos^2\theta-\sin^2\theta)\cdot 1=(\cos\theta-\sin\theta)(\cos\theta+\sin\theta)$$

4. $\frac{\sqrt{2+\sqrt{2}}}{2}$**. Apply the half-angle identity on** $\sin\frac{5\pi}{4}$**. You get** $\sin\frac{5\pi/4}{2}=\pm\sqrt{\frac{1-\cos\frac{5\pi}{4}}{2}}=$

$\pm\sqrt{\frac{1-\left(-\frac{\sqrt{2}}{2}\right)}{2}}=\pm\sqrt{\frac{2+\sqrt{2}}{4}}=\pm\frac{\sqrt{2+\sqrt{2}}}{2}$**. Since** $\frac{5\pi}{8}$ **is in quadrant II, you choose the positive value.**

5. $\sqrt{3}$**. Using the double-angle identity,** $\tan 2x=\frac{2\tan x}{1-\tan^2 x}$**. Since cotangent is the reciprocal of tangent, when** $\cot x=\sqrt{3}$, $\tan x=\frac{1}{\sqrt{3}}=\frac{\sqrt{3}}{3}$**. Substituting into the formula,**

$$\tan 2x=\frac{2\left(\frac{\sqrt{3}}{3}\right)}{1-\left(\frac{\sqrt{3}}{3}\right)^2}=\frac{2\left(\frac{\sqrt{3}}{3}\right)}{1-\frac{1}{3}}=\frac{\frac{2\sqrt{3}}{3}}{\frac{2}{3}}=\frac{\cancel{2}\sqrt{3}}{\cancel{3}}\cdot\frac{\cancel{3}}{\cancel{2}}=\sqrt{3}.$$

6. **45°. Square both sides of the equation, apply the Pythagorean identity, simplify, and apply the double-angle identity:**

$$(\sin x+\cos x)^2=(\sqrt{2})^2\to\sin^2 x+2\sin x\cos x+\cos^2 x=2\to 1+2\sin x\cos x$$
$$=2\to 2\sin x\cos x=1\to\sin 2x=1$$

The sine is equal to 1 when $x=90°$**. If** $2x=90°$**, then** $x=45°$**.**

7 $\frac{\sqrt{2}-\sqrt{6}}{4}$**.** **Applying the sine of the sum of angles using the angles 150° and 45°:**

$$\sin(195°)=\sin(150°+45°)=\sin 150° \cos 45° + \cos 150° \sin 45°$$

$$=\frac{1}{2}\cdot\frac{\sqrt{2}}{2}+\left(-\frac{\sqrt{3}}{2}\right)\cdot\frac{\sqrt{2}}{2}=\frac{\sqrt{2}-\sqrt{6}}{4}$$

8 $\mathbf{\tan x = \tan x}$**.** Apply the two double-angle formulas, simplify the denominator, reduce the fraction, and apply the ratio identity: $\frac{\sin 2x}{\cos 2x+1}=\tan x \to \frac{2\sin x\cos x}{2\cos^2 x-1+1}\to$ $\frac{2\sin x \cos x}{2\cos^2 x}\to\frac{\cancel{2}\sin x\cancel{\cos x}}{\cancel{2}\cos^{\cancel{2}} x}\to\frac{\sin x}{\cos x}=\tan x.$

9 $\mathbf{\tan x = \tan x}$**.** First, apply the sum-of-angles formulas, then evaluate the functions, and, finally, apply the ratio identity:

$$\frac{\sin(x+\pi)}{\cos(x+\pi)}=\frac{\sin x\cos\pi+\cos x\sin\pi}{\cos x\cos\pi-\sin x\sin\pi}=\frac{\sin x(-1)+\cos x(0)}{\cos x(-1)-\sin x(0)}=\frac{-\sin x}{-\cos x}=\tan x$$

10 **0,** $\frac{3\pi}{2}$**.** First, apply the sum-of-angles formula. Then apply the Pythagorean identity and simplify:

$$\cos\left(\frac{x}{3}+\frac{\pi}{4}\right)=\frac{\sqrt{2}}{2}\to\cos\frac{x}{3}\cos\frac{\pi}{4}-\sin\frac{x}{3}\sin\frac{\pi}{4}=\frac{\sqrt{2}}{2}\to\cos\frac{x}{3}\left(\frac{\sqrt{2}}{2}\right)-\sin\frac{x}{3}\left(\frac{\sqrt{2}}{2}\right)=\frac{\sqrt{2}}{2}\to\cos\frac{x}{3}-\sin\frac{x}{3}=1$$

Square both sides, apply the Pythagorean identity, and then apply the double-angle identity:

$$\left(\cos\frac{x}{3}-\sin\frac{x}{3}\right)^2=1^2\to\cos^2\frac{x}{3}-2\cos\frac{x}{3}\sin\frac{x}{3}+\sin^2\frac{x}{3}=1\to 1-2\cos\frac{x}{3}\sin\frac{x}{3}=1$$

$$\to -2\cos\frac{x}{3}\sin\frac{x}{3}=0\to 2\cos\frac{x}{3}\sin\frac{x}{3}=0\to\sin 2\left(\frac{x}{3}\right)=0$$

The sine is equal to 0 at 0 and π. So, setting $2\left(\frac{x}{3}\right)=0$, you have $x=0$. Setting $2\left(\frac{x}{3}\right)=\pi$, you find that $x=\frac{3\pi}{2}$.

11 **120°, 150°.** You can use the product-to-sum formula, or you can recognize that the left side is a version of the sine double-angle formula. Using the second choice, first multiply each side by 2 and then apply the double-angle formula: $\sin x\cos x=-\frac{\sqrt{3}}{2}\to 2\sin x\cos x=$ $-\sqrt{3}\to\sin 2x=-\sqrt{3}$. The sine equals $-\sqrt{3}$ at 240° and 300°. Solving for x: $2x=240°\to x=120°$ and $2x=300°\to x=150°$.

12 $\frac{1}{4}$. **Apply the product-as-sum identity, then simplify the new angles, applying the full period rule to the sine greater than 2π: $\sin\frac{11\pi}{6}\cos\frac{4\pi}{3}=\frac{1}{2}\left(\sin\left(\frac{11\pi}{6}+\frac{4\pi}{3}\right)+\sin\left(\frac{11\pi}{6}-\frac{4\pi}{3}\right)\right)=$**

$$\frac{1}{2}\left(\sin\left(\frac{19\pi}{6}\right)+\sin\left(\frac{3\pi}{6}\right)\right)=\frac{1}{2}\left(\sin\left(\frac{7\pi}{6}\right)+\sin\left(\frac{\pi}{2}\right)\right)$$

Now evaluate the functions and simplify:

$$\frac{1}{2}\left(\sin\left(\frac{7\pi}{6}\right)+\sin\left(\frac{\pi}{2}\right)\right)=\frac{1}{2}\left(-\frac{1}{2}+1\right)=\frac{1}{2}\left(\frac{1}{2}\right)=\frac{1}{4}$$

IN THIS CHAPTER

» Mastering and using the Law of Sines

» Wielding the Law of Cosines

» Utilizing two methods to find the area of triangles

Chapter 14 Getting the Slant on Oblique Triangles

In order to *solve* a triangle, you need to find the measures of all three angles and the lengths of all three sides. You're given as many as three of these pieces of information and you need to find the rest. So far, in this book, you've seen mainly right triangles. In Chapter 8, you find the lengths of missing sides of a right triangle using the Pythagorean Theorem, find missing angles using right-triangle trigonometry, and evaluate trig functions for specific angles. But what happens if you need to solve a triangle that isn't a right triangle?

You can connect any three points in a plane to form a triangle. Then comes the fun of determining all the measures of that triangle. Finding missing angles and sides of *oblique triangles* (acute or obtuse triangles) can be more challenging because they don't have a right angle. And without a right angle, the triangle has no hypotenuse, which means the Pythagorean Theorem is useless. But don't worry; this chapter shows you the way. The Law of Sines and the Law of Cosines are two methods that you can use to solve for missing parts of oblique triangles. The proofs of both laws are long and complicated, and you don't need to concern yourself with them. Instead, use these laws as formulas in which you can plug in information given to you and then use algebra to solve for the missing pieces. Whether you use sines or cosines to solve the triangle, the types of information (sides or angles) given to you and their location on the triangle are factors that help you decide which method is the best to use.

TECHNICAL STUFF

You may be wondering why the Law of Tangents isn't discussed in this chapter. The reason is simple: You can solve every oblique triangle with either the Law of Sines or the Law of Cosines, which are far less complicated than the Law of Tangents.

The techniques presented here have tons of real-world applications, too. You can deal with everything from sailing a boat to putting out a forest fire using triangles. For example, if two forest-fire stations get a call for a fire, they can use the Law of Cosines to figure out which station is closest to the fire.

TIP

Before attempting to solve a triangle, draw a picture that has the sides and angles clearly labeled. This approach helps you visualize which pieces of information you still need. You can use the Law of Sines whenever there's enough information provided in the problem to give you a *pair*: an angle measure and the side length across from it. So if you have the time, try to use the Law of Sines first. If the Law of Sines isn't an option, you'll know because you won't have both a given angle and its partner side. When that happens, the Law of Cosines is there to save the day. The Law of Cosines is specifically needed when you have just the three sides provided with no angles or when you have two sides provided with the angle between them.

REMEMBER

Whether you're using the Law of Sines or the Law of Cosines to solve for missing parts of a triangle, don't haul out your calculator until the very end. Using the calculator too early gives you a greater rounding error in your final answers. So, for example, instead of evaluating the sines of all three angles and using the decimal approximations from the beginning, you should solve the equations and plug the final numeric expression into your calculator all at one time, at the end.

Solving a Triangle with the Law of Sines

You use the *Law of Sines* to find the missing parts of a triangle when you're given any three pieces of information involving at least one angle and the side directly opposite from it. This information comes in three forms.

- **ASA (angle-side-angle):** You're given two angles and the side in between them.
- **AAS (angle-angle-side):** You're given two angles and a non-included side.
- **SSA (side-side-angle):** You're given two sides and a non-included angle.

TECHNICAL STUFF

Following is the formula for the Law of Sines: $\frac{a}{\sin A} = \frac{b}{\sin B} = \frac{c}{\sin C}$.

The small-case letters indicate the sides opposite the angles with the corresponding capital letters. In order to solve for an unknown variable in the Law of Sines, you create a proportion by setting two of the fractions equal to one other and then using cross multiplication.

When using the Law of Sines, take your time and work carefully. Even though you may be tempted to try to solve everything at once, take it one small step at a time. And don't overlook the obvious in order to blindly stick with the formula. If you're given two angles and one side, for instance, finding the third angle is easy because all angles in a triangle must add up to 180°. Fill in the formula with what you know and get crackin'!

In the following sections, you see how to solve a triangle in different situations using the Law of Sines.

REMEMBER

When you solve for an angle by using the Law of Sines, there are instances where a second set of solutions (or none at all) may exist. Those instances are covered in this section as well. (In case you're really curious, these considerations apply only when you're working with a problem where you know two side measures and one angle measure of a triangle.)

When you know two angle measures

In this section, you see two cases where you can use the Law of Sines to solve a triangle: angle–side–angle (ASA) and angle–angle–side (AAS). Whenever you're given two angles, find the third one and work from there. In both of these cases, you find exactly one solution for the triangle in question.

ASA: All-Seems-Appropriate or Angle-Side-Angle?

An *ASA* triangle means that you're given two angles and the side between them in a problem.

EXAMPLE

Q. If $\angle A = 32°$, $\angle B = 47°$, and $c = 21$, as in Figure 14-1, find the missing information. You also could be given $\angle A$, $\angle C$, and b, or $\angle B$, $\angle C$, and a. Figure 14-1 has all the given and unknown parts labeled for you.

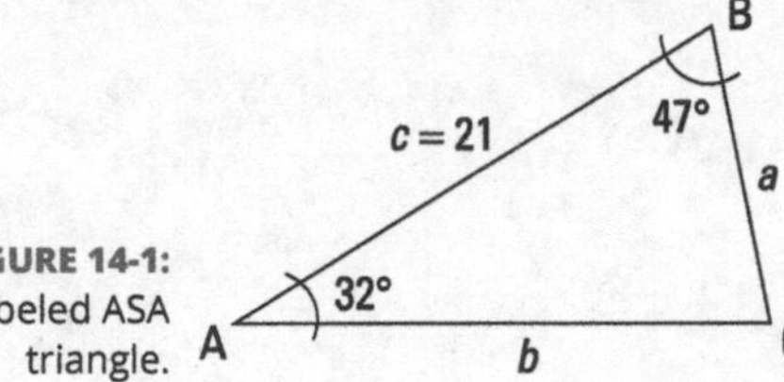

FIGURE 14-1: A labeled ASA triangle.

A. To find the missing information with the Law of Sines, follow these steps:

1. **Determine the measure of the third angle.**

   In triangle ABC, $\angle A + \angle B + \angle C = 180°$. So by plugging in what you know about the angles in this problem, you can solve for the missing angle:

   $$32° + 47° + \angle C = 180°$$
   $$\angle C = 180° - 32° - 47° = 101°$$

2. **Set up the Law of Sines formula, filling in what you know.**

   $$\frac{a}{\sin 32°} = \frac{b}{\sin 47°} = \frac{21}{\sin 101°}$$

3. **Set two of the parts equal to each other and cross-multiply.**

   Use the first and third fractions: $\frac{a}{\sin 32°} = \frac{21}{\sin 101°}$.

   Cross-multiplying, you have $a \sin 101° = 21 \sin 32°$.

4. **Solve for the value of side *a*.**

   $$a = \frac{21 \sin 32°}{\sin 101°} \approx 11.33658963$$

   Round your answer to two decimal places. You have $a = 11.34$.

5. **Repeat Steps 3 and 4 to solve for the other missing side.**

Setting the second and third fractions equal to each other: $\frac{b}{\sin 47°} = \frac{21}{\sin 101°}$.

This equation becomes $b\sin 101° = 21\sin 47°$ when you cross-multiply. Isolate the variable and solve for it:

$$b = \frac{21\sin 47°}{\sin 101°}$$

$$b \approx 15.65$$

6. **State all the parts of the triangle as your final answer.**

Some answers may be approximate, so make sure you maintain the proper signs:

- $\angle A = 32°$, $a \approx 11.34$
- $\angle B = 47°$, $b \approx 15.65$
- $\angle C = 101°$, $c = 21$

YOUR TURN

 Solve the triangle if $B = 46°$, $C = 62°$, and $a = 21$.

 Solve the triangle if $A = 19°$, $C = 100°$, and $b = 4.4$.

AAS: Algebra-Attracts-Students or Angle-Angle-Side?

In many trig problems, you're given two angles and a side that isn't between them. This type of problem is called an *AAS* problem. For example, you may be given $\angle B = 68°$, $\angle C = 29°$, and $b = 15.2$, as shown in Figure 14-2. Notice that if you start at side b and move counterclockwise around the triangle, you come to $\angle C$ and then $\angle B$. This check is a good way to verify whether a triangle is an example of AAS.

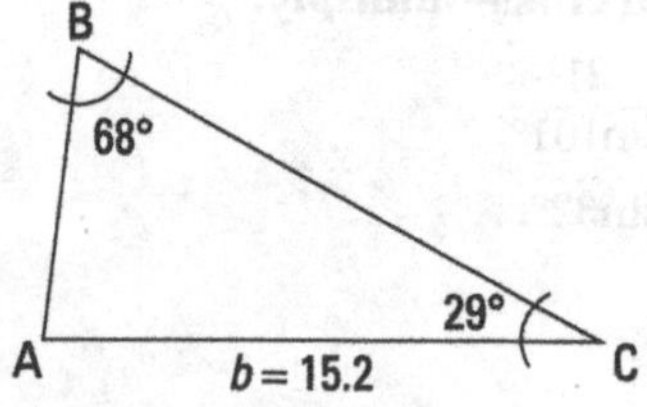

FIGURE 14-2: A labeled AAS triangle.

After you find the third angle, an AAS problem just becomes a special case of ASA. Here are the steps to solve it:

1. **Determine the measure of the third angle.**

   In triangle ABC, $\angle A + 68° + 29° = 180°$. This means $\angle A = 83°$.

2. **Set up the Law of Sines formula, filling in what you know.**

   $$\frac{a}{\sin 83°} = \frac{15.2}{\sin 68°} = \frac{c}{\sin 29°}$$

3. **Set two of the parts equal to each other and then cross-multiply.**

   Choosing the first two fractions: $\frac{a}{\sin 83°} = \frac{15.2}{\sin 68°}$.

   Cross-multiplying, you have $a\sin 68° = 15.2\sin 83°$.

4. **Solve for the missing side.**

   You divide by sin68°, so

   $$a = \frac{15.2\sin 83°}{\sin 68°}$$
   $$a \approx 16.27$$

5. **Repeat Steps 3 and 4 to solve for the other missing side.**

   Setting the second and third fractions equal to each other, you have: $\frac{15.2}{\sin 68°} = \frac{c}{\sin 29°}$.

   Cross-multiply: $15.2\sin 29° = c\sin 68°$.

   Solve for c:

   $$c = \frac{15.2\sin 29°}{\sin 68°}$$
   $$c \approx 7.95$$

6. **State all the parts of the triangle as your final answer.**

   Your final answer sets up as follows:

   1. $\angle A = 83°$, $a \approx 16.27$
   2. $\angle B = 68°$, $b = 15.2$
   3. $\angle C = 29°$, $c \approx 7.95$

YOUR TURN

3 Solve the triangle if $A = 49°$, $B = 21°$, and $a = 5$.

4 Solve the triangle if $A = 110°$, $C = 56°$, and $a = 8$.

When you know two consecutive side lengths

In some trig problems, you may be given two sides of a triangle and an angle that isn't between them, which is the classic case of *SSA*. In this scenario, you may have one solution, two solutions, or no solution.

REMEMBER

Wondering why the number of solutions varies? Recall from geometry that you can't prove that two triangles are congruent using SSA, because these conditions can build you two triangles that aren't the same. Figure 14-3 shows two triangles that fit SSA but aren't congruent.

If you begin with an angle and then continue around to draw the other two sides, you'll find that sometimes you can't make a triangle with those measurements. And sometimes, you can make two different triangles. Unfortunately, the latter means actually solving both of those two different triangles.

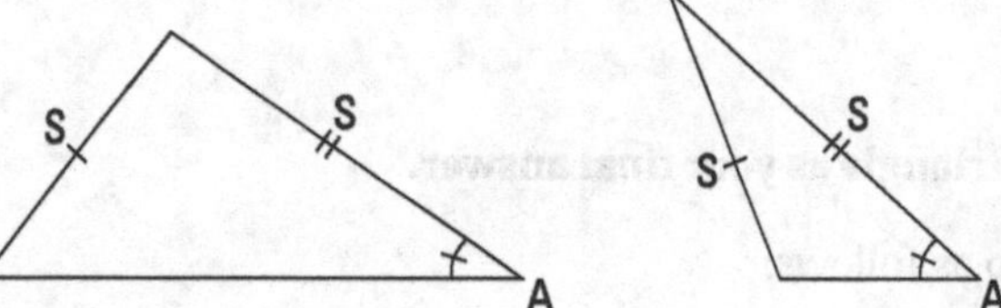

FIGURE 14-3: Non-congruent triangles that follow the SSA format.

Most SSA cases have only one solution, because if you use what you're given to sketch the triangle, there's usually only one way to draw it. When you're faced with an SSA problem, you may be tempted to figure out how many solutions you need to find before you start the solving process. Not so fast! In order to determine the number of possible solutions in an SSA problem, you should start solving first. You'll either arrive at one solution or find that no solution exists. If you find one solution, you can look for the second set of measures. If you get a negative angle in the second set, you'll know that the triangle only has one solution.

TIP

The best approach is to always assume that you'll find two solutions, because remembering all the rules that determine the number of solutions probably will take up far too much time and energy. If you treat every SSA problem as if it has two solutions until you gather enough information to prove otherwise, you'll more easily find all the appropriate solutions.

Doubling the fun: Two solutions

Gaining some experience with solving a triangle that has more than one solution is helpful. The first set of measures that you find in such a situation is always from an acute triangle. The second set of measures is from an obtuse triangle. Remember to always look for two solutions for any problem.

EXAMPLE

Q. Find the other three measures of the triangle ABC when you're given $a = 16$, $c = 20$, and $\angle A = 48°$.

A. Figure 14-4a shows you what the picture may look like. However, couldn't the triangle also look like Figure 14-4b? Both situations follow the constraints of the given information of the triangle. If you start by drawing your picture with the given angle, the side next to the angle has a length of 20, and the side across from the angle is 16 units long. The triangle could be formed two different ways. Angle C could be an acute angle or an obtuse angle; the given information isn't restrictive enough to tell you which one it is. Therefore, you get to find both solutions.

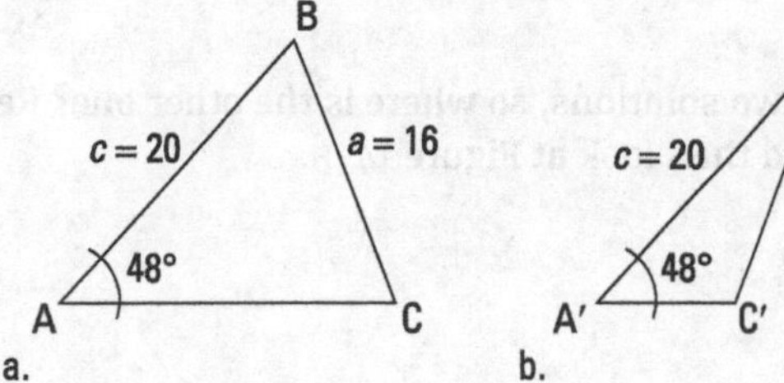

FIGURE 14-4: Two possible representations of an SSA triangle.

Solving this triangle using steps similar to those described for both the ASA and AAS cases gives you the two possible solutions shown in Figure 14-4. Because you have two missing angles, you need to find one of them first, which is why the steps here are different than the other two cases:

1. **Fill in the Law of Sines formula with what you know.**

   The formula here sets up like this: $\frac{16}{\sin 48°} = \frac{b}{\sin B} = \frac{20}{\sin C}$.

2. **Set two fractions equal to each other so that you have only one unknown.**

   If you decide to solve for $\angle C$, you set the first and third fractions equal to each other, so you have:

   $$\frac{16}{\sin 48°} = \frac{20}{\sin C}$$

3. **Cross-multiply and isolate the sine function.**

   This step gives you $16\sin C = 20\sin 48°$. To isolate the sine function, you divide by 16:

   $$\sin C = \frac{20\sin 48°}{16} = \frac{5\sin 48°}{4}$$

4. **Take the inverse sine of both sides.**

$$\sin^{-1}(\sin C) = \sin^{-1}\left(\frac{5\sin 48°}{4}\right)$$

The right-hand side goes right into your handy calculator to give you $\angle C \approx 68.27°$.

5. **Determine the third angle.**

You know that $48° + \angle B + 68.27 = 180°$, so $\angle B \approx 63.73°$.

6. **Plug the final angle back into the Law of Sines formula to find the third side.**

$$\frac{16}{\sin 48°} = \frac{b}{\sin 63.73}$$

This gives you $16\sin 63.73 = b\sin 48°$.

Finally, solving for b:

$$b = \frac{16\sin 63.73}{\sin 48°} \approx 19.31$$

But you're supposed to get two solutions, so where is the other one? Refer to Step 4, where you solved for $\angle C$, and then look at Figure 14-5.

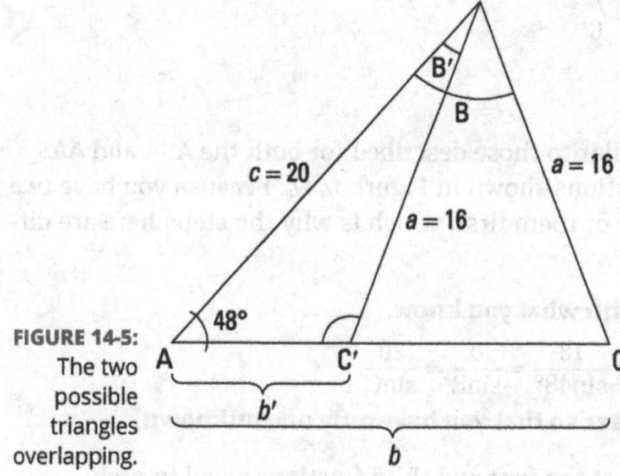

FIGURE 14-5: The two possible triangles overlapping.

REMEMBER

Solving triangle ABC is what you did in the previous steps. Triangle AB'C' has the second set of measures you must look for. A certain trig identity that isn't used in solving or simplifying trig expressions because it isn't helpful for that, is now used for solving triangles. This identity says that $\sin(180° - \theta) = \sin\theta$.

In the case of this triangle, $\sin(180° - 68.27°) = \sin 111.73°$. Checking this out using your calculator, $\sin 68.27° = \sin 111.73° \approx 0.9319$. However, if you plug $\sin^{-1}(0.9319)$ into your calculator to solve for θ, an acute angle is the only solution you get. You find the other angle by subtracting from 180°.

TECHNICAL STUFF

Another way to say this is that sine is positive in both quadrants I and II. So there are two angles that have the same sine — two angles that could fit in the triangle you're solving. You have to check to see if both work or just one of them.

The following steps build on these actions so you can find the second solution for this SSA problem:

1. **Use the trig identity $\sin(180° - \theta) = \sin\theta$ to find the second angle of the second triangle.**

   Because $\angle C \approx 68.27°$, subtract this value from 180° to find that $\angle C' \approx 111.73°$.

2. **Find the measure of the third angle.**

   If $\angle A = 48°$ and $\angle C' \approx 111.73°$, then $\angle B' \approx 20.27°$, because the three angles must add to 180°.

3. **Plug these angle values into the Law of Sines formula.**

   $$\frac{16}{\sin 48°} = \frac{b'}{\sin 20.27°} = \frac{20}{\sin 111.73°}$$

4. **Set two parts equal to each other in the formula.**

   You need to find b', so set the first fraction equal to the second:

   $$\frac{16}{\sin 48°} = \frac{b'}{\sin 20.27°}$$

5. **Cross-multiply to solve for the variable.**

   $$16\sin 20.27° = b'\sin 48°$$

   $$b' = \frac{16\sin 20.27°}{\sin 48°} \approx 7.46$$

6. **List *all* the measures of the two triangles (see the previous numbered steps).**

   Originally, you were given that $a = 16$, $c = 20$, and $\angle A = 48°$. The two solutions are as follows.

   - **First triangle:** $\angle A = 48°$, $a = 16$, $\angle B \approx 63.73°$, $b = 19.31$, $\angle C \approx 68.27°$, $c = 20$
   - **Second triangle:** $\angle A = 48°$, $a = 16$, $\angle B' \approx 20.27°$, $b' \approx 7.46$, $\angle C' \approx 111.73°$, $c = 20$

Arriving at the conclusion: Just one solution

If you don't get an error message in your calculator when attempting to solve a triangle, you know you can find at least one solution. But how do you know if you'll find only one? The answer is, you don't. Keep solving as if a second one exists, and in the end, you will see whether there's only one.

EXAMPLE

Q. Solve the triangle if $a = 19$, $b = 14$, and $\angle A = 35°$. Figure 14-6 shows what this triangle looks like.

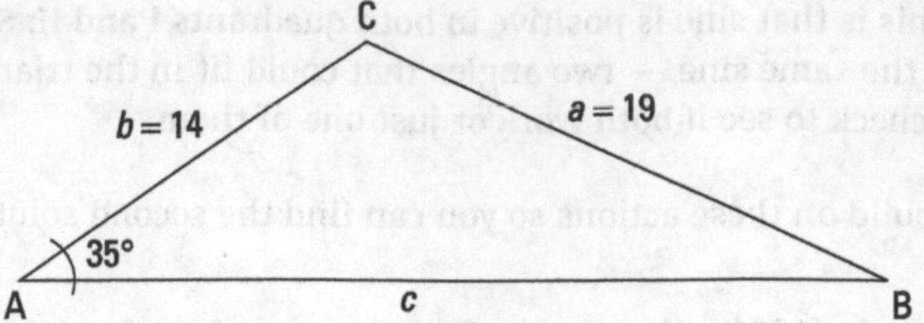

FIGURE 14-6: The setup of an SSA triangle with only one solution set.

Because you know only one of the angles of the triangle, you use the two given sides and the given angle to find one of the missing angles first. That process leads you to the third angle and then the third side.

A. Follow these steps to solve this triangle:

1. **Fill in the Law of Sines formula with what you know.**

   $$\frac{19}{\sin 35°} = \frac{14}{\sin B} = \frac{c}{\sin C}$$

2. **Set two parts of the formula equal to each other.**

TIP

   Because you're given *a*, *b*, and A, solve for ∠B. Setting the first and second fractions equal to one another,

   $$\frac{19}{\sin 35°} = \frac{14}{\sin B}$$

3. **Cross-multiply.**

   $$19\sin B = 14\sin 35°$$

4. **Isolate the sine function.**

   $$\sin B = \frac{14\sin 35°}{19}$$

5. **Take the inverse sine of both sides of the equation.**

   You get $\sin^{-1}(\sin B) = \sin^{-1}\left(\frac{14\sin 35°}{19}\right)$, which simplifies to

   $$B = \sin^{-1}(0.4226352689) \text{ or}$$

   $$B \approx 25.00°$$

6. **Determine the measure of the third angle.**

   With $35° + 25° + \angle B = 180°$, you determine that $\angle C = 120°$.

7. **Use the first and third fractions to solve for c.**

   $$\frac{19}{\sin 35°} = \frac{c}{\sin 120°}$$

8. **Cross-multiply and then isolate the variable to solve.**

   $$19\sin 120° = c\sin 35°$$

   $$c = \frac{19\sin 120°}{\sin 35°}$$

   $$c \approx 28.69$$

9. **Write out all six pieces of information devised from the formula.**

   Your answer sets up as follows:

   $\angle A = 35°,\ a = 19$

   $\angle B = 25°,\ b = 14$

   $\angle C = 120°,\ c \approx 28.69$

10. **Look for a second set of measures.**

    The first thing you did in this example was to find ∠B. You see from Step 5 that B is approximately 25°. If the triangle has two solutions, the measure of B′ is 180° − 25°, or 155°. Then, to find the measure of angle C′, you start with $\angle A + \angle B' + \angle C' = 180°$. This equation becomes $35° + 155° + \angle C' = 180°$, or $\angle C' = -10°$. Angles in triangles can't have negative measures, so this answer tells you that the triangle has only one solution. Don't you feel better knowing that you exhausted the possibilities?

Kind of a pain: No solutions

If a problem gives you an angle and two consecutive sides of a triangle, you may find that the second side won't be long enough to reach the third side of the triangle. In this situation, no solution exists for the problem. However, you may not be able to tell this just by looking at the picture — you really need to solve the problem to know for sure. So begin to solve the triangle just as you do in the previous sections.

EXAMPLE

Q. Solve the triangle where $b = 19$, $\angle A = 35°$, and $a = 10$. Figure 14-7 shows what the picture should look like.

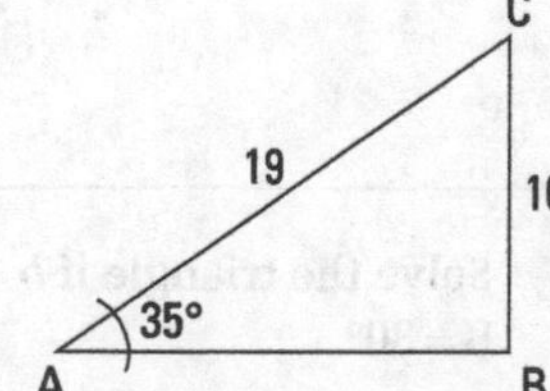

FIGURE 14-7: A triangle with no solution.

If you start solving this triangle using the methods from previous setups, something very interesting happens: Your calculator gives you an error message when you try to find the unknown angle. This error is because the sine of an angle must be between −1 and 1. If you try to take the inverse sine of a number outside this interval, the value for the angle is undefined (meaning it doesn't exist).

A. The following steps illustrate this occurrence:

1. **Fill in the Law of Sines formula with what you know.**

   $$\frac{10}{\sin 35°} = \frac{19}{\sin B} = \frac{c}{\sin C}$$

2. **Set two fractions equal and cross-multiply.**

   You start with the equation, $\frac{10}{\sin 35°} = \frac{19}{\sin B}$, and end up with $10\sin B = 19\sin 35°$.

3. **Isolate the sine function.**

   $$\sin B = \frac{19\sin 35°}{10}$$

4. **Take the inverse sine of both sides to find the missing angle.**

   $$\sin^{-1}(\sin B) = \sin^{-1}\left(\frac{19\sin 35°}{10}\right)$$
   $$B = \sin^{-1}(1.089795229)$$

   You get an error message when you try to plug this into your calculator, which happens because the equation says that $\sin B \approx 1.09$. The sine of an angle can't be larger than 1 or less than −1. Therefore, the measurements given can't form a triangle, meaning that the problem has no solution.

YOUR TURN

 5 Solve the triangle if $b = 8$, $c = 14$, and $C = 37°$.

 6 Solve the triangle if $b = 5$, $c = 12$, and $B = 20°$.

 7 Solve the triangle if $a = 10$, $c = 24$, and $A = 102°$.

 8 Solve the triangle if $b = 8$, $c = 16$, and $B = 30°$.

Conquering a Triangle with the Law of Cosines

You use the *Law of Cosines* formulas to solve a triangle if you're given one of the following situations:

- Two sides and the included angle (SAS)
- All three sides of the triangle (SSS)

REMEMBER

In order to solve for the angles of a triangle by using the Law of Cosines, you first need to find the lengths of all three sides. You have three formulas at your disposal to find missing sides, and three formulas to find missing angles. If a problem gives you all three sides to begin with, you're all set because you can manipulate the side formulas to come up with the angle formulas (you see how in the following section). If a problem gives you two sides and the angle between them, you first find the missing side and then find the missing angles.

TECHNICAL STUFF

To find a missing side of a triangle (a, b, or c), use the following formulas, which comprise the Law of Cosines:

- $a^2 = b^2 + c^2 - 2bc\cos\text{A}$
- $b^2 = a^2 + c^2 - 2ac\cos\text{B}$
- $c^2 = a^2 + b^2 - 2ab\cos\text{C}$

The side formulas are very similar to one another, with only the letters changed around. So if you can remember just two of them, you can change the order to quickly find the other. The following sections put the Law of Cosines formulas into action to solve SSS and SAS triangles.

REMEMBER

When you use the Law of Cosines to solve a triangle, you find only one set of solutions (one triangle), so don't waste any time looking for a second set. With this formula, you're solving SSS and SAS triangles from triangle congruence postulates in geometry. You can use these congruence postulates because they lead to only one triangle every time. (For more on geometry rules, check out *Geometry For Dummies*, by Mark Ryan [Wiley].)

SSS: Finding angles using only sides

The Law of Cosines provides formulas that solve for the lengths of the sides. With some clever manipulations, you can change these formulas to ones that give you the measures of the angles (as shown in the following steps). You really shouldn't worry about memorizing all six formulas. If you remember the formulas to find missing sides using the Law of Cosines, you can use algebra to solve for an angle. For example, here's how to solve for angle A:

1. **Start with the formula.**

   $$a^2 = b^2 + c^2 - 2bc\cos\text{A}$$

2. **Subtract b^2 and c^2 from both sides.**

   $$a^2 - b^2 - c^2 = -2bc\cos\text{A}$$

3. **Divide both sides by $-2bc$.**

$$\frac{a^2-b^2-c^2}{-2bc}=\frac{\cancel{-2bc}\cos A}{\cancel{-2bc}}$$

4. **Simplify and rearrange terms.**

$$\frac{b^2+c^2-a^2}{2bc}=\cos A$$

5. **Take the inverse cosine of both sides.**

$$\cos^{-1}(\cos A)=\cos^{-1}\left(\frac{b^2+c^2-a^2}{2bc}\right)\to A=\cos^{-1}\left(\frac{b^2+c^2-a^2}{2bc}\right)$$

TECHNICAL STUFF

The same process applies to finding angles B and C, so you end up with these formulas for finding angles:

- $A=\cos^{-1}\left(\frac{b^2+c^2-a^2}{2bc}\right)$
- $B=\cos^{-1}\left(\frac{a^2+c^2-b^2}{2ac}\right)$
- $C=\cos^{-1}\left(\frac{a^2+b^2-c^2}{2ab}\right)$

EXAMPLE

Q. Suppose you have three pieces of wood of different lengths. One board is 12 feet long, another is 9 feet long, and the last is 4 feet long. If you want to build a flower box using these pieces of wood, at what angles must you lay down all the pieces so that the sides meet?

A. If each piece of wood is one side of the triangular flower box, you can use the Law of Cosines to solve for the three missing angles.

Let $a=12$, $b=4$, and $c=9$. You can find any of the angles first. Follow these steps to solve:

1. **Decide which angle you want to solve for first and then plug the sides into the formula.**

   Solving for angle A:

$$A=\cos^{-1}\left(\frac{b^2+c^2-a^2}{2bc}\right)=\cos^{-1}\left(\frac{4^2+9^2-12^2}{2(4)(9)}\right)=\cos^{-1}\left(\frac{-47}{72}\right)\approx 130.75°$$

2. **Solve for the other two angles.**

   Angle B:

$$B=\cos^{-1}\left(\frac{12^2+9^2-4^2}{2(12)(9)}\right)=\cos^{-1}\left(\frac{209}{216}\right)\approx 14.63°$$

Angle C:

$$C = \cos^{-1}\left(\frac{12^2+4^2-9^2}{2(12)(4)}\right) = \cos^{-1}\left(\frac{79}{96}\right) \approx 34.62°$$

3. **Check your answers by adding the angles you found.**

   You find that $130.75° + 14.63° + 34.62° = 180°$.

   Note: You could have found angle C by adding A and B together and subtracting from 180°.

   It's always a good idea to use Step 3 as a check after finding the angle measures using the formula.

   Picturing your solutions, the angle across from the 12-foot board ($\angle A$) needs to be 130.75°; the angle across from the 4-foot board ($\angle B$) needs to be 14.63°; and the angle across from the 9-foot board ($\angle C$) needs to be 34.62°. See Figure 14-8.

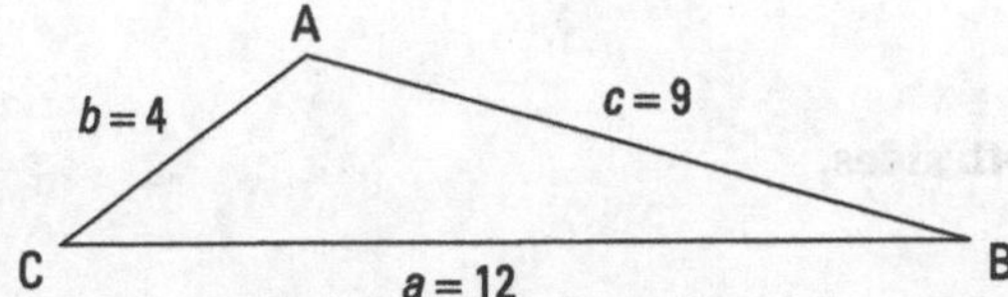

FIGURE 14-8: Determining angles when you know three side lengths.

SAS: Tagging the angle in the middle (and the two sides)

If a problem gives you the lengths of two sides of a triangle and the measure of the angle in between them, you use the Law of Cosines to find the other side (which you need to do first). When you have the third side, you can easily use all the side measures to calculate the remaining angle measures.

EXAMPLE

Q. Solve the triangle if $a = 12$, $b = 23$, and $\angle C = 39°$.

A. You solve for side c first and then solve for $\angle A$ and $\angle B$. Follow these simple steps:

1. **Sketch a picture of the triangle and clearly label all given sides and angles.**

   By drawing a picture, you can make sure that the Law of Cosines is the method you should use to solve the triangle. Figure 14-9 has all the parts labeled.

2. **Decide which side formula you need to use first.**

   Because sides a and b are given, you use the following formula to find side c:

   $$c^2 = a^2 + b^2 - 2ab\cos C$$

3. **Plug the given information into the proper formula.**

   This step gives you this equation: $c^2 = 12^2 + 23^2 - 2(12)(23)\cos 39°$.

If you have a graphing calculator, you can plug in this formula exactly as it's written and then skip directly to Step 6. If you don't have a graphing calculator (meaning you're using a scientific calculator), be very mindful of the order of operations.

WARNING

Following the order of operations when using the Law of Cosines is extremely important. If you try to type the pieces into your calculator all in one step, without the correct use of parentheses, your results could be incorrect. Be sure you're comfortable with your calculator. Some scientific calculators require that you type in the degrees before you hit the trig-function button. If you try to type in an inverse cosine without parentheses to separate the top and bottom of the fraction (see Step 7), your answer will be incorrect as well. The best method is to do all the squaring separately, combine like terms in the numerator and the denominator, divide the fraction, and then take the inverse cosine, as you'll see from this point on.

4. **Square each number and multiply by the cosine separately.**

   You end up with this equation: $c^2 = 144 + 529 - 428.985$.

5. **Combine all the numbers.**

   $c^2 = 244.015$

6. **Find the square root of both sides.**

   $c \approx 15.6$

7. **Find the missing angles.**

   Starting with ∠A, you set up the formula (from the preceding section), plug in the info you know, and solve for A:

   $A = \cos^{-1}\left(\frac{b^2 + c^2 - a^2}{2bc}\right)$, giving you that

   $$A = \cos^{-1}\left(\frac{23^2 + 15.6^2 - 12^2}{2(23)(15.6)}\right) = \cos^{-1}\left(\frac{628.36}{717.6}\right) \approx 28.9$$

REMEMBER

When using your graphing calculator, be sure to use parentheses to separate the numerator and denominator from each other. Put parentheses around the whole numerator *and* the whole denominator.

TIP

You can find the third angle quickly by subtracting the sum of the two known angles from 180°. However, if you're not pressed for time, it's recommended that you use the Law of Cosines to find the third angle because it allows you to check your answer.

   Here's how you find ∠B:

   $B = \cos^{-1}\left(\frac{a^2 + c^2 - b^2}{2ac}\right)$, which gives you that

   $$B = \cos^{-1}\left(\frac{12^2 + 15.6^2 - 23^2}{2(12)(15.6)}\right) = \cos^{-1}\left(\frac{-141.64}{374.4}\right) \approx 112.2°$$

8. **Check to make sure that all angles add to 180°.**

   $39° + 112.2° + 28.9° = 180.1°$

Because of a rounding error, sometimes the angles don't add to 180° exactly. Since measures were rounded to the nearest tenth, an error like this is not unexpected.

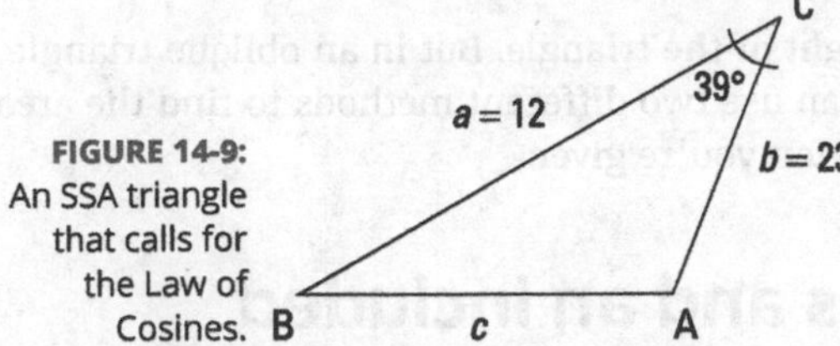

FIGURE 14-9: An SSA triangle that calls for the Law of Cosines.

YOUR TURN

9 **Solve the triangle if $C = 120°$, $a = 6$, and $b = 10$.**

10 **Solve the triangle if $A = 70°$, $b = 6$, and $c = 7$.**

11 **Solve the triangle if $a = 9$, $b = 5$, and $c = 7$.**

12 **Solve the triangle if $a = 7.3$, $b = 9.9$, and $c = 16$.**

Filling in the Triangle by Calculating Area

Geometry provides a nice formula to find the area of triangles: $A=\frac{1}{2}bh$. This formula comes in handy only when you know the base and the height of the triangle. But in an oblique triangle, where is the base? And what is the height? You can use two different methods to find the area of an oblique triangle, depending on the information you're given.

Finding area with two sides and an included angle (for SAS scenarios)

Lucky Mr. Heron (see the next section) got a formula named after him, but the SAS guy to whom this section is dedicated remains nameless.

You use the following formula to find the area when you know two sides of a triangle and the angle between those sides (SAS):

$$A=\frac{1}{2}ab\sin C$$

In the formula, C is the angle between sides a and b.

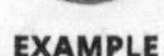

Q. Find the area of a triangle when you know that $a=12$, $b=4$, and $C=34.6°$.

A. Using the formula, you find the area:

$$A=\frac{1}{2}(12)(4)\sin 34.6° \approx 13.64$$

Using Heron's Formula (for SSS scenarios)

TECHNICAL STUFF

You can find the area of a triangle when given only the lengths of all three sides (no angles, in other words) by using a formula called *Heron's Formula*. It says that

$$A=\sqrt{s(s-a)(s-b)(s-c)} \text{ where } s=\frac{1}{2}(a+b+c)$$

The variable s is called the *semi-perimeter* — or half the perimeter.

EXAMPLE

Q. Find the area of the triangle with sides 4, 9, and 12.

A. Follow these steps:

1. **Calculate the semi-perimeter, s.**

   Follow this simple calculation:

   $$s=\frac{1}{2}(4+9+12)=\frac{1}{2}(25)=12.5$$

2. **Plug *s*, *a*, *b*, and *c* into Heron's Formula.**

$$A = \sqrt{12.5(12.5-4)(12.5-9)(12.5-12)} = \sqrt{12.5(8.5)(3.5)(0.5)} \approx 13.64$$

With problems like this, you'll have a choice between the SAS formula and Heron's Formula — whichever is more convenient.

YOUR TURN

13 **Find the area of the triangle where $a = 7$, $c = 17$, and $B = 68°$.**

14 **Find the area of the triangle on the coordinate plane with vertices $(-5,2)$, $(5,6)$, $(4,0)$.**

Making Triangles Work for You

There are many different instances in the real world where you need to know the length of the side of a triangle or the angle formed where two sides meet. When you draw out a picture of the situation you're dealing with, you sometimes discover that the problem has a triangle in it. The problems featured in this section aren't right triangles. They're *oblique triangles*, where you're looking for one missing piece of the puzzle or an area or perimeter.

Here are several opportunities to apply the formulas.

EXAMPLE

Q. **A plane flies for 300 miles in a straight line, makes a 45° turn, and continues for 700 miles. How far is it from its starting point?**

A. **Approximately 936.47 miles. First, draw out a picture like the one shown here. Notice that S is used for the starting point, T for the turning point, and E for the ending point.**

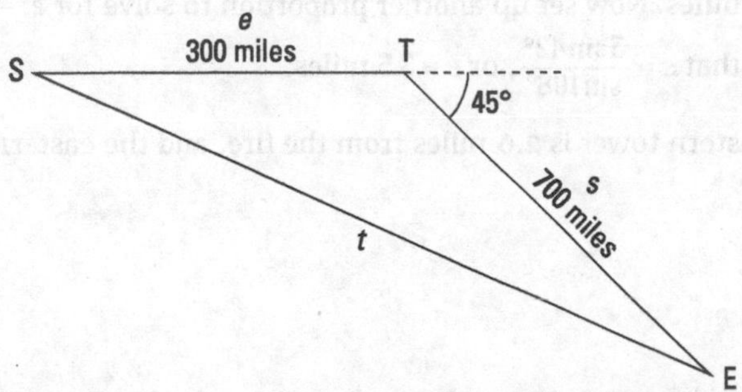

Now that you have the picture, you can figure out whether you need to use the Law of Sines or the Law of Cosines. Because the information given is SAS, you apply the Law of Cosines, using the different variables from the picture. You're solving for the value of t, the length of the side opposite angle T. You have $e = 300$, $s = 700$, and angle T is $180° - 45° = 135°$. Plugging into $t^2 = s^2 + e^2 - 2(s)(e)\cos T$, you have $t^2 = 700^2 + 300^2 - 2(700)(300)\cos 135°$. Using your calculator, you get that $t^2 \approx 876{,}984.8481$ and $t \approx 936.47$. So the distance is approximately 936.47 miles.

EXAMPLE

Q. Two fire towers are exactly 5 miles apart in a forest. Personnel in the towers both spot a forest fire, one at an angle of 30° and the other at an angle of 42°. Which tower is closer?

A. **The western fire tower is closer.** In this case, you need to find two missing pieces of information. You have to find how far both towers are from the fire in order to know which one is closer. First, draw out a figure like the one that follows.

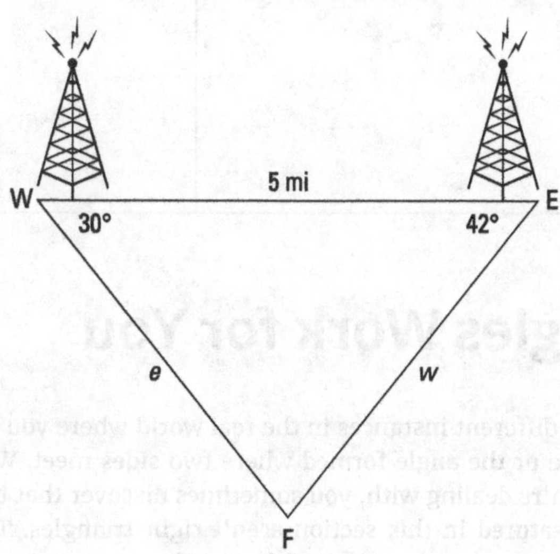

Use W for the western tower, E for the eastern tower, and F for the actual fire. You have a classic case of ASA, so you can use the Law of Sines this time. Knowing two of the angles makes it possible to find the third one easily: F = 108°. Now that you have the third angle, you can use the Law of Sines to form the needed proportions. You have $\frac{w}{\sin W} = \frac{e}{\sin E} = \frac{f}{\sin F}$. Filling in the known values, $\frac{w}{\sin 30°} = \frac{5}{\sin 108°}$. Using the first and third ratios and solving for o, $\frac{w}{\sin 30°} = \frac{e}{\sin 42°} = \frac{5}{\sin 108°}$, you have $w = \frac{5\sin 30°}{\sin 108°}$, or $w \approx 2.6$ miles. Now set up another proportion to solve for e: $\frac{e}{\sin 42°} = \frac{5}{\sin 108°}$, which means that $e = \frac{5\sin 42°}{\sin 108°}$, or $e \approx 3.5$ miles.

The western tower is 2.6 miles from the fire, and the eastern tower is 3.5 miles from the fire.

YOUR TURN

15 Two trains leave a station at the same time on different tracks that have an angle of 100° between them. If the first train is a passenger train that travels 90 miles per hour and the second train is a cargo train that can travel only 50 miles per hour, how far apart are the two trains after three hours?

16 A radio tower is built on top of a hill. The hill makes an angle of 15° with the ground. The tower is 200 feet tall and located 150 feet from the bottom of the hill. If a wire is to connect the top of the tower with the bottom of the hill, how long does the wire need to be?

17 A surveyor stands on one side of a river looking at a flagpole on an island at an angle of 85°. They then walk in a straight line for 100 meters, turn, and look back at the same flagpole at an angle of 40°. Find the distance from their first location to the flagpole.

18 Two scientists stand 350 feet apart, both looking at the same tree somewhere in between them. The first scientist measures an angle of 44° from the ground to the top of the tree. The second scientist measures an angle of 63° from the ground to the top of the tree. How tall is the tree?

Practice Questions Answers and Explanations

1. **A = 72°, *b* ≈ 15.9, *c* ≈ 19.5.** This one is ASA, so you use the Law of Sines to solve it. See the following figure.

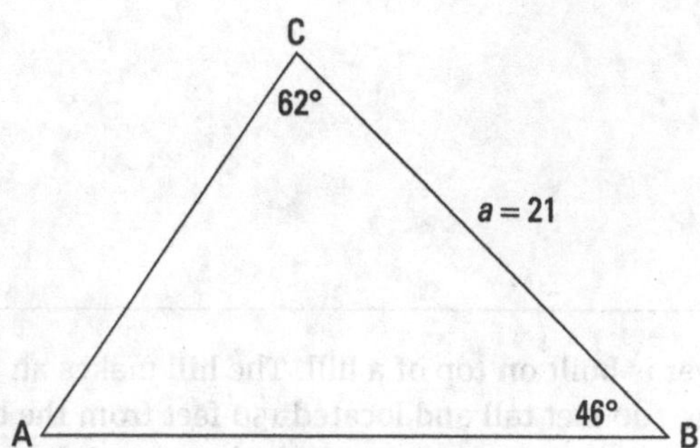

Because you already know two angles, begin by finding the third: $A = 180° - (46° + 62°) = 72°$. Now set up proportions from the Law of Sines to solve for the missing sides. From $\frac{21}{\sin 72°} = \frac{b}{\sin 46°}$, you have $b = \sin 46° \cdot \frac{21}{\sin 72°} \approx 15.9$. And from $\frac{21}{\sin 72°} = \frac{c}{\sin 62°}$, you get that $c \approx 19.5$.

2. **B = 61°, *a* ≈ 1.64, *c* ≈ 4.95.** Draw out the figure first. It's ASA, which means you use the Law of Sines to solve. Start by finding the missing angle: $B = 180° - (19° + 100°) = 61°$. Now set up the first proportion to solve for *a*: $\frac{a}{\sin 19°} = \frac{4.4}{\sin 61°}$; $a = \sin 19° \cdot \frac{4.4}{\sin 61°} \approx 1.64$. Set up another proportion to solve for *c*: $\frac{c}{\sin 100°} = \frac{4.4}{\sin 61°}$; $c = \sin 100° \cdot \frac{4.4}{\sin 61°} \approx 4.95$.

3. **C = 110°, *b* ≈ 2.37, *c* ≈ 6.23.** This AAS case uses the Law of Sines. The missing angle is $C = 180° - (49° + 21°) = 110°$. The first proportion is $\frac{5}{\sin 49°} = \frac{b}{\sin 21°}$, which gets you $b = \sin 21° \cdot \frac{5}{\sin 49°} \approx 2.37$. The second proportion is $\frac{5}{\sin 49°} = \frac{c}{\sin 110°}$, which gives you $c = \sin 110° \cdot \frac{5}{\sin 49°} \approx 6.23$.

4. **B = 14°, *b* ≈ 2.06, *c* ≈ 7.06.** This one involves AAS, so you use the Law of Sines to solve it. First, the missing angle $B = 180° - (110° + 56°) = 14°$. Now set up the proportion, $\frac{8}{\sin 110°} = \frac{c}{\sin 56°}$, to get that $c = \sin 56° \cdot \frac{8}{\sin 110°} \approx 7.06$. Set up another proportion, $\frac{8}{\sin 110°} = \frac{b}{\sin 14°}$, to get that $b = \sin 14° \cdot \frac{8}{\sin 110°} \approx 2.06$.

5. **A ≈ 122.9°, B ≈ 20.1°, *a* ≈ 19.53.** Notice that when you draw this triangle, it's the SSA situation, or the ambiguous case. Always assume there are two answers when you're dealing with these types of problems, until you find out otherwise. Set up the proportion $\frac{14}{\sin 37°} = \frac{8}{\sin B}$. This means that $14 \sin B = 8 \sin 37°$, or $\sin B = \frac{8 \sin 37°}{14} \approx 0.3438942989$. Use inverse sine to get that $B_1 \approx 20.1°$. This is the first quadrant answer. The second quadrant has a second answer:

$B_2 \approx 180° - 20.1° \approx 159.9°$. However, if you look closely, you notice that you start off with $C = 37°$. If you add $159.9° + 37°$, you get over 180°, so you throw this second solution away. Only one triangle satisfies the conditions given. Now that you know $C = 37°$ and (the one and only) $B \approx 20.1°$, it's easy as pi (Get it? Pi!) to find $A \approx 122.9°$. Set up another proportion, $\frac{14}{\sin 37°} = \frac{a}{\sin 122.9°}$, which means that $a = \sin 122.9° \cdot \frac{14}{\sin 37°} \approx 19.53$.

6. **$A_1 \approx 104.8°$, $C_1 \approx 55.2°$, $a_1 \approx 14.13$; $A_2 \approx 35.2°$, $C_2 \approx 124.8°$, $a_2 \approx 8.42$.** Two solutions! How did that happen? Start at the beginning (a *very* good place to start) and use the Law of Sines to set up the proportion $\frac{5}{\sin 20°} = \frac{12}{\sin C}$. By cross-multiplying, you get the equation $5 \sin C = 12 \sin 20°$. Solve for $\sin C$ by dividing the 5: $\sin C = \frac{12 \sin 20°}{5} \approx 0.820848344$. Use the inverse sine function to discover that $C_1 \approx 55.2°$. The second quadrant answer is $C_2 = 180° - 55.2° \approx 124.8°$. If you add 20° to *each* of these answers, you discover that it's possible to make a triangle in both cases (because you haven't exceeded 180°). This sends you on two different paths for two different triangles. The following shows the separate computations.

   If $C_1 \approx 55.2°$, then $A_1 \approx 104.8°$. Next, set up the proportion $\frac{5}{\sin 20°} = \frac{a_1}{\sin 104.8°}$. This gives you $a_1 = \sin 104.8° \cdot \frac{5}{\sin 20°} \approx 14.13$.

   If $C_2 \approx 124.8°$, then $A_2 \approx 35.2°$. Set up another proportion, $\frac{5}{\sin 20°} = \frac{a_2}{\sin 35.2°}$, to then get that $a_2 = \sin 35.2° \cdot \frac{5}{\sin 20°} \approx 8.42$.

7. **No solution.** When you draw this one, you see that you have an ambiguous SSA case. Set up the proportion $\frac{10}{\sin 102°} = \frac{24}{\sin C°}$ using the Law of Sines. Cross-multiply to get $10 \sin C° = 24 \sin 102°$, and then divide the 10 from both sides to get $\sin C° = \frac{24 \sin 102°}{10} \approx 2.35$. That's when the alarms go off.

   Sine can't have a value bigger than 1, so there's no solution.

8. **$A = 60°$, $C = 90°$, $a \approx 13.86$.** Starting with the proportion $\frac{8}{\sin 30°} = \frac{16}{\sin C°}$, you cross-multiply to get $8 \sin C° = 16 \sin 30°$. Dividing, $\sin C° = \frac{16 \sin 30°}{8} = 1$. You don't even need a calculator for this one. An angle of 90° has a sine equaling 1. So $C = 90°$, leaving 60° for angle A. You can use the side proportions for a 30-60-90-degree triangle or just go to the proportion $\frac{8}{\sin 30°} = \frac{a}{\sin 60°}$, which gives you $a = \sin 60° \cdot \frac{8}{\sin 30°} \approx 13.85640646$. The length 13.86 is the approximate value of $8\sqrt{3}$, which finishes the special 30-60-90 right triangle proportions of $\frac{1}{2}n$, $\frac{\sqrt{3}}{2}n$, n.

9. **$c = 14$, $A \approx 21.8°$, $B \approx 38.2°$.** You see that this is a Law of Cosines problem when you draw the triangle (SAS). Find c first: $c^2 = a^2 + b^2 - 2ab\cos C$. Plug in what you know: $c^2 = 6^2 + 10^2 - 2(6)(10)\cos 120°$. Plug this right into your calculator to get that $c^2 = 196$, or $c = 14$. Now find A: $a^2 = b^2 + c^2 - 2bc\cos A$. Plug in what you know: $6^2 = 10^2 + 14^2 - 2(10)(14)\cos A$. Simplify: $36 = 296 - 280\cos A$. Solve for $\cos A$: $-260 = -280\cos A$ gives you $\frac{260}{280} = \cos A$ or $\cos A \approx 0.9285714286$. Now use inverse cosine to get $A \approx 21.8°$, and then use the fact that the sum of the angles of a triangle is 180° to figure out that $B \approx 38.2°$.

10 $a \approx 7.50$, **B ≈ 48.7°, C ≈ 61.3°.** By plugging what you know into the Law of Cosines, $a^2 = b^2 + c^2 - 2bc\cos A$, you get $a^2 = 6^2 + 7^2 - 2(6)(7)\cos 70°$. This simplifies to $a^2 \approx 56.27030796$, or $a \approx 7.50$.

Now switch the substituting in the second law: $b^2 = a^2 + c^2 - 2ac\cos B$; $6^2 = 7.50^2 + 7^2 - 2(7.50)(7)\cos B$. Simplify: $36 = 105.27 - 105\cos B$. Solving for cos B: $-69.27 = -105\cos B$, $\frac{69.27}{105} = \cos B$, resulting in $\cos B \approx 0.6597142857$. Using the inverse operation, you get that B ≈ 48.7°. From there, you can figure out that C ≈ 61.3°.

11 **A ≈ 95.7°, B ≈ 33.6°, C ≈ 50.7°.** You're solving an SSS triangle using the Law of Cosines, so, to find A: $a^2 = b^2 + c^2 - 2bc\cos A$, $81^2 = 5^2 + 7^2 - 2(5)(7)\cos A$, $\cos A = \frac{7}{-70} = -0.1$, $A \approx 95.7°$.

And to find B: $b^2 = a^2 + c^2 - 2ac\cos B$, $5^2 = 9^2 + 7^2 - 2(9)(7)\cos B$, $\cos B = \frac{-105}{-126} \approx 0.83333333$, $B \approx 33.6°$.

Last, but certainly not least, $C = 180° - (95.7° + 33.6°) = 50.7°$.

12 **A ≈ 18.3°, B ≈ 25.3°, C ≈ 136.4°.** To find A: $a^2 = b^2 + c^2 - 2bc\cos A$, $7.3^2 = 9.9^2 + 16^2 - 2(9.9)(16)\cos A$, $\cos A = \frac{-300.72}{-316.8} \approx 0.9492424242$, $A \approx 18.3°$.

And to find B: $b^2 = a^2 + c^2 - 2ac\cos B$, $9.9^2 = 7.3^2 + 16^2 - 2(7.3)(16)\cos B$, $\cos B = \frac{-211.28}{-233.6} \approx 0.9044520548$, $B \approx 25.3°$.

And finally, $C \approx 136.4°$.

13 **About 55.17 square units.** Knowing two sides and the angle between them allows you to use the area formula $A = \frac{1}{2}ab\sin C$. In this case, $A = \frac{1}{2}(7)(17)\sin 68° \approx 55.17$.

14 **About 28 square units.** Since you can use the distance formula to find the length of all three sides, you can then use Heron's Formula, $A = \sqrt{s(s-a)(s-b)(s-c)}$, to compute the area. So you should start by drawing a picture:

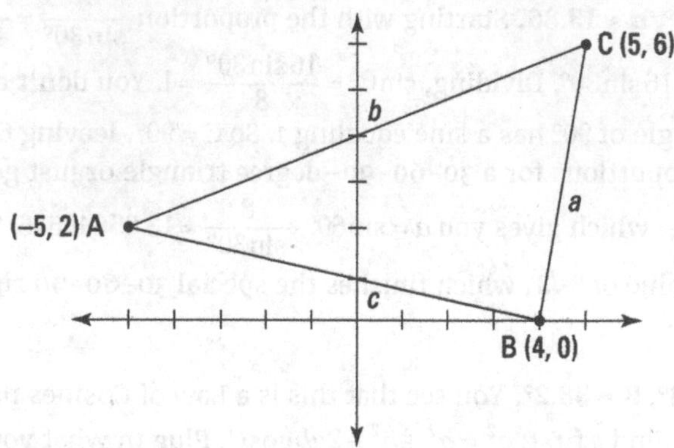

Find all three sides first (for a review of how to find the distance between two points, see Chapter 1):

$$AC = \sqrt{(5-(-5))^2 + (6-2)^2} = \sqrt{10^2 + 4^2} = \sqrt{116}$$

$$AB = \sqrt{(4-(-5))^2 + (0-2)^2} = \sqrt{9^2 + (-2)^2} = \sqrt{85}$$

$$BC = \sqrt{(4-5)^2 + (0-6)^2} = \sqrt{(-1)^2 + (-6)^2} = \sqrt{37}$$

Now that you've found the length of all three sides, use Heron's Formula to find the area.

First, the semi-perimeter is $s = \dfrac{\sqrt{116} + \sqrt{85} + \sqrt{37}}{2} \approx 13.0363.$

Using this in the formula, $A = \sqrt{13.0363(13.0363 - \sqrt{116})(13.0363 - \sqrt{85})}$
$\overline{(13.0363 - \sqrt{37})} \approx 28.05.$

15 **About 330.86 miles.** The two trains depart (F and S) from the same station (T), as shown here:

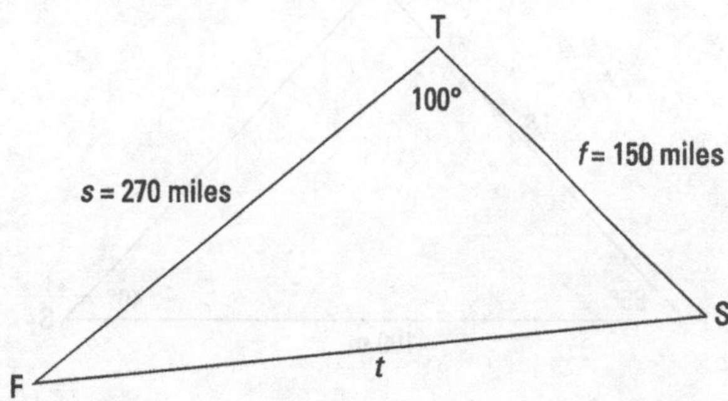

You have to use the Law of Cosines to solve for how far apart the two trains are, t.

To find how far the trains have traveled in three hours, use $d = rt$. For the first train, you get $d = 90 \cdot 3 = 270$ miles; for the second train, $d = 50 \cdot 3 = 150$ miles. Using the equation, $t^2 = f^2 + s^2 - 2fs\cos T$, plug in the values: $t^2 = 150^2 + 270^2 - 2(150)(270)\cos 100°$. This goes right into your calculator to give you $t^2 = 109{,}465.50$, or t is about 330.86 miles.

16 **About 279.3 feet.** This time the picture looks like this:

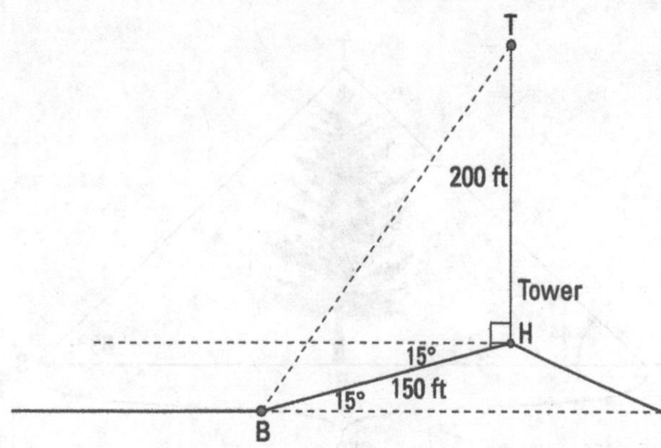

To find the measure of $\angle H$ in the picture, you add a horizontal line that's parallel to the ground. Then, using the facts that alternate interior angles are congruent and that the tower has to be completely vertical (or else you have a leaning tower), you know that $H = 15° + 90° = 105°$.

Now, jump in with the Law of Cosines: $h^2 = t^2 + b^2 - 2tb\cos H$.
Using the known values, $h^2 = 150^2 + 200^2 - 2(150)(200)\cos 105°$.

$h^2 = 78{,}029.14$, giving you $h \approx 279.3$ feet.

17 **About 78.5 meters.** Looking down on the surveyor and the flagpole, here's the picture you use to solve this problem:

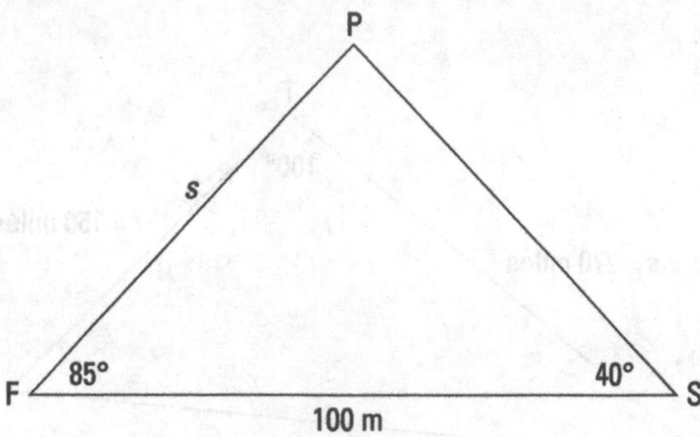

Because you already have two angles, you can find that $P = 55°$ and use the Law of Sines:

$\frac{100}{\sin 55°} = \frac{s}{\sin 40°}$. Solving for s, you have $s = \sin 40° \cdot \frac{100}{\sin 55°} \approx 78.5$ meters.

18 **About 226.53 feet tall.** This problem involves two different triangles: the large triangle with the angles given, and a second triangle drawn with a segment through the tree as one side. You have to figure out the distance from either scientist to the top of the tree (FT or TS in the following illustration) to determine how tall the tree (TB) really is. Here's the drawing of the two scientists and the tree between them:

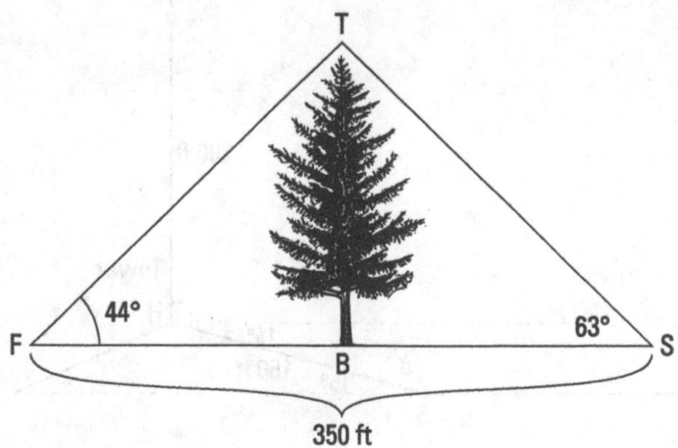

Knowing two angles gets you the third one: $T = 73°$. To determine the length of FT, use the Law of Sines: $\frac{350}{\sin 73°} = \frac{FT}{\sin 63°}$. Solving for FT, $FT = \sin 63° \cdot \frac{350}{\sin 73°} \approx 326.10$ feet.

Now, assuming the tree grows straight up, you have a right triangle FBT in which you know one angle and one side. Go back to Chapter 8, if you need to review the process of finding the missing measures. Knowing that you have the hypotenuse and are looking for the opposite side, you can use the sine function and $\sin 44° = \frac{TB}{FT} = \frac{TB}{326.10}$. Solving for TB: $TB = 326.10 \cdot \sin 44° \approx 226.53$ feet tall. That's one big tree!

If you're ready to test your skills a bit more, take the following chapter quiz that incorporates all the chapter topics.

Whaddya Know? Chapter 14 Quiz

Quiz time! Complete each problem to test your knowledge on the various topics covered in this chapter. You can then find the solutions and explanations in the next section.

1. Solve ΔXYZ if $x = 10$, $y = 20$, and $Z = 35°$.

2. Solve ΔXYZ if $z = 20$, $y = 30$, and $Y = 40°$.

3. Find the area of a triangular garden with sides of 40 feet and 50 feet, where the angle between those sides is $80°$.

4. Solve ΔXYZ if $x = 20$, $X = 40°$, and $Y = 60°$.

5. What are the measures of the three angles of Δ XYZ if $x = 40$, $y = 50$, and $z = 60$?

6. You and your friend are 500 feet apart when you spot a kite that is flying high in the sky between you. The angle from the ground to the kite is $40°$ for you and $60°$ for your friend. How high is the kite?

7. Find the two possible measures of angle Z if $Y = 60°$, $y = 18$, and $z = 20$.

8. Find the area of a triangle whose sides measure 50 feet, 70 feet, and 100 feet.

9. Solve ΔXYZ if $X = 40°$, $Y = 60°$, $z = 20$.

10. You're at the bottom of a staircase leading up to the base of a 100-foot statue. The staircase forms a $10°$ angle with the ground, and the distance from the base of the staircase to the top of the statue is 150 feet. How many feet long is the staircase?

Answers to Chapter 14 Quiz

1. **X ≈ 25.91°, Y ≈ 119.09°, *z* ≈ 13.13.** See the following figure.

   First, apply the Law of Cosines to solve for side *z*.

   $$z^2 = 10^2 + 20^2 - 2(10)(20)\cos 35° = 500 - 400\cos 35° = 500 - 327.6608 = 172.3392$$

   $$z \approx 13.1278 \approx 13.13$$

   Use the Law of Sines to write the proportion and solve for $\sin X$.

   $$\frac{z}{\sin Z} = \frac{x}{\sin X} \rightarrow \frac{13.1278}{\sin 35°} = \frac{10}{\sin X} \rightarrow \sin X = \frac{10(\sin 35°)}{13.1278} = 0.4369$$

   Solve for angle X using the inverse sine: $X = \sin^{-1}(0.4369) = 25.9063 \approx 25.91°$.

   Now find the measure of angle Y by subtracting the sum of angles Z and X from 180.

   $Y° = 180 - (35 + 25.91) = 119.09°$

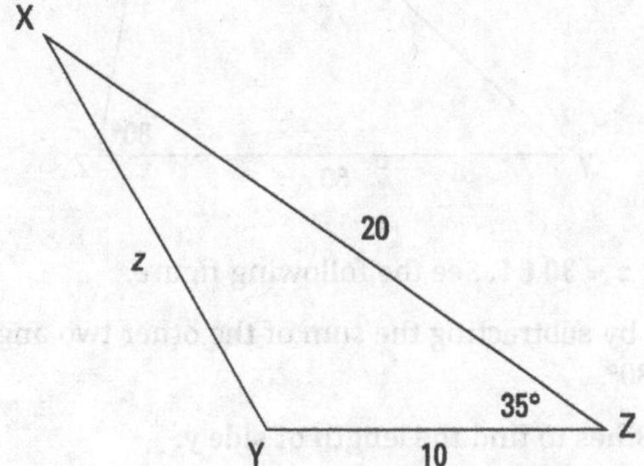

2. **X ≈ 114.63°, Z ≈ 25.37°, *x* ≈ 42.43.** See the following figure.

   Solve for angle Z using the Law of Sines. $\frac{y}{\sin Y} = \frac{z}{\sin Z} \rightarrow \frac{30}{\sin 40°} = \frac{20}{\sin Z} \rightarrow$

   $\sin Z = \frac{20(\sin 40°)}{30} = 0.4285$

   Solve for angle Z using the inverse sine: $Z = \sin^{-1}(0.4285) = 25.3740 \approx 25.37°$.

   Now find the measure of angle X by subtracting the sum of angles Y and Z from 180.

   $X° = 180 - (40 + 25.37) = 114.63°$

   Now use the Law of Sines to solve for side *x*.

   $$\frac{y}{\sin Y} = \frac{z}{\sin X} \rightarrow \frac{30}{\sin 40°} = \frac{x}{\sin 114.63} \rightarrow x = \frac{30(\sin 114.63°)}{\sin 40°} = 42.4254 \approx 42.43$$

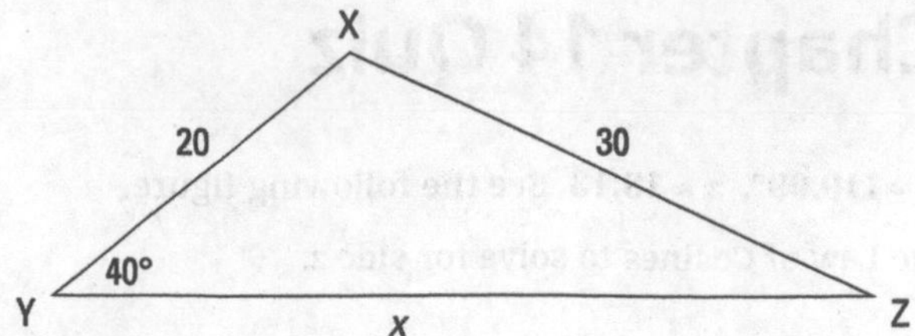

3. **About 984.81 sq. ft.** See the following figure.

Since the given angle is between the two sides, you can use the formula,

$A = \frac{1}{2}(40)(50)\sin 80° = 1000(.984807753) \approx 984.81.$

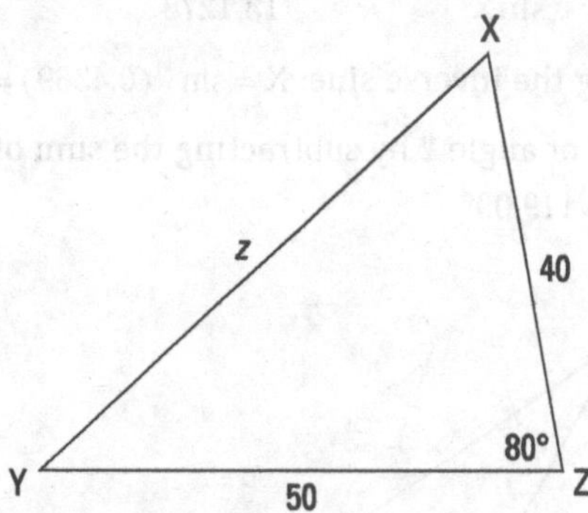

4. **Z ≈ 80°, *y* ≈ 26.95, *z* ≈ 30.64.** See the following figure.

First, find angle Z by subtracting the sum of the other two angles from 180: $180 - (40° + 60°) = 80°$.

Using the Law of Sines to find the length of side *y*:

$$\frac{x}{\sin X} = \frac{y}{\sin Y} \rightarrow \frac{20}{\sin 40°} = \frac{y}{\sin 60°} \rightarrow x = \frac{20(\sin 60°)}{\sin 40°} = 26.9459 \approx 26.95$$

And now, using the Law of Sines to find the length of side *z*:

$$\frac{x}{\sin X} = \frac{z}{\sin Z} \rightarrow \frac{20}{\sin 40°} = \frac{z}{\sin 80°} \rightarrow z = \frac{20(\sin 80°)}{\sin 40°} = 30.6417 \approx 30.64$$

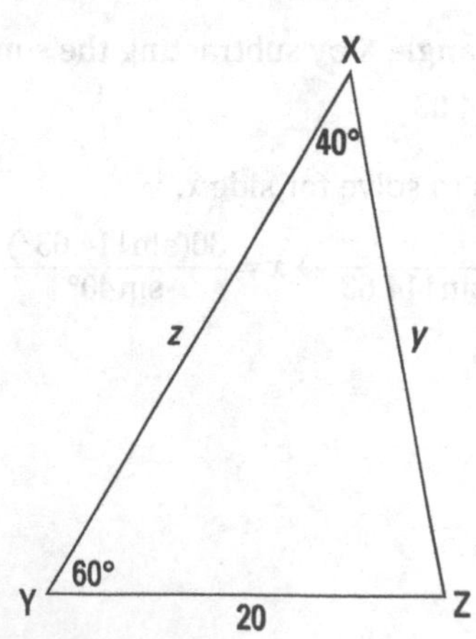

5) **Z ≈ 82.82°, Y ≈ 55.77°, X ≈ 41.41°.** See the following figure.

You can use the Law of Cosines to find the measure of angle Z:

$$60^2 = 40^2 + 50^2 - 2(40)(50)\cos Z \rightarrow -500 = -4000\cos Z \rightarrow \frac{1}{8} = \cos Z$$

Now use the inverse identity to find the angle measure of Z:

$$Z = \cos^{-1}\left(\frac{1}{8}\right) = 82.8192 \approx 82.82°$$

Repeat the process to find the measure of angle Y:

$$50^2 = 40^2 + 60^2 - 2(40)(60)\cos Y \rightarrow -2700 = -4800\cos Y \rightarrow \frac{9}{16} = \cos Y$$

$$Y = \cos^{-1}\left(\frac{9}{16}\right) = 55.7711 \approx 55.77°$$

Now find the measure of angle X by subtracting the sum of angles Z and Y from 180:

$180 - (82.82° + 55.77°) \approx 41.41°$.

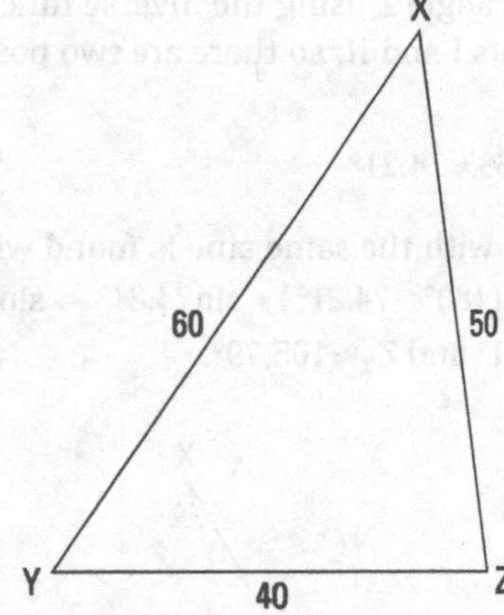

6) **About 282.63 ft.** See the following figure.

First, find the measure of the angle X (the angle of the triangle just under the kite) by subtracting the sum of the other two angles from 180:

$$180 - (40° + 60°) = 80°$$

Use the Law of Sines to find the length of side *y*.

$$\frac{y}{\sin Y} = \frac{x}{\sin X} \rightarrow \frac{y}{\sin 40°} = \frac{500}{\sin 80°} \rightarrow y = \frac{500(\sin 40°)}{\sin 80°} = 326.3518 \approx 326.35$$

Create a right triangle by dropping a line down from the kite. This is a 30-60-90 right triangle with the longer leg opposite the 60-degree angle. That is the height of the kite that you are looking for here. In a 30-60-90-degree triangle, the shorter leg is half the hypotenuse, and the longer leg is $\sqrt{3}$ times the shorter leg.

Half the hypotenuse is $\frac{y}{2} = \frac{326.3518}{2} = 163.1759$. Multiply this by $\sqrt{3}$ and you have $163.1759\sqrt{3} = 282.6290 \approx 282.63$.

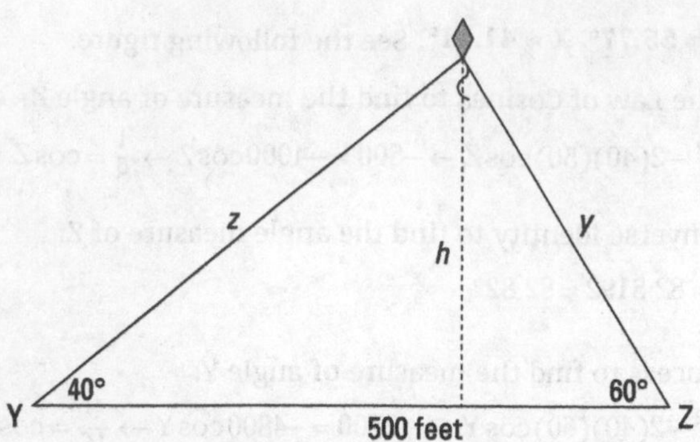

7. **About 74.21°, 105.79°.** See the following figure.

You can use the Law of Sines to solve for the measure of angle Z:

$$\frac{y}{\sin Y}=\frac{z}{\sin Z}\rightarrow\frac{18}{\sin 60°}=\frac{20}{\sin Z}\rightarrow \sin Z=\frac{20(\sin 60°)}{18}=0.9623$$

Solve for the measure of angle Z using the inverse function. But remember that the sine is positive in both quadrants I and II, so there are two possible answers. You find the first one to be:

$$Z=\sin^{-1}(0.9623)=74.2068\approx 74.21°$$

The angle in quadrant II with the same sine is found with the identity $\sin(180°-\theta)=\sin\theta$. So, in this case you have $\sin(180°-74.21°)=\sin 74.21°\rightarrow \sin(105.79°)=\sin 74.21°$. So the two possible values are $Z_1\approx 74.21°$ and $Z_2\approx 105.79°$.

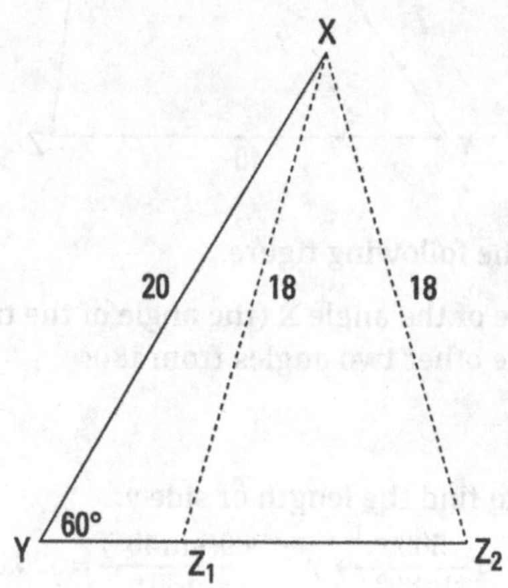

8 **1,624.81 sq. ft.** Using Heron's Formula, first find the semi-perimeter:

$$s = \frac{50 + 70 + 100}{2} = \frac{220}{2} = 110.$$

Inserting the values into the formula, $A = \sqrt{110(110-50)(110-70)(110-100)} =$ $A = \sqrt{110(60)(40)(10)} = 1624.8077.$

9 **Z = 80°, *x* ≈ 13.05, *y* ≈ 17.59.** See the following figure.

First, find the measure of angle Z by subtracting the sum of the other two angles from 180: $180 - (40° + 60°) = 80°$.

Using the Law of Sines to find side *x*:

$$\frac{z}{\sin Z} = \frac{x}{\sin X} \rightarrow \frac{20}{\sin 80°} = \frac{x}{\sin 40°} \rightarrow x = \frac{20(\sin 40°)}{\sin 80°} = 13.0541 \approx 13.05$$

And using the same law to find side *y*:

$$\frac{z}{\sin Z} = \frac{y}{\sin Y} \rightarrow \frac{20}{\sin 80°} = \frac{y}{\sin 60°} \rightarrow y = \frac{20(\sin 60°)}{\sin 80°} = 17.5877 \approx 17.59$$

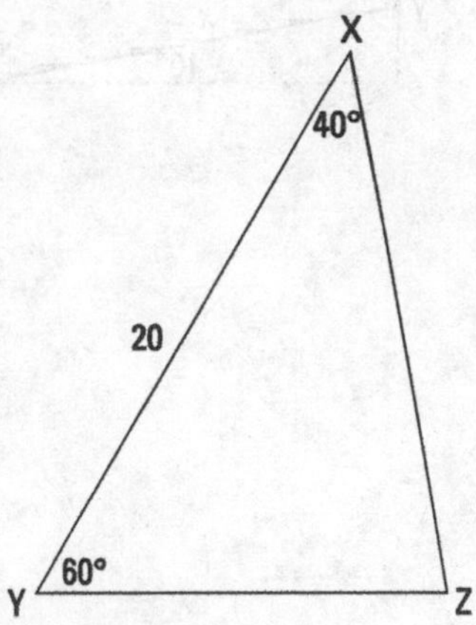

10 **About 95.76 ft.** See the following figure.

First, determine the measure of one of the angles in the triangle XYZ. A right triangle is formed by drawing a horizontal line from the base of the staircase to directly under the statue. The angle under the staircase at Y measures 80 degrees (180 – 90 – 10). This means that the angle Y measures 100 degrees (supplementary to the 80-degree angle).

So now use the Law of Sines to first find angle Z:

$$\frac{z}{\sin Z} = \frac{y}{\sin Y} \rightarrow \frac{100}{\sin Z} = \frac{150}{\sin 100°} \rightarrow \sin Z = \frac{100(\sin 100°)}{150} = 0.656539$$

Using the inverse sine function, $Z = \sin^{-1}(0.656539) = 41.0464°$.

The measure of angle X is then $180 - (100 + 41.0464) = 38.9536°$.

Now use the Law of Sines again to find the length of the staircase, x:

$$\frac{x}{\sin X} = \frac{y}{\sin Y} \rightarrow \frac{x}{\sin 38.9536°} = \frac{150}{\sin 100°} \rightarrow x = \frac{150(\sin 38.9536°)}{\sin 100°} = 95.7584 \approx 95.76$$

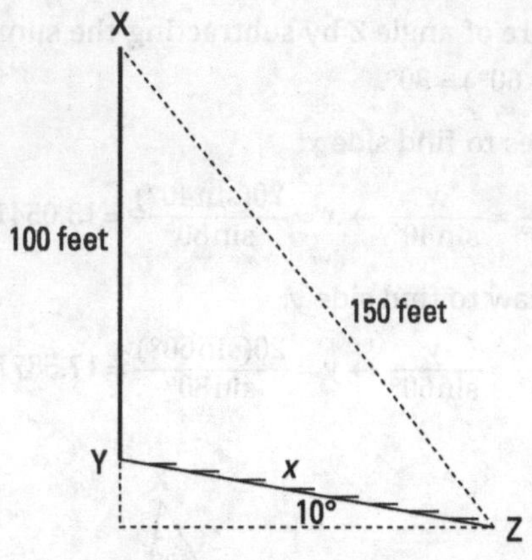

5

Analytic Geometry

IN THIS UNIT . . .

Getting more complex with complex numbers.

Advancing graphing techniques with polar coordinates.

Slicing cones to create conics.

IN THIS CHAPTER

» Pitting real versus imaginary

» Exploring the complex number system

» Plotting complex numbers on a plane

» Solving problems involving complex solutions

Chapter 15 Coordinating with Complex Numbers

Complex numbers can be a most interesting (but frequently neglected) topic in a standard study of pre-calculus. Using complex numbers efficiently can vastly simplify a difficult problem or even allow you to solve a problem that you couldn't solve before.

In early math experiences, you may have been told that you can't find the square root of a negative number. If somewhere in your calculations you stumbled on an answer that required you to take the square root of a negative number, you simply threw that answer out the window. As you advanced in math, you found that you needed complex numbers to explain natural phenomena that real numbers are incapable of explaining. In fact, entire math courses are dedicated to the study of complex numbers and their applications. You won't see that kind of depth here; however, you'll be introduced to the basic topics and enough to allow you to work in those applications.

In this chapter, you find the concepts of and uses for complex numbers. You see where they come from and how you use them (as well as graph them).

Understanding Real versus Imaginary

Courses in Algebra I and II introduce you to the real number system. Pre-calculus can expand your horizons by adding complex numbers to your repertoire. *Complex numbers* are numbers that include both a real *and* an imaginary part; they're widely used for complex analysis, which theorizes functions by using complex numbers as variables (see the following section for more on the complex number system).

If you were previously told to disregard negative roots whenever you found them, here's a quick explanation for why: You can actually take the square root of a negative number, but the square root isn't a real number. However, it does exist! It takes the form of an *imaginary number*.

REMEMBER

Imaginary numbers have the form bi, where b is a real number and i is an imaginary number. The letter i represents $\sqrt{-1}$. A complex number, $a+bi$, can be real or imaginary, depending on the values of a and b. The constant a must also be a real number.

Luckily, you're already familiar with the x-y coordinate plane, which you use to graph functions (such as in Chapter 4). You also can use a complex coordinate plane to graph imaginary numbers. Although these two planes are constructed the same way — two axes perpendicular to one another at the origin — they're very different. For numbers graphed on the x-y plane, the coordinate pairs represent real numbers in the form of variables (x and y). You can show the relationships between these two variables as points on the plane. On the other hand, you use the complex plane simply to plot complex numbers. If you want to graph a real number, all you really need is a real number line. However, if you want to graph a complex number, you need an entire plane so that you can graph both the real and imaginary part.

Enter the *Gauss* or *Argand* coordinate plane. In this plane, pure real numbers in the form $a+0i$ exist completely on the real axis (the horizontal axis), and pure imaginary numbers in the form $0+bi$ exist completely on the imaginary axis (the vertical axis). Figure 15-1a shows the graph of a real number, and Figure 15-1b shows that of an imaginary number.

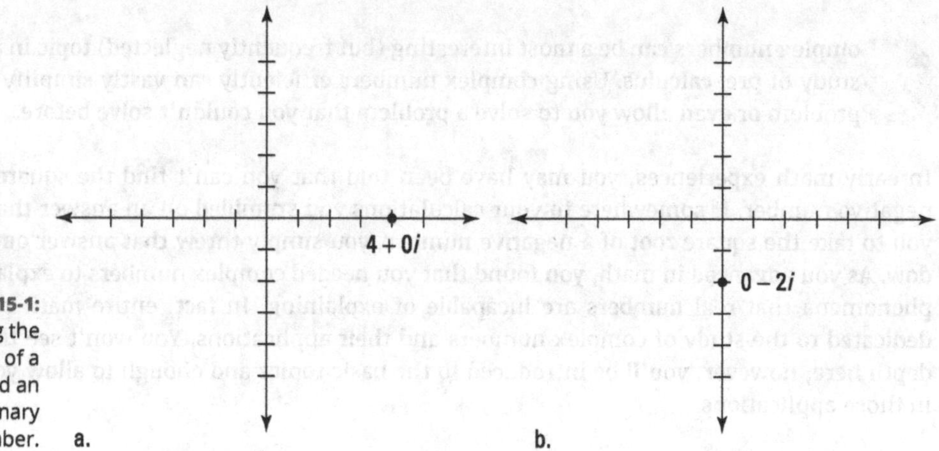

FIGURE 15-1: Comparing the graphs of a real and an imaginary number.

Figure 15-1a shows the graph of the real number 4, and Figure 15-1b shows the graph of the pure imaginary number $-2i$. The 0 in $4+0i$ makes that number *real*, and the 0 in $0-2i$ makes that number *pure imaginary*.

Combining Real and Imaginary: The Complex Number System

The *complex number system* is more complete than the real number system or the pure imaginary numbers in their separate forms. You can use the complex number system to represent real numbers, imaginary numbers, and numbers that have both a real and an imaginary part. In fact, the complex number system is the most comprehensive set of numbers you deal with in pre-calculus.

Simplifying expressions when complex numbers are created

You already know that $\sqrt{-1} = i$. This definition quickly eliminates the bulky and unusable negative-under-a-radical expression. It's also important to be able to simplify more complex (couldn't resist the play-on-words) expressions.

EXAMPLE

Q. Simplify the following:

(a) $\sqrt{-6}$

(b) $\sqrt{-300}$

(c) $\sqrt{36-4(-6)(-2)}$

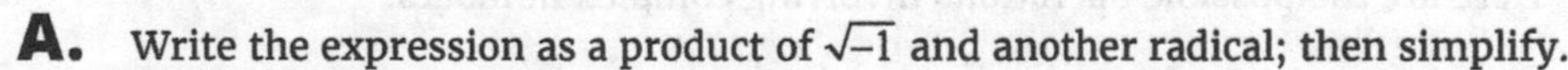

A. Write the expression as a product of $\sqrt{-1}$ and another radical; then simplify.

(a) $\sqrt{-6} = \sqrt{-1} \cdot \sqrt{6} = i\sqrt{6}$

(b) $\sqrt{-300} = \sqrt{-1}\sqrt{300} = \sqrt{-1}\sqrt{100}\sqrt{3} = 10i\sqrt{3}$

(c) $\sqrt{36-4(-6)(-2)} = \sqrt{36-48} = \sqrt{-12} = \sqrt{-1}\sqrt{12} = \sqrt{-1}\sqrt{4}\sqrt{3} = 2i\sqrt{3}$

YOUR TURN

 1 Simplify $\sqrt{-80}$	 2 Simplify $\sqrt{16-4(4)(7)}$

Performing operations with complex numbers

Sometimes you come across situations where you need to operate on real and imaginary numbers together. To do this, you want to write both numbers as complex numbers in order to be able to add, subtract, multiply, or divide them.

Consider the following three types of complex numbers.

- **A real number as a complex number:** $3+0i$

  Notice that the imaginary part of the expression is 0.

- **An imaginary number as a complex number:** $0+2i$

  Notice that the real portion of the expression is 0.

- **A complex number with both a real and an imaginary part:** $1+4i$

  This number can't be described as solely real or solely imaginary — hence the term *complex.*

REMEMBER

You can manipulate complex numbers algebraically just like real numbers to carry out operations. You just have to be careful to keep all the *i*'s straight. You can't combine real parts with imaginary parts by using addition or subtraction, because they're not like terms, so you have to keep them separate. Also, when multiplying complex numbers, the product of two imaginary numbers is a real number; the product of a real and an imaginary number is still imaginary; and the product of two real numbers is real.

Here are the possible operations involving complex numbers.

EXAMPLE

Q. **Perform the following operations:**

(a) Subtract $(3-2i)-(2-6i)$

(b) Multiply $2(3+2i)$

(c) Multiply $2i(3+2i)$

(d) Multiply $(3-2i)(9+4i)$

(e) Divide $\dfrac{1+2i}{3-4i}$

A. **Here are the solutions.**

(a) ***To add and subtract complex numbers,*** **simply combine like terms. In this case, when subtracting, distribute the negative sign first:**

$$(3-2i)-(2-6i)=3-2i-2+6i=(3-2)+(-2i+6i)=1+4i$$

(b) ***To multiply a complex number by a real number,*** **just distribute the real number over both the real and imaginary parts of the complex number:** $2(3+2i)=6+4i$.

(c) ***To multiply a complex number by an imaginary number,*** **first, realize that, as a result of the multiplication, the real part of the complex number becomes imaginary, and the imaginary part becomes real. When you express your final answer, however, you still express the real part first followed by the imaginary part, in the form** $a+bi$. **When** $2i$

multiplies $3+2i$, you get $2i(3+2i)=6i+4i^2$. Since $i=\sqrt{-1}$, $i^2=\left(\sqrt{-1}\right)^2=-1$. Therefore, you really have $6i+4(-1)$. Written in the standard form of a complex number, your answer is $-4+6i$.

(d) *To multiply two complex numbers,* simply follow the FOIL process. In this case, $(3-2i)(9+4i)=27+12i-18i-8i^2=27+12i-18i-8i^2$, which simplifies to $27-6i-8(-1)=27-6i+8=35-6i$.

(e) *To divide complex numbers,* multiply both the numerator and the denominator by the conjugate of the denominator, FOIL the numerator and denominator separately, and then combine like terms. This process is necessary because the imaginary part in the denominator involves a square root (of −1, remember), and the denominator of the fraction shouldn't contain a radical or an imaginary part. In this problem, you're asked to divide $\frac{1+2i}{3-4i}$. The complex conjugate of $3-4i$ is $3+4i$. Follow these steps to finish the problem:

1. **Multiply the numerator and the denominator by the conjugate of the denominator.**

$$\frac{1+2i}{3-4i}\cdot\frac{3+4i}{3+4i}$$

2. **FOIL the numerator and denominator.**

$$\frac{1+2i}{3-4i}\cdot\frac{3+4i}{3+4i}=\frac{3+4i+6i+8i^2}{9+12i-12i-16i^2}=\frac{3+10i+8i^2}{9-16i^2}$$

3. **Substitute −1 for i^2 and simplify.**

$$\frac{3+10i+8(-1)}{9-16(-1)}=\frac{3+10i-8}{9+16}=\frac{-5+10i}{25}$$

4. **Rewrite the result in the standard form of a complex number.**

$$\frac{-5+10i}{25}=\frac{-5}{25}+\frac{10i}{25}=-\frac{1}{5}+\frac{2}{5}i$$

YOUR TURN

3 Add $(2+3i)+(-3-5i)$

4 Subtract $(2+3i)-(-3-5i)$

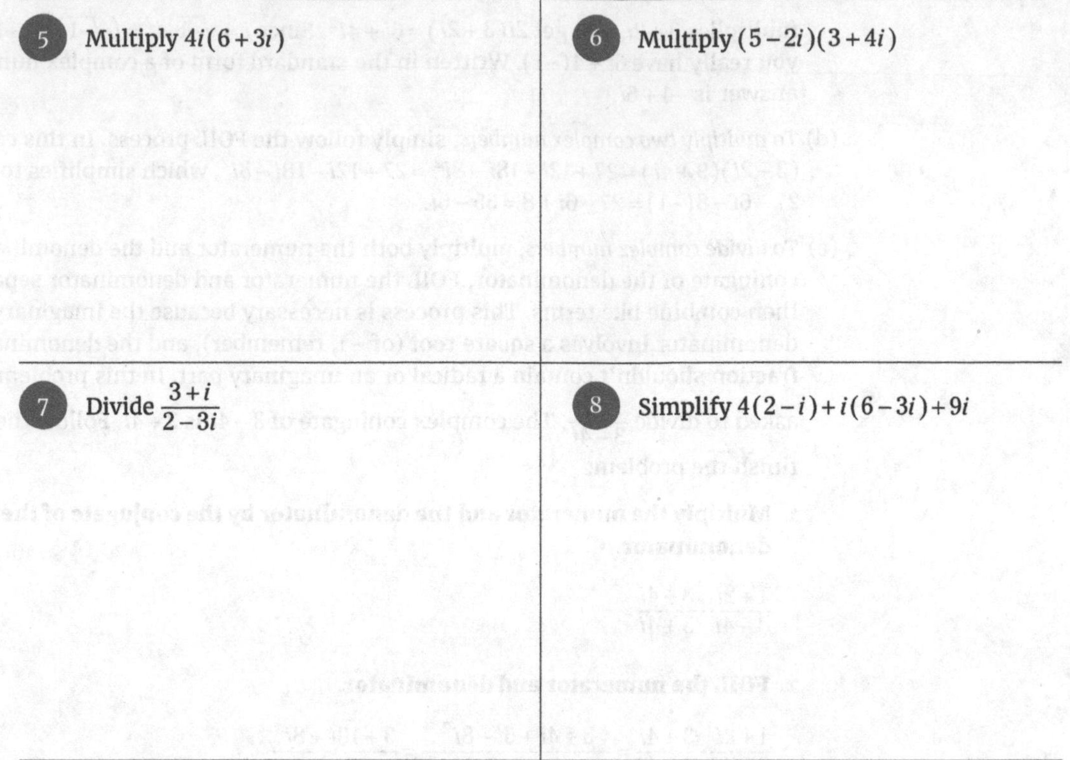

5 Multiply $4i(6+3i)$	6 Multiply $(5-2i)(3+4i)$
7 Divide $\dfrac{3+i}{2-3i}$	8 Simplify $4(2-i)+i(6-3i)+9i$

Graphing Complex Numbers

To graph complex numbers, you simply combine the ideas of the real-number coordinate plane and the Gauss or Argand coordinate plane (which is explained in the earlier section, "Understanding Real versus Imaginary," in this chapter) to create the complex coordinate plane. In other words, you take the real portion of the complex number, a, to represent the x-coordinate, and you take the imaginary portion, b, to represent the y-coordinate.

REMEMBER

Although you graph complex numbers much like any point in the real-number coordinate plane, complex numbers aren't real! The x-coordinate is the only real part of a complex number, so you call the x-axis the *real axis* and the y-axis the *imaginary axis* when graphing in the complex coordinate plane.

Graphing complex numbers gives you a way to visualize them, but a graphed complex number doesn't have the same physical significance as a real-number coordinate pair. For an (x,y)

coordinate, the position of the point on the plane is represented by two numbers. In the complex plane, the value of a single complex number is represented by the position of the point, so each complex number $a+bi$ can be expressed as the ordered pair (a,b).

You can see several examples of graphed complex numbers in Figure 15-2.

Point A: The real part is 2 and the imaginary part is 3, so the complex coordinate is (2,3) where 2 is on the real (or horizontal) axis and 3 is on the imaginary (or vertical) axis. This point is $2+3i$.

Point B: The real part is -1 and the imaginary part is -4; you can draw the point on the complex plane as $(-1,-4)$. This point is $-1-4i$.

Point C: The real part is $\frac{1}{2}$ and the imaginary part is -3, so the complex coordinate is $\left(\frac{1}{2},-3\right)$. This point is $\frac{1}{2}-3i$.

Point D: The real part is -2 and the imaginary part is 1, which means that on the complex plane, the point is $(-2,1)$. This coordinate is $-2+i$.

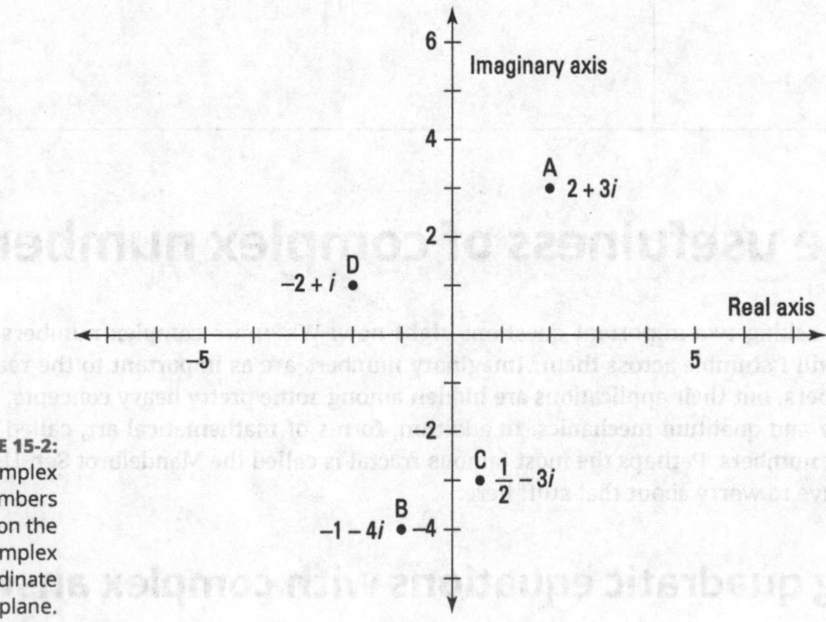

FIGURE 15-2: Complex numbers plotted on the complex coordinate plane.

YOUR TURN

For Questions 9 to 12, plot the following points on the Gauss coordinate plane.

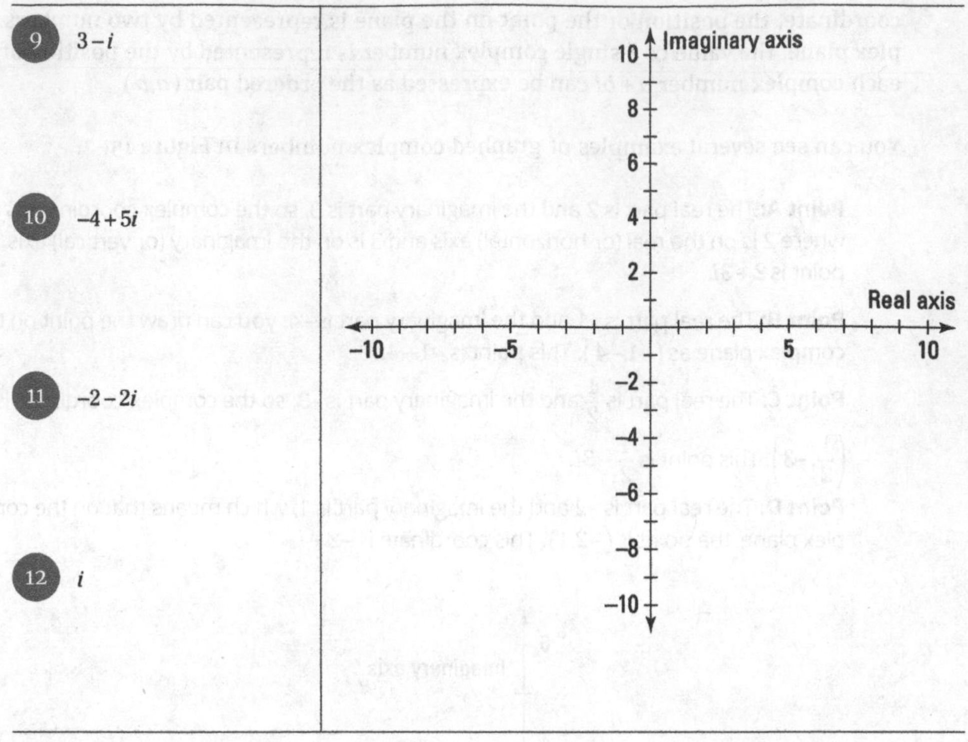

Grasping the usefulness of complex numbers

You may be asking two important questions right now: When are complex numbers useful, and where will I stumble across them? Imaginary numbers are as important to the real world as real numbers, but their applications are hidden among some pretty heavy concepts, such as chaos theory and quantum mechanics. In addition, forms of mathematical art, called *fractals*, use complex numbers. Perhaps the most famous fractal is called the Mandelbrot Set. However, you don't have to worry about that stuff here.

Solving quadratic equations with complex answers

The quadratic formula is such a useful and manageable tool. When a quadratic expression can't be factored to provide solutions in an equation — which, in turn, are extreme values or intercepts or other interesting things — you fall back on the quadratic formula. But what if the numbers don't behave? What if you end up with a negative number under the radical?

EXAMPLE

Q. Solve the quadratic equation for x: $x^2 + x + 1 = 0$.

A. $-\frac{1}{2} \pm \frac{\sqrt{3}}{2}i$. This equation isn't factorable, and using the quadratic formula, you get the following solution to this equation:

$$x = \frac{-b \pm \sqrt{b^2 - 4ac}}{2a} = \frac{-1 \pm \sqrt{1^2 - 4(1)(1)}}{2(1)} = \frac{-1 \pm \sqrt{1-4}}{2} = \frac{-1 \pm \sqrt{-3}}{2}.$$

Now, given $\sqrt{-1} = i$, the value under the radical can be simplified:

$$x = \frac{-1 \pm \sqrt{-3}}{2} = \frac{-1 \pm \sqrt{-1}\sqrt{3}}{2} = \frac{-1 \pm i\sqrt{3}}{2}.$$

Notice that the *discriminant* (the $b^2 - 4ac$ part) is a negative number, which you can't solve with only real numbers. When you first discovered the quadratic formula in algebra, you most likely used it to find real roots only. But because of complex numbers, you don't have to discard this solution. The previous answer is a legitimate complex solution, or a *complex root.* (Perhaps you remember encountering complex roots of quadratics in Algebra II. Check out *Algebra II For Dummies*, by yours truly, [Wiley], for a refresher.)

Writing the solution in the $a + bi$ format, you have:

$$x = \frac{-1 \pm i\sqrt{3}}{2} = \frac{-1}{2} \pm \frac{i\sqrt{3}}{2} = -\frac{1}{2} \pm \frac{\sqrt{3}}{2}i.$$ There are two complex solutions.

YOUR TURN

13 Solve the equation for x: $2x^2 + 4x + 3 = 0$.

14 Solve the equation for x: $x^2 + 25 = 0$.

Analyzing the graphs of quadratics

A quadratic function has the general form $f(x) = ax^2 + bx + c$, where its graph is a parabola that opens upward or downward. It can have one x-intercept, two x-intercepts, or no x-intercepts — all depending on where the vertex is and if it's opening upward or downward.

Another type of parabola that isn't a function is one that opens to the left or right. The equations in this case have the general form $x = ay^2 + by + c$. And with these equations, you can have one y-intercept, two y-intercepts, or no y-intercepts — all depending on where the vertex is and if it's opening left or right.

EXAMPLE

Q. Sketch the graph of the parabola $y = x^2 - 2x + 5$.

A. Since the coefficient of x^2 is positive, you know that the parabola opens upward. You find the vertex by rewriting the equation in the form $y - k = a(x - h)^2$:

$$y = x^2 - 2x + 5 \rightarrow y - 5 = x^2 - 2x \rightarrow y - 4 = x^2 - 2x + 1 \rightarrow y - 4 = (x - 1)^2$$

So the vertex is at (1,4). Opening upward from the vertex (1,4) already tells you that there are no x-intercepts, but, just to confirm this, use the quadratic formula on the equation $0 = x^2 - 2x + 5$ and you have $x = \frac{2 \pm \sqrt{4 - 4(1)(5)}}{2(1)} = \frac{2 \pm \sqrt{4 - 20}}{2} = \frac{2 \pm \sqrt{-16}}{2} = \frac{2 \pm 4i}{2} = 1 \pm 2i$. You have no x-intercepts, but you do have a y-intercept of (0,5). See the sketch of the parabola.

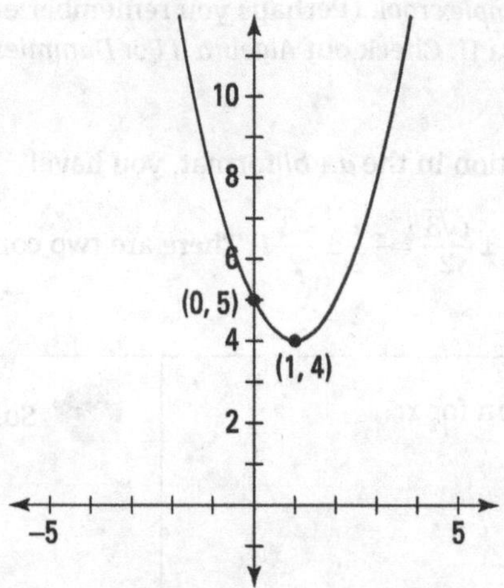

Q. Sketch the graph of the parabola $x = 2y^2 + y + 2$.

EXAMPLE

A. Since the coefficient of $2y^2$ is positive, you know that the parabola opens to the right. You find the vertex by rewriting the equation in the form $x - h = a(y - k)^2$:

$$x = 2y^2 + y + 2 \rightarrow x - 2 = 2\left(y^2 + \frac{1}{2}y\right) \rightarrow x - 2 + \frac{1}{16} = 2\left(y^2 + \frac{1}{2}y + \frac{1}{16}\right) \rightarrow x - \frac{31}{16} = 2\left(y + \frac{1}{4}\right)^2$$

So the vertex is at $\left(\frac{31}{16}, -\frac{1}{4}\right)$. Opening right from the vertex $\left(\frac{31}{16}, -\frac{1}{4}\right)$ already tells you that there are no y-intercepts, but, just to confirm this, use the quadratic formula on the equation $x = 2y^2 + y + 2$ and you have $y = \frac{-1 \pm \sqrt{1 - 4(2)(2)}}{2(2)} = \frac{-1 \pm \sqrt{1 - 16}}{4} = \frac{-1 \pm \sqrt{-15}}{4} = \frac{-1 \pm i\sqrt{15}}{4} = -\frac{1}{4} \pm i\frac{\sqrt{15}}{4}$. You have no y-intercepts, but you do have an x-intercept of (2,0). See the sketch of the parabola.

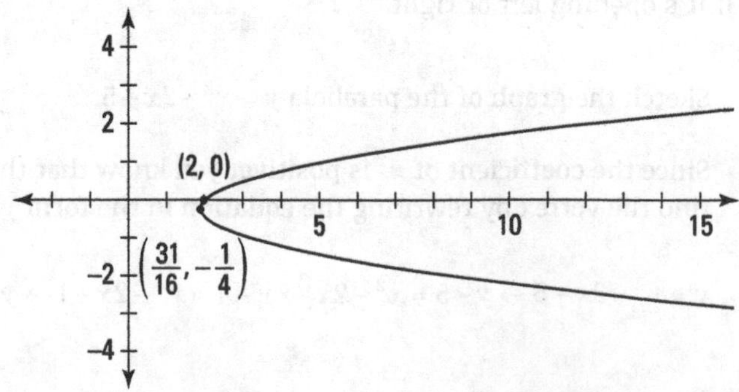

Keeping pace with polynomials

A polynomial function can have all real roots, or it can have no real roots. It can also have some of each type. Even though a complex root doesn't give you a point on the graph, it can tell you something about how the graph of the function is "flattening." The flattening-out is what keeps the graph of the function from crossing the x-axis.

EXAMPLE

Q. Determine any flattening of $y = x^4 + 11x^3 - 11x^2 - 27x - 36$ by finding its complex roots.

A. First, using the Rational Root Theorem, you find choices for roots of $\pm 1, \pm 2, \pm 3, \pm 4, \pm 5, \pm 9, \pm 12, \pm 18, \pm 36$. Whew! Lots of choices. But let's assume you wisely picked +4 and −3 and used synthetic division as follows:

4	1	1	−11	−27	−36
		4	20	36	36
−3	1	5	9	9	
		−3	−6	−9	
	1	2	3		

Now you have the quadratic $x^2 + 2x + 3 = 0$, which doesn't factor and which has two imaginary roots:

$$x = \frac{-2 \pm \sqrt{4 - 4(1)(3)}}{2(1)} = \frac{-2 \pm \sqrt{4 - 12}}{2} = \frac{-2 \pm \sqrt{-8}}{2} = \frac{-2 \pm \sqrt{-1}\sqrt{4}\sqrt{2}}{2} = \frac{-2 \pm 2i\sqrt{2}}{2} = -1 \pm i\sqrt{2}$$

This indicates that there's a flattening of the curve rather than it coming up to cross the x-axis two more times. Look at its graph.

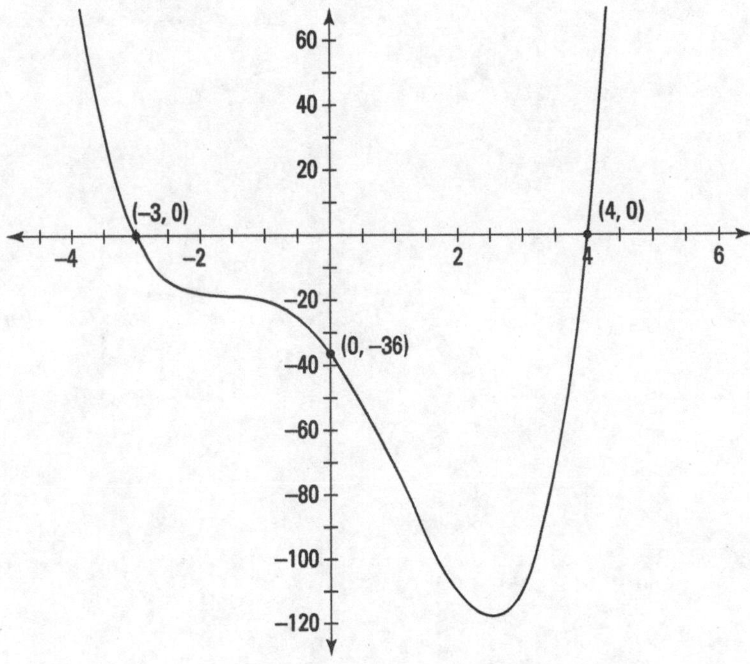

YOUR TURN

15 **Determine the vertex, any intercepts, and the direction in which the parabola opens when given $y = -x^2 - 5x - 11$.**

16 **Determine any intercepts, if there is flattening, and the end-behavior of the polynomial $y = x^4 - 1$.**

Practice Questions Answers and Explanations

1. **$4i\sqrt{5}$.** $\sqrt{-80} = \sqrt{-1}\sqrt{80} = \sqrt{-1}\sqrt{16}\sqrt{5} = 4i\sqrt{5}$

2. **$4i\sqrt{6}$. First, simplify under the radical. Then rewrite the result.**

   $\sqrt{16-4(4)(7)} = \sqrt{16-112} = \sqrt{-96} = \sqrt{-1}\sqrt{96} = \sqrt{-1}\sqrt{16}\sqrt{6} = 4i\sqrt{6}$

3. **$-1-2i$.** Combine the like-terms by adding: $(2+3i)+(-3-5i) = (2+(-3))+(3i+(-5i)) = -1-2i$.

4. **$5+8i$.** First, distribute the negative over the terms in the second parentheses. Then combine like terms by adding: $(2+3i)-(-3-5i) = (2+3i)+(3+5i) = (2+3)+(3i+5i) = 5+8i$.

5. **$-12+24i$.** First, distribute the $4i$ over the two terms: $4i(6+3i) = 24i+12i^2$. Replace the i^2 with -1: $24i+12i^2 = 24i+12(-1) = 24i-12$.

6. **$23+14i$.** Using FOIL, $(5-2i)(3+4i) = 15+20i-6i-8i^2 = 15+14i-8(-1) = 23+14i$.

7. **$\dfrac{3+11i}{13}$.** Multiply the numerator and denominator by the conjugate of the denominator:

   $$\frac{3+i}{2-3i}\cdot\frac{2+3i}{2+3i} = \frac{6+9i+2i+3i^2}{4-9i^2} = \frac{6+11i+3(-1)}{4-9(-1)} = \frac{3+11i}{13}.$$

8. **$11+11i$.** First, distribute; then combine like terms.

   $4(2-i)+i(6-3i)+9i = 8-4i+6i-3i^2+9i = 8+11i-3(-1) = 11+11i$

   For Questions 9 to 12, refer to the following figure. The number i can be written as $0+i$ to indicate the real portion. The letters A, B, C, D corresponding to problems as follows: A – 9, B – 10, C – 11, D – 12.

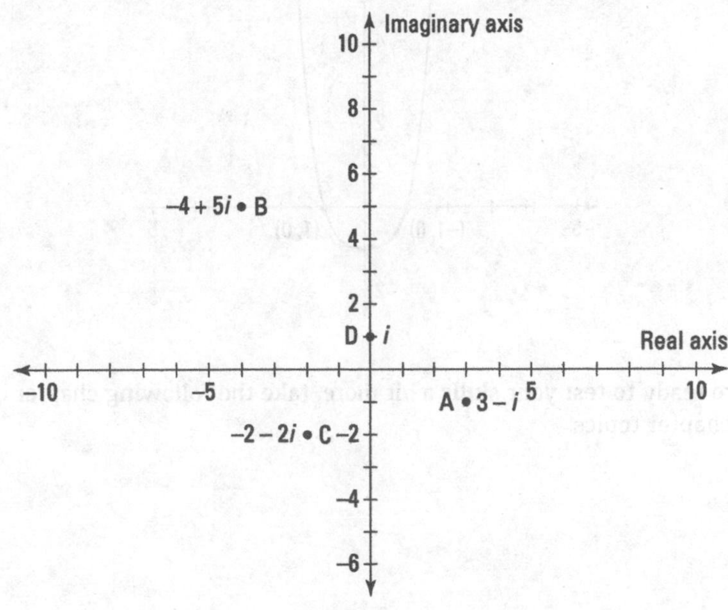

13) $-1\pm\frac{\sqrt{2}}{2}i$. **Applying the quadratic formula, you have**

$$x=\frac{-4\pm\sqrt{16-4(2)(3)}}{2(2)}=\frac{-4\pm\sqrt{-8}}{4}=\frac{-4\pm2\sqrt{-2}}{4}=\frac{-4}{4}\pm\frac{2\sqrt{-2}}{4}=-1\pm\frac{\sqrt{2}}{2}i$$

14) $\pm5i$. **Subtract 25 from both sides and then take their square roots.**

$$x^2+25=0\to x^2=-25\to\sqrt{x^2}=\pm\sqrt{-25}\to x=\pm5i$$

15) $V:\left(-\frac{5}{2},-\frac{19}{4}\right)$; **(0,−11); downward.** To find the vertex, put the equation in the vertex form:

$$y=-x^2-5x-11\to y+11=-1\left(x^2+5x\right)\to y+11-\frac{25}{4}=-1\left(x^2+5x+\frac{25}{4}\right)\to y+\frac{19}{4}=-1\left(x+\frac{5}{2}\right)^2$$

Because of the negative coefficient on the squared term, the parabola opens downward. The only intercept is $(0,-11)$. If you had looked for x-intercepts and solved $0=-x^2-5x-11$, you would have $x=\frac{5\pm\sqrt{25-4(-1)(-11)}}{2(-1)}=\frac{5\pm\sqrt{25-44}}{-2}=\frac{5\pm\sqrt{-19}}{-2}=\frac{5\pm i\sqrt{19}}{-2}$. The two solutions are imaginary numbers.

16) **(−1,0), (1,0); flattening between intercepts;** $y\to\infty$ **as** $x\to\pm\infty$. Factoring the function rule, $y=x^4-1=(x-1)(x+1)\left(x^2+1\right)$. Setting this equal to 0, the four solutions are 1, −1, and $\pm i$. The two imaginary solutions cause the curve to flatten, and the curve opens upward, approaching infinity. See the following graph.

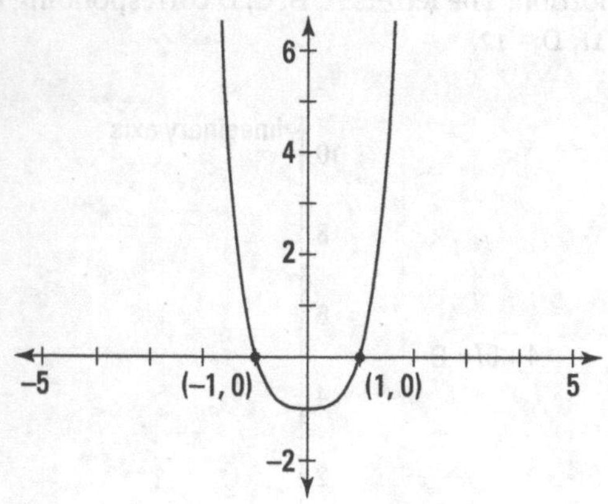

If you're ready to test your skills a bit more, take the following chapter quiz that incorporates all the chapter topics.

Whaddya Know? Chapter 15 Quiz

Quiz time! Complete each problem to test your knowledge on the various topics covered in this chapter. You can then find the solutions and explanations in the next section.

1. Given the graph of points A, B, and C, identify what imaginary numbers they represent.

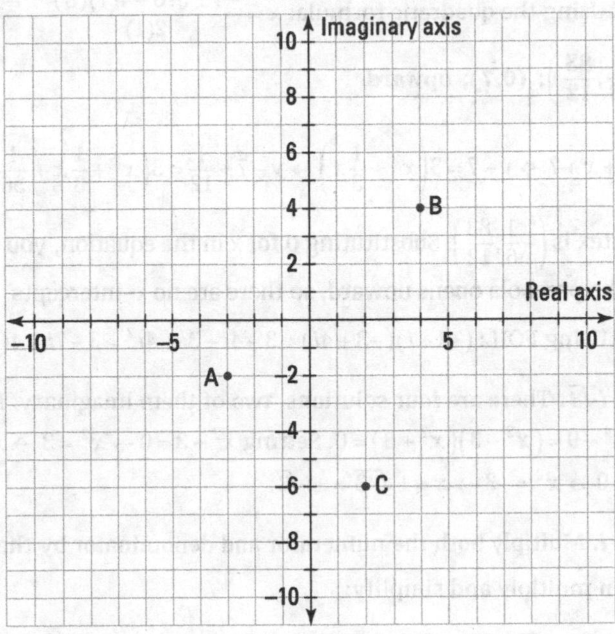

2. Simplify $(3-2i)-4i(3+i)$
3. Solve for x: $x^2+4x+8=0$.
4. Given $y=3x^2+x+7$, determine the vertex, any intercepts, and the direction in which the parabola opens.
5. Multiply $(-1+i)(-3+4i)$
6. Solve for x: $x^4-9=0$.
7. Divide $\dfrac{6-2i}{5+3i}$
8. Given $x=-y^2+y+1$, determine the vertex, any intercepts, and the direction in which the parabola opens.

Answers to Chapter 15 Quiz

1. **A: $-3-2i$; B: $4+4i$; C: $2-6i$**

2. **$7-14i$. First distribute, then simplify and combine like terms:**

   $(3-2i)-4i(3+i)=3-2i-12i-4i^2=3-14i-4(-1)=7-14i$

3. **$-2\pm 2i$. Using the quadratic formula:** $x=\frac{-4\pm\sqrt{16-4(1)(8)}}{2(1)}=\frac{-4\pm\sqrt{-16}}{2}=\frac{-4\pm 4i}{2}=-2\pm 2i$.

4. **$V:\left(-\frac{1}{6},\frac{83}{12}\right)$; (0,7); upward**

   $$y=3x^2+x+7\rightarrow y-7=3\left(x^2+\frac{1}{3}x\right)\rightarrow y-7+\frac{1}{12}=3\left(x^2+\frac{1}{3}x+\frac{1}{36}\right)\rightarrow y-\frac{83}{12}=3\left(x+\frac{1}{6}\right)^2$$

   The vertex is $\left(-\frac{1}{6},\frac{83}{12}\right)$. Substituting 0 for x in the equation, you have the y-intercept of $(0,7)$. The parabola opens upward, so there are no x-intercepts.

5. **$-1-7i$. Using FOIL:** $(-1+i)(-3+4i)=3-4i-3i+4i^2=3-7i+4(-1)=-1-7i$.

6. **$\pm\sqrt{3}$, $\pm i\sqrt{3}$. There are four solutions, two of them imaginary. Factoring, you have:** $x^4-9=\left(x^2-3\right)\left(x^2+3\right)=0$. Setting $x^2-3=0\rightarrow x^2=3\rightarrow x=\pm\sqrt{3}$. When $x^2+3=0\rightarrow x^2=-3\rightarrow x=\pm\sqrt{-3}=\pm i\sqrt{3}$.

7. **$\frac{12}{17}-\frac{14}{17}i$. Multiply both the numerator and denominator by the conjugate of the denominator, then multiply and simplify:**

   $$\frac{6-2i}{5+3i}\cdot\frac{5-3i}{5-3i}=\frac{30-18i-10i+6i^2}{25-9i^2}=\frac{30-28i+6(-1)}{25-9(-1)}=\frac{24-28i}{34}=\frac{12-14i}{17}$$

   The answer, written as a complex number, is $\frac{12}{17}-\frac{14}{17}i$.

8. **$V:\left(\frac{1}{2},\frac{5}{4}\right)$; (1,0); opens left. Putting the equation in the vertex form:**

   $$x=-y^2+y+1\rightarrow x-1=-1\left(y^2-y\right)\rightarrow x-1-\frac{1}{4}=-1\left(y^2-y+\frac{1}{4}\right)\rightarrow x-\frac{5}{4}=-1\left(y-\frac{1}{2}\right)^2$$

   Letting $y=0$, you have $x=1$. The negative coefficient on y makes the parabola open to the left.

IN THIS CHAPTER

» Picturing polar coordinates

» Changing from polar coordinates to rectangular and vice versa

» Graphing polar equations

Chapter 16

Warming Up to Polar Coordinates

Polar coordinates and polar graphs are two interesting but often neglected topics in a standard study of pre-calculus. In this chapter, you'll be introduced to the basics and given enough tools to allow you to work in both applications.

Polar coordinates and graphs do just what their names imply: Coordinates go around the pole (which you'll recognize as the origin from previous graphing), and polar graphing involves equations that are tied to angles and trig functions. A whole new world awaits you.

In this chapter, you move from rectangular coordinates to polar coordinates. You see where they come from and how you use them (as well as graph them).

Plotting around a Pole: Polar Coordinates

Polar coordinates are an extremely useful addition to your mathematics tool kit because they allow you to solve problems that would be extremely difficult if you were to rely on standard x- and y-coordinates. For example, you can use polar coordinates on problems where the relationship between two quantities is most easily described in terms of the angle and distance between them, such as navigation or antenna signals. Instead of relying on the x- and y-axes as reference points, polar coordinates are based on distances along the positive x-axis (the line starting at the origin and continuing in the positive horizontal direction forever). From this line, you measure an angle (which you call theta, or θ) and a length (or radius) along the terminal side of the angle (which you call r). These coordinates replace x- and y-coordinates.

REMEMBER

In polar coordinates, you always write the ordered pair as (r,θ). For instance, a polar coordinate could be $\left(5,\frac{\pi}{6}\right)$ where the radius is 5 and the angle is $\frac{\pi}{6}$, or $(-3,\pi)$ where the radius is 3 and the angle is π. Oops! Did you notice the negative sign on the 3? That just tells you in which direction to graph the radius. More on that later.

In the following sections, you see how to graph points in polar coordinates and how to graph equations as well. You also discover how to change back and forth between Cartesian coordinates and polar coordinates.

Wrapping your brain around the polar coordinate plane

In order to fully grasp how to plot points in the form (r,θ), you need to see what a polar coordinate plane looks like. In Figure 16-1, you can see that the plane is no longer a grid of rectangular coordinates; instead, it's a series of concentric circles around a central point, called the *pole*. The plane appears this way because the polar coordinates are a given radius and a given angle in standard position around the pole. Each circle in the graph represents one radius unit, and each line represents the terminal sides of special angles from the unit circle. (To make finding the angles easier, see Chapter 8.)

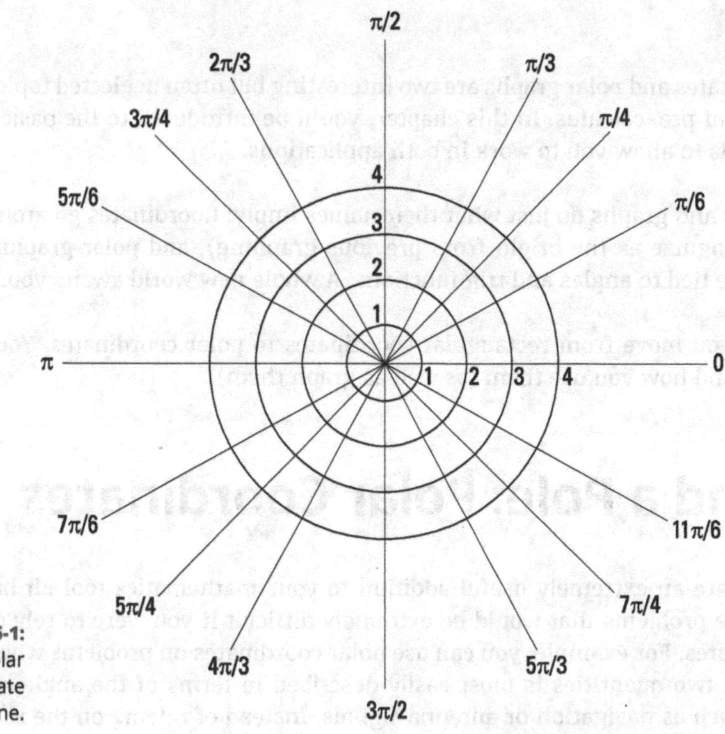

FIGURE 16-1: A polar coordinate plane.

TECHNICAL STUFF

Although θ and r may seem strange as plotting points at first, they're really no more or less strange or useful than x and y. In fact, when you consider a sphere such as Earth, describing points on, above, or below its surface is much more straightforward with polar coordinates on a round coordinate plane.

TIP

Because you write all points on the polar plane as (r,θ), in order to graph a point, it's best to find θ first and then locate r on the line corresponding to θ. This approach allows you to narrow the location of a point to somewhere on one of the lines representing the angles. From there, you can simply count out from the pole the radial distance. If you go the other way and start with r, you may find yourself in a pickle when the problems get more complicated.

EXAMPLE

Q. Plot the point E at $\left(2, \frac{\pi}{3}\right)$.

A. Point E is shown in Figure 16-2. Here's how to graph it.

1. **Locate the angle on the polar coordinate plane.**

   Move counterclockwise from the positive x-axis until you reach the appropriate angle. Refer to Figure 16-1 to find the angle: $\theta = \frac{\pi}{3}$.

2. **Determine where the radius intersects the angle.**

   Because the radius is 2 $(r = 2)$, you start at the pole and move out 2 units along the terminal side of the angle.

3. **Plot the given point.**

   At the intersection of the radius and the angle on the polar coordinate plane, plot a dot and call it a day! Figure 16-2 shows point E on the plane.

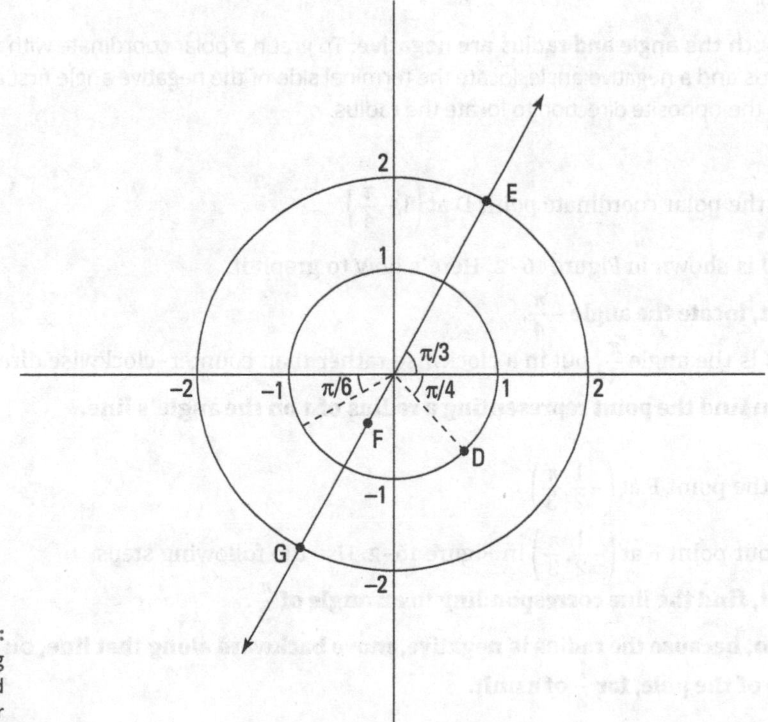

FIGURE 16-2: Visualizing simple and complex polar coordinates.

Polar coordinate pairs can have positive angles or negative angles for values of θ. In addition, they can have positive and negative radii. This concept is probably new to you, as you may have thought that a radius must be positive. When graphing polar coordinates, though, the radius can be negative, which means that you move in the *opposite* direction of the angle's terminal side on the other side of the pole.

REMEMBER

Because polar coordinates are based on angles, unlike Cartesian coordinates, a single point can be represented by many different ordered pairs. Infinitely many values of θ have the same angle in standard position (see Chapter 8), so an infinite number of coordinate pairs describe the same point. Also, a positive and a negative co-terminal angle can describe the same point for the same radius, and because the radius can be either positive or negative, you can express the point with polar coordinates in many ways.

Graphing polar coordinates with negative values

Simple polar coordinates are points where both the radius and the angle are positive. You see the steps for graphing these coordinates in the previous section. But you also must prepare yourself for when you're faced with points that have negative angles and/or radii. The following list shows you how to plot in three situations: when the angle is negative, when the radius is negative, and when both are negative.

TIP

- **When the angle is negative:** Negative angles move in a clockwise direction (see Chapter 8 for more on these angles). Check out Figure 16-2 to see an example point, D.

TIP

- **When the radius is negative:** When graphing a polar coordinate with a negative radius (essentially the *x* value), you move from the pole in the direction opposite the given positive angle (on the same line as the given angle but in the direction opposite from the angle from the pole).

TIP

- **When both the angle and radius are negative:** To graph a polar coordinate with a negative radius and a negative angle, locate the terminal side of the negative angle first and then move in the opposite direction to locate the radius.

EXAMPLE

Q. Locate the polar coordinate point D at $\left(1,-\frac{\pi}{4}\right)$.

A. Point D is shown in Figure 16-2. Here's how to graph it.

1. **First, locate the angle $-\frac{\pi}{4}$.**

   This is the angle $\frac{\pi}{4}$, but in a clockwise rather than counter-clockwise direction.

2. **Then find the point representing a radius of 1 on the angle's line.**

EXAMPLE

Q. Graph the point F at $\left(-\frac{1}{2},\frac{\pi}{3}\right)$.

A. Check out point F at $\left(-\frac{1}{2},\frac{\pi}{3}\right)$ in Figure 16-2. Use the following steps.

1. **First, find the line corresponding to an angle of $\frac{\pi}{3}$.**

2. **Then, because the radius is negative, move backward along that line, on the other side of the pole, for $\frac{1}{2}$ of a unit.**

Q. Graph the point G at $\left(-2, -\frac{5\pi}{3}\right)$.

EXAMPLE

A. You find point G in Figure 16-2. Here are the steps to determining where it is.

1. **First, find the line corresponding to an angle of $-\frac{5\pi}{3}$.**

   It may help to remember that $-\frac{5\pi}{3}$ has the same terminal side as $\frac{\pi}{3}$. (Add 2π to $-\frac{5\pi}{3}$ and you get $\frac{\pi}{3}$ — a coterminal angle.)

2. **Then, using the line for $-\frac{5\pi}{3}$, move 2 units on the other side of the pole, downward and to the left.**

TIP

Indeed, except for the origin, each given point can have the following four types of representations:

- Positive radius, positive angle
- Positive radius, negative angle
- Negative radius, positive angle
- Negative radius, negative angle

Q. Determine the polar representations of point E in Figure 16-2, $\left(2, \frac{\pi}{3}\right)$.

EXAMPLE

A. Point E in Figure 16-2, $\left(2, \frac{\pi}{3}\right)$, can have three other polar coordinate representations with different combinations of signs for the radius and angle:

- $\left(2, -\frac{5\pi}{3}\right)$, because $\frac{\pi}{3}$ and $-\frac{5\pi}{3}$ are different by one rotation.
- $\left(-2, \frac{4\pi}{3}\right)$, because –2 is on the opposite side of the pole and $\frac{4\pi}{3}$ is the angle whose corresponding line is along that of $\frac{\pi}{3}$.
- $\left(-2, -\frac{2\pi}{3}\right)$, because –2 is on the opposite side of the pole and $-\frac{2\pi}{3}$ is half a rotation different from $\frac{\pi}{3}$.

When polar graphing, you can change the coordinate of any point you're given into polar coordinates that are easier to deal with (such as positive radius, positive angle).

For Questions 1 to 4, plot the following points on the polar graph.

YOUR TURN

1 $\left(3, \frac{3\pi}{4}\right)$

2 $\left(1, -\frac{\pi}{3}\right)$

3 $\left(-2, \frac{\pi}{6}\right)$

4 $\left(-4, -\frac{2\pi}{3}\right)$

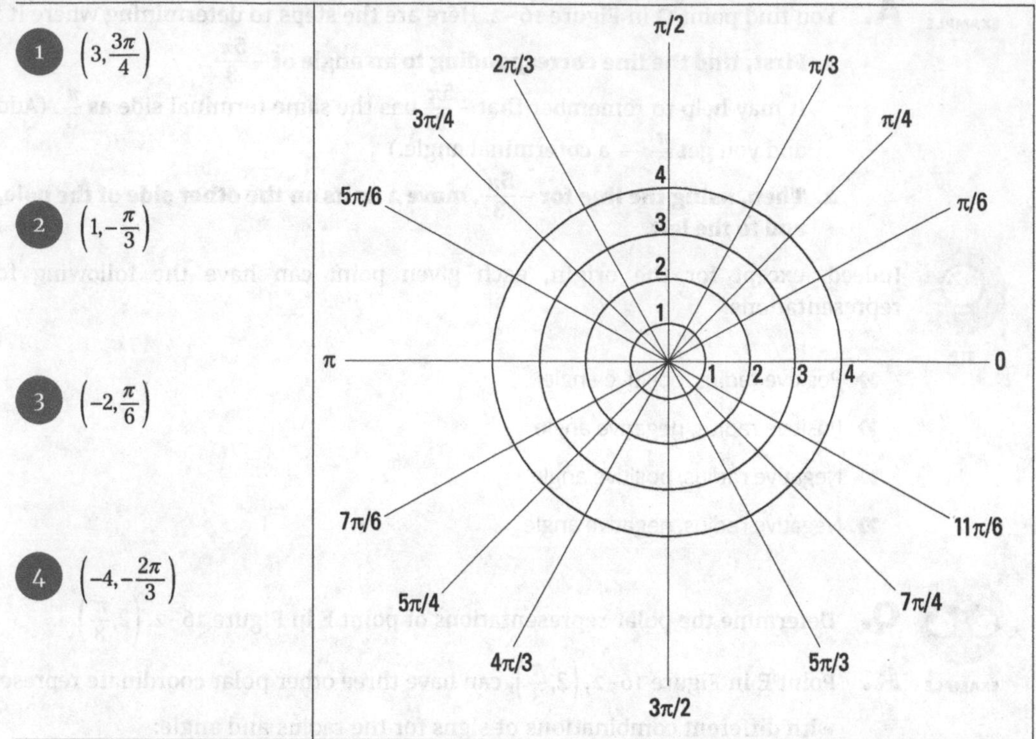

Changing to and from polar coordinates

You can use both polar coordinates and (*x*,*y*) coordinates at any time to describe the same location on the coordinate plane. Sometimes you'll have an easier time using one form rather than the other, and for this reason you'll now see how to navigate between the two. Cartesian coordinates are much better suited for graphs of straight lines or simple curves. Polar coordinates can yield you a variety of pretty, very complex graphs that you can't plot with Cartesian coordinates.

TIP

When changing to and from polar coordinates, your work is often easier if you have all your angle measures in radians. You can make the change by using the conversion factor $180° = \pi$ radians. You may choose, however, to leave your angle measures in degrees, which is fine as long as your calculator is in the right mode.

Devising the changing equations

Examine the point in Figure 16-3, which illustrates a point mapped out in both (*x*,*y*) and (r,θ) coordinates, allowing you to see the relationship between them.

What exactly is the geometric relationship between r, θ, x, and y? Look at how they're labeled on the graph — all parts of the same triangle!

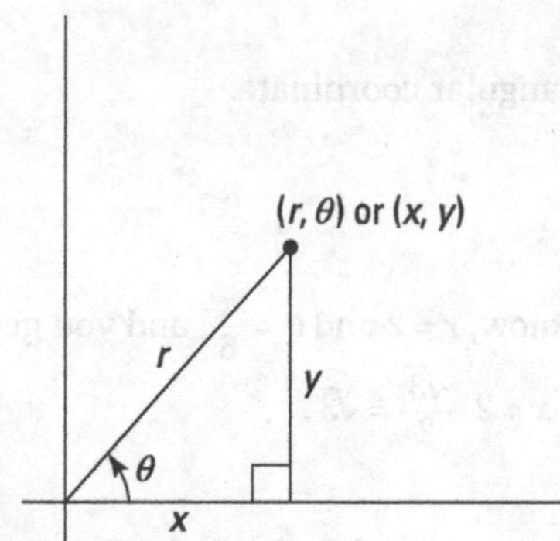

FIGURE 16-3: A polar and (x,y) coordinate mapped in the same plane.

REMEMBER

Using right-triangle trigonometry (see Chapter 8), you know the following facts: $\sin\theta = \frac{y}{r}$ and $\cos\theta = \frac{x}{r}$.

These equations simplify into two very important expressions for x and y in terms of r and θ: $\sin\theta = \frac{y}{r} \rightarrow y = r\sin\theta$ and $\cos\theta = \frac{x}{r} \rightarrow x = r\cos\theta$.

Furthermore, you can use the Pythagorean Theorem in the right triangle of Figure 16-3 to find the radius of the triangle if given x and y:

Since $x^2 + y^2 = r^2$, $r = \sqrt{x^2 + y^2}$.

One final equation allows you to find the angle θ; it derives from the tangent of the angle: $\tan\theta = \frac{y}{x}$.

Solve this equation for θ, and you get the following expression: $\theta = \tan^{-1}\frac{y}{x}$.

WARNING

With respect to the final equation, $\theta = \tan^{-1}\frac{y}{x}$, keep in mind that your calculator always returns a value of tangent that puts θ in the first or fourth quadrant. You need to look at your x- and y-coordinates and decide whether that placement is actually correct for the problem at hand. Your calculator doesn't look for tangent possibilities in the second and third quadrants, but that doesn't mean you don't have to!

REMEMBER

As with degrees (see the earlier section, "Wrapping your brain around the polar coordinate plane"), you can add or subtract 2π to any angle to get a co-terminal angle so you have more than one way to name every point in polar coordinates. In fact, there are infinite ways of naming the same point. For instance, $\left(2,\frac{\pi}{3}\right)$, $\left(2,-\frac{5\pi}{3}\right)$, $\left(-2,\frac{4\pi}{3}\right)$, $\left(2,-\frac{2\pi}{3}\right)$, and $\left(2,\frac{7\pi}{3}\right)$ are several ways of naming the same point.

Putting the equations into action

Together, the four equations for r, θ, x, and y allow you to change (x,y) coordinates into polar (r,θ) coordinates and back again anytime.

Q. Change the polar coordinate $\left(2, \frac{\pi}{6}\right)$ to a rectangular coordinate.

EXAMPLE **A.** Follow these steps:

1. **Find the x value.**

   Since $x = r\cos q$, substitute in what you know, $r = 2$ and $\theta = \frac{\pi}{6}$, and you get $x = 2\cos\frac{\pi}{6}$.

   Substitute in the value of the cosine and $x = 2 \cdot \frac{\sqrt{3}}{2} = \sqrt{3}$.

2. **Find the y value.**

   With $y = r\sin\theta$, substitute what you know to get $y = 2\sin\frac{\pi}{6} = 2 \cdot \frac{1}{2} = 1$.

3. **Express the values from Steps 1 and 2 as a coordinate point.**

   You found that $\left(2, \frac{\pi}{6}\right)$ in the polar graph corresponds to the point $(\sqrt{3}, 1)$ in a rectangular coordinate system. Figure 16-4 shows the two points graphed on their respective coordinate systems. The points $(\sqrt{3}, 1)$ and $\left(2, \frac{\pi}{6}\right)$ are both labeled A. Figure 16-4a is the rectangular graph, and 16-4b is the polar graph.

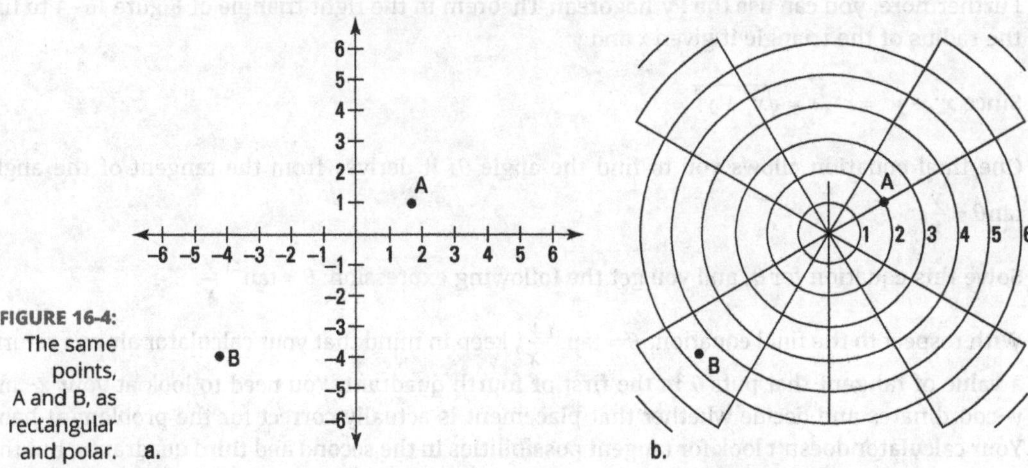

FIGURE 16-4: The same points, A and B, as rectangular and polar.

Now it's time for an example in reverse.

Q. Given the point $(-4, -4)$, find the equivalent polar coordinate.

EXAMPLE **A.** Use the following steps:

1. **Plot the (x,y) point first.**

   Figure 16-4a shows the location of the point, labeled B, in quadrant III.

2. **Find the r value.**

   For this step, you use the Pythagorean Theorem for polar coordinates: $x^2 + y^2 = r^2$. Plug in what you know, $x = -4$ and $y = -4$, to get $(-4)^2 + (-4)^2 = r^2$. Solving for r, you get $r = 4\sqrt{2}$.

3. **Find the value of θ.**

   Use the tangent ratio for polar coordinates: $\theta = \tan^{-1}\frac{y}{x} = \tan^{-1}\frac{-4}{-4} = \tan^{-1}1$. The reference angle for this value is $\theta' = \frac{\pi}{4}$ (see Chapter 8). You know from Figure 16-4 that the point is in the third quadrant, so $\theta = \frac{5\pi}{4}$.

4. **Express the values of Steps 2 and 3 as a polar coordinate.**

   The point in the rectangular system, $(-4,-4)$, is the same as $\left(4\sqrt{2}, \frac{5\pi}{4}\right)$ in the polar system. These points are both labeled B in the Figure 16-4 graphs.

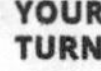

YOUR TURN

For Questions 5 and 6, graph the points on both the rectangular graph and polar graph.

5 $\left(1, \frac{2\pi}{3}\right)$

6 $(3,-3)$

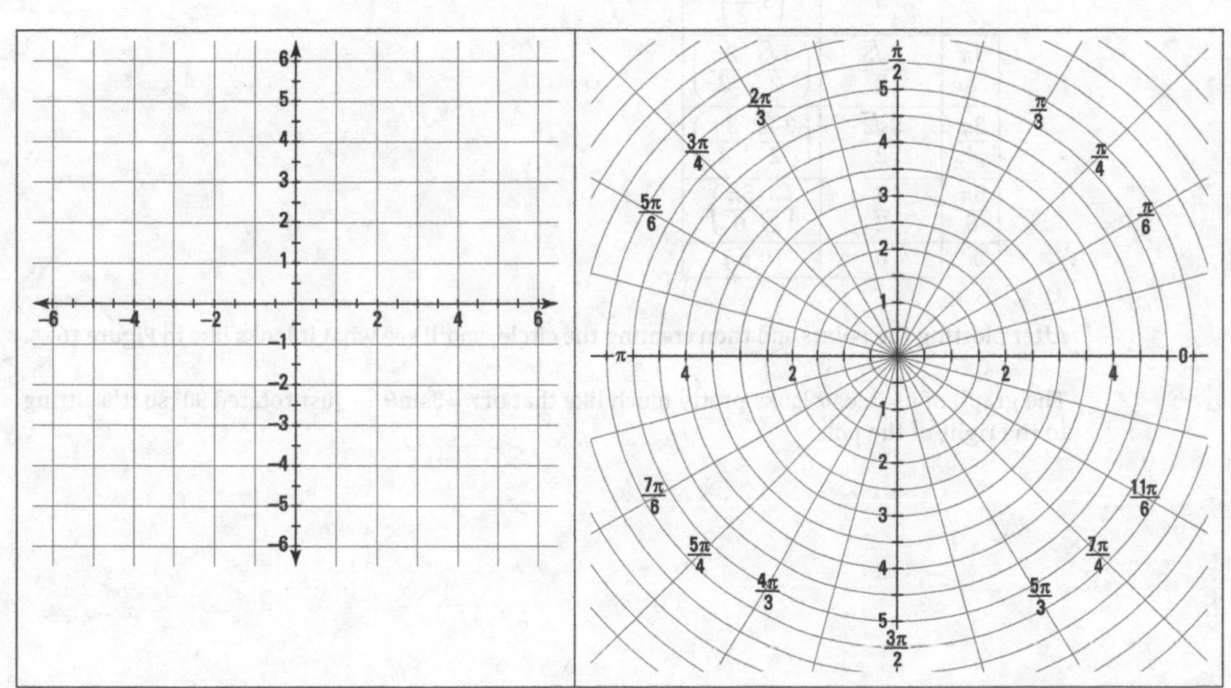

Picturing polar equations

The equation of a circle on the rectangular coordinate system has the form $x^2 + y^2 = r^2$, where r is the radius of the circle. Circles of this form have their center at the origin. The polar equations for a circle are quite different — and much simpler. In fact, there are many rather lovely curves that you can create in a polar graph with very simple statements.

Circling around with two circles

There are two basic equations for the graph of a circle using polar coordinates: $r = a\sin\theta$ and $r = a\cos\theta$. One sits on top of the pole and the other to the side. The multiplier a indicates the diameter of the circle, and the θ represents all the input values in terms of radians.

EXAMPLE

Q. Graph the circle $r = 3\sin\theta$ on a polar graph.

A. First, find a few points on the circle $r = 3\sin\theta$.

θ	$r = 3\sin\theta$	(r,θ)
0	0	$(0,0)$
$\frac{\pi}{6}$	$\frac{3}{2}$	$\left(\frac{3}{2},\frac{\pi}{6}\right)$
$\frac{\pi}{4}$	$\frac{3\sqrt{2}}{2}$	$\left(\frac{3\sqrt{2}}{2},\frac{\pi}{4}\right)$
$\frac{\pi}{3}$	$\frac{3\sqrt{3}}{2}$	$\left(\frac{3\sqrt{3}}{2},\frac{\pi}{3}\right)$
$\frac{\pi}{2}$	3	$\left(3,\frac{\pi}{2}\right)$
$\frac{4\pi}{3}$	$\frac{3\sqrt{3}}{2}$	$\left(\frac{3\sqrt{3}}{2},\frac{2\pi}{3}\right)$
$\frac{3\pi}{4}$	$\frac{3\sqrt{2}}{2}$	$\left(\frac{3\sqrt{2}}{2},\frac{3\pi}{4}\right)$
$\frac{5\pi}{6}$	$\frac{3}{2}$	$\left(\frac{3}{2},\frac{5\pi}{6}\right)$
π	0	$(0,\pi)$

After plotting the points and then creating the circle, you'll see what it looks like in Figure 16-5.

The graph of $r = 3\cos\theta$ looks pretty much like that of $r = 3\sin\theta$ — just rotated 90° so it's sitting to the right of the pole.

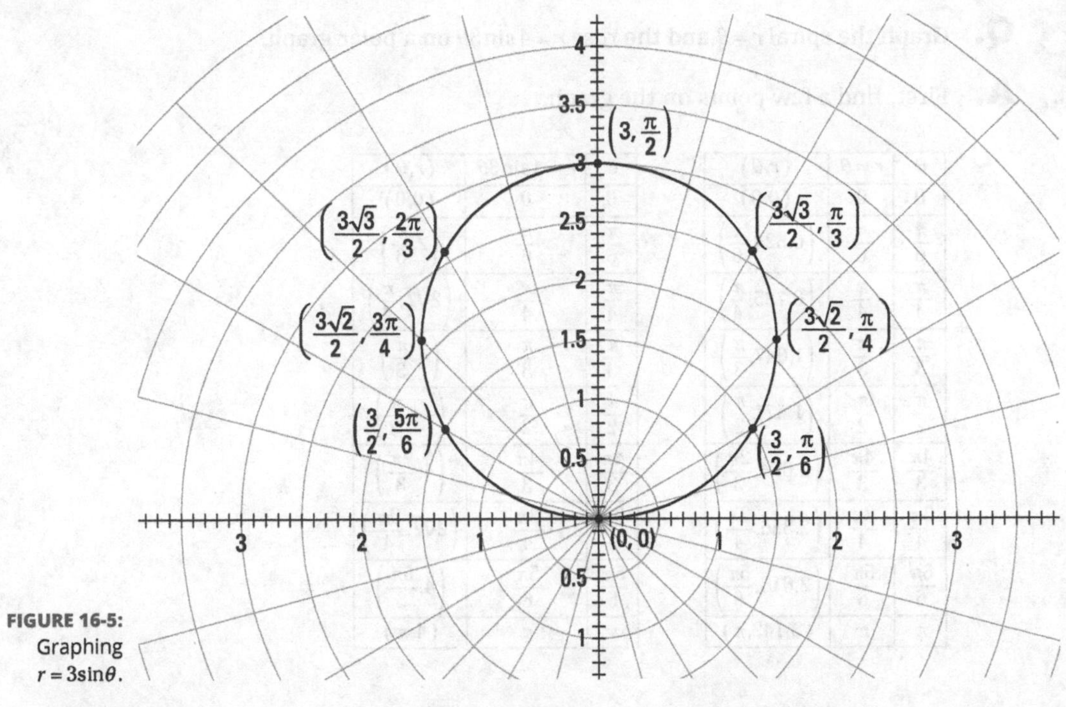

FIGURE 16-5: Graphing $r = 3\sin\theta$.

YOUR TURN

 Graph $r = 3\cos\theta$ on a polar graph.

 Graph $r = -4\sin\theta$ on a polar graph.

Spiraling and planting roses

Another polar graph is the Archimedes spiral. A spiral has the general equation $r = a\theta$. That's all! You multiply the measure of the angle — in radians — times some constant value. θ has to be in radians, because you need real number values in the multiplication. As the value of θ gets greater and greater, the spiral goes out farther and farther.

And a really lovely polar graph is the rose. In polar-speak, a rose can have any number of petals. It all depends on the equation format. The general equations for a rose are either $r = a\sin b\theta$ or $r = a\cos b\theta$. When the value of b is odd, there are b petals. When b is even, there are $2b$ petals.

EXAMPLE

Q. Graph the spiral $r = \theta$ and the rose $r = 4\sin 3\theta$ on a polar graph.

A. First, find a few points on the graph.

θ	$r = \theta$	(r,θ)
0	0	$(0,0)$
$\frac{\pi}{6}$	$\frac{\pi}{6}$	$\left(0.524, \frac{\pi}{6}\right)$
$\frac{\pi}{4}$	$\frac{\pi}{4}$	$\left(0.785, \frac{\pi}{4}\right)$
$\frac{\pi}{3}$	$\frac{\pi}{3}$	$\left(1.047, \frac{\pi}{3}\right)$
$\frac{\pi}{2}$	$\frac{\pi}{2}$	$\left(1.571, \frac{\pi}{2}\right)$
$\frac{4\pi}{3}$	$\frac{4\pi}{3}$	$\left(2.094, \frac{2\pi}{3}\right)$
$\frac{3\pi}{4}$	$\frac{3\pi}{4}$	$\left(2.356, \frac{3\pi}{4}\right)$
$\frac{5\pi}{6}$	$\frac{5\pi}{6}$	$\left(2.618, \frac{5\pi}{6}\right)$
π	π	$(3.142, \pi)$

θ	$r = 4\sin 3\theta$	(r,θ)
0	0	$(0,0)$
$\frac{\pi}{6}$	$\frac{\pi}{6}$	$\left(2, \frac{\pi}{6}\right)$
$\frac{\pi}{4}$	$\frac{\pi}{4}$	$\left(2\sqrt{2}, \frac{\pi}{4}\right)$
$\frac{\pi}{3}$	$\frac{\pi}{3}$	$\left(4, \frac{\pi}{3}\right)$
$\frac{\pi}{2}$	$\frac{\pi}{2}$	$\left(-4, \frac{\pi}{2}\right)$
$\frac{4\pi}{3}$	$\frac{4\pi}{3}$	$\left(4, \frac{2\pi}{3}\right)$
$\frac{3\pi}{4}$	$\frac{3\pi}{4}$	$\left(2\sqrt{2}, \frac{3\pi}{4}\right)$
$\frac{5\pi}{6}$	$\frac{5\pi}{6}$	$\left(4, \frac{5\pi}{6}\right)$
π	π	$(4, \pi)$

Figure 16-6a shows you $r = \theta$, and Figure 16-6b shows $r = 4\sin 3\theta$.

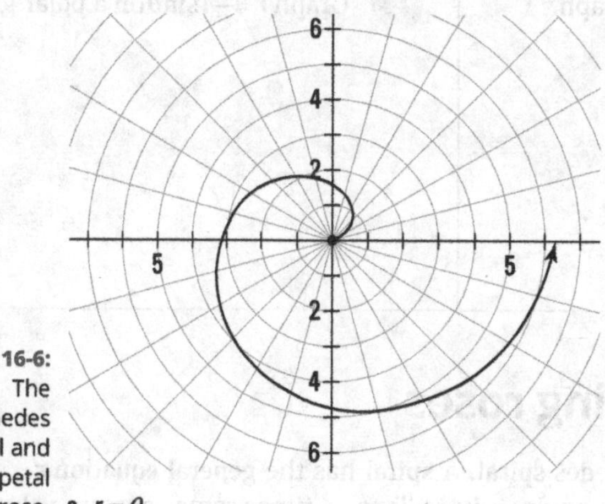

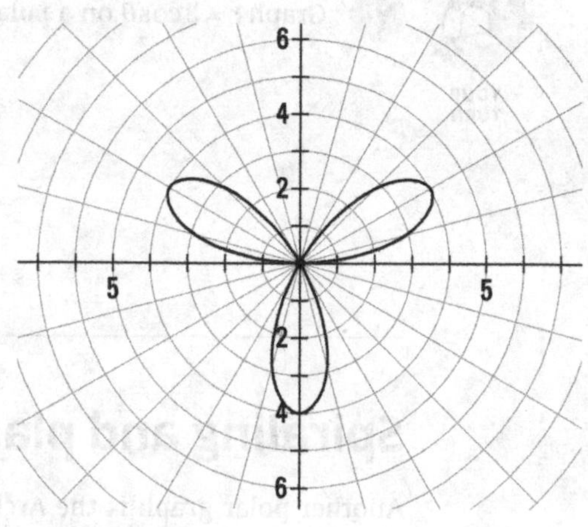

FIGURE 16-6: The Archimedes spiral and three-petal rose. **a.** $r = \theta$ **b.** $r = 4\sin 3\theta$

YOUR TURN

9 Sketch the graph of the spiral $r = 2\theta$.

Sketch the graph of the rose $r = 2\sin 2\theta$.

Practice Questions Answers and Explanations

Refer to the figure for Questions 1 to 4.

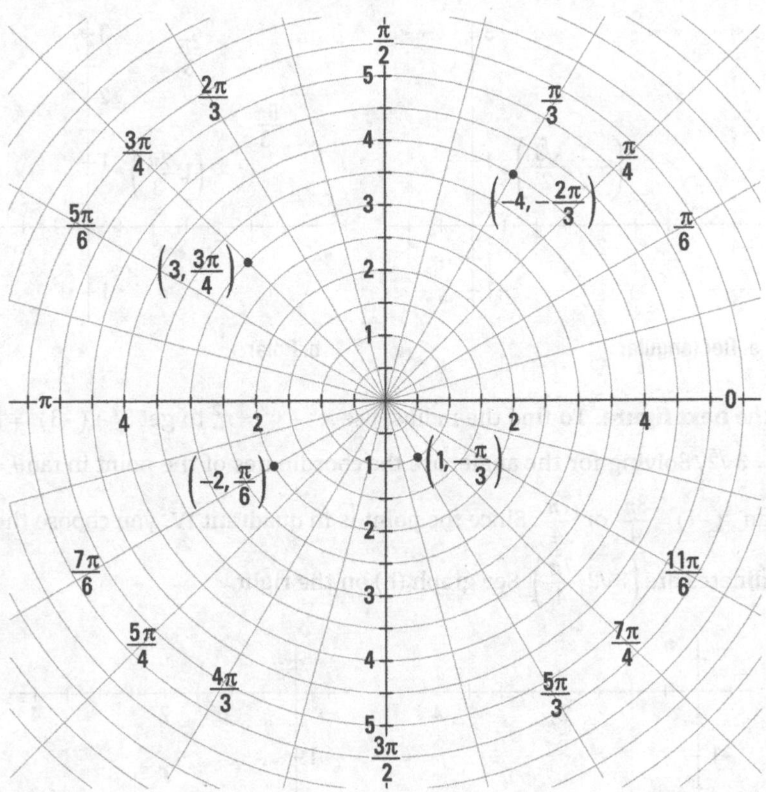

1. **See the figure.** First, find the angle (upper left); then move 3 units out for a radius of 3.
2. **See the figure.** First, find the angle (lower right; $-\frac{\pi}{3}$ has same terminal side as $\frac{5\pi}{3}$). Then move 1 unit out for a radius of 1.
3. **See the figure.** First, find the angle (upper right). Then move 2 units on the opposite side of the center point.
4. **See the figure.** First, find the angle (lower left; $\frac{4\pi}{3}$ has the same terminal side as $-\frac{2\pi}{3}$). Then move 4 units to the opposite side of the center point.

5. **See the next figure.** Using $x = r\cos\theta$, you have $x = 1 \cdot \cos\frac{2\pi}{3} = -\frac{1}{2}$. See graph (a) on the left. Then, using $y = r\sin\theta$, $y = 1 \cdot \sin\frac{2\pi}{3} = \frac{\sqrt{3}}{2}$. So the point on the polar graph is $\left(-\frac{1}{2}, \frac{\sqrt{3}}{2}\right)$. See graph (b), on the right.

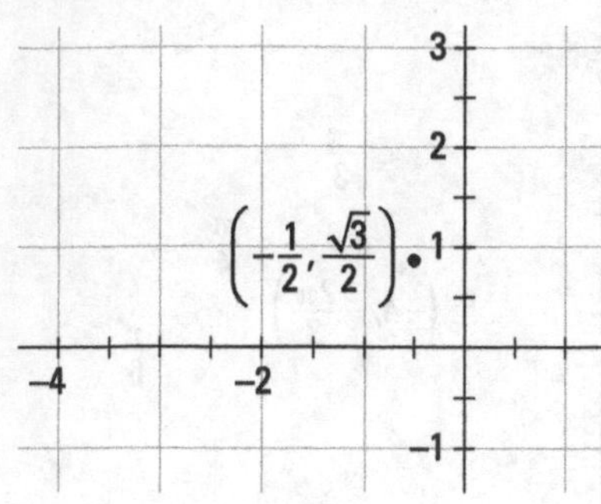

a. Rectangular

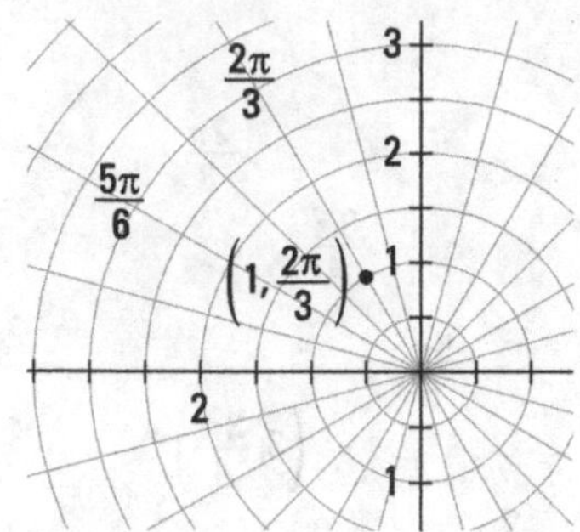

b. Polar

6. **See the next figure.** To find the radius, use $x^2 + y^2 = r^2$ to get $3^2 + (-3)^2 = 18 = r^2$. So $r = 3\sqrt{2}$. Solving for the angle, use the coordinates of the point in $\tan\theta = \frac{y}{x} = \frac{-3}{3} = -1$. Then $\theta = \tan^{-1}(-1) = \frac{3\pi}{4}$ or $\frac{7\pi}{4}$. Since the point is in quadrant IV, you choose the $\frac{7\pi}{4}$, and the polar coordinates are $\left(3\sqrt{2}, \frac{7\pi}{4}\right)$. See graph (b) on the right.

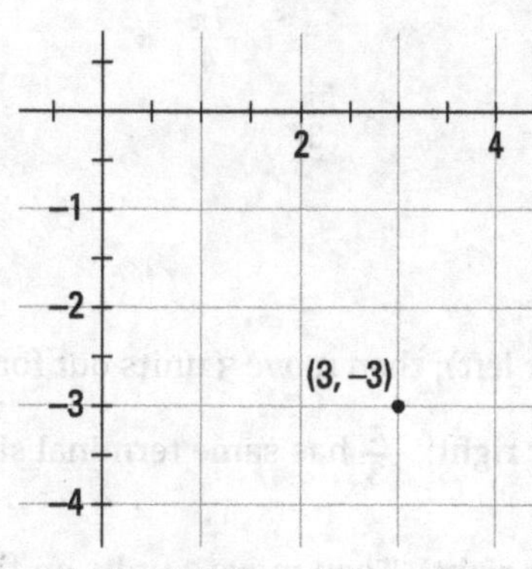

a. Rectangular

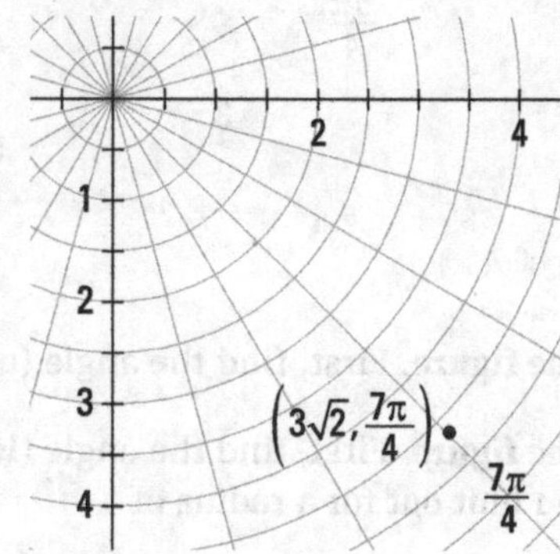

b. Polar

7 **See the next figure.** This is the circle graphed in Figure 16-5, just rotated counterclockwise.

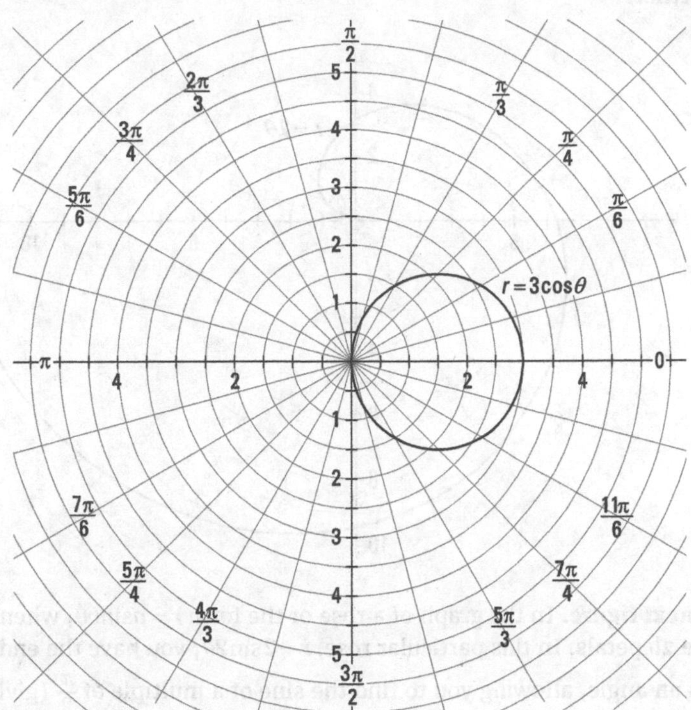

8 **See the next figure.** The multiplier –4 moves the circle graphed in Figure 16-5 on the horizontal axis and increases the diameter.

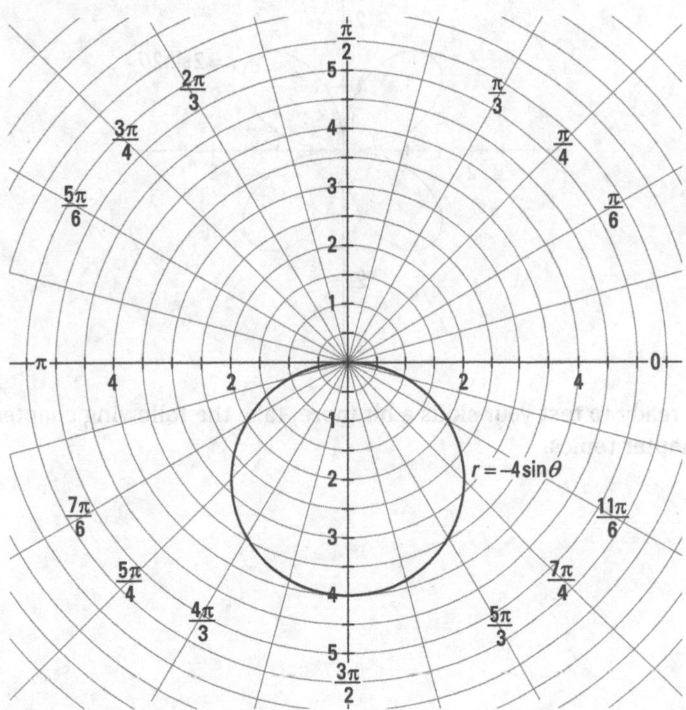

9. **See the next figure.** This is the same shape as $r = \theta$ (see Figure 16-6a). It's just spread 2 times farther.

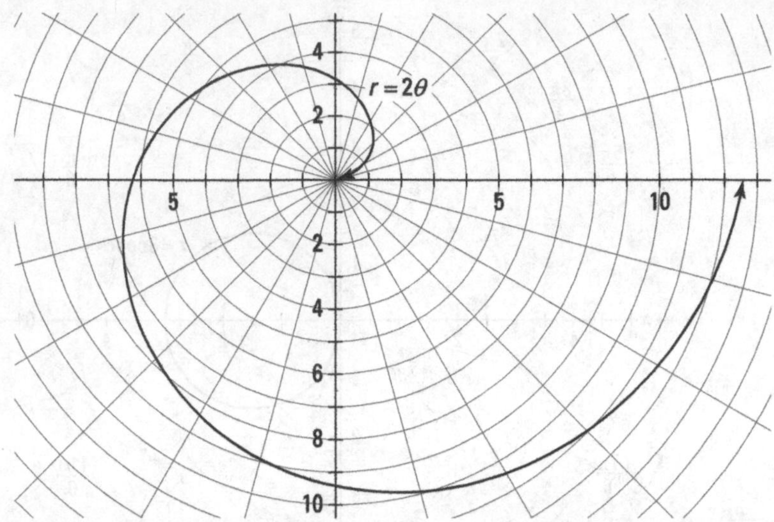

10. **See the next figure.** In the graph of a rose of the form $r = a\sin b\theta$, when the value of b is even, there are $2b$ petals. In this particular rose, $r = 2\sin 2\theta$, you have the ends of the petals whenever θ is an angle, allowing you to find the sine of a multiple of $\frac{\pi}{2}$ (giving you a sine value of 1 or –1).

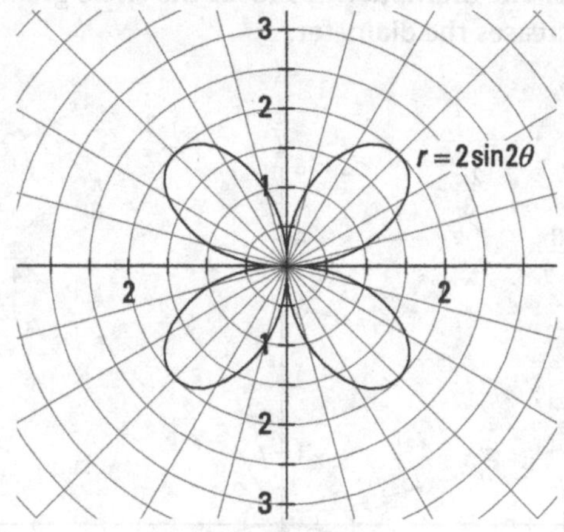

If you're ready to test your skills a bit more, take the following chapter quiz that incorporates all the chapter topics.

Whaddya Know? Chapter 16 Quiz

Quiz time! Complete each problem to test your knowledge on the various topics covered in this chapter. You can then find the solutions and explanations in the next section.

1. Graph the circle $r = 2\cos\theta$.
2. Change the polar coordinates to rectangular coordinates: $\left(-1, \frac{\pi}{4}\right)$.
3. Graph the spiral $r = \frac{1}{2}\theta$.
4. Given the graph of points D, E, and F, identify their polar coordinates.

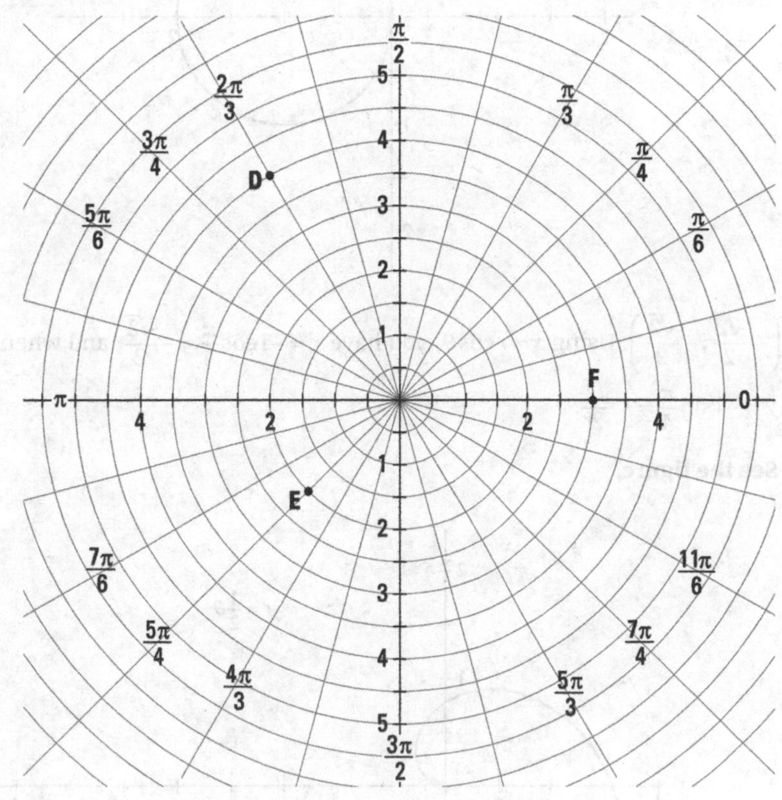

5. Change the rectangular coordinates to polar coordinates: $\left(-1, \sqrt{3}\right)$.
6. Sketch a graph of the rose $r = 4\cos 3\theta$.

Answers to Chapter 16 Quiz

1. **See the figure.**

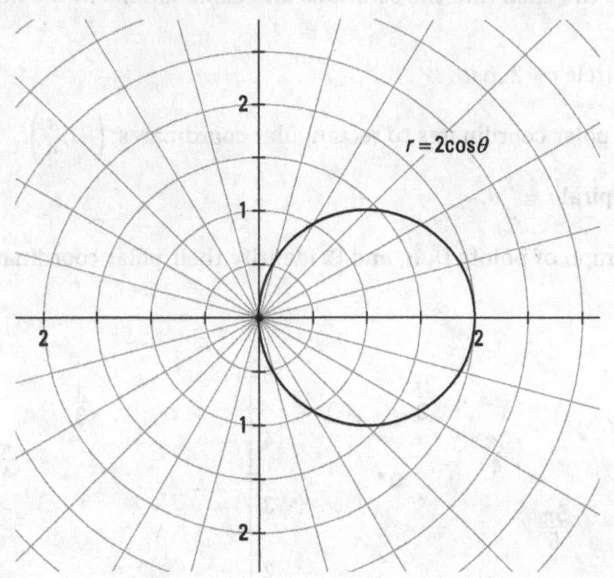

2. $\left(-\frac{\sqrt{2}}{2}, -\frac{\sqrt{2}}{2}\right)$. **Using $x = r\cos\theta$, you have $x = -1\cos\frac{\pi}{4} = -\frac{\sqrt{2}}{2}$, and when $y = r\sin\theta$, you have $y = -1\sin\frac{\pi}{4} = -\frac{\sqrt{2}}{2}$.**

3. **See the figure.**

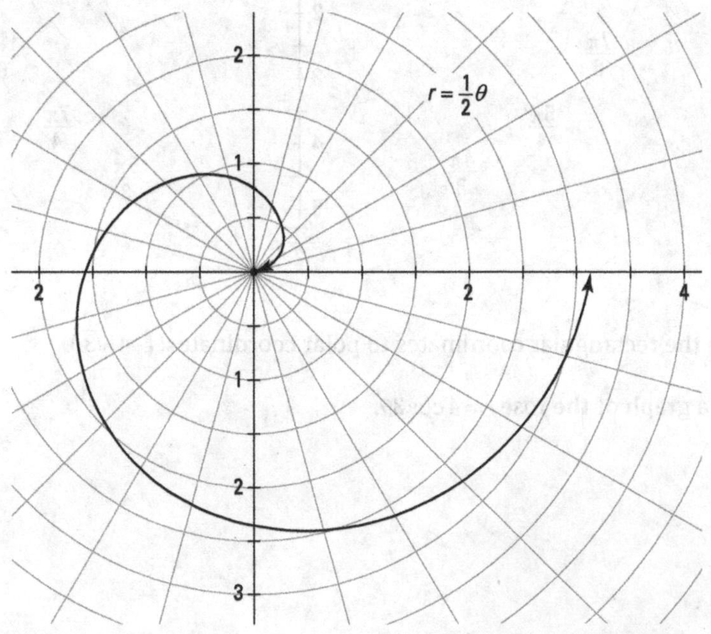

4 D: $\left(4, \frac{2\pi}{3}\right)$; E: $\left(2, \frac{5\pi}{4}\right)$; F: $(3,0)$. These are the simplest coordinates possible for these points. You could also use negative angle measures and negative radii to arrive at the same points.

5 $\left(2, \frac{2\pi}{3}\right)$. First, solve for the radius: $(-1)^2 + \left(\sqrt{3}\right)^2 = r^2 \rightarrow 1+3 = r^2 \rightarrow 4 = r^2 \rightarrow r = 2$. Solving for the angle, $\tan\theta = \frac{y}{x} = \frac{\sqrt{3}}{-1} = -\sqrt{3}$. Determining the angle, $\theta = \tan^{-1}\left(-\sqrt{3}\right)$. This occurs when $\theta = \frac{2\pi}{3}$ or $\theta = \frac{5\pi}{3}$. The point $\left(-1, \sqrt{3}\right)$ is in quadrant II, so $\theta = \frac{2\pi}{3}$. The coordinates of the point are $\left(2, \frac{2\pi}{3}\right)$.

6 **See the figure.** This is the same shape as $r = 4\sin 3\theta$ (see Figure 16-6b). It's just rotated by $\frac{\pi}{6}$.

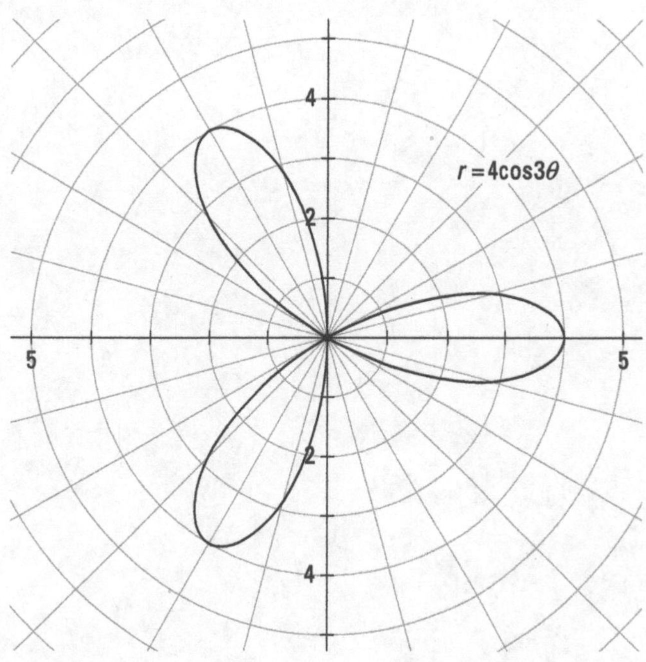

IN THIS CHAPTER

» Spinning around with circles

» Dissecting the parts and graphs of parabolas

» Exploring the ellipse

» Boxing around with hyperbolas

» Writing and graphing conics in two distinct forms

Chapter 17 Relating Conics to Sliced Cones

Astronomers have been looking out into space for a very long time — longer than you've been staring at that caramel-covered ice cream cone. Some of the things that are happening out in space are mysteries; others have shown their true colors to curious observers. One phenomenon that astronomers have discovered and proven is about the movement of bodies in space. They know that the paths of objects moving in space are shaped like one of four conic sections (shapes made from cones): the circle, the parabola, the ellipse, or the hyperbola. Conic sections have evolved into popular ways to describe motion, light, and other natural occurrences in the physical world.

In astronomical terms, an ellipse, for example, describes the path of a planet around the sun. And then a comet, making its way around a planet, may travel so close to the planet's gravity that its path is affected and it gets swung back out into the galaxy. If you were to attach a gigantic pen to the comet, its path would trace out one huge parabola. The movement of objects as they are affected by gravity can often be described using conic sections. For example, you can describe the movement of a ball being thrown up into the air using a conic section. The conic sections you'll find in this chapter have plenty of applications — especially for rocket science!

Conic sections are so named because they're made from two right circular cones (imagine two sugar cones from your favorite ice-cream store). Basically, imagine two right circular cones, touching pointy end to pointy end (the pointy end of a cone is called the *apex* or *vertex*). The conic sections are formed by the intersection of a plane and those two cones. When you slice through the cones with a plane, the intersection of that plane with the cones yields a variety of different curves. The plane is completely arbitrary; where the plane cuts the cone(s) and at what angle are the properties that give you all the different conic sections that are discussed in this chapter.

In this chapter, you'll find each of the conic sections, front to back. You'll see the similarities and the differences between the four conic sections and their applications in mathematics and science. You'll also find the graph of each section and look at its properties. Conic sections are part of the final frontier when it comes to graphing in two dimensions in mathematics, so sit back, relax, and enjoy the ride!

Cone to Cone: Identifying the Four Conic Sections

Each conic section has its own standard form of an equation with x- and y-variables that you can graph on the coordinate plane. You can write the equation of a conic section if you are given key points on the graph, and you can graph the conic section from the equation. You can alter the shape of each of these graphs in various ways, but the general graph shapes still remain true to the type of curve that they are.

Being able to identify which conic section is which by just the equation is important because sometimes that's all you're given (you won't always be told what type of curve you're graphing). Certain key points are common to all conics (vertices, foci, and axes, to name a few), so you start by plotting these key points and then identifying what kind of curve they form.

Picturing the conics — graph form

The whole point of this chapter is to be able to graph conic sections accurately with all the given information. Figure 17-1 illustrates how a plane intersects the cones to create the conic sections, and the following list explains the figure.

- **Circle:** A circle is the set of all points that are a given distance (the radius, *r*) from a given point (the center). To create a circle from the right cones, the plane slice occurs parallel to the base of either cone but doesn't slice through the apex of the cones.
- **Parabola:** A parabola is a curve in which every point is equidistant from one special point (the *focus*) and a line (the *directrix*). A parabola looks a lot like the letter *U,* although it may be upside down or sideways. To form a parabola, the plane slices through parallel to a side of one of the cones (any side, but not one of the bases).
- **Ellipse:** An ellipse is the set of all points where the sum of the distances from two particular points (the *foci*) is constant. You may be more familiar with another term for ellipse, *oval.* In order to get an ellipse from the two right cones, the plane must cut through only one cone, not parallel to the base, and not through the apex.
- **Hyperbola:** A hyperbola is the set of points where the difference of the distances between two given points is constant. The shape of the hyperbola looks visually like two parabolas (although a parabola is very different mathematically) mirroring one another with some space between the vertices. To form a hyperbola, the slice cuts both cones perpendicular to their bases (straight up and down) but not through the apex.

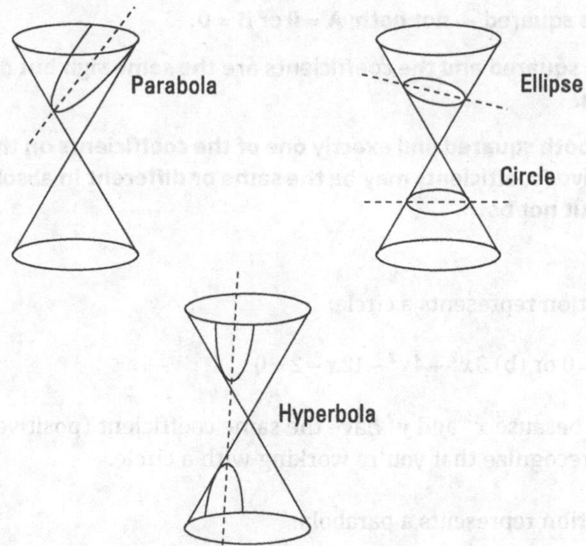

FIGURE 17-1: Cutting cones with a plane to get conic sections.

REMEMBER

Often, sketching a conic is not enough. Each conic section has its own set of properties that you may have to determine to supplement the graph. You can indicate where the center, vertices, major and minor axes, and foci are located. Often, this information is even more important than the actual graph. Besides, knowing all this valuable info helps you sketch the graph more accurately than you could without it.

Writing conic equations — printed form

The equations of conic sections are very important because they not only tell you which conic section you will be graphing, but also give you information on what the graph should look like. The appearance of each conic section has adjustments based on the values of the constants in the equation. Usually these constants are found as the letters *a, b, h,* and *k*. Not every conic has all these constants, but conics that do have them are affected in the same way by changes in the value of the same constant. Conic sections can come in all different shapes and sizes: big, small, fat, skinny, vertical, horizontal, and more. The constants listed here are the culprits of these changes.

TECHNICAL STUFF

An equation has to have x^2 or y^2 or both x^2 and y^2 to create a conic. If neither *x* nor *y* is squared, then the equation is of a line (not considered a conic section for these purposes). None of the variables of a conic section may be raised to any power higher than two.

As briefly mentioned, certain characteristics are unique to each type of conic. In order to recognize which conic you're dealing with, when given an equation in the standard form, refer to the following list. The standard form of a conic is $Ax^2 + By^2 + Cx + Dy + E = 0$.

» **Circle: *x* and *y* are both squared and the coefficients on them are the same, including the sign; A=B.**

- **Parabola: Either x or y is squared — not both; A = 0 or B = 0.**
- **Ellipse: x and y are both squared and the coefficients are the same sign but different numerical values; A ≠ B.**
- **Hyperbola: x and y are both squared and exactly one of the coefficients on the squared terms is negative (coefficients may be the same or different in absolute value); A < 0 or B < 0, but not both.**

EXAMPLE

Q. Determine which equation represents a circle:

(a) $3x^2+3y^2-12x-2=0$ or (b) $3x^2+4y^2-12x-2=0$

A. Choice (a) is the circle because x^2 and y^2 have the same coefficient (positive 3). That info is all you need to recognize that you're working with a circle.

EXAMPLE

Q. Determine which equation represents a parabola:

(a) $y=x^2-4$ or (b) $x=2y^2-3y+10$

A. Actually, these are both parabolas. The equations aren't technically in the standard form, but rewriting them doesn't change the characteristics. In equation (a), you see an x^2 but no y^2, and in equation (b), you see a y^2 but no x^2. Nothing else matters — sign and coefficients change the physical appearance of the parabola (which way it opens or how fat it is) but don't change the fact that it's a parabola.

EXAMPLE

Q. Determine which equation represents an ellipse:

(a) $2x^2-3y^2-12x-2=0$ or (b) $3x^2-9x+2y^2+10y-6=0$

A. Choice (b) is one example of an ellipse. The coefficients on x^2 and y^2 are different, but both are positive. In equation (a) the signs of the coefficients on the squared terms are different.

EXAMPLE

Q. Determine which equation represents a hyperbola:

(a) $4y^2-10y-3x^2=12$ or (b) $x^2-9x+9y^2+10y-6=0$

A. Choice (a) is an example of a hyperbola. This time, the coefficients on x^2 and y^2 are different, but one of them is negative, which is a requirement to get the graph of a hyperbola. The signs on the coefficients in equation (b) are the same — so this is an ellipse!

WARNING

The equations for the four conic sections look very similar to one another, with subtle differences (a plus sign instead of a minus sign, for instance, gives you an entirely different type of conic section).

YOUR TURN

Identify which type of conic section is represented in each of the following equations.

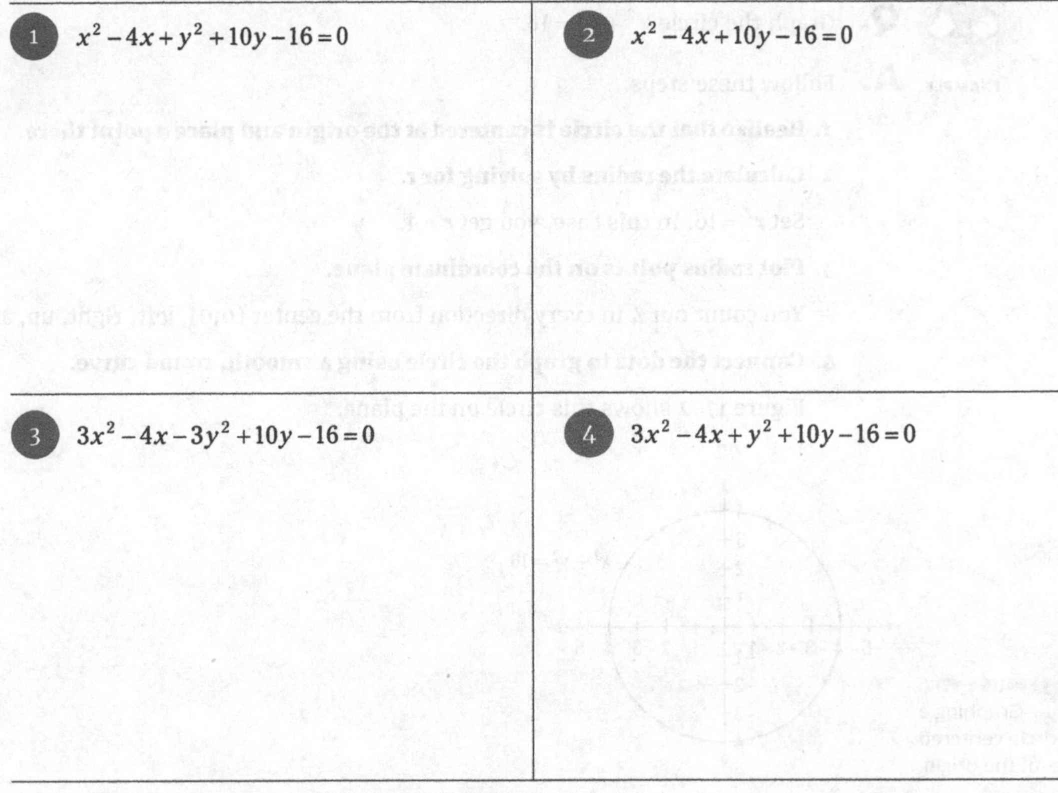

1 $x^2 - 4x + y^2 + 10y - 16 = 0$	2 $x^2 - 4x + 10y - 16 = 0$
3 $3x^2 - 4x - 3y^2 + 10y - 16 = 0$	4 $3x^2 - 4x + y^2 + 10y - 16 = 0$

Going Round and Round: Graphing Circles

Circles are simple to work with. A circle has one center, one radius, and a whole lot of points. In this section, you see how to graph circles on the coordinate plane and figure out from both the graph and the circle's equation where the center lies and what the radius is.

The first thing you need to know in order to graph the equation of a circle is where on a plane the center is located. The equation of a circle that gives you all the necessary information appears as $(x-h)^2 + (y-k)^2 = r^2$. This form is referred to as the *center-radius form* (or standard form) because it gives you both pieces of information at the same time. The h and k represent the center of the circle at point (h,k), and r names the radius. Specifically, h represents the horizontal displacement — how far to the left or to the right of the y-axis the center of the circle is. The variable k represents the vertical displacement — how far above or below the x-axis the center falls. From (h,k), you can count r units (the radius) horizontally in both directions and vertically in both directions to get four different points, all equidistant from the center. Connect these four points with a curve that forms the circle.

Graphing circles at the origin

The simplest circle to graph has its center at the origin (0,0). Because both h and k are zero, you can simplify the standard circle equation to $x^2 + y^2 = r^2$.

Q. Graph the circle $x^2 + y^2 = 16$.

EXAMPLE **A.** Follow these steps:

1. **Realize that the circle is centered at the origin and place a point there.**
2. **Calculate the radius by solving for r.**

   Set $r^2 = 16$. In this case, you get $r = 4$.
3. **Plot radius points on the coordinate plane.**

   You count out 4 in every direction from the center (0,0): left, right, up, and down.
4. **Connect the dots to graph the circle using a smooth, round curve.**

   Figure 17-2 shows this circle on the plane.

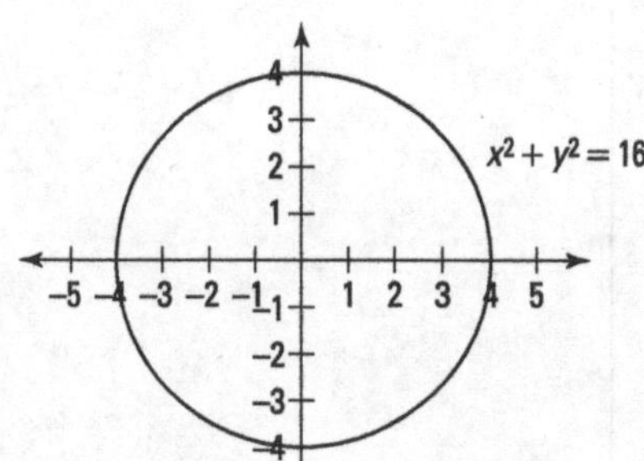

FIGURE 17-2: Graphing a circle centered at the origin.

Graphing circles away from the origin

Graphing a circle anywhere on the coordinate plane is pretty easy when its equation appears in center–radius form. All you do is plot the center of the circle at (h,k), count out from the center r units in the four directions (up, down, left, right), and connect those four points with a circle. Or you can think of it as a transformation involving horizontal and vertical shifts. You can refer to Chapter 4 if you need more information on transformations.

REMEMBER

Don't forget to switch the sign of the h and k from inside the parentheses in the equation. This step is necessary because the h and k are inside the grouping symbols, which means that the shift happens opposite from what you would think.

Q. Graph the equation $(x-3)^2 + (y+1)^2 = 25$.

EXAMPLE **A.** Follow these steps:

1. **Locate the center of the circle from the equation (*h*,*k*).**

   $(x-3)^2$ means that the x-coordinate of the center is positive 3.

   $(y+1)^2$ means that the y-coordinate of the center is -1.

   Place the center of the circle at $(3,-1)$.
2. **Calculate the radius by solving for *r*.**

   Set $r^2 = 25$ and find the square root of both sides to get $r = 5$.

3. **Plot the radius points on the coordinate plane.**

   Count 5 units up, down, left, and right from the center at (3,–1). This step gives you points at (8,–1), (–2,–1), (3,–6), and (3,4).

4. **Connect the dots to the graph of the circle with a round, smooth curve.**

   See Figure 17-3 for a visual representation of this circle.

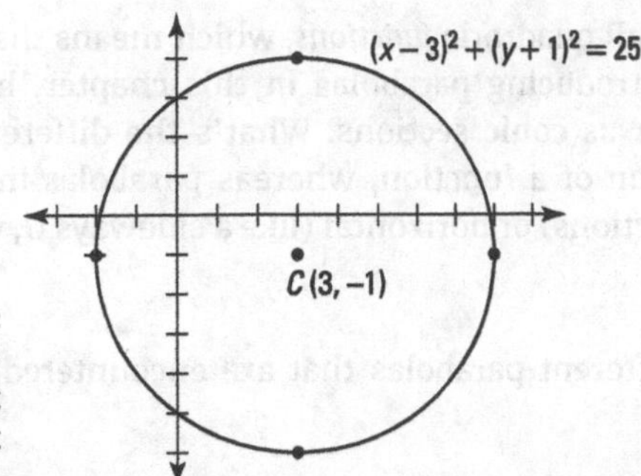

FIGURE 17-3: Graphing a circle not centered at the origin.

Writing in center–radius form

Sometimes the equation of a circle is in center–radius form (and then graphing is a piece of cake), and sometimes you have to manipulate the equation a bit to get it into a form that's easy for you to work with. When a circle doesn't appear in center–radius form, you often have to complete the square in order to find the center.

EXAMPLE

Q. Change the equation $x^2+y^2+4x-6y-23=0$ into the center–radius form.

A. Use the following steps:

1. **Regroup the variables together and move the constant to the right side (by adding 23 to each side):** $x^2+4x+y^2-6y=23$.

2. **Complete the square (see Chapter 5 for more on this) on both x and y.**

   $x^2+4x+y^2-6y=23 \rightarrow x^2+4x+4+y^2-6y+9=23+4+9 \rightarrow x^2+4x+4+y^2-6y+9=36$.

3. **Factor the trinomials:** $\left(x^2+4x+4\right)+\left(y^2-6y+9\right)=36 \rightarrow (x+2)^2+(y-3)^2=36$. The center of the circle is (–2,3), and the radius is 6.

YOUR TURN

5 Graph the circle: $(x-4)^2+y^2=9$.	**6** Graph the circle: $x^2-8x+y^2+12y+27=0$.

Riding the Ups and Downs with Parabolas

Although parabolas look like simple U-shaped curves, there's more to them than first meets the eye. Because the equations of parabolas involve one squared variable (and one variable only), they become a mirror image over their axis of symmetry, just like the quadratic functions from Chapter 3.

The parabolas discussed in Chapter 3 are all quadratic *functions*, which means that they passed the vertical-line test. The purpose of introducing parabolas in this chapter, however, is to discuss them not as functions but rather as conic sections. What's the difference, you ask? Quadratic functions must fit the definition of a function, whereas parabolas in the world of conic sections can be vertical (like the functions) or horizontal (like a sideways U, which doesn't pass the vertical-line test).

In this section, you are introduced to different parabolas that are encountered on a journey through conic sections.

Labeling the parts

Each of the different parabolas in this section has the same general shape; however, the width of the parabolas, their location on the coordinate plane, and which direction they open can vary greatly.

One thing that's true of all parabolas is their symmetry, meaning that you can fold a parabola in half over itself. The line that divides a parabola in half is called the *axis of symmetry*. The *focus* is a point inside (not on) the parabola that lies on the axis of symmetry, and the *directrix* is a line that runs outside the parabola perpendicular to the axis of symmetry. The *vertex* of the parabola is exactly halfway between the focus and the directrix and lies on the curve. Recall from geometry that the distance from any line to a point not on that line is a line segment from the point perpendicular to the line; that's how the distances from points on the parabola to the directrix are measured. A parabola is formed by all the points equidistant from the focus and the directrix. The distance between the vertex and the focus, then, dictates how skinny or how fat the parabola is.

The first thing you must find in order to graph a parabola is where the vertex is located. From there, you can find out whether the parabola goes up and down (a vertical parabola) or sideways (a horizontal parabola). The coefficients of the parabola also tell you which way the parabola opens (toward the positive numbers or the negative numbers).

If you're graphing a vertical parabola, then the vertex is also the maximum or the minimum value of the curve. Calculating the maximum and minimum values has tons of real-world applications for you to dive into. Bigger is usually better, and maximum area is no different. Parabolas are very useful in telling you the maximum (or sometimes minimum) area for rectangles. For example, if you're building a dog run with a preset amount of fencing, you can use parabolas to find the dimensions of the dog run that would yield the maximum area for your dog to exercise in.

Understanding the characteristics of a standard parabola

The squaring of the variables in the equation of the parabola determines how it opens.

- **The x is squared and y is not:** In this case, the axis of symmetry is vertical and the parabola opens up or down. For instance, $y = x^2$ is a vertical parabola; its graph is shown in Figure 17-4a.
- **The y is squared and x is not:** In this case, the axis of symmetry is horizontal and the parabola opens left or right. For example, $x = y^2$ is a horizontal parabola; it's shown in Figure 17-4b.

Both of these parabolas have the vertex located at the origin.

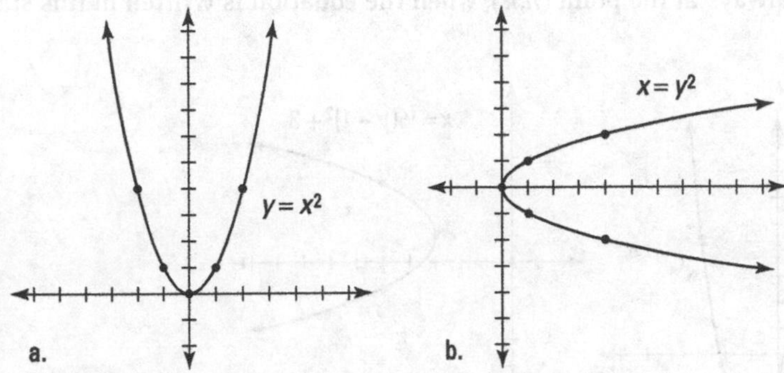

FIGURE 17-4: Vertical and horizontal parabolas based at the origin.

REMEMBER

Be aware of negative coefficients on the squared terms in parabolas. If the parabola is vertical, a negative coefficient on the x^2 (when the equation is written in the standard form) makes the parabola open downward. If the parabola is horizontal, a negative coefficient on y^2 (when the equation is written in the standard form) makes the parabola open to the left.

Plotting the variations: Parabolas all over the plane

Just like with circles and their centers, the vertex of the parabola isn't always at the origin. You need to be comfortable with shifting parabolas around the coordinate plane, too. Certain motions, especially the motion of falling objects, move in a parabolic shape with respect to time. For example, the height of a ball launched up in the air at time t can be described by the equation $h(t) = -16t^2 + 32t$. Finding the vertex of this equation tells you the maximum height of the ball and also when the ball reached that height. Finding the x-intercepts also tells you when the ball will hit the ground.

TECHNICAL STUFF

A vertical parabola written in the form $y = a(x-h)^2 + k$ gives you the following information:

- **A vertical steepening or flattening (designated by the variable a)**
- **The horizontal shift of the graph (designated by the variable h)**
- **The vertical shift of the graph (designated by the variable k)**

Q. Determine the transformations performed on the parabola $y = 2(x-1)^2 - 3$.

EXAMPLE **A.** You find three transformations:

The vertical transformation is determined by the multiplier 2. Every point is stretched vertically by a factor of 2 (see Figure 17-5a for the graph). Therefore, every time you plot a point on the graph, the original height of $y = x^2$ is multiplied by 2.

The horizontal shift is determined by the 1. The vertex is shifted to the right of the origin 1 unit.

The vertical shift is determined by the 3. The vertex is shifted down 3.

Another way to look at the horizontal and vertical shifts is just to recognize that the vertex of the parabola is always at the point (h,k), when the equation is written in this standard form.

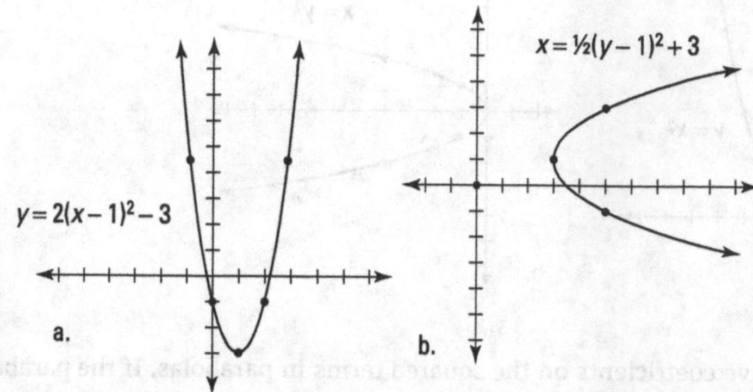

FIGURE 17-5: Graphing transformed horizontal parabolas.

TECHNICAL STUFF

A horizontal parabola written in the form $x = a(y-k)^2 + h$ gives you the following information:

- **A horizontal steepening or flattening (designated by the variable *a*)**
- **The horizontal shift of the graph (designated by the variable *h*)**
- **The vertical shift of the graph (designated by the variable *k*)**

Q. Determine the transformations performed on the parabola $x = \frac{1}{2}(y-1)^2 + 3$.

EXAMPLE **A.** There are three transformations:

The horizontal transformation is determined by the multiplier $\frac{1}{2}$. Every point is flattened horizontally by a factor of $\frac{1}{2}$ (see Figure 17-5b for the graph). Therefore, every time you plot a point on the graph, the original breadth of $y = x^2$ is multiplied by $\frac{1}{2}$.

The horizontal shift is determined by the 3. The vertex is shifted to the right of the origin 3 units.

The vertical shift is determined by the 1. The vertex is shifted up 1.

Again, another way to look at the horizontal and vertical shifts is just to recognize that the vertex of the parabola is always at the point (h,k), when the equation is written in this standard form. You can see this parabola's graph in Figure 17-5b.

The vertex, axis of symmetry, focus, and directrix

In order to graph a parabola correctly, you need to note whether the parabola is horizontal or vertical, because although the variables and constants in the equations for both curves serve the same purpose, their effect on the graphs in the end is slightly different. Adding a constant inside the parentheses of the vertical parabola moves the entire thing horizontally, whereas adding a constant inside the parentheses of a horizontal parabola moves it vertically (see the preceding section for more info). You want to note these differences before you start graphing so that you don't accidentally move your graph in the wrong direction. In the following sections, you see how to find all this information for both vertical and horizontal parabolas.

Finding parts of a vertical parabola

TECHNICAL STUFF

A vertical parabola, $y = a(x-h)^2 + k$, has the following parts or features:

- **Vertex:** (h,k)
- **Axis of symmetry:** $x = h$
- **Focus:** The distance from the vertex to the focus is $\frac{1}{4a}$, where a is the coefficient of the x^2 term. The focus, as a point, is $\left(h, k+\frac{1}{4a}\right)$; it should be directly above or directly below the vertex. It always appears inside the parabola.
- **Directrix:** The equation of the directrix is $y = k - \frac{1}{4a}$. It is the same distance from the vertex along the axis of symmetry as the focus, in the opposite direction. The directrix appears outside the parabola and is perpendicular to the axis of symmetry. Because the axis of symmetry is vertical, the directrix is a horizontal line; thus, it has an equation of the form y = a constant.

The graph in Figure 17-6 is sometimes referred to as the "martini" of parabolas because it looks like a martini glass. The axis of symmetry is the glass stem, the directrix is the base of the glass, and the focus is the olive. You need all those parts to make a good martini *and* a parabola.

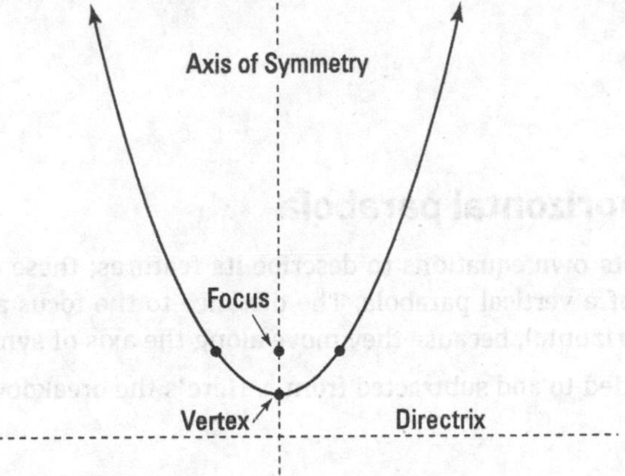

FIGURE 17-6: Identifying the parts of a vertical parabola.

EXAMPLE

Q. **Identify all the parts of the parabola $y=2(x-1)^2-3$ (vertex, axis of symmetry, focus, directrix). Then graph the parabola.**

A. **Working from the form $y=a(x-h)^2+k$, you have that $a=2$, $h=1$, and $k=-3$. With this information, you can identify all the parts of a parabola (vertex, axis of symmetry, focus, and directrix) as points or equations.**

- **Vertex:** The vertex is at (h,k), so for this parabola it's at $(1,-3)$.
- **Axis of symmetry:** The axis of symmetry is the line $x=h$, which for this parabola is $x=1$.
- **Focus:** The focus is the point $\left(h, k+\frac{1}{4a}\right)$, where the distance from the vertex to the focus is $\frac{1}{4a}$. Since $a=2$, the focal distance for this parabola is $\frac{1}{4\cdot 2}=\frac{1}{8}$. With this distance, you can write the focus as the point $\left(1,-3+\frac{1}{8}\right)=\left(1,-2\frac{7}{8}\right)$.
- **Directrix:** To write the equation of the directrix, $y=k-\frac{1}{4a}$, or $y=-3-\frac{1}{8}=-3\frac{1}{8}$.

Now, graph the parabola and label all its parts.

You can see the graph, with all its parts, in Figure 17-7. You should plot at least two other points besides the vertex so that you can better show the vertical transformation. Because the vertical transformation in this equation is a factor of 2, the two points on both sides of the vertex are stretched by a factor of 2. So from the vertex, you plot a point that is to the right 1 and up 2 (instead of up 1). Then you can draw the point at the same height on the other side of the axis of symmetry; the two other points on the graph are at $(2,-1)$ and $(0,-1)$.

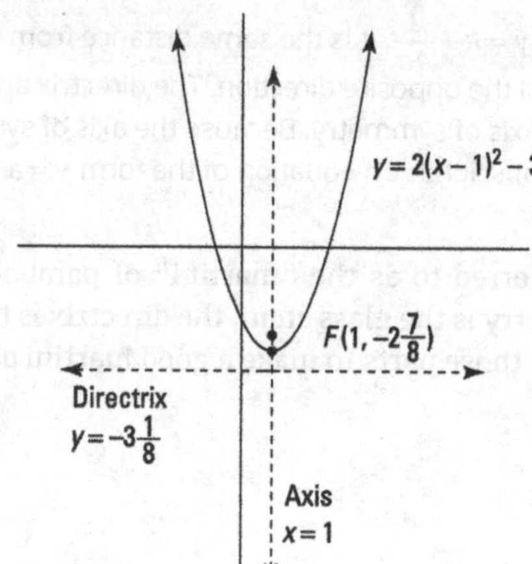

FIGURE 17-7: Finding all the parts of the parabola $y=2(x-1)^2-3$.

Finding parts of a horizontal parabola

TECHNICAL STUFF

A horizontal parabola has its own equations to describe its features; these equations are just a bit different from those of a vertical parabola. The distance to the focus and directrix from the vertex in this case is horizontal, because they move along the axis of symmetry, which is a horizontal line. So $\frac{1}{4a}$ is added to and subtracted from h. Here's the breakdown.

- **Vertex:** (h,k).
- **Axis of symmetry:** $y = k$.
- **Focus:** This is directly to the left or right of the vertex, at the point $\left(h+\frac{1}{4a},k\right)$.
- **Directrix:** The same distance from the vertex as the focus in the opposite direction, at $x = h - \frac{1}{4a}$.

EXAMPLE

Q. **Find the parts of the horizontal parabola $x = \frac{1}{8}(y-1)^2 + 3$. Then graph the parabola.**

A. **Working from the form $x = a(y-k)^2 + h$, you have that $a = \frac{1}{8}$, $h = 3$, and $k = 1$. With this information, you can identify all the parts of a parabola (vertex, axis of symmetry, focus, and directrix) as points or equations.**

- **Vertex:** The vertex is at (h,k), so for this parabola it's at $(3,1)$.
- **Axis of symmetry:** The axis of symmetry is at $y = k$, so in this case, it is at $y = 1$.
- **Focus:** For the given equation, $a = \frac{1}{8}$, and so the focal distance is $\frac{1}{4\left(\frac{1}{8}\right)} = \frac{1}{\frac{1}{2}} = 2$. Add this value to h to find the focus: $(3+2,1)$ or $(5,1)$.
- **Directrix:** Subtract the focal distance that you just found from h to find the equation of the directrix. Because this parabola is horizontal and the axis of symmetry is horizontal, the directrix is vertical. The equation of the directrix is $x = 3 - 2$ or $x = 1$.

Graph the parabola and label its parts.

The focus lies inside the parabola, and the directrix is a vertical line 2 units to the left of the vertex.

Figure 17-8 shows you the graph and has all the parts labeled for you.

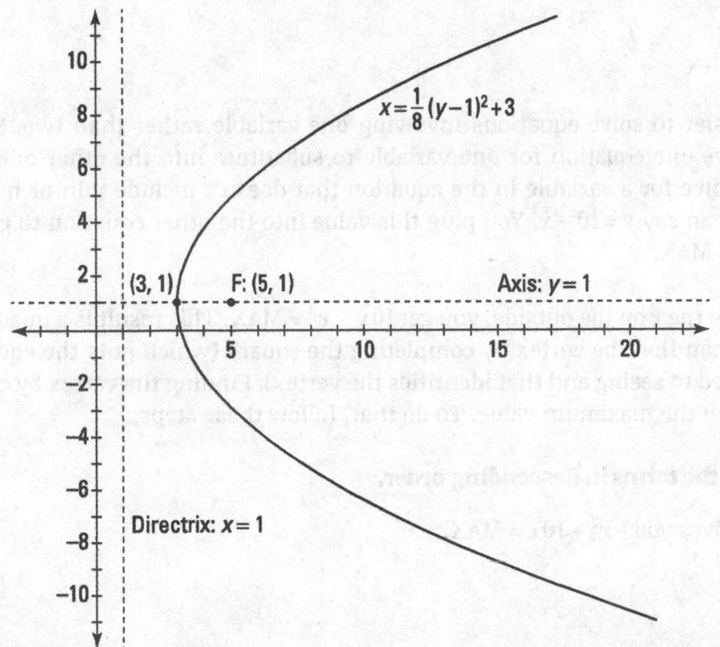

FIGURE 17-8: The graph of a horizontal parabola.

YOUR TURN

 Graph the parabola and identify its various parts/features: $y = -2(x-3)^2 + 5$.

 Graph the parabola and identify its various parts/features: $x = \frac{1}{4}(y-2)^2 - 5$.

Identifying the extremes of vertical parabolas

Vertical parabolas can give an important piece of information: When the parabola opens upward, the vertex is the lowest point on the graph — called the *minimum*. When the parabola opens downward, the vertex is the highest point on the graph — called the *maximum*. Only vertical parabolas can have minimum or maximum values, because horizontal parabolas have no limit on how high or how low they can go. Finding the maximum of a parabola can tell you the maximum height of a ball thrown into the air, the maximum area of a rectangle, the maximum or minimum value of a company's profit, and so on.

EXAMPLE

Q. Find two numbers whose sum is 10 and whose product is a maximum.

A. Letting those two unknown numbers be represented by x and y, you can then write the following math statements about them:

$$x + y = 10$$
$$xy = \text{MAX}$$

TIP

It's usually easier to solve equations involving one variable rather than two. So use substitution and solve one equation for one variable to substitute into the other one. In this case, it's better to solve for a variable in the equation that doesn't include min or max at all. So if $x + y = 10$, you can say $y = 10 - x$. You plug this value into the other equation to get the following: $x(10-x) = \text{MAX}$.

If you distribute the x on the outside, you get $10x - x^2 = \text{MAX}$. This result is a quadratic equation for which you can find the vertex by completing the square (which puts the equation into the form you're used to seeing and that identifies the vertex). Finding the vertex by completing the square gives you the maximum value. To do that, follow these steps:

1. **Rearrange the terms in descending order.**

   This step gives you $-x^2 + 10x = \text{MAX}$.

2. **Factor out the leading term.**

   You now have $-1(x^2 - 10x) = \text{MAX}$.

3. **Complete the square (see Chapter 5 for a reference).**

   This step expands the equation to $-1(x^2 - 10x + 25) = \text{MAX} - 25$. Notice that –1 in front of the parentheses turned the 25 into –25, which is why you must add –25 to the right side as well.

4. **Factor the trinomial inside the parentheses.**

   You get $-1(x-5)^2 = \text{MAX} - 25$.

5. **Move the constant to the other side of the equation.**

   You end up with $-1(x-5)^2 + 25 = \text{MAX}$.

If you think of this in terms of the equation $-1(x-5)^2 + 25 = y$, the vertex of the parabola is $(5,25)$. For more info, see the earlier section, "Plotting the variations: Parabolas all over the plane." Therefore, the number you're looking for (x) is 5, and the maximum product is 25. You can plug 5 in for x to get y in either equation: $5 + y = 10$, gives you $y = 5$. So the two numbers, x and y, are 5 and 5.

Figure 17-9 shows the graph of the maximum function to illustrate that the vertex, in this case, is the maximum point.

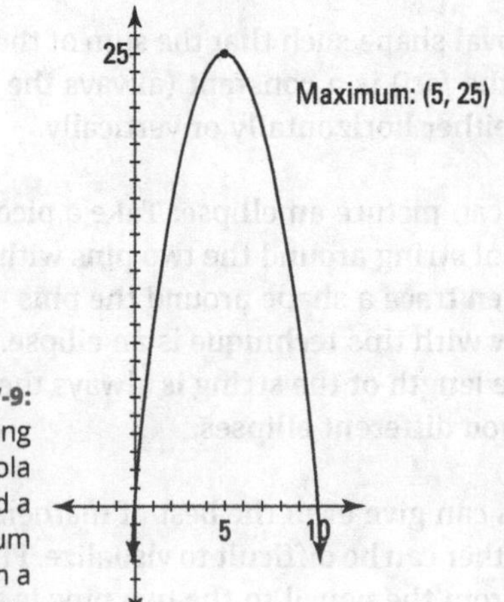

FIGURE 17-9: Graphing a parabola to find a maximum value from a word problem.

TIP

By the way, a graphing calculator can easily find the vertex for this type of question. Even in the table form, you can see from the symmetry of the parabola that the vertex is the highest (or the lowest) point.

YOUR TURN

Find the maximum point on the graph of $y=-\frac{1}{3}(x+1)^2+4$.

You are playing a game on your device where a disgruntled bird flies over a hill to smash into an unwitting pig. The hill is 600 feet high, and the horizontal distance between the bird and pig, at the beginning of the flight, is 100 feet. You realize that the path the bird will take is a parabola! What is the equation of that parabola?

The Fat and the Skinny on the Ellipse

An *ellipse* is a set of points on a plane, creating an oval shape such that the sum of the distances from any point on the curve to two fixed points (the *foci*) is a constant (always the same). An ellipse is basically a circle that has been squished either horizontally or vertically.

Are you more of a visual learner? Here's how you can picture an ellipse: Take a piece of paper and pin it to a corkboard with two pins. Tie a piece of string around the two pins with a little bit of slack. Using a pencil, pull the string taut and then trace a shape around the pins — keeping the string taut the entire time. The shape you draw with this technique is an ellipse. The sums of the distances to the pins, then, is the string. The length of the string is always the same in a given ellipse, and different lengths of string give you different ellipses.

This definition that refers to the sums of distances can give even the best of mathematicians a headache because the idea of adding distances together can be difficult to visualize. Figure 17-10 shows how this works. The sum of the distances from the pencil to the two pins is always the same, no matter where the pencil is — no matter what point on the ellipse.

Labeling ellipses and expressing them with algebra

Graphically speaking, there are two different types of ellipses: horizontal and vertical. A horizontal ellipse spreads left and right — more wide than tall; a vertical one spreads up and down — more tall than wide. Each type of ellipse has the following main parts.

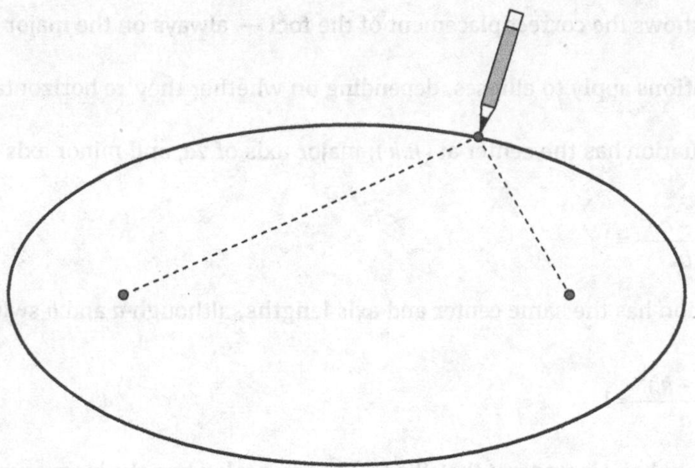

FIGURE 17-10: An ellipse drawn with a pencil.

- **Center:** The point in the middle of the ellipse is called the *center* and is named *(h,k)* just like the vertex of a parabola and the center of a circle. It lies where the two axes cross.
- **Major axis:** The *major axis* is the line that runs through the center of the ellipse the long way. The variable a is the letter used to name the distance from the center to a point on the ellipse on the major axis. The endpoints of the major axis are on the ellipse and are called *vertices.*
- **Minor axis:** The *minor axis* is perpendicular to the major axis and runs through the center the short way. The variable b is the letter used to name the distance to a point on the ellipse from the center on the minor axis. Because the major axis is always longer than the minor one, $a > b$. The endpoints on the minor axis are called *co-vertices.*
- **Foci:** The *foci* are the two points that dictate how fat or how skinny the ellipse is. They are always located on the major axis, on either side of the center, and can be found by the following equation: $a^2 - b^2 = F^2$, where a is the length from the center to a vertex, b is the distance from the center to a co-vertex, and F is the distance from the center to each focus.

Figure 17-11a shows a horizontal ellipse with its parts labeled; Figure 17-11b shows a vertical one. Notice that the length of the major axis is $2a$ and the length of the minor axis is $2b$.

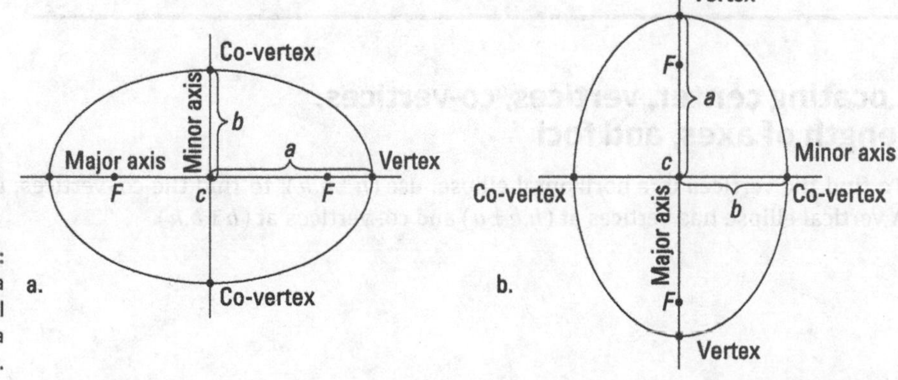

FIGURE 17-11: The parts of a horizontal ellipse and a vertical ellipse.

Figure 17-11 also shows the correct placement of the foci — always on the major axis.

Two types of equations apply to ellipses, depending on whether they're horizontal or vertical.

The horizontal equation has the center at (h,k), major axis of $2a$, and minor axis of $2b$:

$$\frac{(x-h)^2}{a^2}+\frac{(y-k)^2}{b^2}=1$$

The vertical equation has the same center and axis lengths, although a and b switch places:

$$\frac{(x-h)^2}{b^2}+\frac{(y-k)^2}{a^2}=1$$

When the bigger number a is under x, the ellipse is horizontal; when the bigger number is under y, it's vertical.

Identifying the parts from the equation

You have to be prepared not only to graph ellipses but also to name all their parts. If a problem asks you to determine different points or measures of an ellipse, you have to be ready to deal with some interesting square roots and/or decimals. Table 17-1 presents the parts in a handy, at-a-glance format. This section prepares you to graph and to find all the parts of an ellipse.

Table 17-1 Ellipse Parts

	Horizontal Ellipse	Vertical Ellipse
Equation	$\frac{(x-h)^2}{a^2}+\frac{(y-k)^2}{b^2}=1$	$\frac{(x-h)^2}{b^2}+\frac{(y-k)^2}{a^2}=1$
Center	(h,k)	(h,k)
Vertices	$(h\pm a,k)$	$(h,k\pm a)$
Co-vertices	$(h,k\pm b)$	$(h\pm b,k)$
Length of major axis	$2a$	$2a$
Length of minor axis	$2b$	$2b$
Foci where $F^2=a^2-b^2$	$(h\pm F,k)$	$(h,k\pm F)$

Locating center, vertices, co-vertices, length of axes, and foci

To find the vertices in a horizontal ellipse, use $(h\pm a,k)$; to find the co-vertices, use $(h,k\pm b)$. A vertical ellipse has vertices at $(h,k\pm a)$ and co-vertices at $(h\pm b,k)$.

Q. Identify the parts of the ellipse: $\frac{(x-5)^2}{9}+\frac{(y+1)^2}{16}=1.$

EXAMPLE **A.** Use the equation to identify what the variables represent. You see that the ellipse is vertical, because of the larger number 16, so, referring to the form $\frac{(x-h)^2}{b^2}+\frac{(y-k)^2}{a^2}=1$, you find $h=5$, $k=-1$, $a=4$, and $b=3$.

- **Center:** Since $h=5$ and $k=-1$, the center is $(5,-1)$.
- **Vertices:** Using $(h,k\pm a)$, this equation has vertices at $(5,-1\pm 4)$, or $(5,3)$ and $(5,-5)$.
- **Co-vertices:** Using $(h\pm b,k)$, the co-vertices are at $(5\pm 3,-1)$, or $(8,-1)$ and $(2,-1)$.

TECHNICAL STUFF

The major axis in a horizontal ellipse lies on $y=k$; the minor axis lies on $x=h$. The major axis in a vertical ellipse lies on $x=h$; and the minor axis lies on $y=k$. The length of the major axis is $2a$, and the length of the minor axis is $2b$. You can calculate the distance from the center to the foci in an ellipse (either variety) by using the equation $a^2-b^2=F^2$, where F is the distance from the center to each focus. The foci always appear on the major axis at the given distance (F) from the center.

- **Length of major axis:** Using $2a$, the length of the major axis is $2(4)=8$.
- **Length of minor axis:** Using $2b$, the length of the minor axis is $2(3)=6$.
- **Foci:** You find the foci with the equation $16-9=7=F^2$. The focal distance is $\sqrt{7}$. Because the ellipse is vertical, the foci are at $\left(5,-1\pm\sqrt{7}\right)$.

Working with an ellipse in nonstandard form

What if the elliptical equation you're given isn't in standard form? Take a look at the ellipse $3x^2+6x+4y^2-16y-5=0$. Before you do a single thing, determine that the equation is an ellipse because the coefficients on x^2 and y^2 are both positive but not equal. Follow these steps to put the equation in standard form:

1. **Add the constant to the other side.**

   This step gives you $3x^2+6x+4y^2-16y=5$.

2. **Complete the square and factor.**

   You need to factor out two different constants now — the different coefficients for x^2 and y^2. Then complete the squares in the parentheses and add the same amount to both sides of the equation.

$$\begin{aligned}3\left(x^2+2x\right)+4\left(y^2-4y\right)&=5\\3\left(x^2+2x+1\right)+4\left(y^2-4y+4\right)&=5+3+16\\3(x+1)^2+4(y-2)^2&=24\end{aligned}$$

   Note: Adding 1 and 4 inside the parentheses really means adding $3\cdot 1$ and $4\cdot 4$ to each side, because you have to multiply by the coefficient before adding it to the right side.

3. **Divide the equation by the constant on the right to get 1 and then reduce the fractions.**

$$\frac{3(x+1)^2}{24}+\frac{4(y-2)^2}{24}=\frac{24}{24}$$
$$\frac{(x+1)^2}{8}+\frac{(y-2)^2}{6}=1$$

4. **Determine whether the ellipse is horizontal or vertical.**

   Because the bigger number is under *x*, this ellipse is horizontal.

5. **Find the center and the length of the major and minor axes.**

   The center is located at (h,k), or $(-1,2)$.

   Since $a^2=8$, $a=\sqrt{8}=2\sqrt{2}\approx 2.83$.

   Since $b^2=6$, $b=\sqrt{6}\approx 2.45$.

6. **Graph the ellipse to determine the vertices and co-vertices.**

   Go to the center first and mark the point. Because this ellipse is horizontal, *a* moves to the left and right $2\sqrt{2}$ units (about 2.83) from the center and $\sqrt{6}$ units (about 2.45) up and down from the center. Plotting these points locates the vertices of the ellipse.

   Its vertices are at about $(-1\pm 2.83,2)$, and its co-vertices are at about $(-1,2\pm 2.45)$. The major axis lies on $y=2$, and the minor axis lies on $x=-1$. The length of the major axis is $2a=2\cdot 2\sqrt{2}=4\sqrt{2}\approx 5.66$, and the length of the minor axis is $2b=2\sqrt{6}\approx 4.90$.

7. **Plot the foci of the ellipse.**

   You determine the focal distance from the center to the foci in this ellipse with the equation $8-6=2=F^2$, so $F=\sqrt{2}\approx 1.41$. The foci, expressed as points, are located at $(-1\pm\sqrt{2},2)$.

Figure 17-12 shows all the parts of this ellipse.

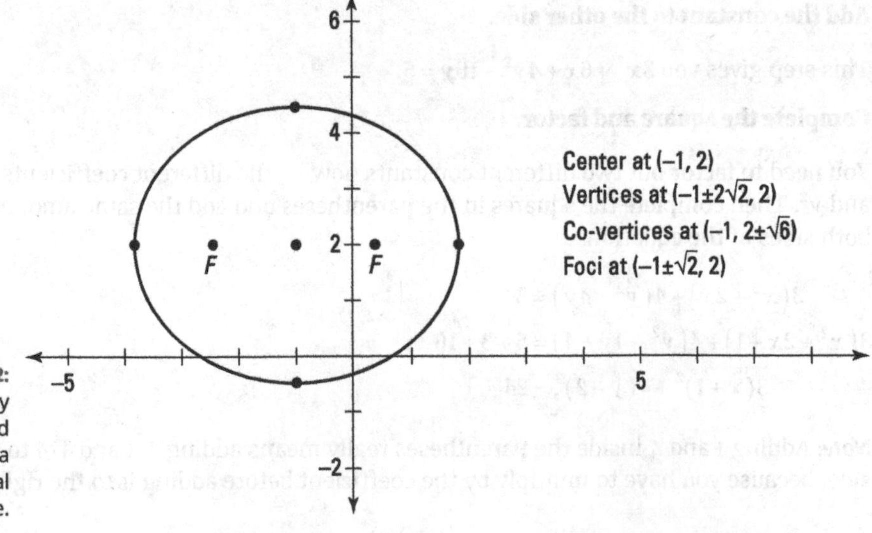

FIGURE 17-12: The many points and parts of a horizontal ellipse.

11 **Graph the ellipse and identify its parts: $\frac{(x+3)^2}{25}+\frac{(y-2)^2}{36}=1$.**

12 **Graph the ellipse and identify its parts: $x^2+16y^2-2x-128y+241=0$.**

Join Two Curves and Get a Hyperbola

Hyperbola literally means "excessive" in Greek, so it's a fitting name. A *hyperbola* is basically twice what you have with a parabola. Think of a hyperbola as two U-shaped curves — each one a perfect mirror image of the other, and each opening away from one another. The vertices of these two curves are a given distance apart, and the curves both open either vertically or horizontally.

The mathematical definition of a hyperbola is the set of all points where the difference in the distance from two fixed points (called the *foci*) is constant. In this section, you discover the ins and outs of the hyperbola, including how to name its parts and graph it.

Visualizing the two types of hyperbolas and their bits and pieces

Similar to ellipses (see the previous section), hyperbolas come in two types: horizontal and vertical.

The equation for a horizontal hyperbola is $\frac{(x-h)^2}{a^2}-\frac{(y-k)^2}{b^2}=1$.

The equation for a vertical hyperbola is $\frac{(y-k)^2}{a^2}-\frac{(x-h)^2}{b^2}=1$.

Notice that *x* and *y* switch places (as well as the *h* and *k* with them) to name horizontal versus vertical, but *a* and *b* stay put. So for hyperbolas, a^2 always comes first, but it isn't necessarily greater. More accurately, *a* is always squared under the positive term (either x^2 or y^2). Basically, to get a hyperbola into standard form, you need to be sure that the positive squared term is first.

The center of a hyperbola is not on the actual curve, but exactly halfway in between the two vertices of the hyperbola. Always plot the center first and then count out from the center to find the vertices, axes, and asymptotes.

A hyperbola has two axes of symmetry. The axis that passes through the center, the vertices of the hyperbola, and the two foci is called the *transverse axis*; the axis that's perpendicular to the transverse axis through the center is called the *conjugate axis*. A horizontal hyperbola has its transverse axis at $y = k$ and its conjugate axis at $x = h$; a vertical hyperbola has its transverse axis at $x = h$ and its conjugate axis at $y = k$.

You can see the two types of hyperbolas in Figure 17-13. Figure 17-13a is a horizontal hyperbola, and Figure 17-13b is a vertical one.

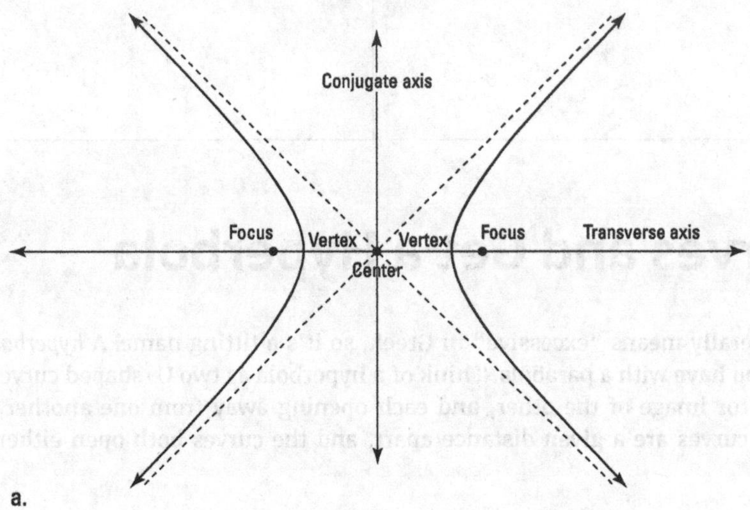

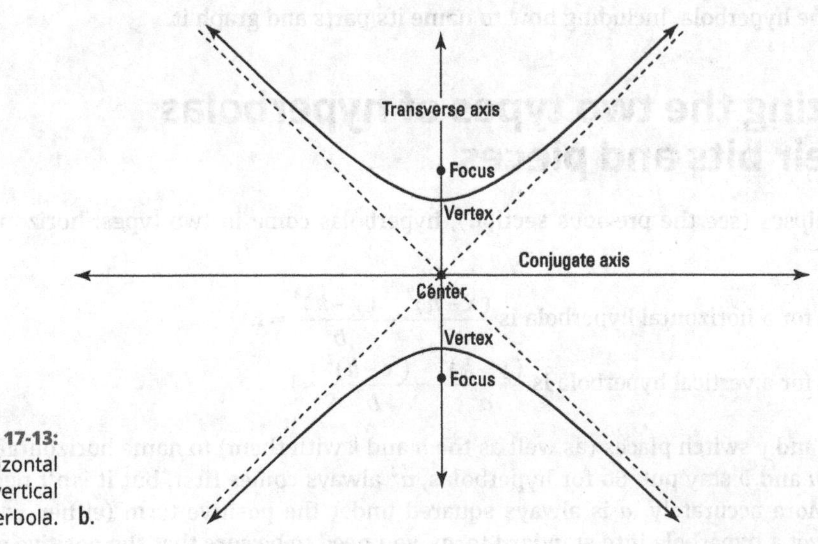

FIGURE 17-13: A horizontal and a vertical hyperbola.

If the hyperbola that you're trying to graph isn't in standard form, then you need to complete the square to get it into standard form. For the steps of completing the square with conic sections, check out the earlier section, "Identifying the extremes of vertical parabolas."

Here are the two standard equations for hyperbolas.

Horizontal hyperbola: $\frac{(x-h)^2}{a^2}-\frac{(y-k)^2}{b^2}=1.$

Vertical hyperbola: $\frac{(y-k)^2}{a^2}-\frac{(x-h)^2}{b^2}=1.$

- **Center:** (h,k)
- **Vertices:**

  A horizontal hyperbola has vertices at $(h \pm a, k)$.

  A vertical hyperbola has vertices at $(h, k \pm a)$.
- **Foci:**

  A horizontal hyperbola has foci $(h \pm F, k)$.

  A vertical hyperbola has foci $(h, k \pm F)$.

  You find the foci of any hyperbola by using the equation $a^2+b^2=F^2$, where F is the distance from the center to the foci along the transverse axis, the same axis that the vertices are on. The distance F moves in the same direction as a.
- **Asymptotes:**

  A horizontal hyperbola's asymptotes have slopes $m=\pm\frac{b}{a}$.

  A vertical hyperbola's asymptotes have slopes $m=\pm\frac{a}{b}$.

  Through the center of the hyperbola and through the corners of a rectangle described in the following bullet run the asymptotes of the hyperbola. These asymptotes help guide your sketch of the curves because the curves don't cross them at any point on the graph.
- **Helpful rectangle:**

  The values of a and b are informative, because they help you form a rectangle that the hyperbola nestles up against. To find the sides of the rectangle, count a or b units right or left of the center and count a or b units up and down from the center. How do you choose the value? You count in vertical directions from the center using the value determined from the number under the y variable, and you count horizontally both to the left and to the right using the value determined under the x variable. These distances from the center to the midpoints of the sides of the rectangle are half the length of the transverse axis and conjugate axis. In a hyperbola, a could be greater than, less than, or equal to b. This rectangle has sides that are parallel to the x- and y-axis (in other words, don't just connect the four points, because they're the midpoints of the sides, not the corners of the rectangle). This rectangle is a useful guide when you graph the hyperbola.

Q. Identify the necessary parts/features of the hyperbola $\frac{(y-3)^2}{16}-\frac{(x+1)^2}{9}=1$ and graph it.

EXAMPLE **A.** This is a vertical hyperbola. Identifying the following parts:

- The center, (h,k), is $(-1,3)$.
- The vertices, $(h,k\pm a)$, are at $(-1,3\pm 4)$, or $(-1,7)$ and $(-1,-1)$.
- The foci, $(h,k\pm F)$, are $(-1,3\pm 5)$, or $(-1,8)$ and $(-1,-2)$. The points are created using the equation $a^2+b^2=F^2$, where F is the distance from the center to the foci along the transverse axis, the same axis that the vertices are on. The distance F moves in the same direction as a. Continuing, $16+9=F^2$, or $25=F^2$. Taking the root of both sides gives you $5=F$.

To graph the hyperbola, you take all the information that you've found and put it to work. Follow these steps:

1. **Mark the center, $(-1,3)$.**
2. **From the center in Step 1, find points on the transverse and conjugate axes.**

   Go up and down the transverse axis a distance of 4 (because 4 is under y) and mark the points, and then go right and left 3 (because 3 is under x) and mark the points. The points you marked as a (on the transverse axis) are your vertices, at $(-1,7)$ and $(-1,-1)$.

3. **Use these points to draw a rectangle that will help guide the shape of your hyperbola.**

   These points are the midpoints of the sides of your rectangle. Because you went up and down 4, the height of your rectangle is 8; going left and right 3 gives you a width of 6.

4. **Draw diagonal lines through the center and the corners of the rectangle that extend beyond the rectangle.**

   This step gives you two lines that will be your asymptotes.

5. **Sketch the curves.**

   Beginning at each vertex separately, draw the curves that hug the asymptotes more closely the farther away from the vertices the curve gets.

   The graph approaches the asymptotes but never actually touches them.

   Figure 17-14 shows the finished hyperbola. You also see the foci marked at $(-1,8)$ and $(-1,-2)$.

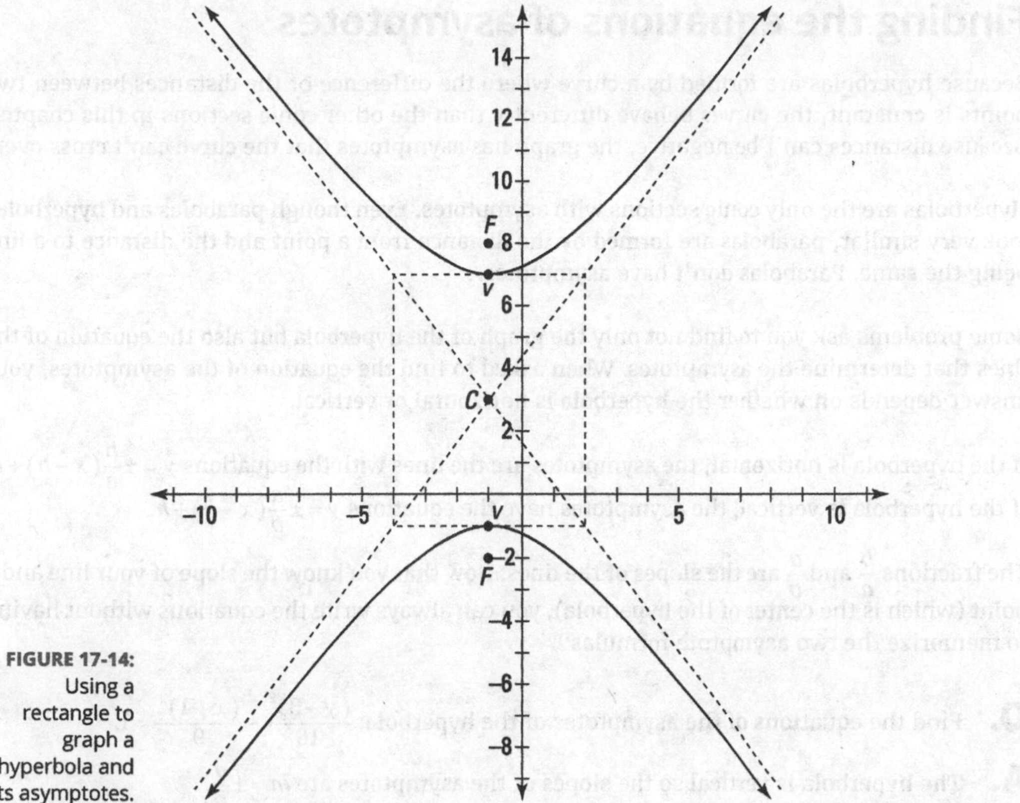

FIGURE 17-14: Using a rectangle to graph a hyperbola and its asymptotes.

YOUR TURN

13 **Graph the hyperbola and identify its parts:** $\frac{(x-3)^2}{25}-\frac{(y-1)^2}{9}=1.$

14 **Graph the hyperbola and identify its parts:** $\frac{(y-3)^2}{4}-\frac{(x+5)^2}{16}=1.$

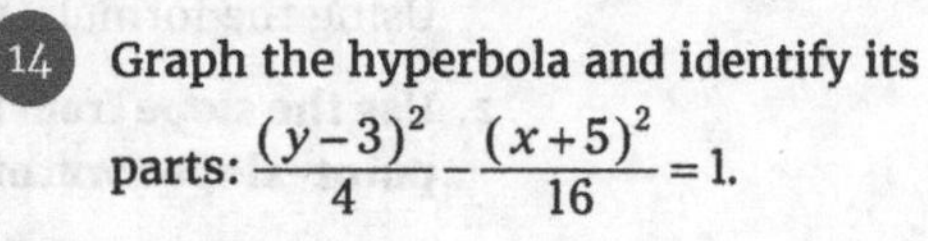

Finding the equations of asymptotes

Because hyperbolas are formed by a curve where the difference of the distances between two points is constant, the curves behave differently than the other conic sections in this chapter. Because distances can't be negative, the graph has asymptotes that the curve can't cross over.

REMEMBER

Hyperbolas are the only conic sections with asymptotes. Even though parabolas and hyperbolas look very similar, parabolas are formed by the distance from a point and the distance to a line being the same. Parabolas don't have asymptotes.

Some problems ask you to find not only the graph of the hyperbola but also the equation of the lines that determine the asymptotes. When asked to find the equation of the asymptotes, your answer depends on whether the hyperbola is horizontal or vertical.

TECHNICAL STUFF

If the hyperbola is horizontal, the asymptotes are the lines with the equations $y = \pm\frac{b}{a}(x-h)+k$. If the hyperbola is vertical, the asymptotes have the equations $y = \pm\frac{a}{b}(x-h)+k$.

The fractions $\frac{b}{a}$ and $\frac{a}{b}$ are the slopes of the lines. Now that you know the slope of your line and a point (which is the center of the hyperbola), you can always write the equations without having to memorize the two asymptote formulas.

EXAMPLE

Q. Find the equations of the asymptotes of the hyperbola: $\frac{(y-3)^2}{16} - \frac{(x+1)^2}{9} = 1$.

A. The hyperbola is vertical so the slopes of the asymptotes are $m = \pm\frac{a}{b}$.

1. **Find the slopes of the asymptotes.**

   Using the formula, the slopes of the asymptotes are $\pm\frac{4}{3}$.

2. **Use the slope from Step 1 and the center of the hyperbola as the point to find the point-slope form of the equation.**

   Remember that the equation of a line with slope m through point (x_1, y_1) is $y - y_1 = m(x - x_1)$. Therefore, since the slopes are $\pm\frac{4}{3}$ and the point is $(-1, 3)$, the equations of the lines are $y - 3 = \pm\frac{4}{3}(x+1)$.

3. **Solve for y to find the equations in slope-intercept form.**

   You do each asymptote separately here.

- Distribute $\frac{4}{3}$ on the right to get $y - 3 = \frac{4}{3}x + \frac{4}{3}$, and then add 3 to both sides to get $y = \frac{4}{3}x + \frac{13}{3}$.
- Distribute $-\frac{4}{3}$ on the right to get $y - 3 = -\frac{4}{3}x - \frac{4}{3}$, and then add 3 to both sides to get $y = -\frac{4}{3}x + \frac{5}{3}$.

YOUR TURN

15 **Find the equations of the asymptotes to the hyperbola: $\frac{(x-3)^2}{25}-\frac{(y-1)^2}{9}=1$.**

16 **Find the equations of the asymptotes to the hyperbola: $\frac{(y-3)^2}{4}-\frac{(x+5)^2}{16}=1$.**

Expressing Conics Outside the Realm of Cartesian Coordinates

Up to this point, the graphs of conics have been constructed using rectangular coordinates (x,y). There are also two other ways to graph conics.

- **In parametric form:** *Parametric form* is a fancy way of saying a form in which you can deal with conics that aren't easily expressed as the graph of a function $y = f(x)$. Parametric equations are usually used to describe the motion or velocity of an object with respect to time. Using parametric equations allows you to evaluate both *x* and *y* as dependent variables, as opposed to *x* being independent and *y* dependent on *x*.
- **In polar form:** As you can see in Chapter 16, in *polar form* every point on the polar graph is (r,θ).

The following sections show you how to graph conics in these forms.

Graphing conic sections in parametric form

Parametric form defines both the *x*- and the *y*-variables of conic sections in terms of a third, arbitrary variable, called the *parameter*, which is usually represented by *t*; you can find both *x* and *y* by plugging *t* into the parametric equations. As *t* changes, so do *x* and *y*, which means that *y* is no longer dependent on *x* but is dependent on *t*.

Why switch to this form? Consider, for example, an object moving in a plane during a specific time interval. If a problem asks you to describe the path of the object and its location at any certain time, you need three variables:

- Time *t*, which usually is the parameter
- The coordinates *(x,y)* of the object at time *t*

TECHNICAL STUFF

The x_t variable gives the horizontal movement of an object as t changes; the y_t variable gives the vertical movement of an object over time.

EXAMPLE

Q. Graph the following equations. The equations given here define both x and y for the same parameter, t, and the inequality statement defines the parameter in a set interval:

$$x = 2t - 1x^2$$
$$y = t^2 - 3t + 1$$
$$1 < t \le 5$$

Time t exists only between 1 and 5 seconds for this problem.

A. If you're asked to graph these equations, you can do it in one of two ways. The first method is the plug and chug: Set up a chart and pick t values from the given interval in order to figure out what x and y should be, and then graph these points like normal. Table 17-2 shows the results of this process. *Note:* The value for $t = 1$ is included in the chart, even though the parameter isn't defined there. You need to see what it would've been, because you graph the point where $t = 1$ with an open circle to show what happens to the function arbitrarily close to 1. Be sure to make that point an open circle on your graph.

Table 17-2 Plug and Chug *t* Values from the Interval

Variable			Interval Time		
t value	1	2	3	4	5
x value	1	3	5	7	9
y value	−1	−1	1	5	11

The points (x,y) are plotted in Figure 17-15.

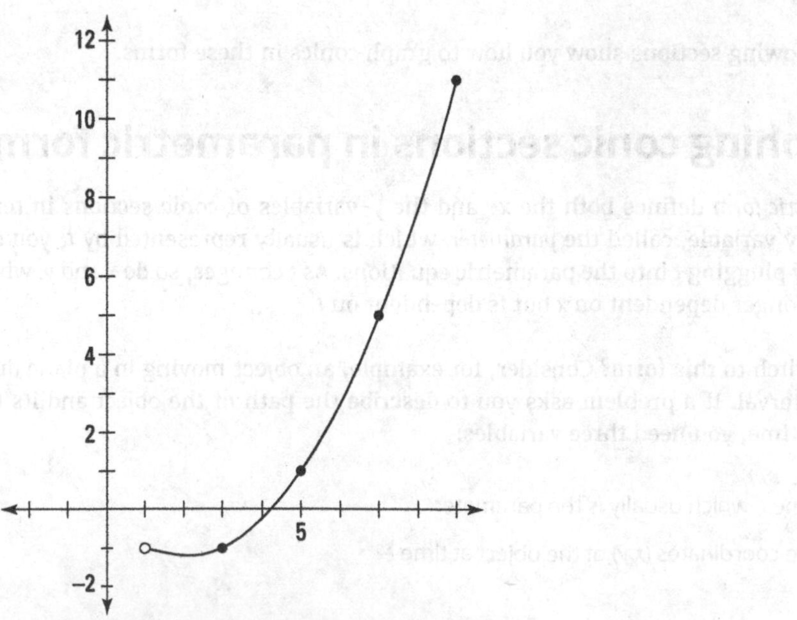

FIGURE 17-15: Graphing a parametric curve.

The other way to graph a parametric curve is to solve one equation for the parameter and then substitute that equation into the other equation. You should pick the simplest equation to solve and start there.

EXAMPLE

Q. Solve the linear equation $x = 2t - 1$ for t.

A. Use the following steps:

1. **Solve the simplest equation.**

   For the chosen equation, you get $t = \frac{x+1}{2}$.

2. **Plug the solved equation into the other equation.**

   For this step, you get $y = \left(\frac{x+1}{2}\right)^2 - 3\left(\frac{x+1}{2}\right) + 1$.

3. **Simplify this equation.**

   This simplifies to $y = \frac{x^2}{4} - x - \frac{1}{4}$.

   Because this step gives you an equation in terms of *x* and *y*, you can graph the points on the coordinate plane just like you always do. The only problem is that you don't draw the entire graph, because you have to look at a specific interval of *t*.

4. **Substitute the endpoints of the *t* interval into the *x* function to know where the graph starts and stops.**

   This is shown in Table 17-2. When $t = 1$, $x = 1$, and when $t = 5$, $x = 9$.

Figure 17-15 shows the parametric curve from this example (for both methods). You end up with a parabola, but you can also write parametric equations for ellipses, circles, and hyperbolas using this same method.

TIP

If you have a graphing calculator, you can set it to parametric mode to graph. When you go into your graphing utility, you get two equations — one is "$x =$" and the other is "$y =$". Input both equations exactly as they're given, and the calculator will do the work for you!

The equations of conic sections on the polar coordinate plane

Graphing conic sections on the polar plane (see Chapter 16) is based on equations that depend on a special value known as *eccentricity*. Eccentricity is a numerical value that describes the overall shape of a conic section. The value of a conic's eccentricity can tell you what type of conic section the equation describes, as well as how fat or skinny it is. When graphing equations using polar coordinates, you may have trouble telling which conic section you should be graphing based solely on the equation (unlike graphing in Cartesian coordinates, where each conic section has its own unique equation). Therefore, you can use the eccentricity of a conic section to find out exactly which type of curve you should be graphing.

TECHNICAL STUFF

Here are the general equations that allow you to put conic sections in polar coordinate form, where (r,θ) is the coordinate of a point on the curve in polar form. Recall from Chapter 16 that r is the radius, and θ is the angle in standard position on the polar coordinate plane.

$$r=\frac{ke}{1-e\cos\theta} \text{ or } r=\frac{ke}{1-e\sin\theta}$$

$$r=\frac{-ke}{1-e\cos\theta} \text{ or } r=\frac{-ke}{1-e\sin\theta}$$

When graphing conic sections in polar form, you can plug in various values of θ to get the graph of the curve. In the equations shown here, k is a constant value, θ takes the place of time, and e is the eccentricity. The variable e determines the conic section:

- If $0<e<1$, the conic section is an ellipse.
- If $e=1$, the conic section is a parabola.
- If $e>1$, the conic section is a hyperbola.

EXAMPLE

Q. Graph $r=\dfrac{2}{4-\cos\theta}$.

A. First, realize that as it's shown, it doesn't exactly fit the form of any of the equations shown for the conic sections. It doesn't fit because all the denominators of the conic sections begin with 1, and this equation begins with 4. Have no fear; you can perform mathematical maneuvers so you can tell what k and e are!

First, multiply both the numerator and denominator by $\frac{1}{4}$, distributing over the denominator: $r=\dfrac{\frac{1}{4}}{\frac{1}{4}}\cdot\dfrac{2}{4-\cos\theta}=\dfrac{\frac{1}{2}}{1-\frac{1}{4}\cos\theta}$.

Fitting this equation to the form $r=\dfrac{ke}{1-e\cos\theta}$, you see that the eccentricity is $\frac{1}{4}$. Since the numerator is the product of k and the eccentricity, k must be equal to 2. You could rewrite the equation as $r=\dfrac{2\left(\frac{1}{4}\right)}{1-\frac{1}{4}\cos\theta}$. Since the eccentricity is $\frac{1}{4}$, you know that this is an ellipse.

In order to graph the polar function of the ellipse, you can plug in values of θ and solve for r. Then plot the coordinates of (r,θ) on the polar coordinate plane to get the graph. For the graph of the equation, $r=\dfrac{2}{4-\cos\theta}$, you can plug in 0, $\frac{\pi}{2}$, π, and $\frac{3\pi}{2}$, and find r.

- $r(0)$: The cosine of 0 is 1, so $r(0)=\frac{2}{3}$.
- $r\left(\frac{\pi}{2}\right)$: The cosine of $\frac{\pi}{2}$ is 0, so $r\left(\frac{\pi}{2}\right)=\frac{1}{2}$.
- $r(\pi)$: The cosine of π is -1, so $r(\pi)=\frac{2}{5}$.
- $r\left(\frac{3\pi}{2}\right)$: The cosine of $\frac{3\pi}{2}$ is 0, so $r\left(\frac{3\pi}{2}\right)=\frac{1}{2}$.

These four points are enough to give you a rough sketch of the graph. You can see the graph of the example ellipse in Figure 17-16.

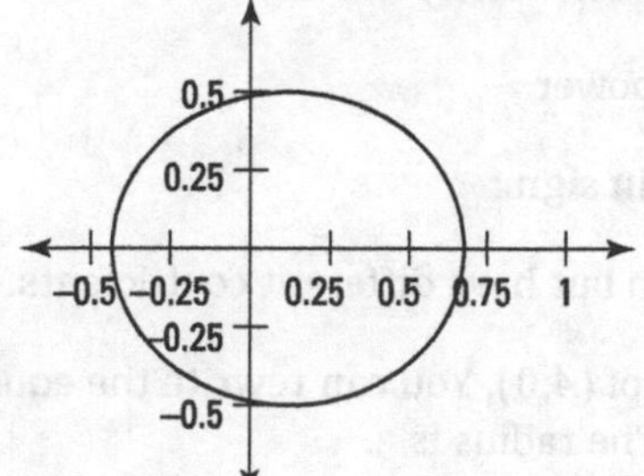

FIGURE 17-16: The graph of an ellipse in polar coordinates.

YOUR TURN

17 Write the parametric equations $x = 2t + 1$, $y = t^2 - 3t + 1$, and $1 < t \leq 5$ in rectangular form.

18 Graph the equation $r = \dfrac{2}{4 - \cos\theta}$.

Practice Questions Answers and Explanations

1. **Circle.** The coefficients of the two squared terms are exactly the same.
2. **Parabola.** Only one term is raised to the second power.
3. **Hyperbola.** The two squared terms are opposite in sign.
4. **Ellipse.** The two squared terms are the same sign but have different coefficients.
5. **See the following figure.** The circle has a center of $(4,0)$. You can rewrite the equation to create the standard form: $(x-4)^2+(y-0)^2=9$. The radius is 3.

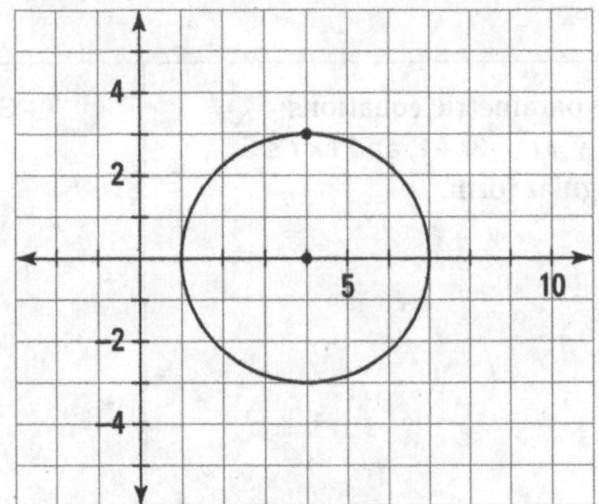

6. **See the following figure.** First, change the equation to the standard form.

$$x^2-8x+y^2+12y+27=0 \rightarrow \left(x^2-8x+16\right)+\left(y^2+12y+36\right)=-27+16+36$$

$$\rightarrow (x-4)^2+(y+6)^2=25$$

The center of the circle is $(4,-6)$, and the radius is 5.

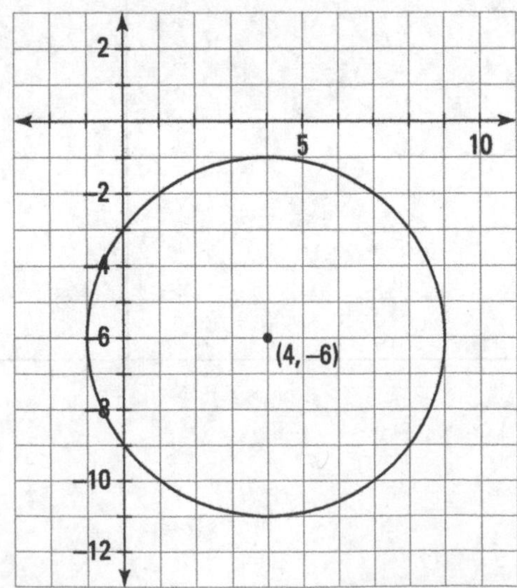

7. **See the following figure.** The vertex is at $(3,5)$. It opens downward and is steepened by the 2 multiplier. The focus is at $\left(3,4\frac{7}{8}\right)$, and the directrix is $y=5\frac{1}{8}$. The axis of symmetry is $x=3$.

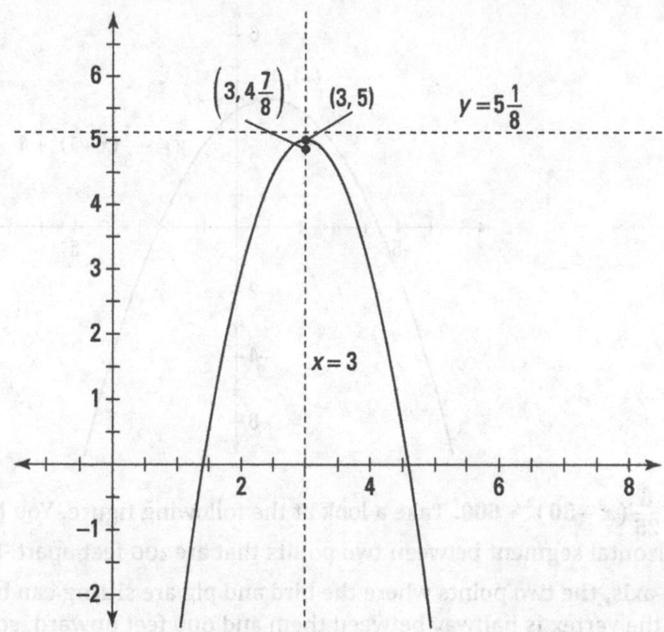

8. **See the following figure.** The vertex is at $(-5,2)$. It opens to the right and is flattened by the $\frac{1}{4}$ multiplier. The focus is at $(-4,2)$, and the directrix is $x=-6$. The axis of symmetry is $y=2$.

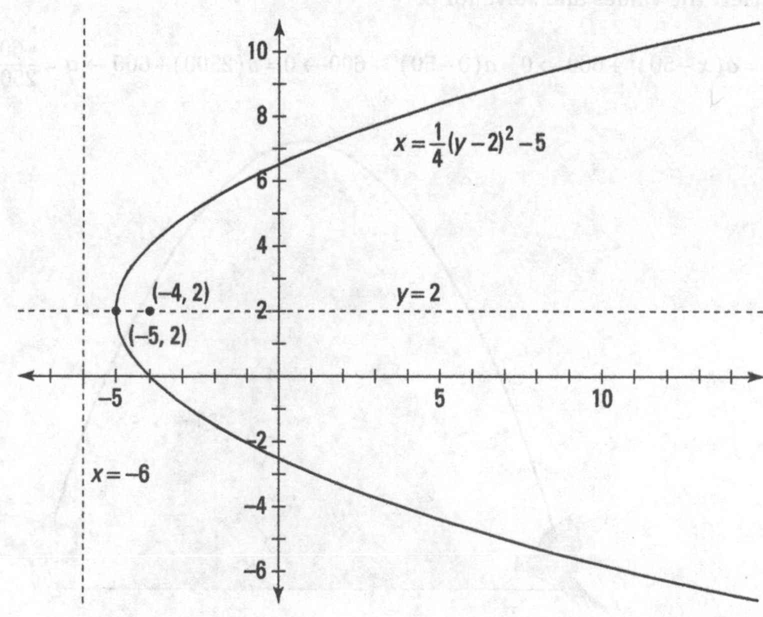

9 $(-1,4)$. **This is a parabola that opens downward, so the maximum point is the vertex. See the graph.**

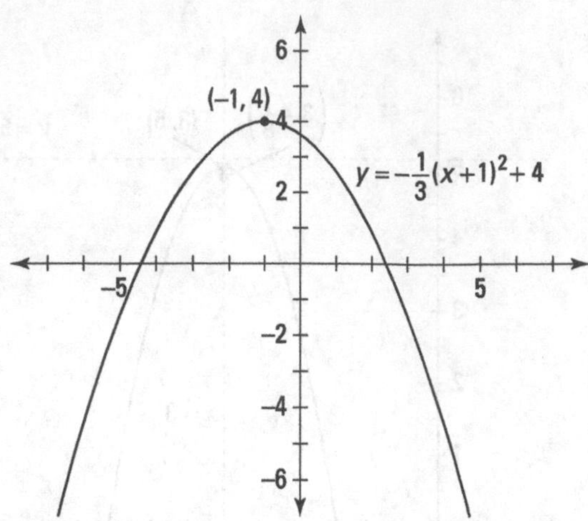

10 $y=-\frac{6}{25}(x-50)^2+600$. **Take a look at the following figure. You have created a parabola with a horizontal segment between two points that are 100 feet apart. Letting the segment lie on the x-axis, the two points where the bird and pig are sitting can be (0,0) and (100,0). In that case, the vertex is halfway between them and 600 feet upward, so the coordinates of the vertex are (50,600). The parabola opens downward. Inserting all this information into the standard form of a vertical parabola, you have $y=a(x-50)^2+600$. You just need to solve for a. You can use either point on the horizontal segment, but (0,0) would be the easiest. Insert the values and solve for a.**

$$y=a(x-50)^2+600 \rightarrow 0=a(0-50)^2+600 \rightarrow 0=a(2500)+600 \rightarrow a=\frac{-600}{2500}=-\frac{6}{25}$$

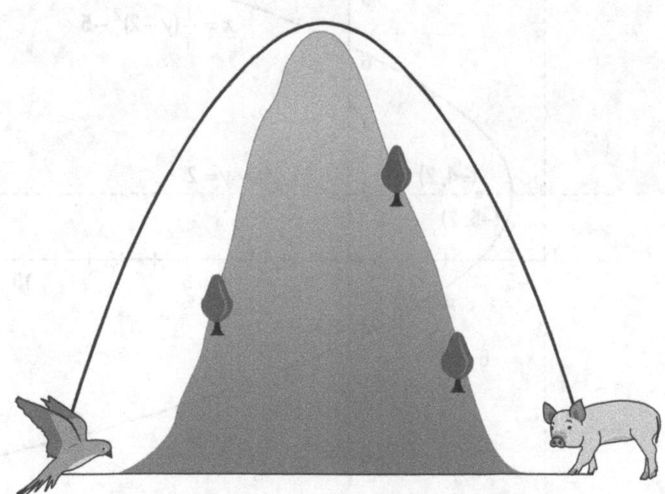

(11) **See the following figure.** The center is at $(-3,2)$. The vertices are $(-3,8)$ and $(-3,-4)$; the co-vertices are $(-8,2)$ and $(2,2)$. The foci are $(-3,2+\sqrt{11})$ and $(-3,2-\sqrt{11})$.

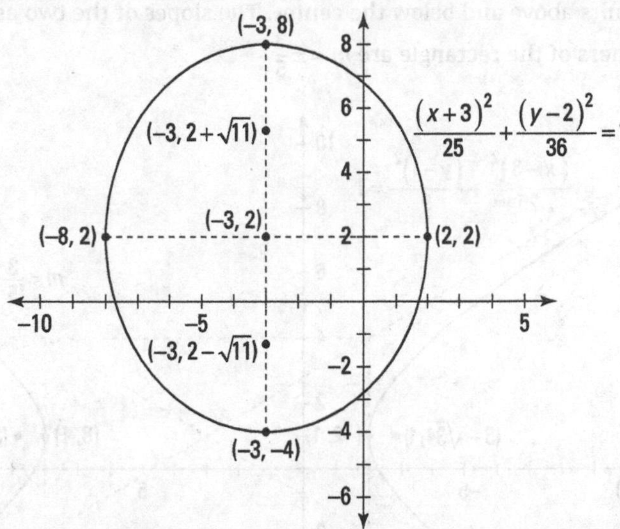

(12) **See the following figure.** First, write the equation in the standard form.

$$x^2+16y^2-2x-128y+241=0 \rightarrow (x^2-2x+1)+16(y^2-8y+16)=-241+1+256$$

$$\rightarrow (x-1)^2+16(y-4)^2=16 \rightarrow \frac{(x-1)^2}{16}+\frac{(y-4)^2}{1}=1$$

You find the center at $(1,4)$. The vertices are $(-3,4)$ and $(5,4)$; the co-vertices are $(1,5)$ and $(1,3)$. The foci are $(1+\sqrt{15},4)$ and $(1-\sqrt{15},4)$.

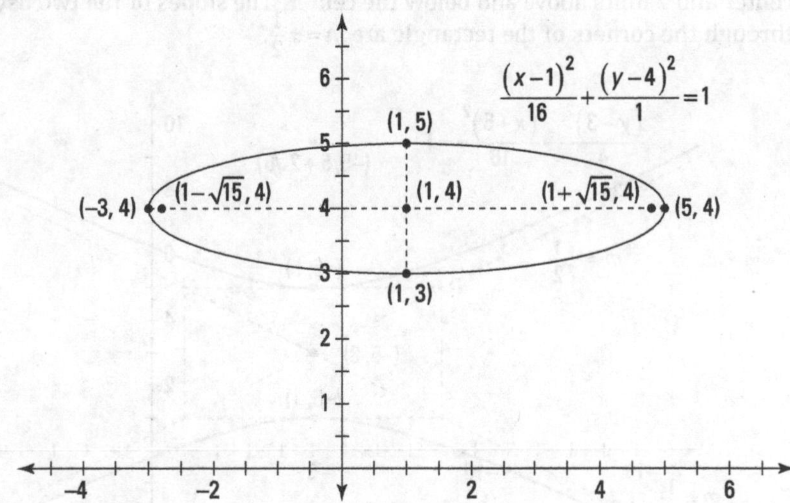

13 **See the following figure.** The hyperbola opens left and right. The center is at (3,1). The vertices are at (−2,1) and (8,1). The foci are at $\left(3+\sqrt{34},1\right)$ and $\left(3-\sqrt{34},1\right)$. The helpful rectangle has its sides on the lines $x=-2$, $x=8$, $y=4$, $y=-2$, which are 5 units right and left of the center and 3 units above and below the center. The slopes of the two asymptotes passing through the corners of the rectangle are $m=\pm\frac{3}{5}$.

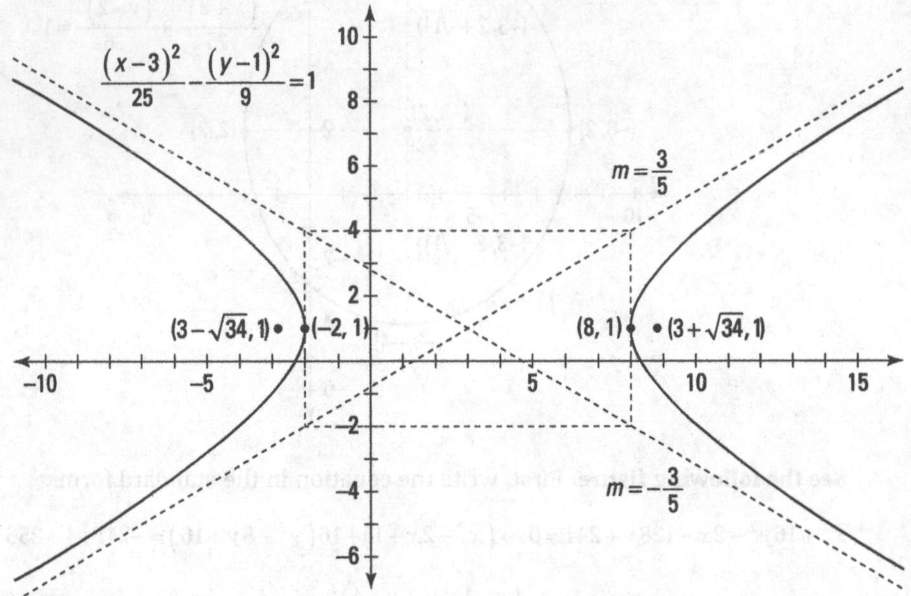

14 **See the following figure.** The hyperbola opens up and down. The center is at (−5,3). The vertices are at (−5,5) and (−5,1). The foci are at $\left(-5,5+2\sqrt{5}\right)$ and $\left(-5,1-2\sqrt{5}\right)$. The helpful rectangle has its sides on the lines $x=-9$, $x=-1$, $y=5$, $y=1$, which are 4 units right and left of the center and 2 units above and below the center. The slopes of the two asymptotes passing through the corners of the rectangle are $m=\pm\frac{1}{2}$.

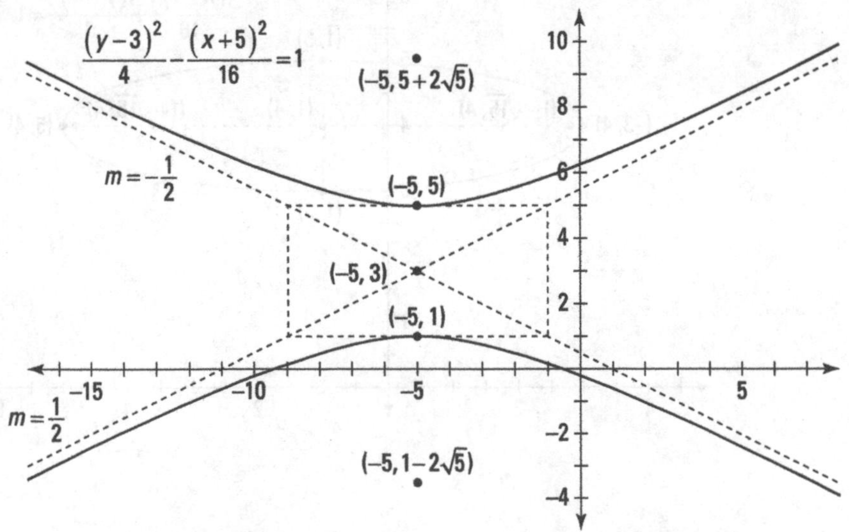

15. $\mathbf{y = \frac{3}{5}x - \frac{4}{5}}$, $\mathbf{y = -\frac{3}{5}x + \frac{14}{5}}$. The slopes of the asymptotes are $m = \pm\frac{3}{5}$. Substituting into the equation $y = \pm m(x-h)+k$, where h and k are the coordinates of the center of the hyperbola, you have $y = \pm\frac{3}{5}(x-3)+1$. Simplifying for the positive slope, $y = \frac{3}{5}(x-3)+1 = \frac{3}{5}x - \frac{9}{5} + 1 = \frac{3}{5}x - \frac{4}{5}$. And, when the slope is negative, $y = -\frac{3}{5}(x-3)+1 = -\frac{3}{5}x + \frac{9}{5} + 1 = -\frac{3}{5}x + \frac{14}{5}$.

16. $\mathbf{y = \frac{1}{2}x + \frac{11}{2}}$, $\mathbf{y = -\frac{1}{2}x + \frac{1}{2}}$. The slopes of the asymptotes are $m = \pm\frac{1}{2}$. Substituting into the equation $y = \pm m(x-h)+k$, where h and k are the coordinates of the center of the hyperbola, you have $y = \pm\frac{1}{2}(x+5)+3$. Simplifying for the positive slope, $y = \frac{1}{2}(x+5)+3 = \frac{1}{2}x + \frac{5}{2} + 3 = \frac{1}{2}x + \frac{11}{2}$. And, when the slope is negative, $y = -\frac{1}{2}(x+5)+3 = -\frac{1}{2}x - \frac{5}{2} + 3 = -\frac{1}{2}x + \frac{1}{2}$.

17. $\mathbf{y = \frac{1}{4}x^2 - 2x + \frac{11}{4}}$. First, solve the equation that's linear for t: $t = \frac{x-1}{2}$. Then substitute this value into the other equation for t: $y = \left(\frac{x-1}{2}\right)^2 - 3\left(\frac{x-1}{2}\right) + 1$. Simplify this equation to get $y = \frac{1}{4}x^2 - 2x + \frac{11}{4}$. This is a parabola. Since $1 < t \le 5$, the graph will only exist for $3 < x \le 11$.

18. **See the following figure.** First, notice that the equation as shown doesn't fit exactly into any of the equations given in this section. All those denominators begin with 1, and this equation begins with 4! So factor out the 4 from the denominator to get $r = \frac{2}{4\left(1 - \frac{1}{4}\cos\theta\right)}$ or $r = \frac{2 \cdot \frac{1}{4}}{1 - \frac{1}{4}\cos\theta}$.

Notice that this makes e the same in the numerator and denominator, $\frac{1}{4}$, and that k is 2. Now that you know e is $\frac{1}{4}$, that tells you the equation is an ellipse. Plugging in values gives you points $\theta = 0, r = \frac{2}{3}$; $\theta = \frac{\pi}{2}, r = \frac{1}{2}$; $\theta = \pi, r = \frac{2}{5}$; $\theta = \frac{3\pi}{2}, r = \frac{1}{2}$. These help you with the shape of the figure.

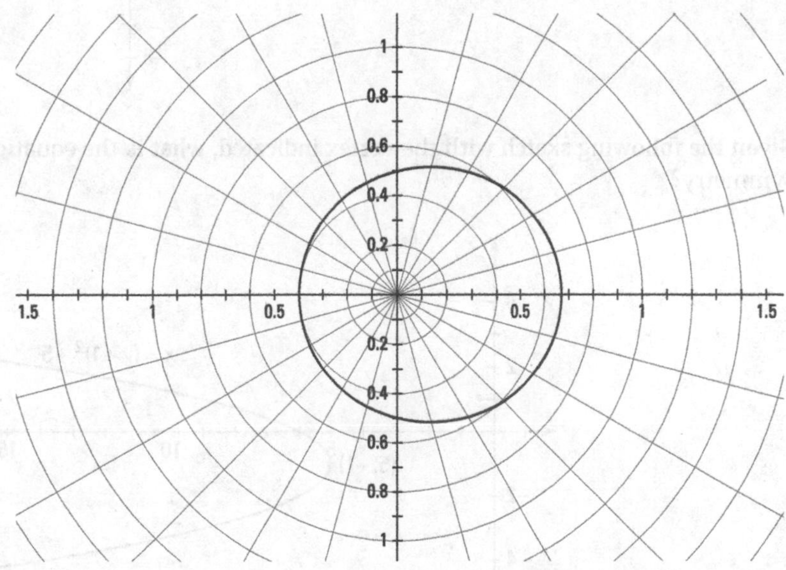

If you're ready to test your skills a bit more, take the following chapter quiz that incorporates all the chapter topics.

Whaddya Know? Chapter 17 Quiz

Quiz time! Complete each problem to test your knowledge on the various topics covered in this chapter. You can then find the solutions and explanations in the next section.

1. **Eliminate the parameter and find an equation in x and y whose graph contains the curve defined by the parametric equations $x = t^2 + 2$, $y = 1 - t$, and $t \geq 0$.**

2. **Identify which conic is represented by each equation.**

   a) $y = -3x + x^2 + 11$

   b) $y^2 = -3x + x^2 + 11$

   c) $y^2 = -3x - x^2 + 11$

   d) $y^2 = -3x - 3x^2 + 11$

3. **Given the following sketch, what is the diameter of the circle?**

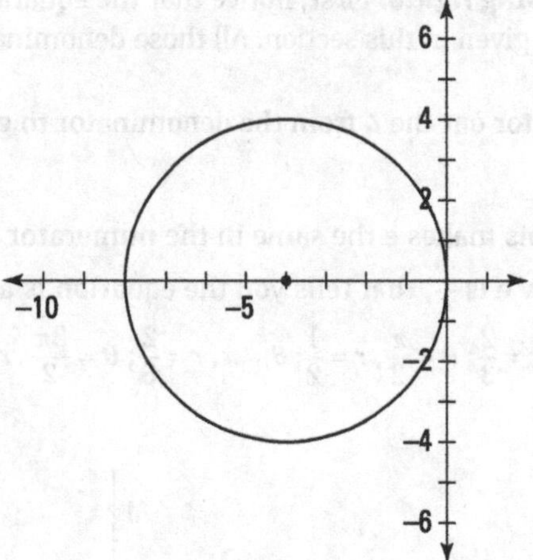

4. **Given the following sketch with the vertex indicated, what is the equation of the axis of symmetry?**

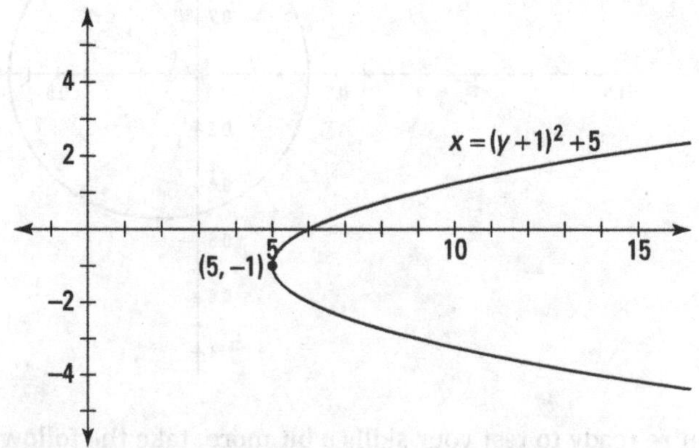

5 Given the following sketch, what is the vertex?

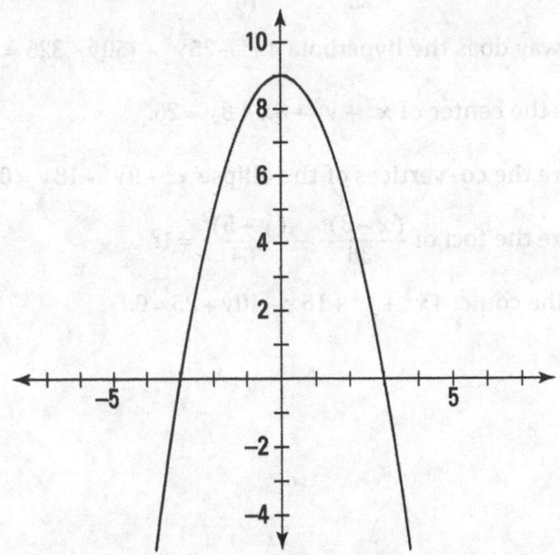

6 Given the following sketch, what is the length of the major axis?

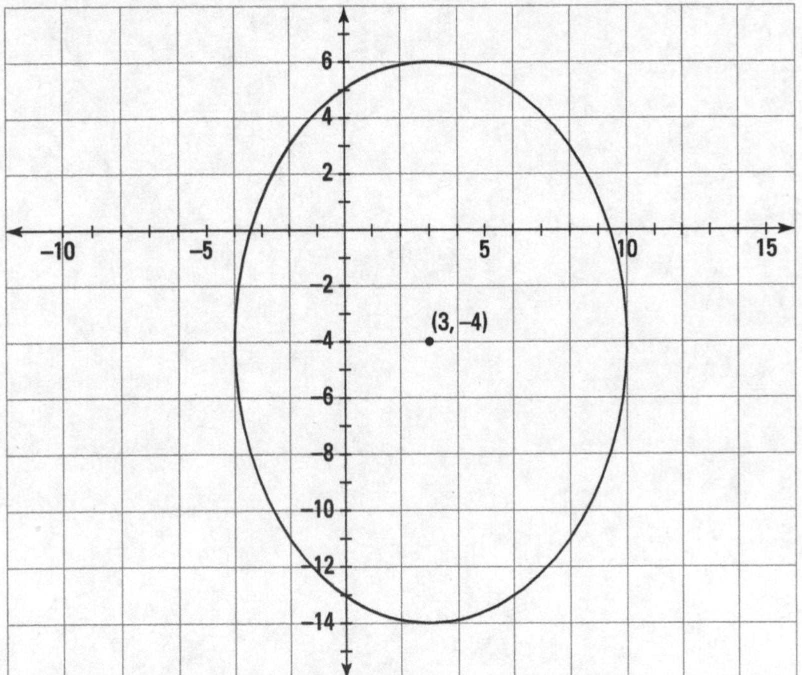

7 What is the center of the circle $2x^2 + 2y^2 - 4x + 12y + 9 = 0$?

8 What is the vertex of $y^2 - 8y - 2x - 4 = 0$?

9. What are the foci of $\frac{(x-1)^2}{25}+\frac{(y+1)^2}{16}=1$?

10. Which way does the hyperbola $4x^2-25y^2+150y-325=0$ open?

11. What is the center of $x^2+y^2+6x+8y=26$.

12. What are the co-vertices of the ellipse $x^2+9y^2-18y=0$?

13. What are the foci of $\frac{(x-3)^2}{36}-\frac{(y+5)^2}{64}=1$?

14. Graph the conic: $4x^2+y^2+16x-10y+25=0$.

Answers to Chapter 17 Quiz

Refer to the graph for answers 1, 7, 8, and 9.

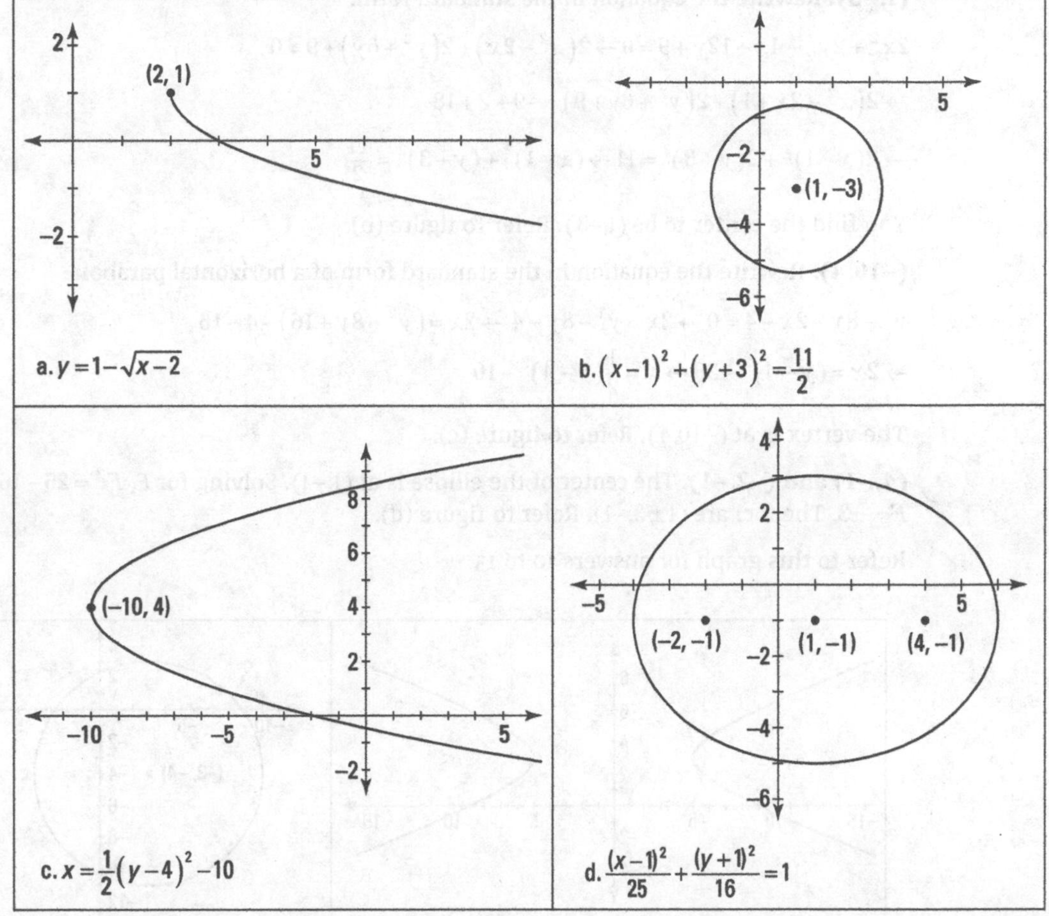

1. $x = y^2 - 2y + 3$. **First, solve for t in the second equation and substitute into the first equation:** $y=1-t \to t=1-y$ and then $x=t^2+2 \to x=(1-y)^2+2 \to x=1-2y+y^2+2 \to x=y^2-2y+3$. This is the part of a right-opening parabola where y-values $y \leq 1$. Refer to figure (a).

2. **a) parabola; b) hyperbola; c) circle; d) ellipse:**

   a) There is only one squared term.

   b) The coefficients of the squared terms have opposite signs.

   c) The coefficients of the squared terms are the same number.

   d) The coefficients of the squared terms are different but have the same sign.

3. **8.** The center is at (−4,0), and the point (0,0) is on the circle. That gives you a radius of 4 and a diameter of 8.

4. $y = -1$. **The axis of symmetry goes through the vertex, and the parabola reflects over that line.**

5 **(0,9).** The vertex of a vertical parabola is the highest or lowest point. In this case, it's the highest point and lies on the y-axis.

6 **20.** The major axis is the longer of the two axes in an ellipse. The major axis goes through the center and, in this case, goes from (3,–14) to (3,6).

7 **(1,–3).** Rewrite the equation in the standard form.

$$2x^2+2y^2-4x+12y+9=0\rightarrow 2\left(x^2-2x\right)+2\left(y^2+6y\right)+9=0$$

$$\rightarrow 2\left(x^2-2x+1\right)+2\left(y^2+6y+9\right)=-9+2+18$$

$$\rightarrow 2(x-1)^2+2(y+3)^2=11\rightarrow (x-1)^2+(y+3)^2=\frac{11}{2}$$

You find the center to be (1,–3). Refer to figure (b).

8 **(–10, 4).** Rewrite the equation in the standard form of a horizontal parabola.

$$y^2-8y-2x-4=0\rightarrow 2x=y^2-8y-4\rightarrow 2x=\left(y^2-8y+16\right)-4-16$$

$$\rightarrow 2x=(y-4)^2-20\rightarrow x=\frac{1}{2}(y-4)^2-10$$

The vertex is at (–10,4). Refer to figure (c).

9 **(4,–1) and (–2,–1).** The center of the ellipse is at (1,–1). Solving for F, $F^2=25-16=9$, so $F=\pm 3$. The foci are $(1\pm 3,-1)$. Refer to figure (d).

Refer to this graph for answers 10 to 13.

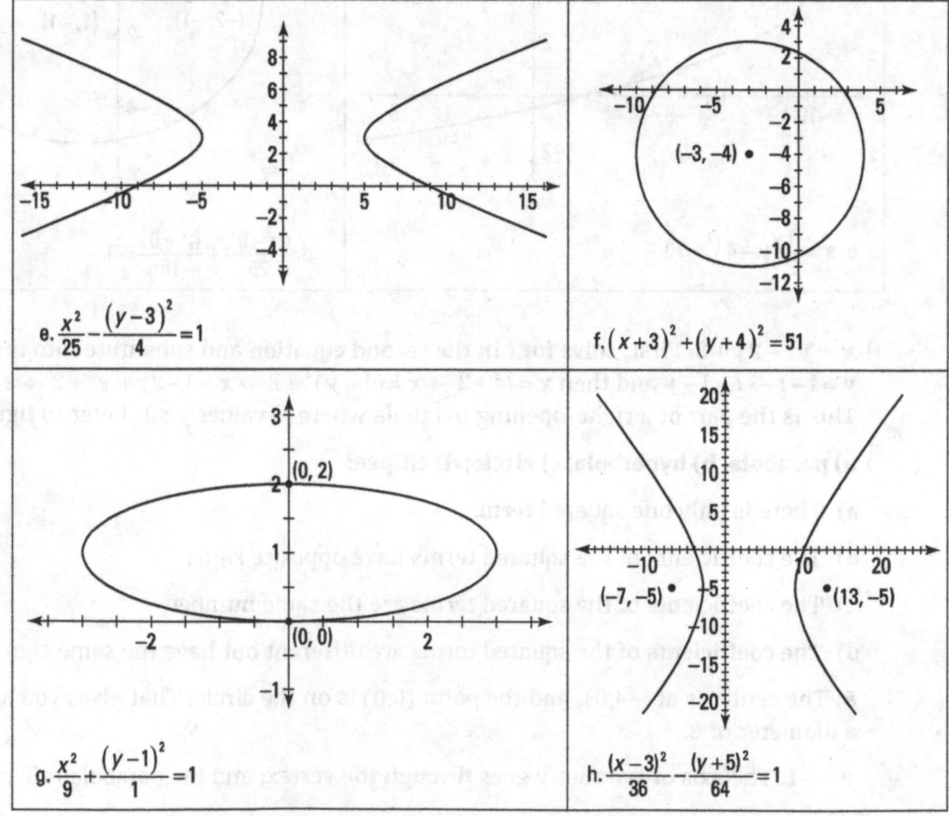

(10) **Left and right.** Writing the hyperbola's equation in standard form,

$$4x^2-25y^2+150y-325=0 \to 4x^2-\left(25y^2-150y\right)=325$$

$$\to 4x^2-25\left(y^2-6y\right)=325 \to 4x^2-25\left(y^2-6y+9\right)=325-225$$

$$\to 4x^2-25(y-3)^2=100 \to \frac{4x^2}{100}-\frac{25(y-3)^2}{100}=1 \to \frac{x^2}{25}-\frac{(y-3)^2}{4}=1$$

This is the standard form for a hyperbola opening left and right. Refer to figure (e).

(11) **(–3,–4).** Writing the equation of the circle in standard form,

$$x^2+y^2+6x+8y=26 \to \left(x^2+6x\right)+\left(y^2+8y\right)=26$$

$$\to \left(x^2+6x+9\right)+\left(y^2+8y+16\right)=26+25 \to (x+3)^2+(y+4)^2=51$$

The center is at (–3,–4). Refer to figure (f).

(12) **(0,2) and (0,0).** First, write the ellipse equation in standard form.

$$x^2+9y^2-18y=0 \to x^2+9\left(y^2-2y\right)=0 \to x^2+9\left(y^2-2y+1\right)=9$$

$$\to x^2+9(y-1)^2=9 \to \frac{x^2}{9}+\frac{9(y-1)^2}{9}=\frac{9}{9} \to \frac{x^2}{9}+\frac{(y-1)^2}{1}=1$$

This is a horizontal ellipse with its center at (0,1). The co-vertices are 1 unit above and below the center. Refer to figure (g).

(13) **(13,–5) and (–7,–5).** This hyperbola opens left and right. The center is at (3,–5), and the foci are at $(3\pm F,-5)$. To solve for F, $F^2=a^2+b^2=36+64=100$, so the foci are at $(3\pm 10,-5)$. Refer to figure (h).

(14) **See the following graph.** First, write the ellipse in its standard form.

$$4x^2+y^2+16x-10y+25=0 \to 4\left(x^2+4x\right)+\left(y^2-10y\right)+25=0$$

$$\to 4\left(x^2+4x+4\right)+\left(y^2-10y+25\right)+25=-25+16+25$$

$$\to 4(x+2)^2+(y-5)^2=-25+16+25 \to 4(x+2)^2+(y-5)^2=16$$

$$\to \frac{4(x+2)^2}{16}+\frac{(y-5)^2}{16}=\frac{16}{16} \to \frac{(x+2)^2}{4}+\frac{(y-5)^2}{16}=1$$

This is a vertical ellipse with its center at (–2,5). The vertices are $(-2,5\pm 4)$, and the co-vertices are at $(-2\pm 2,5)$.

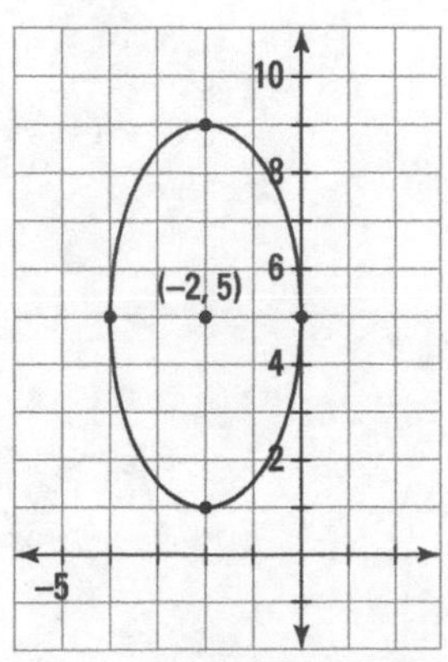

6

Systems, Sequences, and Series

IN THIS UNIT . . .

Dealing with systems of equations.

Making matrices cooperate.

Coordinating sequences and series.

Tackling the Binomial Theorem.

IN THIS CHAPTER

» Showing the why before the how

» Taking on two-equation systems with substitution and elimination

» Breaking down systems with more than two equations

» Graphing systems of inequalities

Chapter 18 Streamlining Systems of Equations

Solving equations: Why do you do it? You need an answer to a question about the height of a kite. You need the answer to the question about the length and width of a garden. You need an answer about an answer! Sometimes you need one equation to solve a problem, and sometimes you need more than one equation. These instances are covered here.

Solving equations: What do you do? When you have one variable and one equation, you can almost always solve the equation for the variable. When a problem has two variables, however, you need at least two equations to solve and get a unique solution; this set of equations is called a *system*. When you have three variables, you need at least three equations in the system to get a single answer. Basically, for every variable present, you need a separate unique equation if you want to solve the system for its special answer.

For an equation with three variables, an infinite number of values for two variables would work for that particular equation. Why? Because you can pick any two numbers to plug in for two of the variables in order to find the value of the third one that makes the equation true. If you add another equation into the system, the solutions to the first equation now have to *also* work in the second equation, which makes for fewer solutions to choose from. The set of solutions (usually *x, y, z,* and so on) must work when plugged into each and every equation in the system.

Of course, the bigger the system of equations becomes, the longer it may take to solve it algebraically. Therefore, certain systems are easier to solve in special ways, which is why math books usually show several methods. In this chapter, you'll see where particular methods are preferable to others. You want to be comfortable with as many of these methods as possible so that you can choose the best and quickest route.

As if systems of equations weren't enough on their own, in this chapter you also find systems of inequalities. The systems of inequalities require that you use a graph to determine the solution. The solution to a system of inequalities is shown as a shaded region on a graph.

A Primer on Your System-Solving Options

Before you can know which method of solving a system of equations is best, you need to be familiar with the possibilities. Your system-solving options are as follows:

- If the system has two variables, then algebraically solving the system or graphing the equations involving the variables are two options.
- When working algebraically, if the system has only two or three variables, then substitution or elimination is often your best plan of attack.
- If the system has four or more variables, a good choice is to use matrices on linear systems. You find lots of information on matrices in Chapter 19.
- You can still use substitution or elimination on large systems of equations, but it's often a lot of extra work.

TIP

A note to the calculator-savvy out there: You may be tempted to just plug the numbers into a calculator, not ask any questions, and move on. If you're lucky enough to have a graphing calculator and are allowed to use it, you could let the calculator do the math for you. However, because you're reading this book, I show you how those calculators get the answers they do. And there are some problems that calculators can't help you with, so you do need to know these techniques.

REMEMBER

No matter which system-solving method you use, check the answers you get, because even the best mathematicians sometimes make mistakes. The more variables and equations you have in a system, the more likely you are to make a misstep. And if you make a mistake in calculations somewhere, it can affect more than one part of the answer, because one variable usually is dependent on another. Always verify!

Graphing Equations to Find Solutions

Graphing equations can be helpful to get a better perspective of what is being asked and what the possible answer or answers might be. You see how to graph lines and parabolas and polynomials and circles and many, many others in the earlier chapters. Now you can put those graphing skills to use to find intersections of the curves, which represent common points, which in turn represent solutions to systems of equations.

Common solutions of linear equations

When two different lines intersect, they have only one point in common.

EXAMPLE

Q. Use a graph to find the common solution of the lines $y = 3x + 2$ and $x - 3y = 2$.

A. The equation $y = 3x + 2$ is in slope-intercept form, so just graph the intercept $(0, 2)$, then count 1 unit to the right and 3 units up to find a second point. Draw the line through the points. For the equation $x - 3y = 2$, just find the two easiest points possible and draw a line through them. For the x-intercept, $y = 0$, and that gives you $x = 2$. You have the point $(2, 0)$. For a second point, let $x = 5$, giving you $5 - 3y = 2 \rightarrow -3y = -3 \rightarrow y = 1$. This tells you that the point $(5, 1)$ is on that line. Take a look at the graphs of the lines in Figure 18-1. Can you find the common point?

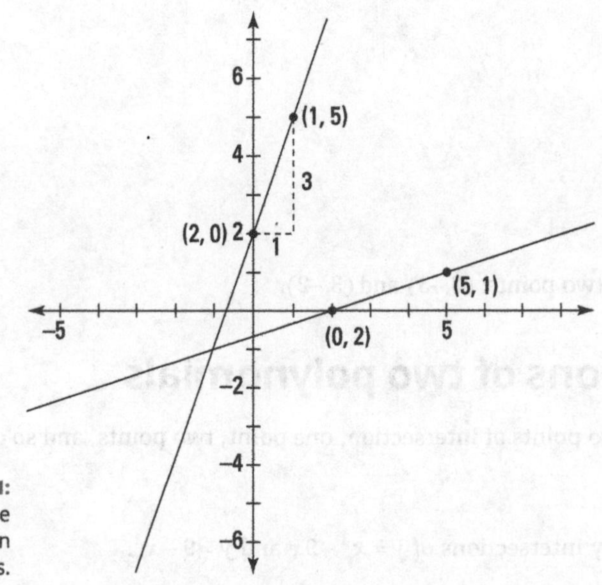

FIGURE 18-1: The intersection of two lines.

Yes, the point of intersection is at $(-1,-1)$. This is also called the *solution of the system* of the two equations.

Common solutions of a linear and a quadratic equation

When a line and a parabola are graphed, you'll see two points of intersection, one point of intersection, or no points of intersection at all.

EXAMPLE

Q. Use a graph to find any intersections of $y = x - 5$ and $y = (x - 2)^2 - 3$.

A. The line has a y-intercept at (0,5) and a slope of −1. The parabola has its vertex at (2,−3) and opens upward. Their graphs are shown in Figure 18-2. Can you spot any points of intersection?

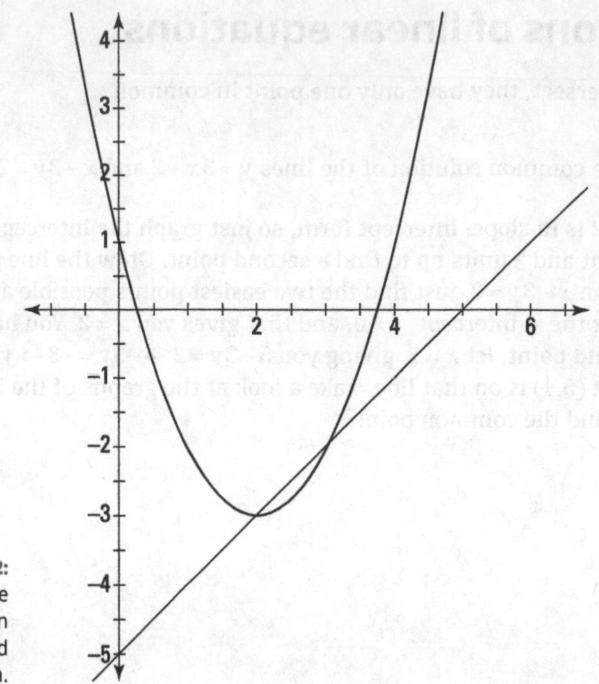

FIGURE 18-2: The intersection of a line and parabola.

Yes. They intersect in two points: (2,–3) and (3,–2).

Common solutions of two polynomials

Two polynomials can have no points of intersection, one point, two points, and so on, depending on their degree.

EXAMPLE

Q. Use a graph to find any intersections of $y = x^3 - 9x$ and $y = 9 - x^2$.

A. The cubic polynomial has x-intercepts at (0,0), (–3,0), and (3,0). The parabola shares two of those intercepts: (–3,0) and (3,0). Their third point of intersection is a little more difficult to determine from the graph, but it is at (–1,8). You see all these points in Figure 18-3.

This last example was meant to show you how challenging it can be to solve systems of equations using graphs. There are other methods that are much more efficient — and deal with fractional and decimal solutions more accurately.

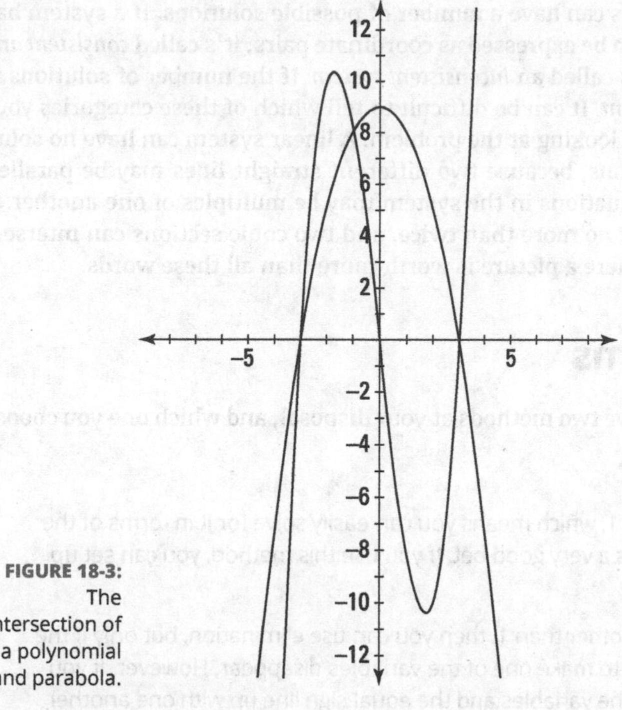

FIGURE 18-3: The intersection of a polynomial and parabola.

YOUR TURN

Use a graph to find the intersections of $y = x^3$ and $y = x$.

Use a graph to find the intersections of $y = -x^2 + 6x - 3$ and $y = 2x - 3$.

Algebraic Solutions of Two-Equation Systems

When you solve systems with two variables and two equations, the equations can be linear or nonlinear. Linear equations are usually expressed in the form $Ax + By = C$, where A, B, and C are real numbers.

Nonlinear equations can include circles, other conics, polynomials, and exponential, logarithmic, or rational functions. Elimination usually doesn't work in nonlinear systems, because there can be terms that aren't alike and therefore can't be added together. In that case, you resort to substitution to solve the nonlinear system.

Like many algebra problems, systems can have a number of possible solutions. If a system has one or more unique solutions that can be expressed as coordinate pairs, it's called *consistent and independent*. If it has no solution, it's called an *inconsistent system*. If the number of solutions is infinite, the system is called *dependent*. It can be difficult to tell which of these categories your system of equations falls into just by looking at the problem. A linear system can have no solution, one solution, or infinite solutions, because two different straight lines may be parallel, may intersect in one point, or the equations in the system may be multiples of one another. A line and a conic section can intersect no more than twice, and two conic sections can intersect a maximum of four times. This is where a picture is worth more than all these words.

Solving linear systems

When solving linear systems, you have two methods at your disposal, and which one you choose depends on the problem:

» If the coefficient of any variable is 1, which means you can easily solve for it in terms of the other variable, then substitution is a very good bet. If you use this method, you can set up each equation any way you want.

» If all the coefficients are anything other than 1, then you can use elimination, but only if the equations can be added together to make one of the variables disappear. However, if you use this method, be sure that all the variables and the equal sign line up with one another before you add the equations together.

The substitution method

In the *substitution method*, you use one equation to solve for one of the variables and then substitute that expression into the other equation to solve for the other variable. To begin using the easiest way, look for a variable with a coefficient of 1 to solve for. You just have to add or subtract terms in order to move everything else to the other side of the equal sign. With a coefficient of 1, you won't have to divide by a number when you're solving, which means you won't have to deal with fractions.

EXAMPLE

Q. Suppose you're managing a theater and you need to know how many adults and children are in attendance at a show. The auditorium is sold out and contains a mixture of adults and children. The tickets cost $23 per adult and $15 per child. If the auditorium has 250 seats and the total ticket revenue for the event is $4,846, how many adults and how many children are in attendance?

A. To solve the problem with the substitution method, follow these steps:

1. **Express the word problem as a system of equations.**

   Let a represent the number of adult tickets sold and c represent the number of child tickets sold. If the auditorium has 250 seats and was sold out, the sum of a and c must be 250.

   The ticket prices also lead you to the revenue (or money made) from the event. The adult ticket price times the number of adults present lets you know how much money you made from the adults. The same with the child tickets. The sum of $23 times a and $15 times c must be $4,846.

Here's how you write this system of equations:

$$\begin{cases} a + c = 250 \\ 23a + 15c = 4{,}846 \end{cases}$$

2. **Solve for one of the variables.**

   Pick a variable with a coefficient of 1 if you can, because solving for this variable will be easy. In this example, you can choose to solve for either a or c in the first equation. If solving for a, subtract c from both sides: $a = 250 - c$.

3. **Substitute the solved variable into the other equation.**

   You take what a is equal to, $250 - c$, and substitute it into the other equation for a.

   The second equation now says $23(250 - c) + 15c = 4{,}846$.

4. **Solve for the unknown variable.**

   You distribute the number 23:

   $5{,}750 - 23c + 15c = 4{,}846$

   And then you simplify:

   $-8c = -904$

   Dividing by -8, you get $c = 113$. A total of 113 children attended the event.

5. **Substitute the value of the unknown variable into one of the original equations to solve for the other unknown variable.**

   TIP

   You don't have to substitute into one of the original equations, but it's best if you do. Your answer is more likely to be accurate.

   When you plug 113 into the first equation for c, you get $a + 113 = 250$. Solving this equation, you get $a = 137$. You sold a total of 137 adult tickets.

6. **Check your solution.**

   When you plug a and c into the original equations, you should get two true statements. Does $137 + 113 = 250$? Yes. Does $23(137) + 15(113) = 4{,}846$? Indeed.

Process of elimination

If solving a system of two equations with the substitution method isn't the best choice because none of the coefficients is 1 or the system involves fractions, the elimination method is your next best option. In the *elimination method*, you eliminate one of the variables by doing some creative adding of the two equations.

When using elimination, you often have to multiply one or both equations by constants in order to eliminate a variable. (Remember that in order for one variable to be eliminated, the coefficients of the variable must be opposites.)

Q. Solve the system using the process of elimination.

EXAMPLE

$$\begin{cases} 10x + 12y = 5 \\ \frac{1}{3}x + \frac{4}{5}y = \frac{5}{6} \end{cases}$$

A. Use the following steps:

1. **Rewrite the equations, if necessary, to make like variables line up underneath each other.**

   The order of the variables doesn't matter; just make sure that like terms line up with like terms from top to bottom. The equations in this system have the variables x and y lined up already:

   $$\begin{cases} 10x + 12y = 5 \\ \frac{1}{3}x + \frac{4}{5}y = \frac{5}{6} \end{cases}$$

2. **Multiply the equations by constants so that one set of variables has opposite coefficients.**

   Decide which variable you want to eliminate.

   Say you decide to eliminate the x-variables; first, you have to find their least common multiple. What number do 10 and $\frac{1}{3}$ both go into evenly? The answer is 30. But one of the new coefficients has to be negative so that when you add the equations, the terms cancel out (that's why it's called elimination!). Multiply the top equation by -3 and the bottom equation by 90. (Be sure to distribute this number to each term — even on the other side of the equal sign.) Doing this step gives you the following equations:

   $$\begin{cases} -30x - 36y = -15 \\ 30x + 72y = 75 \end{cases}$$

3. **Add the two equations.**

   You now have $36y = 60$.

4. **Solve for the unknown variable that remains.**

   Dividing by 36 gives you $y = \frac{5}{3}$.

5. **Substitute the value of the found variable into either equation.**

   Using the first original equation: $10x + 12\left(\frac{5}{3}\right) = 5$.

6. **Solve for the final unknown variable.**

   You end up with $x = -\frac{3}{2}$.

7. **Check your solutions.**

Always verify your answer by plugging the solutions back into the original system. These check out!

$$10\left(-\frac{3}{2}\right)+12\left(\frac{5}{3}\right)=-15+20=5$$

It works! Now check the other equation:

$$\frac{1}{3}\left(-\frac{3}{2}\right)+\frac{4}{5}\left(\frac{5}{3}\right)=-\frac{1}{2}+\frac{4}{3}=-\frac{3}{6}+\frac{8}{6}=\frac{5}{6}$$

Because the set of values is a solution to both equations, the solution to the system is correct.

YOUR TURN

3 **If the sum of two numbers is 14 and their difference is 2, find the numbers using substitution.**

---

4 **Solve the system using elimination:**

$$\begin{cases} 2x-3y=5 \\ 4x+5y=-1 \end{cases}$$

Working nonlinear systems

In a *nonlinear system* of equations, at least one equation has a graph that isn't a straight line. You can always write a linear equation in the form $Ax + By = C$ (where A, B, and C are real numbers); a nonlinear equation is represented by any other form — depending on what type function you have. Examples of nonlinear equations include, but are not limited to, any conic section, polynomial, rational function, exponential, or logarithmic (all of which are discussed in this book). The nonlinear systems you see in pre-calculus have two equations with two variables, because the three-dimensional systems are a tad more difficult to solve. Because you're really working with a system with two equations and two variables, you have the same two methods at your disposal as you do with linear systems: substitution and elimination. Usually, substitution is your best bet. Unless the variable you want to eliminate is raised to the same power or has the same type of function operator in both equations, elimination won't work.

When one system equation is nonlinear

If one equation in a system is nonlinear, your first thought before solving should be, "Bingo! Substitution method!" (or something to that effect). In this situation, you can solve for one variable in the linear equation and substitute this expression into the nonlinear equation. And any time you can solve for one variable easily, you can substitute that expression into the other equation to solve for the other variable.

EXAMPLE

Q. Solve this system: $\begin{cases} x - 4y = 3 \\ xy = 6 \end{cases}$

A. The system consists of a line and a rational function. Making a quick sketch is often helpful, because it gives you a preview as to whether you may have no intersection, one intersection, or two intersections of the graphs. Figure 18-4a shows the two functions graphed. Use the following steps:

1. **Solve the linear equation for one variable.**

   In this system, the top equation is linear. If you solve for *x*, you get $x = 3 + 4y$.

2. **Substitute the value of the variable into the nonlinear equation.**

   When you plug $3 + 4y$ into the second equation for *x*, you get $(3 + 4y)y = 6$.

3. **Solve the nonlinear equation for the variable.**

   When you distribute the *y*, you get $4y^2 + 3y = 6$. Because this equation is quadratic (refer to Chapter 5), you must get 0 on one side, so subtract the 6 from both sides to get $4y^2 + 3y - 6 = 0$. This doesn't factor, so you have to use the quadratic formula to solve this equation for *y*:

   $$y = \frac{-3 \pm \sqrt{3^2 - 4(4)(-6)}}{2(4)} = \frac{-3 \pm \sqrt{9 + 96}}{8} = \frac{-3 \pm \sqrt{105}}{8}$$

4. **Substitute the solution(s) into either equation to solve for the other variable.**

   Because you find two solutions for *y*, you have to substitute them both to get two different coordinate pairs. Here's what happens when you substitute into the linear equation:

$$x = 3 + 4\left(\frac{-3 \pm \sqrt{105}}{8}\right) = 3 + \frac{-3 \pm \sqrt{105}}{2} = \frac{6}{2} + \frac{-3 \pm \sqrt{105}}{2} = \frac{3 \pm \sqrt{105}}{2}$$

Pairing up the x and y values:

When $y = \frac{-3 + \sqrt{105}}{8}$, $x = \frac{3 + \sqrt{105}}{2}$.

And when $y = \frac{-3 - \sqrt{105}}{8}$, $x = \frac{3 - \sqrt{105}}{2}$.

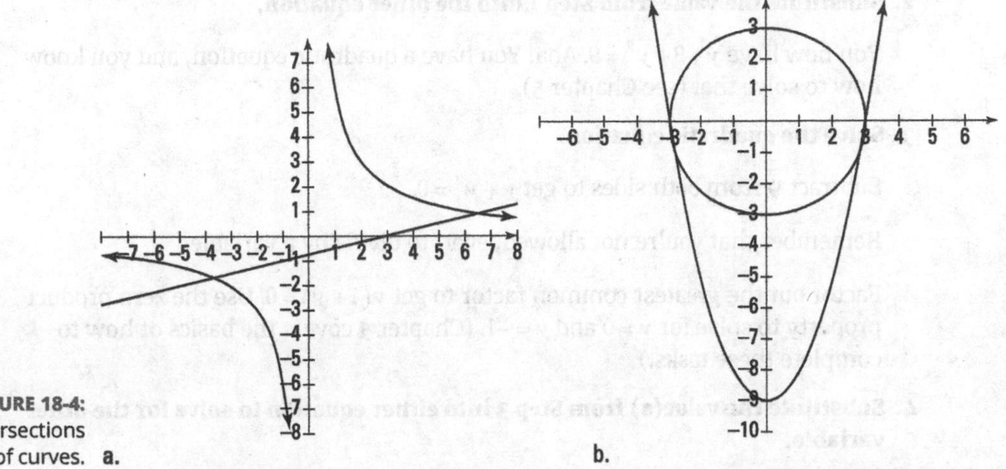

FIGURE 18-4: Intersections of curves.

Putting these solutions into your graph, the intersections are at about (6.62,0.91) and (−3.62,−1.66). Check out these points in Figure 18-4a.

When both system equations are nonlinear

If both of the equations in a system are nonlinear, well, you just have to get more creative to find the solutions. Unless one variable is raised to the same power in both equations, elimination is out of the question. Solving for one of the variables in either equation isn't necessarily easy, but it can usually be done. After you solve for a variable, plug this expression into the other equation and solve for the other variable just as you did before. Unlike linear systems, many operations may be involved in the simplification or solving of these equations. Just remember to keep your order of operations in mind at each step.

REMEMBER

When both equations in a system are conic sections, you'll never find more than four solutions (unless the two equations describe the same conic section, in which case the system has an infinite number of solutions — and therefore is a dependent system). Four is the limit because conic sections are all very smooth curves with no sharp corners or crazy bends, so two different conic sections can't intersect more than four times.

Q. Solve the following system: $\begin{cases} x^2 + y^2 = 9 \\ y = x^2 - 9 \end{cases}$

EXAMPLE

A. Again, a quick sketch is helpful. Check out Figure 18-4b for the graphs of the circle and parabola. Then follow these steps to find the solutions:

1. **Solve for x^2 or y^2 in one of the given equations.**

   The second equation is attractive because all you have to do is add 9 to both sides to get $y + 9 = x^2$.

2. **Substitute the value from Step 1 into the other equation.**

   You now have $y + 9 + y^2 = 9$. Aha! You have a quadratic equation, and you know how to solve that (see Chapter 5).

3. **Solve the quadratic equation.**

   Subtract 9 from both sides to get $y + y^2 = 0$.

WARNING

   Remember that you're not allowed, ever, to divide by a variable.

   Factor out the greatest common factor to get $y(1 + y) = 0$. Use the zero product property to solve for $y = 0$ and $y = -1$. (Chapter 5 covers the basics of how to complete these tasks.)

4. **Substitute the value(s) from Step 3 into either equation to solve for the other variable.**

   Using the equation $y + 9 = x^2$ from Step 1: when y is 0, $9 = x^2$, so $x = \pm 3$; and when y is -1, $8 = x^2$, so $x = \pm\sqrt{8} = \pm 2\sqrt{2}$.

   Your answers are $(-3,0)$, $(3,0)$, $(-2\sqrt{2},-1)$, and $(2\sqrt{2},-1)$.

   This solution set represents the intersections of the circle and the parabola given by the equations in the system. You see the curves and their intersections in Figure 18-4b.

Q. Solve $\begin{cases} \dfrac{14}{x+3} + \dfrac{7}{4-y} = 9 \\ \dfrac{21}{x+3} - \dfrac{3}{4-y} = 0 \end{cases}$

EXAMPLE

A. First, rewrite the system by letting $u = \dfrac{1}{x+3}$ and $v = \dfrac{1}{4-y}$ and getting $\begin{cases} 14u + 7v = 9 \\ 21u - 3v = 0 \end{cases}$. Now, multiply the first equation by 3 and the second equation by 7: $\begin{cases} 42u + 21v = 27 \\ 147u - 21v = 0 \end{cases}$. Adding these two equations gets you $189u = 27$, which means that $u = \dfrac{1}{7}$. Remember, though, that you're looking for x, not u. So, going back to $u = \dfrac{1}{x+3}$, write it as $\dfrac{1}{7} = \dfrac{1}{x+3}$ and you get that $x = 4$. Now you can use that to get $y = 3$, starting with the top equation in u and v: $14u + 7v = 9$ becomes $14\left(\dfrac{1}{7}\right) + 7v = 9$ or $2 + 7v = 9$. The value of v is 1, so go back to $v = \dfrac{1}{4-y}$ and solve for y: $\dfrac{1}{1} = \dfrac{1}{4-y}$ gives you that $1 = 4 - y$ or $y = 3$.

Because this is a rational expression, also be sure to always check your solution to see whether it's extraneous. In other words, if $x = 4$ or $y = 3$, do you get 0 in the denominator of either given equation? In this case, the answer is no — so this solution is legit!

YOUR TURN

5 Solve the following system: $\begin{cases} x^2 - y = 1 \\ x + y = 5 \end{cases}$

6 Solve the following system: $\begin{cases} \dfrac{12}{x+1} - \dfrac{12}{y-1} = 8 \\ \dfrac{6}{x+1} + \dfrac{6}{y-1} = -2 \end{cases}$

Solving Systems with More Than Two Equations

Larger systems of linear equations involve more than two equations that go along with more than two variables. These larger systems can be written in the form $Ax + By + Cz + \ldots = K$ where all coefficients (and K) are constants. These linear systems can have many variables, and they may have one unique solution. Three variables need three equations to find a unique solution, four variables need four equations, ten variables would have to have ten equations, and so on. For these types of linear systems, the solutions you can find vary widely:

- You may find no solution.
- You may find one unique solution.
- You may come across infinitely many solutions.

The number of solutions you find depends on how the equations interact with one another. Because linear systems of three variables describe equations of planes, not lines (as two-variable equations do), the solution to the system depends on how the planes lie in three-dimensional space relative to one another. Unfortunately, just like in the systems of equations with two variables, you can't tell how many solutions the system has without doing the problem. Treat each problem as if it has one solution, and if it doesn't, you will arrive at a statement that is either never true (no solutions) or always true (which means the system has infinitely many solutions).

REMEMBER

Typically, you must use the elimination method more than once to solve systems with more than two variables and two equations (see the earlier section, "Process of elimination").

EXAMPLE

Q. Solve the following system:

$$\begin{cases} x+2y+3z=-7 \\ 2x-3y-5z=9 \\ -6x-8y+z=-22 \end{cases}$$

A. To find the solution(s), follow these steps:

1. **Look at the coefficients of all the variables and decide which variable is easiest to eliminate.**

   With elimination, you want to find the least common multiple (LCM) for one of the variables, so go with the one that's the easiest. In this case, both x and z are candidates, because they each appear with a coefficient of 1. The choice here will be x.

2. **Set apart two of the equations and eliminate one variable.**

   Looking at the first two equations, you have to multiply the first by -2 and add it to the second equation. Doing this, you get:

$$\begin{array}{rcrcrcr} \cancel{-2x} & - & 4y & - & 6z & = & 14 \\ \cancel{2x} & - & 3y & - & 5z & = & 9 \\ \hline & - & 7y & - & 11z & = & 23 \end{array}$$

3. **Set apart another two equations and eliminate the *same variable*.**

   The first and the third equations allow you to easily eliminate x again. Multiply the top equation by 6 and add it to the third equation to get the following:

$$\begin{array}{rcrcrcr} \cancel{6x} & + & 12y & + & 18z & = & -42 \\ \cancel{-6x} & - & 8y & + & z & = & -22 \\ \hline & & 4y & + & 19z & = & -64 \end{array}$$

4. **Repeat the elimination process with your two new equations.**

   You now have these two equations with two variables:

$$-7y-11z=23$$
$$4y+19z=-64$$

You need to eliminate one of these variables. Eliminate the y-variable by multiplying the top equation by 4 and the bottom by 7 and then adding the results. Here's what that step gives you:

$$\begin{array}{rcrcr} -\cancel{28y} & - & 44z & = & 92 \\ \cancel{28y} & + & 133z & = & -448 \\ \hline & & 89z & = & -356 \end{array}$$

5. **Solve the final equation for the variable that remains.**

   If $89z = -356$, then $z = -4$.

6. **Substitute the value of the solved variable into one of the equations that has two variables to solve for another one.**

   Using the equation $-7y - 11z = 23$, you substitute to get $-7y - 11(-4) = 23$, which simplifies to $-7y + 44 = 23$. Now finish the job: $-7y = -21$, $y = 3$.

7. **Substitute the two values you now have into one of the original equations to solve for the last variable.**

   Using the first equation in the original system, the substitution gives you $x + 2(3) + 3(-4) = -7$. Simplify to get your final answer: $x = -1$.

   The solutions to this equation are $x = -1$, $y = 3$, and $z = -4$. You should now check your answer by substituting these three values back into the original three equations.

   This process is called *back substitution* because you literally solve for one variable and then work your way backward to solve for the others (you see this process again later when solving matrices). In this last example, you go from the solution for one variable in one equation to two variables in two equations to the last step with three variables in three equations. Always move from the simpler to the more complicated.

YOUR TURN

Solve the system:

$$\begin{cases} 3x - 2y = 17 \\ x - 2z = 1 \\ 3y + 2z = 1 \end{cases}$$

Solve the system:

$$\begin{cases} 3a + b + c + d = 0 \\ 4a + 5b + 2c = 15 \\ 4a + 2b + 5d = -10 \\ -5a + 3b - d = 8 \end{cases}$$

Decomposing Partial Fractions

No. This isn't about dead bodies and forensic science. The process called *decomposition of fractions* takes one fraction and expresses it as the sum or difference of two other fractions. There are many reasons why you'd want to do this. In calculus, this process is useful before you integrate a function. Because integration is so much easier when the degree of a rational function is 1 in the denominator, partial fraction decomposition is a useful tool for you.

The process of decomposing partial fractions requires you to separate the fraction into two (or sometimes more) separate fractions with unknowns (usually *A*, *B*, *C*, and so on) standing in as placeholders in the numerators. Then you can set up a system of equations to solve for these unknowns.

EXAMPLE

Q. Use partial fraction decomposition to write the fraction as the sum of two fractions:

$$\frac{11x+21}{2x^2+9x-18}$$

A. Use the following steps:

1. **Factor the denominator (see Chapter 5) and rewrite the fraction as a sum with *A* over one factor and *B* over the other.**

   You do this because you want to break the fraction into two. The process unfolds as follows:

   $$\frac{11x+21}{2x^2+9x-18}=\frac{11x+21}{(2x-3)(x+6)}=\frac{A}{2x-3}+\frac{B}{x+6}$$

2. **Multiply the equation by the factored denominator $(2x-3)(x+6)$ and reduce.**

   $$\frac{11x+21}{\cancel{(2x-3)(x+6)}}\cdot\cancel{(2x-3)(x+6)}=\frac{A}{\cancel{2x-3}}\cdot\cancel{(2x-3)}(x+6)+\frac{B}{\cancel{x+6}}\cdot(2x-3)\cancel{(x+6)}$$

   This expression simplifies to $11x+21=A(x+6)+B(2x-3)$.

3. **Distribute A and B.**

   This step gives you $11x+21=Ax+6A+2Bx-3B$.

4. **On the right side of the equation, group all terms with an *x* together and all terms without it together.**

   Rearranging gives you $11x+21=Ax+2Bx+6A-3B$.

5. **Factor out the *x* from the terms on the right side.**

   You now have $11x+21=(A+2B)x+6A-3B$.

6. **Create a system from the equation by pairing up terms.**

   For an equation to work, everything must be in balance. Because of this fact, the coefficients of *x* must be equal and the constants must be equal. If the coefficient of x is 11 on the left and $A+2B$ on the right, you can say that $11=A+2B$ is one equation. Constants are the terms with no variable, and in this case, the constant on the left is 21. On the right side, $6A-3B$ is the constant (because no variable is attached) and so $21=6A-3B$.

7. **Solve the system, using either substitution or elimination (see the earlier sections of this chapter).**

   Using elimination in this system, you work with the following equations:

   $$\begin{cases} A+2B=11 \\ 6A-3B=21 \end{cases}$$

   You can multiply the top equation by −6 and then add to eliminate and solve. You find that $A=5$ and $B=5$.

8. **Write the solution as the sum of two fractions.**

   $$\frac{11x+21}{2x^2+9x-18}=\frac{11x+21}{(2x-3)(x+6)}=\frac{5}{2x-3}+\frac{3}{x+6}$$

YOUR TURN

9 Find the constants A and B: $\dfrac{x-38}{x^2+x-12}=\dfrac{A}{x+4}+\dfrac{B}{x-3}$.

---

10 Perform fraction decomposition on $\dfrac{8x-16}{x^2-6x+5}$.

Surveying Systems of Inequalities

In a *system of inequalities*, you usually see more than one inequality with more than one variable. Your first exposure to systems of linear inequalities probably dealt with the straight line representing the real number system or a system of linear equations and the areas representing the solution. Now, you get to expand your study to systems of nonlinear inequalities.

In these systems of inequalities, at least one inequality isn't linear. The only way to solve a system of inequalities is to graph the solution. Fortunately, these graphs look very similar to the graphs you have already seen. And for the most part, these inequalities probably resemble the parent functions from Chapter 3 and conic sections from Chapter 17. The lines and curves you graph are either solid or dashed, depending on whether there's an "or equal to," and you get to color (or shade) where the solutions lie!

EXAMPLE

Q. Find the solution of the system of inequalities:

$$\begin{cases} x^2+y^2\le 25 \\ y\ge -x^2+5 \end{cases}$$

A. To solve this system of inequalities:

1. **Graph the system of related equations.**

   Change the inequality signs to equal signs. The fact that these expressions are inequalities and not equations doesn't change the general shape of the graph at all. Therefore, you can graph these inequalities just as you would graph them if they were equations.

   The top equation of this example is a circle centered at the origin, and the radius is 5. The second equation is a down-facing parabola that is shifted vertically 5 units.

   Because both of the inequality signs in this example include the equality line underneath (the first one is "less than or equal to" and the second is "greater than or equal to"), both curves should be solid. Refer to Figure 18-5.

   REMEMBER

   If the inequality symbol says "strictly greater than: >" or "strictly less than: <", then the curve (or line) should be dashed.

2. **Pick one test point that isn't on a boundary curve and plug it into the equations to see if you get true or false statements. The point that you pick as a solution must work in every equation.**

   When you can, use the origin, (0,0). If you plug this point into the inequality for the circle, you get $0^2+0^2\le 25$. This statement is true because $0\le 25$, so you shade inside the circle.

   Now plug the same point into the parabola to get $0\ge -0^2+5$, but because 0 isn't greater than 5, this statement is false. You shade outside the parabola.

   REMEMBER

   The solution of this system of inequalities is where the shading overlaps.

   In this case, the overlap is everything that's inside the circle but not inside (or underneath) the parabola. See Figure 18-5 for the final graph.

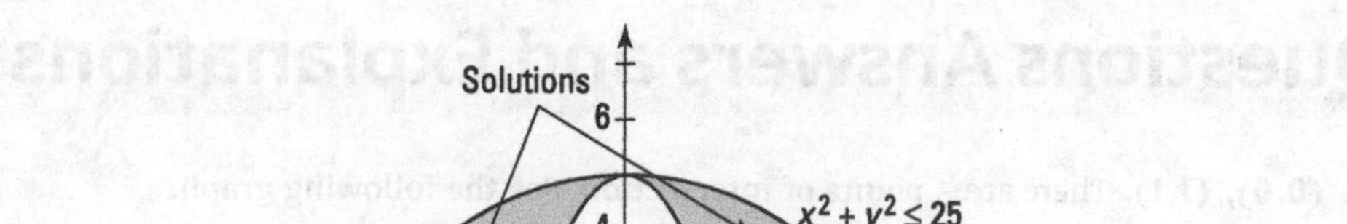

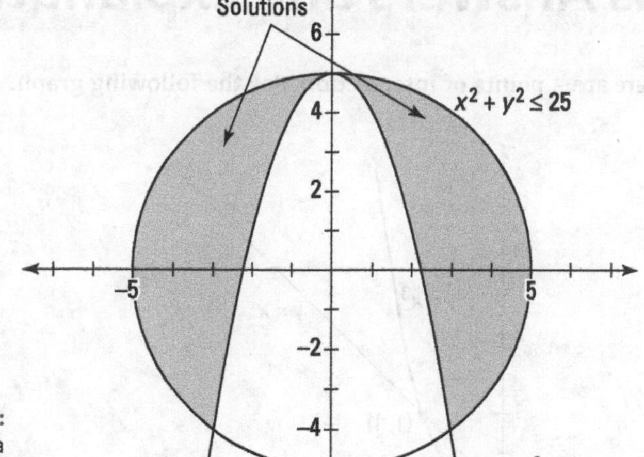

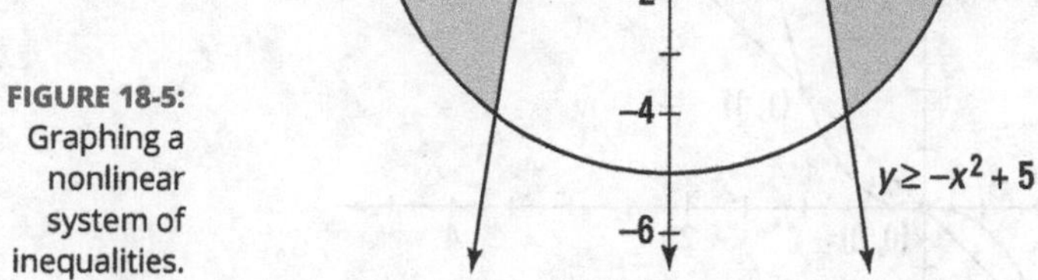

FIGURE 18-5: Graphing a nonlinear system of inequalities.

YOUR TURN

11 Solve the system of inequalities: $\begin{cases} x^2 - y > 2 \\ x - y < 4 \end{cases}$

---

12 Solve the system of inequalities: $\begin{cases} x^2 + y^2 \ge 9 \\ x^2 + (y-3)^2 \ge 9 \end{cases}$

Practice Questions Answers and Explanations

1. **(−1,−1), (0,0), (1,1). There are 3 points of intersection. See the following graph.**

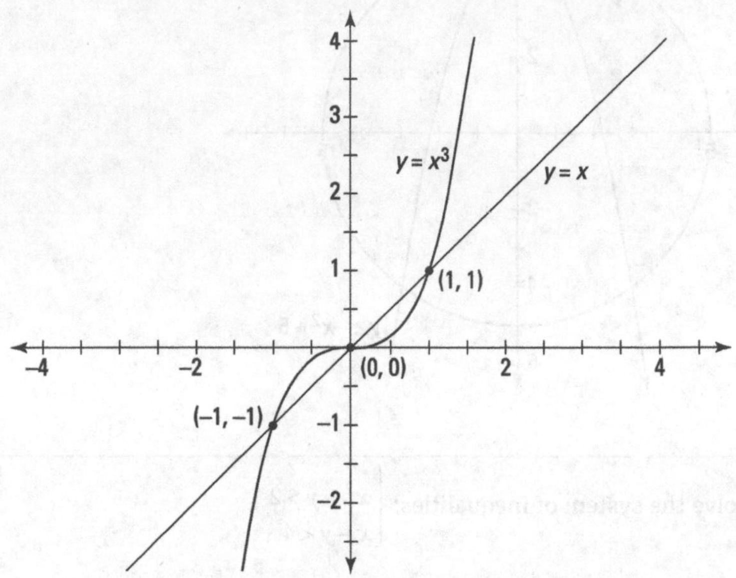

2. **(4,5), (0,−3). There are 2 points of intersection. See the following graph.**

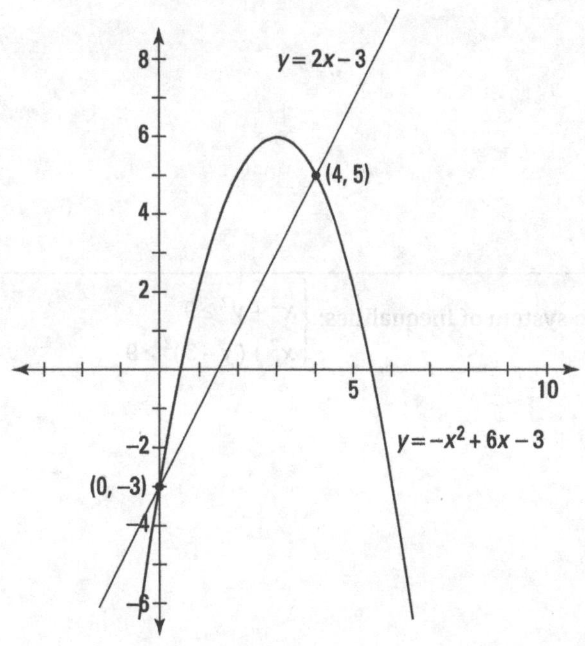

3. **6, 8.** Let the numbers be x and y. Now write the two equations representing their relationships. When the sum is 14 you have $x+y=14$, and when the difference is 2 you have $x-y=2$. Solve for x in the second equation: $x-y=2 \rightarrow x=2+y$. Now substitute the equivalence of x into the first equation and simplify: $x+y=14 \rightarrow (2+y)+y=14 \rightarrow 2+2y=14 \rightarrow 2y=12 \rightarrow y=6$. Solve for x using $x=2+y$ and $y=6$. You get $x=2+y \rightarrow x=2+6 \rightarrow x=8$.

4. $x=1$, $y=-1$. Notice that the y terms have opposite signs. This is an easier choice, because you don't have to multiply through by a negative. The coefficients are 3 and 5; their least common multiple is 15, so you have to multiply the top equation by 5 and the bottom equation by 3. This gives you

$$\begin{cases} 10x-15y=25 \\ 12x+15y=-3 \end{cases}$$

Adding these two equations together gives you $22x=22$, which gives you the solution $x=1$. You then substitute this value back into one of the two original equations to solve for y. In this example, $y=-1$.

5. $x=2$, $y=3$ **or** $x=-3$, $y=8$. In this problem, you have a parabola and a line, so there is the possibility of two solutions. First, solve the linear equation for a variable, like x in the second equation: $x=5-y$. Now substitute this into the first equation: $(5-y)^2-y=1$. Multiply out and combine like terms: $(5-y)^2-y=1 \rightarrow (25-10y+y^2)-y=1 \rightarrow 25-11y+y^2=1$. Then set the equation equal to 0: $24-11y+y^2=0$. This factors to $(3-y)(8-y)=0$, which, when you use the zero product property, gets you two solutions for y: $y=3$ and $y=8$. Substitute them, one at a time, into the original linear equation $x+y=5$ to get the solutions for x. First: If $y=3$, then $x+3=5$ and $x=2$. Now do the same thing for $y=8$:$x+8=5$ $x=-3$. So the other solution is $x=-3$, $y=8$.

6. $x=5$, $y=-1$. If you let $u=\frac{1}{x+1}$ and $v=\frac{1}{y-1}$, you can rewrite the system as $\begin{cases} 12u-12v=8 \\ 6u+6v=-2 \end{cases}$.

Now, multiply the second equation by 2 and you get $\begin{cases} 12u-12v=8 \\ 12u+12v=-4 \end{cases}$. Add them to get $24u=4$, or $u=\frac{1}{6}$. Work your way backward from there: $u=\frac{1}{x+1}$ means $\frac{1}{6}=\frac{1}{x+1}$ or $x=5$. Using the adjusted first equation, $12u-12v=8$, you have $12\left(\frac{1}{6}\right)-12v=8$ or $2-12v=8$, which becomes $-12v=6$ or $v=-\frac{1}{2}$. Since $v=\frac{1}{y-1}$, you write $\frac{1}{-2}=\frac{1}{y-1}$, and when $-2=y-1$, you get $y=-1$.

7. $x=5$, $y=-1$, $z=2$. Each equation has just two of the three variables. The second and third equations have z variables with opposite coefficients, so add them together to get

$$\begin{array}{rcrcrcr} x & & & - & 2z & = & 1 \\ & & 3y & + & 2z & = & 1 \\ \hline x & + & 3y & & & = & 2 \end{array}$$

Use this new equation and the first equation in the next step. Multiply the new equation through by -3, and you have $\begin{cases} 3x-2y=17 \\ -3x-9y=-6 \end{cases}$. Adding the two equations together, you get $-11y=11$ or $y=-1$. Substituting back into the original first equation, you get $3x-2(-1)=17$ or $x=5$. Finally, going back to the original second equation, let $x=5$ and you get $5-2z=1$ or $z=2$.

8 $a = 1$, $b = 3$, $c = -2$, $d = -4$. Look for a good candidate to eliminate first. There's no d term in the second equation, and the d terms are opposites in the first and last. There's where you can start.

Add the first and last equations together, and add –5 times the first equation to the third equation. Your new system, without the d term, is:

$$\begin{matrix}\text{original second equation}\\ \text{first plus last equation}\\ -5\text{ times first plus third}\end{matrix}\begin{cases}4a+5b+2c=15\\ -2a+4b+c=8\\ -11a-3b-5c=-10\end{cases}$$

The next good candidate for elimination is the c term. Use multiples of the second term in the new system to do the elimination.

$$\begin{matrix}-2\text{ times second plus first}\\ 5\text{ times second plus third}\end{matrix}\begin{cases}8a-3b=-1\\ -21a+17b=30\end{cases}$$

Solving for a or b will require multiplying both equations by some number to make the coefficients the opposite. If you multiply the first new equation by 17 and the second by 3, you get:

$$\begin{cases}136a-51b=-17\\ -63a+51b=90\end{cases}$$

Adding the equations together, you have $73a = 73$ or $a = 1$. Back-substituting this into $8a - 3b = -1$, you get $b = 3$. Then, using $-2a + 4b + c = 8$ and the two values you've found, you get that $c = -2$. And, finally, if $3a + b + c + d = 0$, you have $3(1) + 3 + (-2) + d = 0$, giving you $d = -4$.

9 $A = 6$, $B = -5$. The decomposition process has been started for you, showing the factorization of the denominator of the fraction being decomposed. The first thing you should do is multiply every term in the equation by the factored denominator:

$$\frac{x-38}{\cancel{x^2+x-12}}\cdot\cancel{(x+4)(x-3)}=\frac{A}{\cancel{x+4}}\cdot\cancel{(x+4)}(x-3)+\frac{B}{\cancel{x-3}}\cdot(x+4)\cancel{(x-3)}$$

After reducing the fractions, you have $x - 38 = A(x-3) + B(x+4)$. Multiply everything out to get $x - 38 = Ax - 3A + Bx + 4B$. Collect the like terms on the right: $x - 38 = A(x-3) + B(x+4)$ or $x - 38 = Ax - 3A + Bx + 4B$. Now factor out the x on the right: $x - 38 = (A+B)x - 3A + 4B$. The coefficients of the x terms are equal, which gives you one equation: $1 = A + B$. The constants are also equal, which gives you a second equation: $-38 = -3A + 4B$. Solving this new system of equations, $\begin{cases}A+B=1\\ -3A+4B=-38\end{cases}$, you first multiply the top equation by 3 and then add the two equations together to get $7B = -35$. This gives you that $B = -5$. Substituting this into the top equation, $A + (-5) = 1$, gives you that $A = 6$.

(10) **$A = 6$, $B = 2$.** First, factor the denominator and break the expression into the sum of two fractions: $\frac{8x-16}{x^2-6x+5} = \frac{8x-16}{(x-5)(x-1)} = \frac{A}{x-5} + \frac{B}{x-1}$. Multiply through by the common denominator, distribute the terms, and factor out the variable on the right:

$$\frac{8x-16}{\cancel{(x-5)(x-1)}} \cdot \cancel{(x-5)(x-1)} = \frac{A}{\cancel{x-5}} \cdot \cancel{(x-5)}(x-1) + \frac{B}{\cancel{x-1}} \cdot (x-5)\cancel{(x-1)}$$

$$\rightarrow 8x-16 = A(x-1)+B(x-5) \rightarrow 8x-16 = Ax-A+Bx-5B \rightarrow 8x-16 = (A+B)x-A-5B$$

The coefficient of x, $A+B$, is 8, and the constants, $-A-5B$, are equal to -16. Solving the system of equations $\begin{cases} A+B=8 \\ -A-5B=-16 \end{cases}$, you can add the two equations together to get $-4B=-8$ or $B=2$. Since $A+B=8$, you have $A+2=8$ or $A=6$.

(11) **See the following graph.** Sketch the graph of $\begin{cases} x^2-y>2 \\ x-y<4 \end{cases}$. The first inequality represents the region below the parabola, $y = x^2-2$. The second inequality represents the region above the line, $y = x-4$. Graph them both on the same graph.

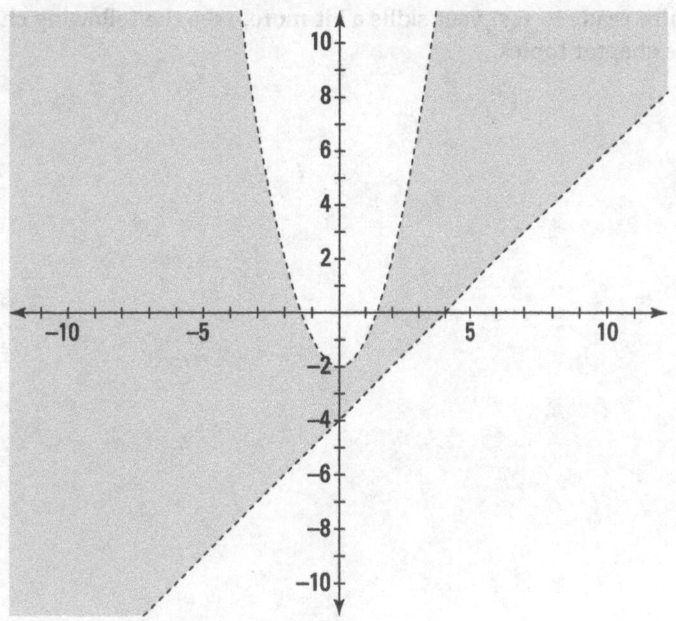

(12) **See the following graph.** Both of these inequalities describe regions bounded by circles. The answer is the area outside of both these two circles. If you don't recognize them as such, turn to Chapter 17 to read up on conic sections. Graph them both on the same graph.

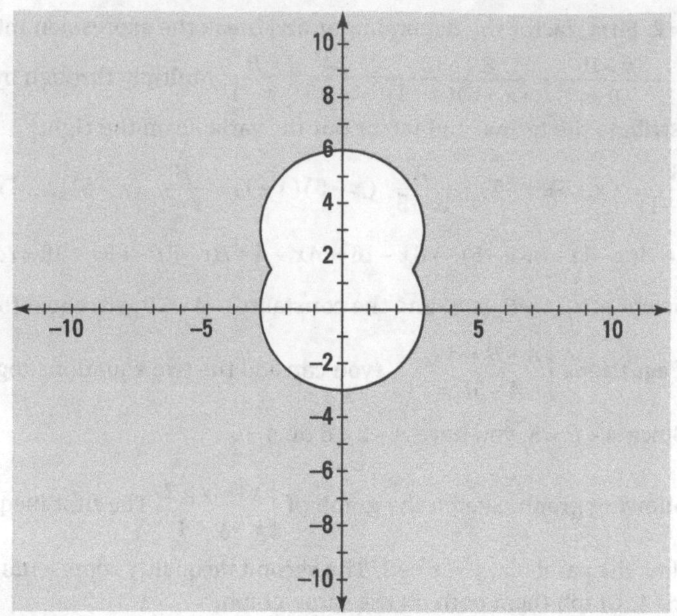

If you're ready to test your skills a bit more, take the following chapter quiz that incorporates all the chapter topics.

Whaddya Know? Chapter 18 Quiz

Quiz time! Complete each problem to test your knowledge on the various topics covered in this chapter. You can then find the solutions and explanations in the next section.

1 **Solve the system of equations:**

$$\begin{cases} x^2 + y^2 = 25 \\ y = 5 - x^2 \end{cases}$$

2 **Solve the system of equations using elimination:**

$$\begin{cases} 5x - 3y = 14 \\ 19x + 6y = 1 \end{cases}$$

3 **A recipe calls for three times as much flour as sugar. To make 10 loaves of bread, the total number of cups, combined, of flour and sugar is 40 cups. How much of each ingredient is needed to make the 10 loaves?**

4 **Solve the system of equations:** $\begin{cases} x + 2y - 3z = 15 \\ 2x - y - z = 10 \\ x + 3y + z = 4 \end{cases}$

5 **Use a graph to find the intersection of $y = 4$ and $3x + 2y + 4 = 0$.**

6 **Solve the system of inequalities:** $\begin{cases} y \geq x^2 + 4x + 1 \\ y < 7 - x \end{cases}$

7 **Find the points of intersection using substitution:**

$$\begin{cases} y = x^2 - 6x + 11 \\ x + 2y = 8 \end{cases}$$

8 **Perform fraction decomposition on** $\dfrac{14x}{3x^2 - 5x - 2}$.

9 **Solve the system of equations:**

$$\begin{cases} \dfrac{1}{x+3} + \dfrac{6}{y} = 4 \\ \dfrac{4}{x+3} - \dfrac{6}{y} = 1 \end{cases}$$

Answers to Chapter 18 Quiz

1. **(0,5), (3,–4), (–3,–4).** Solve the second equation for x^2: $y = 5 - x^2 \rightarrow x^2 = 5 - y$. Now substitute the equivalence into the first equation: $x^2 + y^2 = 25 \rightarrow (5 - y) + y^2 = 25 \rightarrow y^2 - y - 20 = 0$. Factor and solve for *y*: $y^2 - y - 20 = (y-5)(y+4) = 0 \rightarrow y = 5$ or $y = -4$. Substitute back into the original second equation to solve for *x*. When $y = 5$, $y = 5 - x^2 \rightarrow 5 = 5 - x^2 \rightarrow x^2 = 0 \rightarrow x = 0$. When $y = -4$, $y = 5 - x^2 \rightarrow -4 = 5 - x^2 \rightarrow x^2 = 9 \rightarrow x = \pm 3$.

2. **$x = 1$, $y = -3$.** Multiply the first equation through by 2 and add the two equations together.

$$\begin{array}{rcrcr} 10x & - & 6y & = & 28 \\ 19x & + & 6y & = & 1 \\ \hline 29x & & & = & 29 \end{array}$$

You have $x = 1$. Substitute this back into the original first equation to solve for *y*: $5x - 3y = 14 \rightarrow 5(1) - 3y = 14 \rightarrow 5 - 3y = 14 \rightarrow -3y = 9 \rightarrow y = -3$.

3. **30 cups flour, 10 cups sugar.** Let *f* represent the number of cups of flour and *g* represent the number of cups of sugar. You can say that $f = 3g$ and $f + g = 40$. Replace the *f* in the second equation with $3g$, and you get $3g + g = 40 \rightarrow 4g = 40 \rightarrow g = 10$ cups of sugar. So, if the total number of cups is 40, then you need 30 cups of flour.

4. **$x = 4$, $y = 1$, $z = -3$.** The *z* variable will be handy to aim at eliminating. Add the first equation to 3 times the third equation, and add the second equation to the third equation.

$$\begin{array}{rcrcrcr} x & + & 2y & - & 3z & = & 15 \\ 3x & + & 9y & + & 3z & = & 12 \\ \hline 4x & + & 11y & & & = & 27 \end{array} \qquad \begin{array}{rcrcrcr} 2x & - & y & - & z & = & 10 \\ x & + & 3y & + & z & = & 4 \\ \hline 3x & + & 2y & & & = & 14 \end{array}$$

Next, use the sums to eliminate the *x* terms by adding 3 times the first result to –4 times the second result.

$$\begin{array}{rcrcr} 12x & + & 33y & = & 81 \\ -12x & - & 8y & = & -56 \\ \hline & & 25y & = & 25 \end{array}$$

You have that $y = 1$. Substitute this back into $3x + 2y = 14$ to get $3x + 2(1) = 14 \rightarrow 3x = 12 \rightarrow x = 4$. Now go back to the third original equation to solve for *z*: $x + 3y + z = 4 \rightarrow 4 + 3(1) + z = 4 \rightarrow 7 + z = 4 \rightarrow z = -3$.

5 **(−4, 4).** See the following graph.

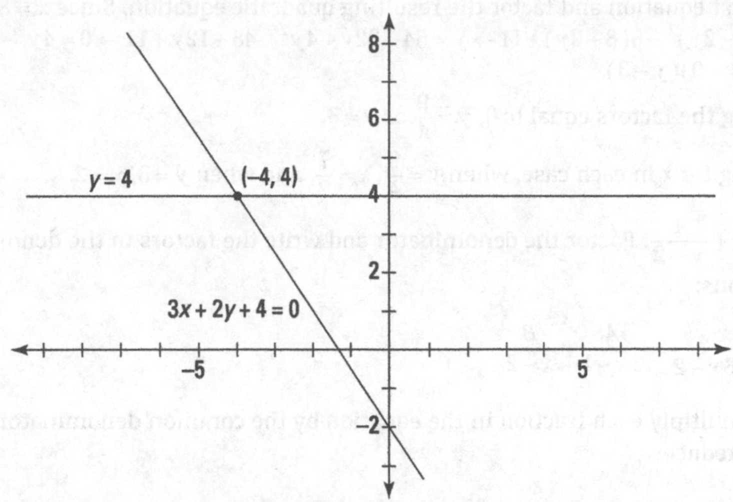

6 **See the following graph.** The parabola is drawn with a solid curve. The line is drawn with a dashed line. Shade above the parabola and below the line.

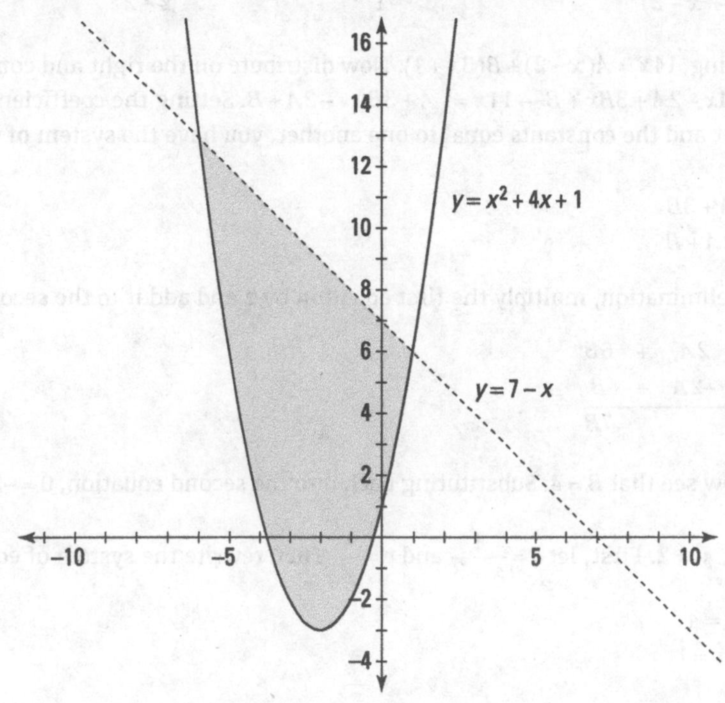

7 $\left(\frac{7}{2}, \frac{9}{4}\right)$, **(2, 3).** Solve for x in the second equation and substitute that equivalence into the first equation and factor the resulting quadratic equation. Since $x = 8 - 2y$, $y = (8-2y)^2 - 6(8-2y) + 11 \to y = 64 - 32y + 4y^2 - 48 + 12y + 11 \to 0 = 4y^2 - 21y + 27 \to$ $0 = (4y-9)(y-3)$.

Setting the factors equal to 0, $y = \frac{9}{4}$ or $y = 3$.

Solving for x in each case, when $y = \frac{9}{4}$, $x = \frac{7}{2}$, and when $y = 3$, $x = 2$.

8 $\frac{2}{3x+1} + \frac{4}{x-2}$. Factor the denominator and write the factors in the denominators of two fractions:

$$\frac{14x}{3x^2 - 5x - 2} = \frac{A}{3x+1} + \frac{B}{x-2}$$

Now multiply each fraction in the equation by the common denominator of the fractions. Then reduce:

$$\frac{14x}{3x^2 - 5x - 2} \cdot (3x+1)(x-2) = \frac{A}{3x+1} \cdot (3x+1)(x-2) + \frac{B}{x-2} \cdot (3x+1)(x-2)$$

$$\frac{14x}{\cancel{(3x+1)(x-2)}} \cdot \cancel{(3x+1)(x-2)} = \frac{A}{\cancel{3x+1}} \cdot \cancel{(3x+1)}(x-2) + \frac{B}{\cancel{x-2}} \cdot (3x+1)\cancel{(x-2)}$$

Rewriting, $14x = A(x-2) + B(3x+1)$. Now distribute on the right and combine the like terms: $14x = Ax - 2A + 3Bx + B \to 14x = (A + 3B)x - 2A + B$. Setting the coefficients equal to one another and the constants equal to one another, you have the system of equations.

$$\begin{cases} 14 = A + 3B \\ 0 = -2A + B \end{cases}$$

Using elimination, multiply the first equation by 2 and add it to the second.

$$\begin{array}{rcrcr} 28 & = & 2A & + & 6B \\ 0 & = & -2A & + & B \\ \hline 28 & = & & & 7B \end{array}$$

You now see that $B = 4$. Substituting back into the second equation, $0 = -2A + 4$ or $A = 2$.

9 $x = -2$, $y = 2$. First, let $u = \frac{1}{x+3}$ and $v = \frac{1}{y}$. Then rewrite the system of equations as:

$$\begin{cases} u + 6v = 4 \\ 4u - 6v = 1 \end{cases}$$

Add the two equations together to get $5u = 5$ or $u = 1$. Substitute that back into the first equation to get $1 + 6v = 4 \to 6v = 3 \to v = \frac{1}{2}$. Now go back to the original substitutions and solve for x and y. When $u = 1$ in $u = \frac{1}{x+3}$, you have $1 = \frac{1}{x+3} \to x + 3 = 1 \to x = -2$. And when $v = \frac{1}{2}$ in $v = \frac{1}{y}$, you have $\frac{1}{2} = \frac{1}{y} \to y = 2$.

IN THIS CHAPTER

» Forming and operating on matrices

» Creating matrices from equations

» Putting matrices into simpler forms to solve systems of equations

Chapter 19
Making Matrices Work

A matrix is a rectangular array that displays numbers or variables, called *elements*. Matrices are most helpful when solving systems of equations — especially when there are three or more equations involved. The mathematical procedures performed on matrices allow you to do operations on the equations involved very quickly and efficiently. And, the best part: these procedures can also be performed on calculators and in computer programs!

Introducing Matrices: The Basics

In Chapter 18, you see how to solve systems of two or more equations by using substitution or elimination. But these methods get very messy when the size of a system rises above three equations. Not to worry; whenever you have four or more equations to solve simultaneously, matrices are a great option.

A *matrix* is a rectangular array of elements arranged in rows and columns. You use matrices to organize complicated data. Say, for example, you want to keep track of sales records in your store. Matrices help you do that, because they can separate the sales by day in columns, while organizing different types of items sold by row.

After you get comfortable with what matrices are and how they're important, you can start adding, subtracting, and multiplying them by scalars and each other. Operating on matrices is useful when you need to add, subtract, or multiply large groups of data in an organized fashion. (*Note*: Matrix division doesn't exist; you multiply by an inverse instead.) For a more detailed explanation of how matrices work, refer to *Finite Math For Dummies* (Wiley). This section shows you how to perform these operations.

REMEMBER

One thing to always remember when working with matrices is the order of operations, which is the same across all math applications: First, do any multiplication and then do the addition/subtraction.

You express the *dimensions*, sometimes called *order*, of a matrix as the number of rows by the number of columns. For example, if matrix M is 3×2, it has three rows and two columns.

Applying basic operations to matrices

Operating on matrices works very much like operating on multiple terms within parentheses; you just have more terms in the "parentheses" to work with. Just like with operations on numbers, a certain order is involved with operating on matrices. Multiplication comes before addition and/or subtraction. When multiplying by a *scalar*, a constant that multiplies a quantity (which changes its size, or scale), each and every element of the matrix gets multiplied. The following sections show you how to perform some of the basic operations on matrices: addition, subtraction, and multiplication.

When adding or subtracting matrices, you just add or subtract their corresponding elements. It's as simple as that. When adding or subtracting, you just apply that operation to each set of corresponding elements.

REMEMBER

Note, however, that you can add or subtract matrices only if their dimensions are exactly the same. To add or subtract matrices, you add or subtract their corresponding elements; if the dimensions aren't exactly the same, then the terms won't line up.

EXAMPLE

Q. Given matrices A and B, find their sum and difference.

$$A = \begin{bmatrix} -5 & 1 & -3 \\ 6 & 0 & 2 \\ 2 & 6 & 1 \end{bmatrix} \quad B = \begin{bmatrix} 2 & 4 & 5 \\ -8 & 10 & 3 \\ -2 & -3 & -9 \end{bmatrix}$$

A. Add and subtract the corresponding elements

$$A + B = \begin{bmatrix} -5+2 & 1+4 & -3+5 \\ 6+(-8) & 0+10 & 2+3 \\ 2+(-2) & 6+(-3) & 1+(-9) \end{bmatrix} = \begin{bmatrix} -3 & 5 & 2 \\ -2 & 10 & 5 \\ 0 & 3 & -8 \end{bmatrix}$$

$$A - B = \begin{bmatrix} -5-2 & 1-4 & -3-5 \\ 6-(-8) & 0-10 & 2-3 \\ 2-(-2) & 6-(-3) & 1-(-9) \end{bmatrix} = \begin{bmatrix} -7 & -3 & -8 \\ 14 & -10 & -1 \\ 4 & 9 & 10 \end{bmatrix}$$

When you multiply a matrix by a scalar, you're just multiplying by a constant. To do that, you multiply each element inside the matrix by the constant on the outside.

EXAMPLE

Q. Using the same matrix A from the previous example, find 3A.

A. Multiply each term of matrix A by 3.

$$3A = 3\begin{bmatrix} -5 & 1 & -3 \\ 6 & 0 & 2 \\ 2 & 6 & 1 \end{bmatrix} = \begin{bmatrix} 3(-5) & 3\cdot 1 & 3(-3) \\ 3\cdot 6 & 3\cdot 0 & 3\cdot 2 \\ 3\cdot 2 & 3\cdot 6 & 3\cdot 1 \end{bmatrix} = \begin{bmatrix} -15 & 3 & -9 \\ 18 & 0 & 6 \\ 6 & 18 & 3 \end{bmatrix}$$

Suppose a problem asks you to combine the operations of multiplication and addition or subtraction. In this case, you simply multiply each matrix by the scalar separately and then add or subtract the results.

EXAMPLE

Q. Find 3C − 2D using the following two matrices: $C = \begin{bmatrix} 3 & -4 & 0 \\ 2 & 6 & -1 \end{bmatrix}$ $D = \begin{bmatrix} 8 & -10 & 4 \\ 2 & -6 & 9 \end{bmatrix}$.

A. Use the following steps:

1. **Insert the matrices into the problem.**

$$3C - 2D = 3\begin{bmatrix} 3 & -4 & 0 \\ 2 & 6 & -1 \end{bmatrix} - 2\begin{bmatrix} 8 & -10 & 4 \\ 2 & -6 & 9 \end{bmatrix}$$

2. **Multiply the scalars into the matrices.**

$$= \begin{bmatrix} 9 & -12 & 0 \\ 6 & 18 & -3 \end{bmatrix} - \begin{bmatrix} 16 & -20 & 8 \\ 4 & -12 & 18 \end{bmatrix}$$

3. **Complete the problem by subtracting the matrices.**

After subtracting, here is your final answer:

$$= \begin{bmatrix} -7 & 8 & -8 \\ 2 & 30 & -21 \end{bmatrix}$$

YOUR TURN

Use matrices J and K to perform the operations indicated:

$$J = \begin{bmatrix} 3 & -8 \\ 2 & 7 \\ 5 & -9 \end{bmatrix} \text{ and } K = \begin{bmatrix} 0 & 5 \\ -2 & 3 \\ 6 & 6 \end{bmatrix}$$

1 J + 2K

2 3J − 4K

Multiplying matrices by each other

Multiplying matrices is very useful when solving systems of equations because you can multiply a matrix by its inverse (don't worry, you'll see how to find that) on both sides of the equal sign to eventually get the variable matrix on one side and the solution to the system on the other.

Multiplying two matrices together is not as simple as multiplying by a scalar. And you can't just multiply the corresponding elements together.

TECHNICAL STUFF

If you want to multiply matrix A times matrix B, AB, the number of columns in A must match the number of rows in B. Each element in the first row of A is multiplied by each corresponding element from the first column of B, and then all these products are added together to give you the element in the first row, first column of AB. To find the value in the first row, second column position, multiply each element in the first row of A by each element in the second column of B and then add them all together. In the end, after all the multiplication and addition are finished, your new matrix should have the same number of rows as A and the same number of columns as B.

For example, when you multiply a matrix with three rows and two columns by a matrix with two rows and four columns, you end up with a matrix of three rows and four columns.

REMEMBER

If matrix A has dimensions $m \times n$ and matrix B has dimensions $n \times p$, then AB is an $m \times p$ matrix.

EXAMPLE

Q. Multiply the following two matrices:

$$A = \begin{bmatrix} 5 & -6 \\ -3 & 9 \\ 2 & 4 \end{bmatrix} \quad B = \begin{bmatrix} -2 & 4 & 8 & -5 \\ 1 & 3 & -4 & -2 \end{bmatrix}$$

A. First, check to make sure that you can multiply the two matrices. Matrix A is 3×2 and B is 2×4, so you can multiply them to get a 3×4 matrix as an answer. Now you can proceed to multiply every row of the first matrix times every column of the second.

The products and sums are given in the matrix. You start by multiplying each term in the first row of matrix A by the sequential terms in the columns of matrix B. Note that multiplying row one by column one and adding them together gives you row one, column one's answer.

$$AB = \begin{bmatrix} 5(-2)+(-6)(1) & 5(4)+(-6)(3) & 5(8)+(-6)(-4) & 5(-5)+(-6)(-2) \\ -3(-2)+9(1) & -3(4)+9(3) & -3(8)+9(-4) & -3(-5)+9(-2) \\ 2(-2)+4(1) & 2(4)+4(3) & 2(8)+4(-4) & 2(-5)+4(-2) \end{bmatrix}$$

Simplifying each element: $AB = \begin{bmatrix} -16 & 2 & 64 & -13 \\ 15 & 15 & -60 & -3 \\ 0 & 20 & 0 & -18 \end{bmatrix}$

YOUR TURN

3 Given $C = \begin{bmatrix} 1 & 4 \\ 0 & 6 \end{bmatrix}$ and $D = \begin{bmatrix} -3 & 1 \\ 2 & -2 \end{bmatrix}$, find CD.

4 Given $E = \begin{bmatrix} -1 & 2 \\ 3 & -4 \\ 5 & -6 \\ -7 & 0 \end{bmatrix}$ and $F = \begin{bmatrix} 1 & -2 & -3 \\ 7 & 3 & 5 \end{bmatrix}$, find EF.

Simplifying Matrices to Ease the Solving Process

In a system of linear equations, where each equation is in the form $Ax + By + Cz + \cdots = K$, the coefficients of this system can be represented in a matrix, called the *coefficient matrix*. If all the variables line up with one another vertically, then the first column of the coefficient matrix is dedicated to all the coefficients of the first variable, the second row is for the second variable, and so on. Each row then represents the coefficients of each variable in order as they appear in the system of equations. Through a couple of different processes, you can use the coefficient matrix in order to find the solutions to the system of equations.

Solving a system of equations using a matrix is a great method, especially for larger systems (with more variables and more equations). These methods work for systems of all sizes, so you have to choose which method is appropriate for which problem. The following sections break down the available processes.

Writing a system in matrix form

You can write any linear system of equations as a matrix. Write the equations with the variables in the same order.

EXAMPLE

Q. Express this system in matrix form: $\begin{cases} x+2y+3z=-7 \\ 2x-3y-5z=9 \\ -6x-8y+z=-22 \end{cases}$

A. Follow three simple steps:

1. **Write all the coefficients in one matrix first (this is called a *coefficient matrix*). If any variable doesn't appear in one of the equations, enter its coefficient as 0.**
2. **Write the variables in a column matrix (called the *variable matrix*). Show this as a matrix multiplying the coefficient matrix.**
3. **Write the constants in a column matrix (*constant matrix*). Put this matrix on the other side of an = sign.**

   The setup appears as follows: $\begin{bmatrix} 1 & 2 & 3 \\ 2 & -3 & -5 \\ -6 & -8 & 1 \end{bmatrix}\begin{bmatrix} x \\ y \\ z \end{bmatrix}=\begin{bmatrix} -7 \\ 9 \\ -22 \end{bmatrix}$

If you want to go directly to using your calculator, go on to the section, "Multiplying a matrix by its inverse," later in this chapter.

YOUR TURN

5 Express the system in matrix form: $\begin{cases} 2x-2y+4z=11 \\ 3x-3z-4y=-9 \\ -x+4y+z=2 \end{cases}$

6 Express the system in matrix form: $\begin{cases} x+3z=-7 \\ 2x-3y=9 \\ 3y+z=2 \end{cases}$

Reduced row-echelon form

You can use the *reduced row-echelon form* of a matrix to solve for the solutions of a system of equations. This form is very specific and allows you to read solutions right from the matrices.

Changing a matrix into reduced row-echelon form is beneficial because this form of a matrix is unique to each matrix (and that unique matrix could give you the solutions to your system of equations).

TECHNICAL STUFF

Reduced row-echelon form shows a matrix with a very specific set of requirements. To be considered to be in reduced row-echelon form, a matrix must meet *all* the following requirements:

- All rows containing all 0s are at the bottom of the matrix.
- All leading coefficients (first nonzero coefficient of a row) are 1.
- Any element above or below a leading coefficient is 0.
- The leading coefficient of any row is always to the left of the leading coefficient below it.

The matrices shown here are in *reduced* row-echelon form:

$$\begin{bmatrix}1&0&0&0\\0&1&0&0\\0&0&1&0\\0&0&0&1\end{bmatrix}\begin{bmatrix}1&0&0&0\\0&1&0&0\\0&0&1&0\\0&0&0&0\end{bmatrix}\begin{bmatrix}1&0&0&0\\0&0&1&0\\0&0&0&1\\0&0&0&0\end{bmatrix}$$

Augmented form

An alternative to writing a matrix in row-echelon or reduced row-echelon form is what's known as *augmented form*, where the coefficient matrix and the solution matrix are written in the same matrix, separated in each row by colons or a bar. This setup makes using elementary row operations to solve a matrix much simpler because you only have one matrix on your plate at a time (as opposed to three!).

Using the augmented form cuts down on the amount that you have to write. And when you're attempting to solve a system of equations that requires many steps, you'll be thankful to be writing less! Then you can use elementary row operations just as before to get the solution to your system.

Consider this matrix equation:

$$\begin{bmatrix}1&2&3\\2&-3&-5\\-6&-8&1\end{bmatrix}\begin{bmatrix}x\\y\\z\end{bmatrix}=\begin{bmatrix}-7\\9\\-22\end{bmatrix}$$

Written in augmented form, it looks like this:

$$\left[\begin{array}{ccc|c}1&2&3&-7\\2&-3&-5&9\\-6&-8&1&-22\end{array}\right]$$

In the next section you see how to perform row operations on this augmented matrix, just like a 4×3 matrix. You reduce the 3×3 coefficient matrix on the left side to reduced row-echelon form and obtain the solution while doing so.

The row-echelon form of a matrix comes in handy for solving systems of equations. You'll see how to get a matrix into row-echelon form using *elementary row operations*. You can use any of these operations to get a matrix into reduced row-echelon form:

- Multiply each element in a single row by a constant.
- Interchange two rows.
- Add two rows together or multiples of rows together.

Using these elementary row operations, you can rewrite any matrix so that the solutions to the system that the matrix represents become apparent.

Making Matrices Work for You

When you're comfortable with changing the appearances of matrices (to get them into augmented and reduced row-echelon, form, for instance), you are ready to tackle matrices and really start solving the more challenging systems. Hopefully, for really large systems (four or more variables), you'll have the aid of a graphing calculator. Computer programs can also be very helpful with matrices and can solve systems of equations in a variety of ways. The three ways you find to solve systems using matrices in this section are Gaussian elimination, matrix inverses, and Cramer's Rule.

Gaussian elimination is probably the best method to use if you don't have a graphing calculator or computer program to help you. If you do have these tools, then you can use either of them to find the inverse of any matrix, and in that case the inverse operation is the best plan. If the system only has two or three variables and you don't have a graphing calculator to help you, then Cramer's Rule is a good way to go.

Using Gaussian elimination to solve systems

Gaussian elimination requires the use of the elementary row operations and uses an augmented matrix.

The goals of Gaussian elimination are to make the upper-left corner element a 1, use elementary row operations to get 0s in all positions underneath that first 1, get 1s for leading coefficients in every row diagonally from the upper-left to the lower-right corner, and get 0s beneath all leading coefficients. Basically, you eliminate all variables in the last row except for one, all variables except for two in the equation above that one, and so on to the top equation, which has all the variables. Then you can use back substitution to solve for one variable at a time by plugging the values you know into the equations from the bottom up.

Elementary operations for Gaussian elimination are stated here with the notation used.

TECHNICAL STUFF

You can perform three operations on matrices in order to eliminate variables in a system of linear equations.

- **You can multiply any row by a constant:** $-2R_3 \to R_3$ says to multiply row three by –2 to give you a new row three.
- **You can switch any two rows:** $R_1 \leftrightarrow R_2$ swaps rows one and two.
- **You can add two rows together:** $R_1 + R_2 \to R_2$ adds rows one and two and writes the result in row two.
- **You can add the multiple of one row to another row:** $-3R_1 + R_3 \to R_3$ multiplies row one by –3, adds the product to row three, and puts the result in row three.

Putting the matrix in *row-echelon form* allows you to find the value of the last variable and then back-substitute to find the rest of the variables. If you continue the row operations and create the *reduced row-echelon form*, you can read the values of the variables right off the right column of the completed matrix.

Using row-echelon form to solve a system of equations

You've decided to take the plunge and use matrices to solve a system of equations.

EXAMPLE

Q. Solve the system of equations using Gaussian elimination: $\begin{cases} x+2y+3z=-7 \\ 2x-3y-5z=9 \\ -6x-8y+z=-22 \end{cases}$

A. First, write the system using the augmented matrix: $\left[\begin{array}{ccc|c} 1 & 2 & 3 & -7 \\ 2 & -3 & -5 & 9 \\ -6 & -8 & 1 & -22 \end{array}\right]$

Now take a look at the steps used in Gaussian elimination to rewrite this matrix:

1. **Position a 1 in the upper-left corner.**

   You already have it!

2. **Create 0s underneath the 1 in the first column.**

   You need to use the combo of two matrix operations together here. Here's what you should ask: "What do I need to add to row two to make a 2 become a 0?" The answer is –2.

   You can do this step by multiplying the first row by −2 and adding the resulting elements to the second row.

$$\left[\begin{array}{ccc|c} 1 & 2 & 3 & -7 \\ 2 & -3 & -5 & 9 \\ -6 & -8 & 1 & -22 \end{array}\right] -2R_1 + R_2 \to R_2 \left[\begin{array}{ccc|c} 1 & 2 & 3 & -7 \\ 0 & -7 & -11 & 23 \\ -6 & -8 & 1 & -22 \end{array}\right]$$

3. In the third row, get a 0 under the 1.

To do this step, you need to multiply row one by 6 and add it to row three.

$$\left[\begin{array}{ccc|c} 1 & 2 & 3 & -7 \\ 0 & -7 & -11 & 23 \\ -6 & -8 & 1 & -22 \end{array}\right] 6R_1 + R_3 \to R_3 \left[\begin{array}{ccc|c} 1 & 2 & 3 & -7 \\ 0 & -7 & -11 & 23 \\ 0 & 4 & 19 & -64 \end{array}\right]$$

4. Get a 1 in the second row, second column.

To do this step, you need to multiply the row by a constant; in other words, multiply row two by the reciprocal of –7.

$$\left[\begin{array}{ccc|c} 1 & 2 & 3 & -7 \\ 0 & -7 & -11 & 23 \\ 0 & 4 & 19 & -64 \end{array}\right] -\frac{1}{7}R_2 \to R_2 \left[\begin{array}{ccc|c} 1 & 2 & 3 & -7 \\ 0 & 1 & \frac{11}{7} & -\frac{23}{7} \\ 0 & 4 & 19 & -64 \end{array}\right]$$

5. Get a 0 under the 1 you created in row two.

Multiply row two by –4 and add it to row three to create a new row three.

$$\left[\begin{array}{ccc|c} 1 & 2 & 3 & -7 \\ 0 & 1 & \frac{11}{7} & -\frac{23}{7} \\ 0 & 4 & 19 & -64 \end{array}\right] -4R_2 + R_3 \to R_3 \left[\begin{array}{ccc|c} 1 & 2 & 3 & -7 \\ 0 & 1 & \frac{11}{7} & -\frac{23}{7} \\ 0 & 0 & \frac{89}{7} & -\frac{356}{7} \end{array}\right]$$

6. Get another 1, this time in the third row, third column.

Multiply the third row by the reciprocal of $\frac{89}{7}$ to get a 1.

$$\left[\begin{array}{ccc|c} 1 & 2 & 3 & -7 \\ 0 & 1 & \frac{11}{7} & -\frac{23}{7} \\ 0 & 0 & \frac{89}{7} & -\frac{356}{7} \end{array}\right] \frac{7}{89}R_3 \to R_3 \left[\begin{array}{ccc|c} 1 & 2 & 3 & -7 \\ 0 & 1 & \frac{11}{7} & -\frac{23}{7} \\ 0 & 0 & 1 & -4 \end{array}\right]$$

You now have a matrix in *row-echelon form*, which allows you to find the solutions when you use back substitution.

Referring to the last matrix, the original system of equations forming the augmented matrix was:

$$\begin{cases} x + 2y + 3z = -7 \\ 2x - 3y - 5z = 9 \\ -6x - 8y + z = -22 \end{cases}$$

This system has been transformed into:

$$\begin{cases} x + 2y + 3z = -7 \\ \quad y + \frac{11}{7}z = -\frac{23}{7} \\ \qquad z = -4 \end{cases}$$

You easily read that $z = -4$. Substituting this into the second equation, you have:

$$y + \frac{11}{7}(-4) = -\frac{23}{7}$$
$$y - \frac{44}{7} = -\frac{23}{7}$$
$$y = \frac{21}{7} = 3$$

Then, substituting the values for y and z into the first equation:

$$x + 2(3) + 3(-4) = -7$$
$$x = -1$$

The solution of the system is $x = -1$, $y = 3$, $z = -4$.

Applying the reduced row-echelon form

However, if you want to continue with the matrix operations and get this matrix into *reduced row-echelon form* to find the solutions and be able to read them right off the matrix, follow these steps:

1. **Starting with the row-echelon form, create a 0 above the 1 in row two.**

   Multiply row two by –2 and add it to row one.

$$\left[\begin{array}{ccc|c} 1 & 2 & 3 & -7 \\ 0 & 1 & \frac{11}{7} & -\frac{23}{7} \\ 0 & 0 & 1 & -4 \end{array}\right] -2R_2 + R_1 \to R_1 \left[\begin{array}{ccc|c} 1 & 0 & -\frac{1}{7} & -\frac{3}{7} \\ 0 & 1 & \frac{11}{7} & -\frac{23}{7} \\ 0 & 0 & 1 & -4 \end{array}\right]$$

2. **Create a 0 in row one, column three.**

   To do this, multiply row three by $\frac{1}{7}$ and add it to row one.

$$\left[\begin{array}{ccc|c} 1 & 0 & -\frac{1}{7} & \frac{3}{7} \\ 0 & 1 & \frac{11}{7} & -\frac{23}{7} \\ 0 & 0 & 1 & -4 \end{array}\right] \frac{1}{7}R_3 + R_1 \to R_1 \left[\begin{array}{ccc|c} 1 & 0 & 0 & -1 \\ 0 & 1 & \frac{11}{7} & -\frac{23}{7} \\ 0 & 0 & 1 & -4 \end{array}\right]$$

3. **Get a 0 in row two, column three.**

   Multiply row three by $-\frac{11}{7}$ and add it to row two.

$$\left[\begin{array}{ccc|c} 1 & 0 & 0 & -1 \\ 0 & 1 & \frac{11}{7} & \frac{23}{7} \\ 0 & 0 & 1 & -4 \end{array}\right] -\frac{11}{7}R_3 + R_2 \to R_2 \left[\begin{array}{ccc|c} 1 & 0 & 0 & -1 \\ 0 & 1 & 0 & 3 \\ 0 & 0 & 1 & -4 \end{array}\right]$$

This matrix, in reduced row-echelon form, shows the solution to the system right there in the rightmost column: $x = -1$, $y = 3$, $z = -4$.

YOUR TURN

7 Solve the system of equations using the row-echelon form: $\begin{cases} 3x-2y+6z=7 \\ x-2y-z=-2 \\ -3x+10y+11z=18 \end{cases}$

8 Solve the system of equations using the reduced row-echelon form: $\begin{cases} x-y+2z=8 \\ 3x \quad +2z=7 \\ x-3y-z=8 \end{cases}$

Arriving at infinite solutions or no solution

So far, all the systems shown have had a unique solution. This isn't always the case. Sometimes there are infinite solutions, and other times there's no solution at all. The two systems shown next are examples of just such cases. The first two equations in each system are the same; it's the constant in the third equation that makes the difference.

$$\text{I:}\begin{cases} x+2y-3z=8 \\ 4x-y+z=15 \\ 6x+3y-5z=31 \end{cases} \qquad \text{II:}\begin{cases} x+2y-3z=8 \\ 4x-y+z=15 \\ 6x+3y-5z=23 \end{cases}$$

Working through these two systems using an augmented matrix and row operations, you end up with the following matrices; the last two steps are shown here:

$$\text{I:}\left[\begin{array}{ccc|c} 1 & 2 & -3 & 8 \\ 0 & -9 & 13 & -17 \\ 0 & -9 & 13 & -17 \end{array}\right] -1R_2+R_3 \rightarrow R_3 \left[\begin{array}{ccc|c} 1 & 2 & -3 & 8 \\ 0 & -9 & 13 & -17 \\ 0 & 0 & 0 & 0 \end{array}\right]$$

$$\text{II:}\left[\begin{array}{ccc|c} 1 & 2 & -3 & 8 \\ 0 & -9 & 13 & -17 \\ 0 & -9 & 13 & -25 \end{array}\right] -1R_2+R_3 \rightarrow R_3 \left[\begin{array}{ccc|c} 1 & 2 & -3 & 8 \\ 0 & -9 & 13 & -17 \\ 0 & 0 & 0 & -8 \end{array}\right]$$

The last row of the first system ends up with all zeros. The associated equation with this row is $0=0$, which is always true. This indicates that the last equation was some linear combination of the other two equations. There are an infinite number of solutions to the system. For example, $(4,-1,-2)$, $(5,12,7)$, and $(6,25,16)$ all work. And there are infinitely more solutions, all in the form $\left(\frac{z+38}{9}, \frac{13z+17}{9}, z\right)$. Just pick a value for z, and the other coordinates can be determined. Where did the coordinates come from? Just use the equations from the matrix. The second line reads $-9y+13z=-17$. Solving for y, you get $y=\frac{13z+17}{9}$. Plug that into the equation formed from the first line, $x+2y-3z=8$, and you find that $x=\frac{z+38}{9}$.

The last row of the second system has a false statement. Putting it in equation form, it reads that $0=-8$. That statement indicates that there is no solution to the system.

YOUR TURN

9 Given the matrix $\begin{bmatrix} 1 & 3 & 4 & 7 \\ 0 & 1 & 2 & 5 \\ 0 & 0 & 0 & 0 \end{bmatrix}$, write the infinite number of solutions (x, y, z) in terms of the variable z.

10 Given the matrix $\begin{bmatrix} 1 & -2 & 3 & 4 \\ 0 & 1 & 1 & 9 \\ 0 & 0 & 0 & 0 \end{bmatrix}$, what are the values of x and y when $z=-2$?

REMEMBER

Perhaps one of the most recognizable matrices is the *identity matrix*, which has 1s along the diagonal from the upper-left to the lower-right corner and has 0s everywhere else. It is a square matrix in reduced row-echelon form and stands for the identity element of multiplication in the world of matrices.

The identity matrix is an important structure in solving systems of equations because, if you can manipulate the coefficient matrix to look like the identity matrix (using matrix operations), then the solution to the system is right there in front of you.

Multiplying a matrix by its inverse

You can use matrices in yet another way to solve a system of equations. This method is based on the simple idea that if you have a coefficient tied to a variable on one side of an equation, you can multiply by the coefficient's inverse to make that coefficient go away and leave you with just the variable. For example, if $3x = 12$, how would you solve the equation? You'd divide both sides by 3, which is the same thing as multiplying by $\frac{1}{3}$, to get $x = 4$. So it goes with matrices.

In variable form, an inverse function is written as $f^{-1}(x)$, where f^{-1} is the inverse of the function f. You name an inverse matrix similarly; the inverse of matrix A is A^{-1}. If A, B, and C are matrices in the matrix equation $AB = C$, and you want to solve for B, how do you do that? Just multiply by the inverse of matrix A, which is written $A^{-1}[AB] = A^{-1}C$.

The simplified version is $B = A^{-1}C$.

Using this property of inverse matrices, you can find the solution of a system of equations.

Finding the inverse of a matrix

First off, you need to know that only square matrices have inverses — in other words, the number of rows must be equal to the number of columns. And even then, not every square matrix has an inverse. If the *determinant* of a matrix is not 0, then the matrix has an inverse. See the section, "Using determinants: Cramer's Rule," for more on determinants.

When a matrix has an inverse, you have several ways to find it, depending on how big the matrix is. If the matrix is a 2×2 matrix, then you can use a simple formula to find the inverse. However, for anything larger than 2×2, you can either use a graphing calculator or computer program, or solve for the inverse using row operations — somewhat like solving a system of equations (but different). To solve for an identity using this method, you augment your original, invertible matrix with the identity matrix and use elementary row operations to get the identity matrix where your original matrix once was.

With that said, here's how you find an inverse of a 2×2 matrix:

If matrix A is the 2×2 matrix $\begin{bmatrix} a & b \\ c & d \end{bmatrix}$, its inverse is $\frac{1}{ad-bc}\begin{bmatrix} d & -b \\ -c & a \end{bmatrix}$.

Simply follow this format to find the inverse of a 2×2 matrix.

Using an inverse to solve a system

Armed with a system of equations and the knowledge of how to use inverse matrices (see the previous section), you can follow a series of simple steps to arrive at a solution to the system, again using the trusty old matrix.

EXAMPLE

Q. Solve the system that follows by using an inverse matrix:

$$\begin{cases} 4x + 3y = -13 \\ -10x - 2y = 5 \end{cases}$$

A. These steps show you the way:

1. **Write the system as a matrix equation.**

   Show the coefficient matrix, variable matrix, and constant matrix.

   $$\begin{bmatrix} 4 & 3 \\ -10 & -2 \end{bmatrix}\begin{bmatrix} x \\ y \end{bmatrix} = \begin{bmatrix} -13 \\ 5 \end{bmatrix}$$

2. **Create the inverse matrix out of the matrix equation.**

   Use the inverse formula, $\frac{1}{ad - bc}\begin{bmatrix} d & -b \\ -c & a \end{bmatrix}$.

   In this case, $a = 4$, $b = 3$, $c = -10$, and $d = -2$. The inverse matrix is

   $$\frac{1}{4(-2) - 3(-10)}\begin{bmatrix} -2 & -3 \\ 10 & 4 \end{bmatrix} = \frac{1}{22}\begin{bmatrix} -2 & -3 \\ 10 & 4 \end{bmatrix}$$

   You could multiply each element by $\frac{1}{22}$ at this point, but it creates messy fractions. Instead, you can just wait and multiply by that scalar at the end of the process.

3. **Multiply the matrix equation by the inverse.**

   You now have the following equation:

   $$\frac{1}{22}\begin{bmatrix} -2 & -3 \\ 10 & 4 \end{bmatrix}\begin{bmatrix} 4 & 3 \\ -10 & -2 \end{bmatrix}\begin{bmatrix} x \\ y \end{bmatrix} = \frac{1}{22}\begin{bmatrix} -2 & -3 \\ 10 & 4 \end{bmatrix}\begin{bmatrix} -13 \\ 5 \end{bmatrix}$$

4. **Cancel the matrix on the left and multiply the matrices on the right (see the section, "Multiplying matrices by each other").**

   An inverse matrix times its matrix cancels out.

   $$\cancel{\frac{1}{22}\begin{bmatrix} -2 & -3 \\ 10 & 4 \end{bmatrix}\begin{bmatrix} 4 & 3 \\ -10 & -2 \end{bmatrix}}\begin{bmatrix} x \\ y \end{bmatrix} = \frac{1}{22}\begin{bmatrix} -2 & -3 \\ 10 & 4 \end{bmatrix}\begin{bmatrix} -13 \\ 5 \end{bmatrix}$$

   $$\begin{bmatrix} x \\ y \end{bmatrix} = \frac{1}{22}\begin{bmatrix} -2(-13) + (-3)(5) \\ 10(-13) + 4(5) \end{bmatrix}$$

   $$\begin{bmatrix} x \\ y \end{bmatrix} = \frac{1}{22}\begin{bmatrix} 11 \\ -110 \end{bmatrix}$$

5. **Multiply the scalar to solve the system.**

   You finish with the x and y values: $\begin{bmatrix} x \\ y \end{bmatrix} = \frac{1}{22}\begin{bmatrix} 11 \\ -110 \end{bmatrix} = \begin{bmatrix} \frac{1}{2} \\ -5 \end{bmatrix}$

   The solution is $x = \frac{1}{2}$, $y = -5$.

YOUR TURN

11 Find the inverse of the matrix: $A = \begin{bmatrix} 4 & -1 \\ 2 & 3 \end{bmatrix}$.

12 Solve the system of equations using an inverse matrix: $\begin{cases} 4x - y = -10 \\ 2x + 3y = 16 \end{cases}$.

Using determinants: Cramer's Rule

The final method that you find here for solving systems of equations was conceived by Gabriel Cramer and named after him. As with much of what this chapter covers, a graphing calculator enables you to bypass much of the legwork and makes life a lot easier. However, you may run across an occasion or two where Cramer's Rule is needed and useful. So here goes!

First, a quick example should make all the notations used in Cramer's Rule a bit more understandable.

EXAMPLE

Q. Use Cramer's Rule to solve the system of equations: $\begin{cases} x + 3y = 6 \\ 2x - 5y = 1 \end{cases}$.

A. The following steps will get you to the solution.

1. **Write the determinant of the coefficient matrix.**

   The coefficient matrix is $\begin{bmatrix} 1 & 3 \\ 2 & -5 \end{bmatrix}$, and its determinant is $\begin{vmatrix} 1 & 3 \\ 2 & -5 \end{vmatrix}$. Note the change from brackets to vertical bars.

2. **Find the value, *d*, of the determinant.**

   The determinant of a 2×2 matrix $\begin{vmatrix} a_1 & b_1 \\ a_2 & b_2 \end{vmatrix}$ is $d = a_1b_2 - b_1a_2$. So $\begin{vmatrix} 1 & 3 \\ 2 & -5 \end{vmatrix} = 1(-5) - 3(2) = -11$.

3. **Find the value of the variable *x* by replacing the two *x* coefficients in the determinant with the constants, evaluating the new determinant, and dividing by *d*.**

   $$x = \frac{\begin{vmatrix} 6 & 3 \\ 1 & -5 \end{vmatrix}}{-11} = \frac{6(-5) - 3(1)}{-11} = \frac{-33}{-11} = 3$$

4. **Find the value of the variable *y* by replacing the two *y* coefficients in the determinant with the constants, evaluating the new determinant, and dividing by *d*.**

   $$y = \frac{\begin{vmatrix} 1 & 6 \\ 2 & 1 \end{vmatrix}}{-11} = \frac{1(1) - 6(2)}{-11} = \frac{-11}{-11} = 1$$

   The solution of the system is $x = 3$, $y = 1$.

Now, on to the more general description.

TECHNICAL STUFF

Cramer's Rule says that if the determinant of a coefficient matrix |A| is not 0, then the solutions to a system of linear equations can be found as follows:

If a linear system of equations,

$$\begin{array}{ccccccc} a_1x_1 & + & b_1x_2 & +c_1x_3 & + & \cdots & =k_1 \\ a_2x_1 & + & b_2x_2 & +c_2x_3 & + & \cdots & =k_2 \\ a_3x_1 & + & b_3x_2 & +c_3x_3 & + & \cdots & =k_3 \\ \vdots & & \vdots & \vdots & & & \vdots \end{array}$$

has a coefficient determinant, d, and constant matrix K,

$$d=\begin{vmatrix} a_1 & b_1 & c_1 & \cdots \\ a_2 & b_2 & c_2 & \cdots \\ a_3 & b_3 & c_3 & \cdots \\ \vdots & \vdots & \vdots & \end{vmatrix} \quad K=\begin{bmatrix} k_1 \\ k_2 \\ k_3 \\ \vdots \end{bmatrix}$$

then each variable is equal to the determinant formed by replacing its coefficients with the constant matrix values and dividing that determinant by d.

So $x_1=\dfrac{\begin{vmatrix} k_1 & b_1 & c_1 & \cdots \\ k_2 & b_2 & c_2 & \cdots \\ k_3 & b_3 & c_3 & \cdots \\ \vdots & \vdots & \vdots & \end{vmatrix}}{d}$, and so on for the other variables.

The determinant of a 2×2 matrix is simply the difference between the cross-products. The determinant of a 3×3 matrix is a bit more complicated. Consider the general 3×3 matrix,

$$A=\begin{bmatrix} a_1 & b_1 & c_1 \\ a_2 & b_2 & c_2 \\ a_3 & b_3 & c_3 \end{bmatrix}$$

You find the determinant of the matrix by following these steps:

1. **Rewrite the first two columns immediately after the third column.** $\left|\begin{array}{ccc|cc} a_1 & b_1 & c_1 & a_1 & b_1 \\ a_2 & b_2 & c_2 & a_2 & b_2 \\ a_3 & b_3 & c_3 & a_3 & b_3 \end{array}\right|$

2. **Draw three diagonal lines from the upper-left to the lower-right corner and three diagonal lines from the lower-left to the upper-right corner (see Figure 19-1).**

3. **Multiply down the three diagonals from left to right, and up the other three.**

   The determinant of the 3×3 matrix is $(a_1b_2c_3+b_1c_2a_3+c_1a_2b_3)-(a_3b_2c_1+b_3c_2a_1+c_3a_2b_1)$.

$$|A| = \begin{vmatrix} a_1 & b_1 & c_1 \\ a_2 & b_2 & c_2 \\ a_3 & b_3 & c_3 \end{vmatrix} \begin{matrix} a_1 & b_1 \\ a_2 & b_2 \\ a_3 & b_3 \end{matrix}$$

FIGURE 19-1: Computing the determinant of a 3×3 matrix.

EXAMPLE

Q. Find the determinant of this 3×3 matrix:

$$\begin{bmatrix} 1 & 2 & 3 \\ 2 & -3 & -5 \\ -6 & -8 & 1 \end{bmatrix}$$

A. The determinant is found *using diagonals*:

$$\begin{vmatrix} 1 & 2 & 3 \\ 2 & -3 & -5 \\ -6 & -8 & 1 \end{vmatrix} \begin{matrix} 1 & 2 \\ 2 & -3 \\ -6 & -8 \end{matrix}$$

$$= (1(-3)(1) + 2(-5)(-6) + 3(2)(-8)) - (3(-3)(-6) + 1(-5)(-8) + 2(2)(1))$$

$$= (-3 + 60 - 48) - (54 + 40 + 4) = 9 - 98 = -89$$

After you find the determinant of the coefficient matrix (either by hand or with a technological device), replace the first column of the coefficient matrix with the constant matrix from the other side of the equal sign and find the determinant of that new matrix. Then replace the second column of the coefficient matrix with the constant matrix and find the determinant of that matrix. Continue this process until you have replaced each column and found each new determinant. The values of the respective variables are equal to the determinant of the new matrix (when you replaced the respective column) divided by the determinant of the coefficient matrix.

You can't use Cramer's Rule when the matrix isn't square or when the determinant of the coefficient matrix is 0, because you can't divide by 0. Cramer's Rule is most useful for a 2×2 or 3×3 system of linear equations.

YOUR TURN

13 Find the determinant of the matrix:

$$\begin{vmatrix} 3 & 2 & 1 \\ 1 & -1 & -2 \\ 3 & 1 & 0 \end{vmatrix}.$$

14 Solve the system using Cramer's Rule:

$$\begin{cases} 2x - 3y = 12. \\ -x + 2y = 9 \end{cases}$$

Practice Questions Answers and Explanations

1. $\begin{bmatrix} 3 & 2 \\ -2 & 13 \\ 17 & 3 \end{bmatrix}$. $J+2K=\begin{bmatrix} 3 & -8 \\ 2 & 7 \\ 5 & -9 \end{bmatrix}+2\begin{bmatrix} 0 & 5 \\ -2 & 3 \\ 6 & 6 \end{bmatrix}=\begin{bmatrix} 3 & -8 \\ 2 & 7 \\ 5 & -9 \end{bmatrix}+\begin{bmatrix} 0 & 10 \\ -4 & 6 \\ 12 & 12 \end{bmatrix}=\begin{bmatrix} 3 & 2 \\ -2 & 13 \\ 17 & 3 \end{bmatrix}$

2. $\begin{bmatrix} 9 & -44 \\ 14 & 9 \\ -9 & -51 \end{bmatrix}$. $3J-4K=3\begin{bmatrix} 3 & -8 \\ 2 & 7 \\ 5 & -9 \end{bmatrix}-4\begin{bmatrix} 0 & 5 \\ -2 & 3 \\ 6 & 6 \end{bmatrix}=\begin{bmatrix} 9 & -24 \\ 6 & 21 \\ 15 & -27 \end{bmatrix}-\begin{bmatrix} 0 & 20 \\ -8 & 12 \\ 24 & 24 \end{bmatrix}=\begin{bmatrix} 9 & -44 \\ 14 & 9 \\ -9 & -51 \end{bmatrix}$

3. $\begin{bmatrix} 5 & -7 \\ 12 & -12 \end{bmatrix}$. $CD=\begin{bmatrix} 1 & 4 \\ 0 & 6 \end{bmatrix}\cdot\begin{bmatrix} -3 & 1 \\ 2 & -2 \end{bmatrix}=\begin{bmatrix} 1(-3)+4(2) & 1(1)+4(-2) \\ 0(-3)+6(2) & 0(1)+6(-2) \end{bmatrix}=\begin{bmatrix} 5 & -7 \\ 12 & -12 \end{bmatrix}$

4. $\begin{bmatrix} 13 & 8 & 13 \\ -25 & -18 & -29 \\ -37 & -28 & -45 \\ -7 & 14 & 21 \end{bmatrix}$. $EF=\begin{bmatrix} -1(1)+2(7) & -1(-2)+2(3) & -1(-3)+2(5) \\ 3(1)+(-4)(7) & 3(-2)+(-4)(3) & 3(-3)+(-4)(5) \\ 5(1)+(-6)(7) & 5(-2)+(-6)(3) & 5(-3)+(-6)(5) \\ -7(1)+0(7) & -7(-2)+0(3) & -7(-3)+0(5) \end{bmatrix}=\begin{bmatrix} 13 & 8 & 13 \\ -25 & -18 & -29 \\ -37 & -28 & -45 \\ -7 & 14 & 21 \end{bmatrix}$

5. $\begin{bmatrix} 2 & -2 & 4 \\ 3 & -4 & -3 \\ -1 & 4 & 1 \end{bmatrix}\begin{bmatrix} x \\ y \\ z \end{bmatrix}=\begin{bmatrix} 11 \\ -9 \\ 2 \end{bmatrix}$. **You start by putting the terms in the second equation in the correct order:**

$$\begin{cases} 2x-2y+4z=11 \\ 3x-3z-4y=-9 \\ -x+4y+z=2 \end{cases} \rightarrow \begin{cases} 2x-2y+4z=11 \\ 3x-4y-3z=-9 \\ -x+4y+z=2 \end{cases}$$

6. $\begin{bmatrix} 1 & 0 & 3 \\ 2 & -3 & 0 \\ 0 & 3 & 1 \end{bmatrix}\begin{bmatrix} x \\ y \\ z \end{bmatrix}=\begin{bmatrix} -7 \\ 9 \\ 2 \end{bmatrix}$. **Replace the missing variable terms with 0s:**

$$\begin{cases} x+3z=-7 \\ 2x-3y=9 \\ 3y+z=2 \end{cases} \rightarrow \begin{cases} x+0+3z=-7 \\ 2x-3y+0=9 \\ 0+3y+z=2 \end{cases}$$

7. $x=1$, $y=1$, $z=1$. **Set up the system as an augmented matrix:**

$$\left[\begin{array}{ccc|c} 3 & -2 & 6 & 7 \\ 1 & -2 & -1 & -2 \\ -3 & 10 & 11 & 18 \end{array}\right]$$

Reverse rows; $r_1 \leftrightarrow r_2$ gets a 1 in the upper-left corner:

$$= \left[\begin{array}{ccc|c} 1 & -2 & -1 & -2 \\ 13 & -2 & 6 & 7 \\ -3 & 10 & 11 & 18 \end{array}\right]$$

Add a multiple of row three to row two; $-3r_1 + r_2 \rightarrow r_2$ gets a 0 under the 1 in the second row:

$$= \left[\begin{array}{ccc|c} 1 & -2 & -1 & -2 \\ 0 & 4 & 9 & 13 \\ -3 & 10 & 11 & 18 \end{array}\right]$$

Add a multiple of row one to row three; $3r_1 + r_3 \rightarrow r_3$ gets a 0 under the 1 in the third row:

$$= \left[\begin{array}{ccc|c} 1 & -2 & -1 & -2 \\ 0 & 4 & 9 & 13 \\ 0 & 4 & 8 & 12 \end{array}\right]$$

Next, to avoid fractions, switch rows two and three, and then multiply the new row two by $\frac{1}{4}$:

$$r_2 \leftrightarrow r_3 = \left[\begin{array}{ccc|c} 1 & -2 & -1 & -2 \\ 0 & 4 & 8 & 12 \\ 0 & 4 & 9 & 13 \end{array}\right] \text{ and then } \frac{1}{4}r_2 \rightarrow r_2 = \left[\begin{array}{ccc|c} 1 & -2 & -1 & -2 \\ 0 & 1 & 2 & 3 \\ 0 & 4 & 9 & 13 \end{array}\right]$$

Add a multiple of row two to row three; $-4r_2 + r_3 \rightarrow r_3$ gets a 0 under the 1 you just created on the main diagonal:

$$= \left[\begin{array}{ccc|c} 1 & -2 & -1 & -2 \\ 0 & 1 & 2 & 3 \\ 0 & 0 & 1 & 1 \end{array}\right]$$

Reading the corresponding equation from the third row, you have that $z = 1$. Back-substituting, you use the second row to get that $y + 2(1) = 3$ or $y = 1$. Back-substitute again using the first row, and you have $x - 2(1) - 1(1) = -2$ or $x = 1$.

8 $\boldsymbol{x = 1, y = -3, z = 2}$. Set up the system as an augmented matrix:

$$\left[\begin{array}{ccc|c} 1 & -1 & 2 & 8 \\ 3 & 0 & 2 & 7 \\ 1 & -3 & -1 & 8 \end{array}\right]$$

Perform the operations indicated to create 0s below the 1 in row one, column one:

$$\left[\begin{array}{ccc|c} 1 & -1 & 2 & 8 \\ 3 & 0 & 2 & 7 \\ 1 & -3 & -1 & 8 \end{array}\right] \begin{array}{l} -3r_1 + r_2 \rightarrow r_2 \\ -1r_1 + r_3 \rightarrow r_3 \end{array} \left[\begin{array}{ccc|c} 1 & -1 & 2 & 8 \\ 0 & 3 & -4 & -17 \\ 0 & -2 & -3 & 0 \end{array}\right]$$

Perform the operation indicated to create a 1 in row two, column 2:

$$\left[\begin{array}{ccc|c} 1 & -1 & 2 & 8 \\ 0 & 3 & -4 & -17 \\ 0 & -2 & -3 & 0 \end{array}\right] \frac{1}{3}r_2 \to r_2 \left[\begin{array}{ccc|c} 1 & -1 & 2 & 8 \\ 0 & 1 & -\frac{4}{3} & -\frac{17}{3} \\ 0 & -2 & -3 & 0 \end{array}\right]$$

Perform the operations indicated to create 0s above and below the 1 in row two, column two:

$$\left[\begin{array}{ccc|c} 1 & -1 & 2 & 8 \\ 0 & 1 & -\frac{4}{3} & -\frac{17}{3} \\ 0 & -2 & -3 & 0 \end{array}\right] \begin{array}{c} r_2 + r_1 \to r_1 \\ 2r_2 + r_3 \to r_3 \end{array} \left[\begin{array}{ccc|c} 1 & 0 & \frac{2}{3} & \frac{7}{3} \\ 0 & 1 & -\frac{4}{3} & -\frac{17}{3} \\ 0 & 0 & -\frac{17}{3} & -\frac{34}{3} \end{array}\right]$$

Perform the operation indicated to create a 1 in row three, column three:

$$\left[\begin{array}{ccc|c} 1 & 0 & \frac{2}{3} & \frac{7}{3} \\ 0 & 1 & -\frac{4}{3} & -\frac{17}{3} \\ 0 & 0 & -\frac{17}{3} & -\frac{34}{3} \end{array}\right] -\frac{3}{17}r_3 \to r_3 \left[\begin{array}{ccc|c} 1 & 0 & \frac{2}{3} & \frac{7}{3} \\ 0 & 1 & -\frac{4}{3} & -\frac{17}{3} \\ 0 & 0 & 1 & 2 \end{array}\right]$$

Perform the operations indicated to create 0s above the 1 in row three, column three:

$$\left[\begin{array}{ccc|c} 1 & 0 & \frac{2}{3} & \frac{7}{3} \\ 0 & 1 & -\frac{4}{3} & -\frac{17}{3} \\ 0 & 0 & 1 & 2 \end{array}\right] \begin{array}{c} -\frac{2}{3}r_3 + r_1 \to r_1 \\ \frac{4}{3}r_3 + r_2 \to r_2 \end{array} \left[\begin{array}{ccc|c} 1 & 0 & 0 & 1 \\ 0 & 1 & 0 & -3 \\ 0 & 0 & 1 & 2 \end{array}\right]$$

9. **$(2z-8, 5-2z, z)$.** From the second row, you have $y+2z=5 \to y=5-2z$. Then, using the first row, $x+3y+4z=7$. Substitute in what you found for y: $x+3(5-2z)+4z = 7 \to x+15-6z+4z=7 \to x=2z-8$.

10. **$x=32$, $y=11$.** From the second row, you have $y+z=9 \to y=9-z$. If $z=-2$, $y=9-(-2)=11$. Then, using the first row, $x-2y+3z=4 \to x-2(11)+3(-2)=4 \to x=22+6+4=32$.

11. **$\frac{1}{14}\begin{bmatrix} 3 & 1 \\ -2 & 4 \end{bmatrix}$.** Using the formula $A^{-1} = \frac{1}{ad-bc}\begin{bmatrix} d & -b \\ -c & a \end{bmatrix}$, you get:

$$A^{-1} = \frac{1}{4(3)-(-1)(2)}\begin{bmatrix} 3 & 1 \\ -2 & 4 \end{bmatrix} = \frac{1}{14}\begin{bmatrix} 3 & 1 \\ -2 & 4 \end{bmatrix}$$

12. **$x=-1$, $y=6$.** First, write the system as a matrix equation: $\begin{bmatrix} 4 & -1 \\ 2 & 3 \end{bmatrix}\begin{bmatrix} x \\ y \end{bmatrix} = \begin{bmatrix} -10 \\ 16 \end{bmatrix}$. Now, use the inverse matrix from problem 11. Multiply both sides of the matrix equation by the inverse:

$$\frac{1}{14}\begin{bmatrix} 3 & 1 \\ -2 & 4 \end{bmatrix}\begin{bmatrix} 4 & -1 \\ 2 & 3 \end{bmatrix}\begin{bmatrix} x \\ y \end{bmatrix} = \frac{1}{14}\begin{bmatrix} 3 & 1 \\ -2 & 4 \end{bmatrix}\begin{bmatrix} -10 \\ 16 \end{bmatrix}\frac{1}{14}\begin{bmatrix} 3 & 1 \\ -2 & 4 \end{bmatrix}\begin{bmatrix} 4 & -1 \\ 2 & 3 \end{bmatrix}$$

This gives you $\begin{bmatrix} x \\ y \end{bmatrix} = \frac{1}{14}\begin{bmatrix} 3 & 1 \\ -2 & 4 \end{bmatrix}\begin{bmatrix} -10 \\ 16 \end{bmatrix}$.

Multiply the two matrices on the right, and then multiply by the scalar:

$$\begin{bmatrix} x \\ y \end{bmatrix} = \frac{1}{14}\begin{bmatrix} -14 \\ 84 \end{bmatrix} = \begin{bmatrix} -1 \\ 6 \end{bmatrix}$$

Your solutions from top to bottom are $x = -1$ and $y = 6$.

13 **−2.** Copy the first three columns to the right. Then find the sum of the products of the left-downward diagonals minus the sum of the products of the right-upward diagonals.

$$\begin{vmatrix} 3 & 2 & 1 \\ 1 & -1 & -2 \\ 3 & 1 & 0 \end{vmatrix} \begin{matrix} 3 & 2 \\ 1 & -1 \\ 3 & 1 \end{matrix} \rightarrow (0 + (-12) + 1) - (-3 + (-6) + 0) = -11 - (-9) = -2$$

14 **$x = 51$, $y = 30$.** First, find the value of the determinant: $\begin{vmatrix} 2 & -3 \\ -1 & 2 \end{vmatrix} = 4 - (3) = 1$. The value of x is found by replacing the x-coefficients in the equations with the constants, finding the determinant, and dividing by the 1:

$$x = \frac{\begin{vmatrix} 12 & -3 \\ 9 & 2 \end{vmatrix}}{1} = 24 - (-27) = 51$$

The value of y is found by replacing the y-coefficients in the equations with the constants, finding the determinant, and dividing by the 1: $y = \frac{\begin{vmatrix} 2 & 12 \\ -1 & 9 \end{vmatrix}}{1} = 18 - (-12) = 30$.

If you're ready to test your skills a bit more, take the following chapter quiz that incorporates all the chapter topics.

Whaddya Know? Chapter 19 Quiz

Quiz time! Complete each problem to test your knowledge on the various topics covered in this chapter. You can then find the solutions and explanations in the next section.

1 **Solve the system of equations using Gaussian elimination:**

$$\begin{cases} x-2y+3z=13 \\ 2x+y-z=4 \\ x-y+4z=11 \end{cases}$$

2 **Given** $A=\begin{bmatrix} 2 & 3 & -1 \\ 4 & -2 & 5 \end{bmatrix}$ **and** $B=\begin{bmatrix} 0 & -6 & -3 \\ 2 & 1 & 1 \end{bmatrix}$, **find 2A + 3B.**

3 **Find the value of the determinant:** $\begin{vmatrix} 3 & -2 \\ 1 & 4 \end{vmatrix}$.

4 **Write the infinite solutions (x,y,z) in terms of z for** $\left[\begin{array}{ccc|c} 1 & -1 & 4 & 3 \\ 0 & 1 & -5 & 6 \\ 0 & 0 & 0 & 0 \end{array}\right]$.

5 **Given** $C=\begin{bmatrix} 1 & 0 & 2 \\ 2 & -3 & 1 \\ 0 & 1 & -1 \end{bmatrix}$ **and** $D=\begin{bmatrix} 4 & 3 \\ 1 & 1 \\ 0 & -1 \end{bmatrix}$, **find CD.**

6 **Use the reduced row-echelon form to solve the system of equations:**

$$\begin{cases} 3x-y+2z=2 \\ x+y+z=3 \\ -4x+y-z=2 \end{cases}$$

7 **Use Cramer's Rule to solve the system of equations:**

$$\begin{cases} x-3y=13 \\ 2x-y=6 \end{cases}$$

8 **Find the inverse of A:** $A=\begin{bmatrix} 3 & -2 \\ 1 & 4 \end{bmatrix}$.

Answers to Chapter 19 Quiz

1. **$x = 4,\ y = -3,\ z = 1.$**

   Set up the system as an augmented matrix: $\left[\begin{array}{ccc|c} 1 & -2 & 3 & 13 \\ 2 & 1 & -1 & 4 \\ 1 & -1 & 4 & 11 \end{array}\right]$.

   Perform the operations indicated to create 0s below the 1 in row one, column one:

   $$\left[\begin{array}{ccc|c} 1 & -2 & 3 & 13 \\ 2 & 1 & -1 & 4 \\ 1 & -1 & 4 & 11 \end{array}\right] \begin{array}{c} -2r_1 + r_2 \to r_2 \\ -1r_1 + r_3 \to r_3 \end{array} \left[\begin{array}{ccc|c} 1 & -2 & 3 & 13 \\ 0 & 5 & -7 & -22 \\ 0 & 1 & 1 & -2 \end{array}\right]$$

   Switch rows two and three to get the 1 into row two, column two:

   $$\left[\begin{array}{ccc|c} 1 & -2 & 3 & 13 \\ 0 & 5 & -7 & -22 \\ 0 & 1 & 1 & -2 \end{array}\right] r_2 \leftrightarrow r_3 \left[\begin{array}{ccc|c} 1 & -2 & 3 & 13 \\ 0 & 1 & 1 & -2 \\ 0 & 5 & -7 & -22 \end{array}\right]$$

   Perform the operations indicated to create 0s above and below the 1 in row two, column two:

   $$\left[\begin{array}{ccc|c} 1 & -2 & 3 & 13 \\ 0 & 1 & 1 & -2 \\ 0 & 5 & -7 & -22 \end{array}\right] \begin{array}{c} 2r_2 + r_1 \to r_1 \\ -5r_2 + r_3 \to r_3 \end{array} \left[\begin{array}{ccc|c} 1 & 0 & 5 & 9 \\ 0 & 1 & 1 & -2 \\ 0 & 0 & -12 & -12 \end{array}\right]$$

   Multiply the terms in row three by $-\frac{1}{12}$:

   $$\left[\begin{array}{ccc|c} 1 & 0 & 5 & 9 \\ 0 & 1 & 1 & -2 \\ 0 & 0 & -12 & -12 \end{array}\right] -\frac{1}{12}r_3 \to r_3 \left[\begin{array}{ccc|c} 1 & 0 & 5 & 9 \\ 0 & 1 & 1 & -2 \\ 0 & 0 & 1 & 1 \end{array}\right]$$

   From row three, you have $z = 1$. From row two, you have $y + z = -2$. Insert the value of z to find $y + 1 = -2 \to y = -3$. And, with row one, $x + 5z = 9 \to x + 5(1) = 9 \to x = 4$.

2. $\begin{bmatrix} \mathbf{4} & \mathbf{-12} & \mathbf{-11} \\ \mathbf{14} & \mathbf{-1} & \mathbf{13} \end{bmatrix}$

   $$2A + 3B = 2\begin{bmatrix} 2 & 3 & -1 \\ 4 & -2 & 5 \end{bmatrix} + 3\begin{bmatrix} 0 & -6 & -3 \\ 2 & 1 & 1 \end{bmatrix} = \begin{bmatrix} 4 & 6 & -2 \\ 8 & -4 & 10 \end{bmatrix} + \begin{bmatrix} 0 & -18 & -9 \\ 6 & 3 & 3 \end{bmatrix} = \begin{bmatrix} 4 & -12 & -11 \\ 14 & -1 & 13 \end{bmatrix}$$

3. **14**

   $$\begin{vmatrix} 3 & -2 \\ 1 & 4 \end{vmatrix} = 3(4) - (-2)(1) = 12 + 2 = 14$$

4. **$(z + 9,\ 5z + 6,\ z)$**

   From row two, $y - 5z = 6 \to y = 5z + 6$. And from row one: $x - y + 4z = 3 \to x - (5z + 6) + 4z = 3 \to x - 5z - 6 + 4z = 3 \to x = z + 9$

(5) $\begin{bmatrix} \mathbf{4} & \mathbf{1} \\ \mathbf{5} & \mathbf{2} \\ \mathbf{1} & \mathbf{2} \end{bmatrix}$

$$CD = \begin{bmatrix} 1 & 0 & 2 \\ 2 & -3 & 1 \\ 0 & 1 & -1 \end{bmatrix} \cdot \begin{bmatrix} 4 & 3 \\ 1 & 1 \\ 0 & -1 \end{bmatrix} = \begin{bmatrix} 1(4)+0(1)+2(0) & 1(3)+0(1)+2(-1) \\ 2(4)+(-3)(1)+1(0) & 2(3)+(-3)(1)+1(-1) \\ 0(4)+1(1)+(-1)(0) & 0(3)+1(1)+(-1)(-1) \end{bmatrix} = \begin{bmatrix} 4 & 1 \\ 5 & 2 \\ 1 & 2 \end{bmatrix}$$

(6) $\boldsymbol{x = -1, y = 1, z = 3}$

Set up the system as an augmented matrix: $\left[\begin{array}{ccc|c} 3 & -1 & 2 & 2 \\ 1 & 1 & 1 & 3 \\ -4 & 1 & -1 & 2 \end{array}\right]$.

Switch rows one and two so there's a 1 in row one, column one.

$$\left[\begin{array}{ccc|c} 3 & -1 & 2 & 2 \\ 1 & 1 & 1 & 3 \\ -4 & 1 & -1 & 2 \end{array}\right] r_1 \leftrightarrow r_2 \left[\begin{array}{ccc|c} 1 & 1 & 1 & 3 \\ 3 & -1 & 2 & 2 \\ -4 & 1 & -1 & 2 \end{array}\right]$$

Perform the operations indicated to create 0s below the 1 in row one, column one:

$$\left[\begin{array}{ccc|c} 1 & 1 & 1 & 3 \\ 3 & -1 & 2 & 2 \\ -4 & 1 & -1 & 2 \end{array}\right] \begin{array}{l} -3r_1 + r_2 \to r_2 \\ 4r_1 + r_3 \to r_3 \end{array} \left[\begin{array}{ccc|c} 1 & 1 & 1 & 3 \\ 0 & -4 & -1 & -7 \\ 0 & 5 & 3 & 14 \end{array}\right]$$

Multiply row two through by $-\frac{1}{4}$:

$$\left[\begin{array}{ccc|c} 1 & 1 & 1 & 3 \\ 0 & -4 & -1 & -7 \\ 0 & 5 & 3 & 14 \end{array}\right] -\frac{1}{4}r_2 \to r_2 \left[\begin{array}{ccc|c} 1 & 1 & 1 & 3 \\ 0 & 1 & \frac{1}{4} & \frac{7}{4} \\ 0 & 5 & 3 & 14 \end{array}\right]$$

Perform the operations indicated to create 0s above and below the 1 in row two, column two:

$$\left[\begin{array}{ccc|c} 1 & 1 & 1 & 3 \\ 0 & 1 & \frac{1}{4} & \frac{7}{4} \\ 0 & 5 & 3 & 14 \end{array}\right] \begin{array}{l} -1r_2 + r_1 \to r_1 \\ -5r_2 + r_3 \to r_3 \end{array} \left[\begin{array}{ccc|c} 1 & 0 & \frac{3}{4} & \frac{5}{4} \\ 0 & 1 & \frac{1}{4} & \frac{7}{4} \\ 0 & 0 & \frac{7}{4} & \frac{21}{4} \end{array}\right]$$

Multiply row three through by $\frac{4}{7}$:

$$\left[\begin{array}{ccc|c} 1 & 0 & \frac{3}{4} & \frac{5}{4} \\ 0 & 1 & \frac{1}{4} & \frac{7}{4} \\ 0 & 0 & \frac{7}{4} & \frac{21}{4} \end{array}\right] \frac{4}{7}r_3 \to r_3 \left[\begin{array}{ccc|c} 1 & 0 & \frac{3}{4} & \frac{5}{4} \\ 0 & 1 & \frac{1}{4} & \frac{7}{4} \\ 0 & 0 & 1 & 3 \end{array}\right]$$

Perform the operations indicated to create 0s above and below the 1 in row three, column three:

$$\left[\begin{array}{cc c|c} 1 & 0 & \frac{3}{4} & \frac{5}{4} \\ 0 & 1 & \frac{1}{4} & \frac{7}{4} \\ 0 & 0 & 1 & 3 \end{array}\right] \begin{array}{l} -\frac{3}{4}r_3 + r_1 \to r_1 \\ -\frac{1}{4}r_3 + r_2 \to r_2 \end{array} \left[\begin{array}{ccc|c} 1 & 0 & 0 & -1 \\ 0 & 1 & 0 & 1 \\ 0 & 0 & 1 & 3 \end{array}\right]$$

7. $\boldsymbol{x = 1, y = -4}$. **First, find the value of the determinant:** $\begin{vmatrix} 1 & -3 \\ 2 & -1 \end{vmatrix} = 1(-1) - (-3)(2) = -1 + 6 = 5$. Then,

$$x = \frac{\begin{vmatrix} 13 & -3 \\ 6 & -1 \end{vmatrix}}{5} = \frac{-13-(-18)}{5} = \frac{5}{5} = 1 \text{ and } y = \frac{\begin{vmatrix} 1 & 13 \\ 2 & 6 \end{vmatrix}}{5} = \frac{6-26}{5} = \frac{-20}{5} = -4.$$

8. $\boldsymbol{\frac{1}{14}\begin{bmatrix} 4 & 2 \\ -1 & 3 \end{bmatrix}}$

Using $\dfrac{\begin{bmatrix} d & -b \\ -c & a \end{bmatrix}}{ad-bc}$, $A^{-1} = \dfrac{\begin{bmatrix} 4 & 2 \\ -1 & 3 \end{bmatrix}}{3(4)-(-2)(1)} = \dfrac{\begin{bmatrix} 4 & 2 \\ -1 & 3 \end{bmatrix}}{12+2} = \dfrac{1}{14}\begin{bmatrix} 4 & 2 \\ -1 & 3 \end{bmatrix}$.

IN THIS CHAPTER

» Exploring the terms and formulas of sequences

» Grasping arithmetic and geometric sequences

» Summing sequences to create a series

Chapter 20
Sequences and Series

You can breathe a sigh of relief: In this chapter, you get to put aside your graph paper and use even more mathematical topics from the real world. The real-world applications from previous chapters are useful only to special areas of interest. This chapter is different because the applications are useful to everyone. No matter who you are or what you do, you should understand the value of your belongings. For example, you may want to know your car's worth after a number of years and what your credit card or loan balance will be if you don't pay it on time. The focus here takes math out of the classroom and into the fresh air.

- **Sequences:** This application of pre-calculus helps you understand patterns. You can see how patterns develop, for example, in how much your car depreciates, how credit-card interest builds, and how scientists estimate the growth of bacterial populations.
- **Series:** Series help you understand the sum of a sequence of numbers, such as annuities, the total distance a ball bounces, how many seats are in a theater, and so on.

This chapter dives into these topics and debunks the myth that math isn't useful in the real world.

Speaking Sequentially: Grasping the General Method

A *sequence* is basically an ordered list of objects (numbers, letters, and so forth) following some sort of pattern. This pattern can usually be described by a general rule that allows you to determine any of the numbers in this list without having to find *all* the numbers in between.

A sequence can be infinite, meaning it can continue in the same pattern forever. A sequence's mathematical definition is a function defined over the set of positive integers, usually written in the following form: $\{a_n\} = a_1, a_2, a_3, \ldots, a_n, \ldots$

The $\{a_n\}$ notation represents the general term for the entire set of numbers. Each a_n is called a *term of the sequence;* a_1 is the first term, a_2 is the second term, and so on. The a_n is the nth term, meaning it can be any term you need it to be. The values of n are always positive integers.

In the real world, sequences are helpful when describing a quantity that increases or decreases with time — financial interest, debt, sales, populations, and asset depreciation or appreciation, to name a few. All quantities that change with time based on a certain percentage follow a pattern that can be described using a sequence. Depending on the rule for a particular sequence, you can multiply the initial value of an object by a certain percentage to find a new value after a certain length of time. Repeating this process reveals the general pattern and the change in value for the object.

Determining a sequence's terms

TECHNICAL STUFF

The general formula or rule used to determine the terms of a sequence is usually referred to as the *general term* and is denoted with something like $\{a_n\}$. This rule involves the letter n, which is the number of the term (the first term would be $n = 1$, and the 20th term would be $n = 20$). You can find any term of a sequence by plugging the number of the term you want into the expression representing the general term. The formula associated with the general term gives you specific instructions on what to do with the value you're plugging in. If you're given a few terms of a sequence, you can use these terms to find the general term for the sequence. If you're given the general term (complete with n as the variable), you can find any term by plugging in the number of the term you want for n.

REMEMBER

Unless otherwise noted, the first term of any sequence $\{a_n\}$ begins with $n = 1$. The next n always goes up by 1.

EXAMPLE

Q. Find the first three terms of $a_n = (-1)^{n-1} \cdot (n^2)$.

A. Use the following steps:

1. **Find a_1 first by plugging in 1 wherever you see n.**

$$a_1 = (-1)^{1-1} \cdot (1^2) = (-1)^0 \cdot (1) = 1 \cdot 1 = 1$$

2. **Continue plugging in consecutive integers for n.**

This process gives you terms two and three:

$$a_2 = (-1)^{2-1} \cdot (2^2) = (-1)^1 \cdot (4) = -1 \cdot 4 = -4$$

$$a_3 = (-1)^{3-1} \cdot (3^2) = (-1)^2 \cdot (9) = 1 \cdot 9 = 9$$

The first three terms of this sequence are: 1, −4, 9. If you continued on further, you'd find it to be an *alternating sequence,* where every other term is negative.

YOUR TURN

1 **Find the first three terms of the sequence $a_n = 3^{n-1}$.**	**2** **Find terms 100, 101, and 102 of the sequence $a_n = \frac{n-1}{2n+2}$.**

Working in reverse: Forming an expression from terms

If you know the first few terms of a sequence, you can usually write a mathematical expression for the general term. To write the formula for this term, you must look for a pattern in the first few terms of the sequence, which demonstrates logical thinking (and we all want to be logical thinkers, right?). The formula you write must work for every integer value of n, starting with $n = 1$.

Sometimes this calculation is an easy task, and sometimes it's less apparent and more complicated. Sequences involving fractions and/or exponents tend to be more complicated and less obvious in their patterns. The easy ones to write include addition, subtraction, multiplication, or division by integers.

EXAMPLE

Q. Find the general term for the nth term of the sequence $\frac{2}{3}, \frac{3}{5}, \frac{4}{7}, \frac{5}{9}, \frac{6}{11}, \ldots$

A. You should look at the numerator and the denominator separately:

- The numerators begin with 2 and increase by one each time. This sequence is described by $b_n = n + 1$.
- The denominators start with 3 and increase by two each time. This sequence is described by $c_n = 2n + 1$.

Therefore, the terms in this sequence can be expressed using $\frac{b_n}{c_n}$: $a_n = \frac{n+1}{2n+1}$.

To double-check your formula and ensure that the answers work, plug in 1, 2, 3, and so on to make sure you get the original numbers from the given sequence.

$n = 1,\ a_1 = \frac{1+1}{2(1)+1} = \frac{2}{3} \qquad n = 2,\ a_2 = \frac{2+1}{2(2)+1} = \frac{3}{5}$

$n = 3,\ a_3 = \frac{3+1}{2(3)+1} = \frac{4}{7} \qquad n = 4,\ a_4 = \frac{4+1}{2(4)+1} = \frac{5}{9}$

$n = 5,\ a_5 = \frac{5+1}{2(5)+1} = \frac{6}{11}$

They all work! And, of course, you really want to know the value of the hundredth term, so you use the formula and get: $n = 100,\ a_{100} = \frac{100+1}{2(100)+1} = \frac{101}{201}$.

YOUR TURN

 Find the general term for the nth term of the sequence $32, 8, 2, \frac{1}{2}, \ldots$

 Find the general term for the nth term of the sequence $1, 4, 7, 10, \ldots$

Recognizing recursive formulas: A type of general sequence

A *recursive sequence* is a sequence in which each term depends on the term before it. To find any term in a recursive sequence, you use the given term (at least one term, usually the first, is given for the problem) and the formula in the general term, allowing you to find the other terms.

You can recognize recursive sequences because the given formula typically has a_n (the nth or general term of the sequence) as well as a_{n-1} (the term before the nth term of the sequence).

EXAMPLE

Q. Find the general term for the Fibonacci sequence.

A. The most famous recursive sequence is the Fibonacci sequence, in which each term after the second term is defined as the sum of the two terms before it. The first term of this sequence is 1, and the second term is 1 also. The formula for the Fibonacci sequence includes the first two terms and then a_{n-1} (the term before the nth term of the sequence); $a_n = a_{n-2} + a_{n-1}$ for $n \geq 3$.

So, if you were asked to find the next three terms of this sequence, beginning with 1, 1, you'd use the formula as follows:

$$a_3 = a_{3-2} + a_{3-1} = a_1 + a_2 = 1 + 1 = 2$$
$$a_4 = a_{4-2} + a_{4-1} = a_2 + a_3 = 1 + 2 = 3$$
$$a_5 = a_{5-2} + a_{5-1} = a_3 + a_4 = 2 + 3 = 5$$

The first ten terms of this sequence are 1, 1, 2, 3, 5, 8, 13, 21, 34, 55. This sequence is very famous because many things in the natural world follow the pattern of the Fibonacci sequence. For example, the florets in the head of a sunflower form two oppositely directed spirals, 55 of them clockwise and 34 counterclockwise. Seeds of coneflowers and sunflowers have also been observed to follow the same pattern as the Fibonacci sequence. And the Golden Ratio is found in consecutive terms of the Fibonacci sequence. When dividing the a_n term by the a_{n-1} term, the larger n is, the more decimal places of the Golden Ratio are found.

EXAMPLE

Q. Find the recursive formula for the sequence 5, 10, 20, 40, 80, ...

A. Each term after the first term is twice the previous term, so you can say that $a_1 = 5$ and $a_{n+1} = 2a_n$ for $n \geq 1$.

Checking, you have:

$a_1 = 5$

$a_2 = 2(a_1) = 2(5) = 10$

$a_3 = 2(a_2) = 2(10) = 20$

$a_4 = 2(a_3) = 2(20) = 40$

$a_5 = 2(a_4) = 2(40) = 80$

$a_6 = 2(a_5) = 2(80) = 160, \ \ldots$

YOUR TURN

5 Find the recursive formula for the sequence 15, 215, 415, 615, 815, ...

6 Find the recursive formula for the sequence 2, 4, 7, 11, 16, ...

Finding the Difference between Terms: Arithmetic Sequences

One of the most common types of sequences is called an *arithmetic sequence.* In an arithmetic sequence, each term differs from the one before it by the same number, called the *common difference.* To determine whether a sequence is arithmetic, you subtract each term by its preceding term; if the difference between each set of terms is the same, the sequence is arithmetic.

Arithmetic sequences have the same format for the *n*th or general term: $a_n = a_1 + (n-1)d$, where a_1 is the first term and d is the common difference.

Q. What are the first ten terms of the arithmetic sequence, where $a_n = -6 + (n-1)2$?

EXAMPLE **A.** You start with $n = 1$ and end with $n = 10$, to create the list $-6, -4, -2, 0, 2, 4, 6, 8, 10, 12$. You can see the difference of 2 between the terms just standing up and shouting at you.

Q. Find the general term for the arithmetic sequence 41, 36, 31, 26, 21, ...

EXAMPLE **A.** Each term is 5 less than the previous term, so the common difference $d = -5$. The first term is 41, so the general term is $a_n = 41 + (n-1)(-5)$. You may also prefer to simplify this general term by distributing the difference and combining like terms: $a_n = 41 + (n-1)(-5) \rightarrow a_n = 41 - 5n + 5 \rightarrow a_n = 46 - 5n$.

You may encounter two other types of arithmetic-sequence problems: one where you're given a list of consecutive terms, and one where you're given two terms that are not consecutive (but you're told which terms they are). In the next two sections, you see how to handle each of these.

Using consecutive terms to find another

If you're given two consecutive terms of an arithmetic sequence, the common difference between these terms is not too far away. And you can write the terms of the sequence for as long as you want. But what if you need to find the 100th or 1,000th term of a sequence? Yikes!

Q. Given the arithmetic sequence $-7, -4, -1, 2, 5, \ldots$, you want to find the 55th term.

EXAMPLE **A.** You can continue the pattern begun by the first few terms 50 more times. However, that process would be very time-consuming and not very effective to find terms that come later in the sequence.

Instead, create a formula to find any term of the arithmetic sequence. Finding the formula for the *n*th term of an arithmetic sequence is easy as long as you know the first term and the common difference. See the following steps:

1. **Find the common difference.**

   To find the common difference, simply subtract one term from the one after it. In this case, you see the first term subtracted from the second term: $-4 - (-7) = 3$. The common difference is 3. You'll get this same result, no matter which pair of terms you choose.

2. **Plug a_1 and d into the formula to write the specific general term for the given sequence.**

   Start with: $a_n = a_1 + (n-1)d$

   Then plug in what you know: The first term of the sequence is -7, and the common difference is 3: $a_n = -7 + (n-1)3 = -7 + 3n - 3 = 3n - 10$.

3. **Plug in the number of the term you're trying to find for *n*.**

   To find the 55th term, plug 55 in for n into the formula for a_n. Using $a_n = 3n - 10$, you find $a_{55} = 3(55) - 10 = 165 - 10 = 155$.

Using any two terms

At times you'll need to find the general term for the *n*th term of an arithmetic sequence without knowing the first term or the common difference. In this case, you're given two terms (not necessarily consecutive), and you use this information to find a_1 and *d*. Your steps are still the same: Find the common difference, write the specific rule for the given sequence, and then find the term you're looking for.

EXAMPLE

Q. Find the general term of an arithmetic sequence, where $a_4 = -23$ and $a_{22} = 40$.

A. Follow these steps:

1. Find the common difference.

You have to be more creative in finding the common difference for these types of problems.

a. Use the formula $a_n = a_1 + (n-1)d$ to set up two equations that use the given information.

For the first equation, you know that when $n = 4$, $a_n = -23$:
$-23 = a_1 + (4-1)d \rightarrow -23 = a_1 + 3d$.

For the second equation, you know when $n = 22$, $a_n = 40$:
$40 = a_1 + (22-1)d \rightarrow 40 = a_1 + 21d$.

b. Set up a system of equations (see Chapter 18) and solve for *d*.

The system looks like this:

$$\begin{cases} a_1 + 3d = -23 \\ a_1 + 21d = 40 \end{cases}$$

You can use elimination or substitution to solve the system (you can refer to Chapter 18). Elimination works nicely because you can multiply the top equation by -1 and add the two together to get $18d = 63$. From this, you get that $d = 3.5$.

2. Find the value of a_1.

Plug *d* into one of the equations in Step 1b to solve for a_1. You can plug 3.5 back into either equation; using the top equation: $a_1 + 3(3.5) = -23 \rightarrow a_1 + 10.5 = -23 \rightarrow a_1 = -33.5$.

3. Write the general term.

Replacing the values you found in $a_n = a_1 + (n-1)d$, you have $a_n = -33.5 + (n-1)3.5$, which simplifies to $a_n = 3.5n - 37$. Using this, you can find any other term in the sequence. For example, the 20th term is $a_{20} = 3.5(20) - 37 = 70 - 37 = 33$.

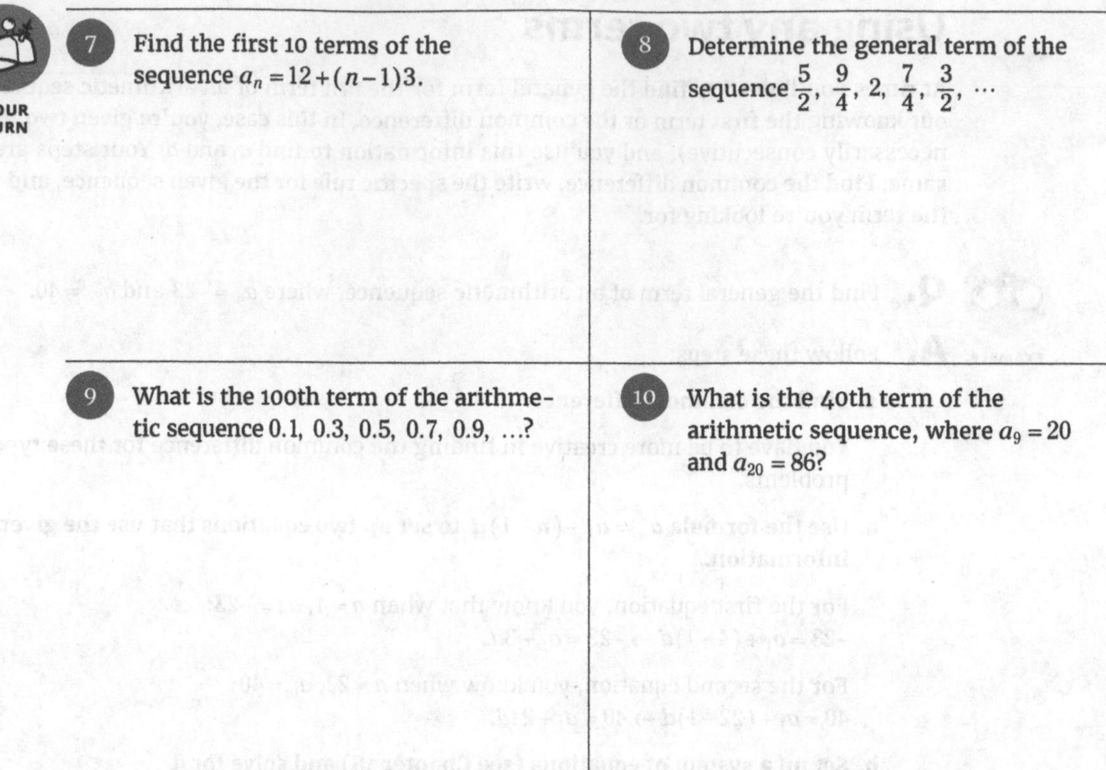

YOUR TURN

7. Find the first 10 terms of the sequence $a_n = 12 + (n-1)3$.

8. Determine the general term of the sequence $\frac{5}{2}, \frac{9}{4}, 2, \frac{7}{4}, \frac{3}{2}, \ldots$

9. What is the 100th term of the arithmetic sequence 0.1, 0.3, 0.5, 0.7, 0.9, …?

10. What is the 40th term of the arithmetic sequence, where $a_9 = 20$ and $a_{20} = 86$?

Looking at Ratios and Consecutive Paired Terms: Geometric Sequences

A *geometric sequence* is one in which consecutive terms have a common ratio. In other words, if you divide any term by the term before it, the quotient, denoted by the letter *r*, is the same.

Certain objects, such as cars, depreciate with time. And, often, this depreciation rate is constant. You can describe this depreciation using a geometric sequence. The common ratio is always the rate as a percent (sometimes called APR, which stands for annual percentage rate). Finding the value of the car at any time, as long as you know its original value, is fairly easy to do. The following sections show you how to identify the terms and expressions of geometric sequences, which allow you to apply the sequences to real-world situations (such as trading in your car!).

Embracing some basic concepts

In the following work with geometric sequences, you find how to determine a term in the sequence as well as how to write the formula for the specific sequence when you're not given the formula. But first, here are some general ideas to remember.

REMEMBER

Denote the first term of a geometric sequence as g_1. To find the second term of a geometric sequence, multiply the first term by the common ratio, r. You follow this pattern infinitely to find any term of a geometric sequence:

$$\{g_n\} = g_1, g_1 \cdot r, g_1 \cdot r^2, g_1 \cdot r^3, \ldots, g \cdot r^{n-1}, \ldots$$

TECHNICAL STUFF

More simply put, the formula for the nth term of a geometric sequence is $g_n = g_1 r^{n-1}$.

EXAMPLE

Q. In the formula, $5, \frac{5}{2}, \frac{5}{4}, \frac{5}{8}, \frac{5}{16}, \ldots$ write the general term of the sequence.

A. The first term is 5. The second term is $\frac{1}{2}$ the first term, the third term is $\frac{1}{2}$ the second term, and so on. So the general term is written: $g_n = 5\left(\frac{1}{2}\right)^{n-1}$.

The steps for dealing with geometric sequences are remarkably similar to those in the arithmetic sequence sections. You find the common ratio (not the difference!), you write the specific formula for the given sequence, and then you find the term you're looking for.

EXAMPLE

Q. Find the 15th term of the geometric sequence 2, 4, 8, 16, 32.

A. To find the 15th term, follow these steps:

1. **Find the common ratio.**

   In this sequence, each consecutive term is twice the previous term. If you can't see the common difference by looking at the sequence, divide any term by the term before it. For example, divide the fourth term by the third: $\frac{16}{8} = 2$. So $r = 2$.

2. **Find the general term for the given sequence.**

   The first term, $g_1 = 2$, and the ratio $r = 2$. So the general term for this sequence is $g_n = 2 \cdot 2^{n-1}$, which simplifies (using the rules of exponents) to $g_n = 2^1 \cdot 2^{n-1} = 2^n$.

3. **Find the term you're looking for.**

   Since $g_n = 2^n$, $g_{15} = 2^{15} = 32{,}768$.

Going out of order: Dealing with nonconsecutive terms

If you know any two nonconsecutive terms of a geometric sequence, you can use this information to find the general term of the sequence as well as any specified term.

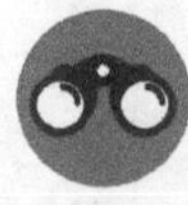

EXAMPLE

Q. If the 5th term of a geometric sequence is 64 and the 10th term is 2, what is the 15th term?

A. Just follow these steps:

1. **Determine the value of *r*.**

   You can use the general geometric formula $g_n = g_1 r^{n-1}$ to create a system of two equations:

   $$g_5 = g_1 \cdot r^{5-1} = 64 \text{ and } g_{10} = g_1 \cdot r^{10-1} = 2, \text{ or } \begin{cases} g_1 r^4 = 64 \\ g_1 r^9 = 2 \end{cases}$$

You can use substitution to solve the first equation for g_1 (see Chapter 18 for more on this method of solving systems):

$r^4 g_1 = 64$ becomes $g_1 = \frac{64}{r^4}$.

Plug this expression in for g_1 in the second equation: $r^9\left(\frac{64}{r^4}\right) = 2.$

Simplifying the equation and solving for r by taking the fifth root of each side:

$r^9\left(\frac{64}{r^4}\right) = 2 \rightarrow r^5(64) = 2 \rightarrow r^5 = \frac{2}{64} = \frac{1}{32} \rightarrow r = \frac{1}{2}.$

2. Write the specific formula for the given sequence.

a. Plug r into one of the equations to find g_1.

This step gives you $r^4 g_1 = 64 \rightarrow \left(\frac{1}{2}\right)^4 g_1 = 64 \rightarrow \frac{1}{16} g_1 = 64 \rightarrow \frac{16}{1} \cdot \frac{1}{16} g_1 = \frac{16}{1} \cdot 64$ $\rightarrow g_1 = 1{,}024.$

b. Plug g_1 and r into the general term $g_n = g_1 r^{n-1}$.

$g_n = 1{,}024\left(\frac{1}{2}\right)^{n-1}$

3. Find the term you're looking for.

In this case, you want to find the 15th term ($n = 15$):

$g_{15} = 1{,}024\left(\frac{1}{2}\right)^{15-1} = 1{,}024\left(\frac{1}{2}\right)^{14} = 1{,}024\left(\frac{1}{16{,}384}\right) = \frac{1{,}024}{16{,}384} = \frac{1}{16}$

TECHNICAL STUFF

The annual depreciation of a car's value is approximately 30 percent. Every year, the car is actually worth 70 percent of its value from the year before. If g_1 represents the value of a car when it was new and n represents the number of years that have passed, then $g_n = g_1(0.70)^n$ when $n \geq 0$. Notice that this sequence starts with $n = 0$ rather than $n = 1$, which is okay because this allows for the original value to be the price when bought.

YOUR TURN

11 Find the first 10 terms of the geometric sequence, where $g_1 = 3$ and $r = -2$.

12 Find the general term of the sequence

$5, \frac{5}{2}, \frac{5}{4}, \frac{5}{8}, \frac{5}{16}, \ldots$

What is the general term of the geometric sequence, where $g_7 = 384$ and $g_{10} = 3072$?

If the annual depreciation of a boat's value is approximately 20%, then what is its value after 10 years if you bought it for $125,000?

Creating a Series: Summing Terms of a Sequence

A *series* is the sum of the terms in a sequence. Except for situations where you can actually determine the sum of an infinite series, you are usually asked to find the sum of a certain number of terms (the first 12, for example). Summing a sequence is especially helpful in calculus when you begin discussing integration, which is used to find the areas under curves. Finding the area of a rectangle is easy, but because curves aren't straight, finding the area under them isn't as easy. You can find the area under a curve by breaking up the region in question into very small rectangles and adding them together. This process is somewhat like finding the sum of the terms in a sequence.

Reviewing general summation notation

The sum of the first *k* terms of a sequence is referred to as the *kth partial sum*. They're called partial sums because you only need to find the sum of a certain number of terms — no infinite series here! You may use partial sums when you want to find the area under a curve (graph) between two certain values of *x*.

REMEMBER

Don't let the use of a different variable here confuse you. Switching to *k* for the number of the term, alerts you to the fact that you're dealing with a partial sum. Remember that a variable stands in for an unknown, so it really can be any variable you want — even those Greek variables that are used in the trig chapters.

TECHNICAL STUFF

The notation of the *k*th partial sum of a sequence is as follows: $\sum_{n=1}^{k} a_n = a_1 + a_2 + a_3 + \cdots + a_k$

You read this equation as "the kth partial sum of a_n is . . .", where $n=1$ is the *lower limit* of the sum and k is the *upper limit* of the sum. To find the kth partial sum, you begin by plugging the lower limit into the general term and continue in order, plugging in integers until you reach the upper limit of the sum. At that point, you simply add all the terms to find the sum.

Q. Find the fifth partial sum of $a_n=n^3-4n+2$.

EXAMPLE **A.** Writing this as $\sum_{n=1}^{5}(n^3-4n+2)$, follow these steps:

1. **Plug all values of n (starting with 1 and ending with k) into the formula.**

   Because you want to find the fifth partial sum, plug in 1, 2, 3, 4, and 5:

   $$a_1=1^3-4(1)+2=1-4+2=-1$$
   $$a_2=2^3-4(2)+2=8-8+2=2$$
   $$a_3=3^3-4(3)+2=27-12+2=17$$
   $$a_4=4^3-4(4)+2=64-16+2=50$$
   $$a_5=5^3-4(5)+2=125-20+2=107$$

2. **Add all the values from a_1 to a_k to determine the sum.**

   You find: $\sum_{n=1}^{5}(n^3-4n+2)=-1+2+17+50+107=175$.

Summing an arithmetic sequence

The kth partial sum of an arithmetic sequence still calls for you to add the first k terms. But with an arithmetic sequence, you do have a formula to use instead of plugging in each of the values for n. The kth partial sum of an arithmetic series is $S_k=\sum_{n=1}^{k}a_n=\frac{k}{2}(a_1+a_k)$. You simply plug the lower and upper limits into the formula for a_n to find a_1 and a_k.

One real-world application of an arithmetic sum involves stadium seating.

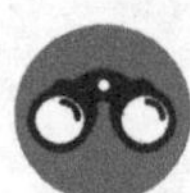

Q. A stadium has 35 rows of seats; there are 20 seats in the first row, 21 seats in the second row, 22 seats in the third row, and so on. How many seats do all 35 rows contain?

EXAMPLE **A.** Follow these steps to find out:

1. **Find the first term of the sequence.**

   The first term of this sequence (or the number of seats in the first row) is given: 20.

2. **Find the kth term of the sequence.**

   Because the stadium has 35 rows, find a_{35}. Use the formula for the nth term of an arithmetic sequence (see the earlier section, "Finding the Difference between Terms: Arithmetic Sequences"). The first term is 20, and each row has one more seat than the row before it, so $d=1$. Plug these values into the formula:

   $$a_{35}=a_1+(35-1)d=20+34\cdot1=54$$

   Note: This gives you the number of seats in the 35th row, not the answer to how many seats the stadium contains.

3. Use the formula for the kth partial sum of an arithmetic sequence to find the sum.

$$S_{35}=\frac{35}{2}(a_1+a_{35})=\frac{35}{2}(20+54)=\frac{35}{2}(74)=1{,}295$$

Seeing how a geometric sequence adds up

As well as the arithmetic sequence and its special formula for a sum, you also have a formula for the sum of a geometric sequence — well, actually, there are two formulas.

The first formula computes a finite sum (comparable to a kth partial sum from the previous section), and it too has an upper limit and a lower limit. The common ratio of partial sums of this type has no specific restrictions.

The second type of geometric sum is called an *infinite* geometric sum, and the common ratio for this type is very specific (it *must* be strictly between −1 and 1). This type of geometric sequence is very helpful if you drop a ball and count how far it travels up and down, and then up and down, until it finally starts rolling. By definition, a geometric series continues infinitely, for as long as you want to keep plugging in values for n. However, in this special type of geometric series, no matter how long you plug in values for n, the sum never gets larger than a certain value. This type of series has a specific formula to find the infinite sum. In mathematical terms, you say that some geometric sequences (ones with a common ratio between −1 and 1) have a limit to their sequence of partial sums. In other words, the partial sum comes closer and closer to a particular number without ever actually reaching it. You call this number the *sum of the sequence*, as opposed to the kth partial sum you find in previous sections in this chapter.

Stop right there: Determining the partial sum of a finite geometric sequence

You can find a partial sum of a geometric sequence by using the following formula. The k represents the number of terms being added, g_1 is the first term, and r is the ratio.

$$S_k=\sum_{n=1}^{k}g_1\cdot r^{n-1}=g_1\left(\frac{1-r^k}{1-r}\right)$$

EXAMPLE

Q. Find the sum of the first seven terms: $\sum_{n=1}^{7}9\left(-\frac{1}{3}\right)^{n-1}$.

A. Follow these steps:

1. Find g_1 and r.

The sequence formula is in standard form, so you can just read off the values of g_1 and r from the problem statement. You see that $g_1=9$ and $r=-\frac{1}{3}$.

If the format hadn't been exactly in the general form, you could solve for g_1 by replacing the n with 1, and then you solve for r by finding the first two terms and dividing the second term by the first term.

2. **Plug g_1, r, and k into the sum formula.**

$$\sum_{n=1}^{7} 9\left(-\frac{1}{3}\right)^{n-1} = 9\left(\frac{1-\left(-\frac{1}{3}\right)^7}{1-\left(-\frac{1}{3}\right)}\right).$$ Simplifying,

$$9\left(\frac{1-\left(-\frac{1}{3}\right)^7}{1-\left(-\frac{1}{3}\right)}\right) = 9\left(\frac{1-\left(-\frac{1}{2{,}187}\right)}{1+\frac{1}{3}}\right) = 9\left(\frac{1+\frac{1}{2{,}187}}{1+\frac{1}{3}}\right) = 9\left(\frac{\frac{2{,}188}{2{,}187}}{\frac{4}{3}}\right) = 9\cdot\frac{2{,}188}{2{,}187}\cdot\frac{3}{4} = \frac{547}{81}$$

Geometric summation problems take a bit of work when they involve fractions, so make sure to find a common denominator, invert, and multiply when necessary. Or you can use a calculator and then reconvert to a fraction. Just be careful to use correct parentheses when entering the numbers.

To geometry and beyond: Finding the value of an infinite sum

Finding the value of an infinite sum in a geometric sequence is actually quite simple — as long as you keep your fractions and decimals straight. If r lies outside the range $-1 < r < 1$, g_n grows without bound, so there's no limit on how large the absolute value of g_n ($|g_n|$) can get — so the sum also keeps growing. If $|r| < 1$, for every value of n, $|r^n|$ continues to decrease infinitely until it becomes arbitrarily close to 0. This decrease occurs because when you multiply a fraction between –1 and 1 by itself, the absolute value of that fraction continues to get smaller until it becomes so small that you hardly notice it. Therefore, the term r^k almost disappears completely in the finite geometric sum formula: $S_k = \sum_{n=1}^{k} g_1 \cdot r^{n-1} = g_1\left(\frac{1-r^k}{1-r}\right)$.

So if the r^k disappears — gets closer and closer to 0 — the finite formula changes to the following and allows you to find the sum of an infinite geometric series: $S_k = \sum_{n=1}^{\infty} g_1 \cdot r^{n-1} = g_1\left(\frac{1-0}{1-r}\right) = \frac{g_1}{1-r}$

EXAMPLE

Q. Find: $\sum_{n=1}^{\infty} 4\left(\frac{2}{5}\right)^{n-1}$

A. Use the new formula and follow these steps:

1. **Find g_1 and r.**

   Reading from the general term, $g_1 = 4$ and $r = \frac{2}{5}$.

2. **Plug g_1 and r into the formula to find the infinite sum.**

   $$\sum_{n=1}^{\infty} 4\left(\frac{2}{5}\right)^{n-1} = \frac{4}{1-\frac{2}{5}} = \frac{4}{\frac{3}{5}} = 4\cdot\frac{5}{3} = \frac{20}{3}$$

Rewriting in general term format

When using the formula for the sum of a finite or infinite geometric series, you need the first term and the ratio. But what if the given expression isn't in this form?

EXAMPLE

Q. Find the sum of the first ten terms of the following: $\sum_{n=1}^{10} 9\left(\frac{2}{3}\right)^n$.

A. The general term isn't in standard form, because the exponent is n instead of $n-1$. You could perform some algebraic processes and put the expression in the correct format, or you can just get down-and-dirty and make use of what you have to find what you need.

All you have to do is find the first two terms from the expression given.

When $n=1$, $9\left(\frac{2}{3}\right)^1 = 6$, and when $n=2$, $9\left(\frac{2}{3}\right)^2 = 9\left(\frac{4}{9}\right) = 4$. The first term is $g_1 = 6$. And the ratio, r, is what you get when you divide the second term by the first term: $r = \frac{4}{6} = \frac{2}{3}$.

Using the formula for the first ten terms,

$$\sum_{n=1}^{10} 6\cdot\left(\frac{2}{3}\right)^{n-1} = 6\left(\frac{1-\left(\frac{2}{3}\right)^{10}}{1-\frac{2}{3}}\right) = 6\left(\frac{1-\frac{1,024}{59,049}}{1-\frac{2}{3}}\right) = 6\left(\frac{\frac{58,025}{59,049}}{\frac{1}{3}}\right) = 6\cdot\frac{58,025}{59,049}\cdot\frac{3}{1} = \frac{116,050}{6,561}$$

TECHNICAL STUFF

Repeating decimals also can be expressed as infinite sums. Consider the number 0.555555.... You can write this number as $0.5+0.05+0.005+0.0005+\ldots$, and so on forever. The first term of this sequence is 0.5; to find r, divide the second term by the first term; $\frac{0.05}{0.5} = 0.1$. Plug these values into the infinite sum formula:

$$\sum_{n=1}^{\infty} 0.5(0.1)^{n-1} = \frac{0.5}{1-0.1} = \frac{0.5}{0.9} = \frac{5}{9}$$

REMEMBER

This sum is finite only if r lies strictly between -1 and 1.

YOUR TURN

15 Find the partial sum of the series $\sum_{n=1}^{17}(3+2n)$.

16 Find the sum of the first 8 terms of the series $\sum_{n=1}^{8} 2\left(\frac{1}{2}\right)^{k-1}$.

17 Find the sum of the series $\sum_{n=1}^{\infty}\left(\frac{1}{4}\right)^{k-1}$.

18 Find the sum of the first 10 terms of the series $\sum_{n=1}^{10} 2^k$.

Practice Questions Answers and Explanations

1 **1, 3, 9.** Given $a_n = 3^{n-1}$: $a_1 = 3^{1-1} = 3^0 = 1$, $a_2 = 3^{2-1} = 3^1 = 3$, $a_3 = 3^{3-1} = 3^2 = 9$.

2 $\mathbf{\frac{99}{202}, \frac{25}{51}, \frac{101}{206}}$. Given $a_n = \frac{n-1}{2n+2}$: $a_{100} = \frac{100-1}{2(100)+2} = \frac{99}{202}$, $a_{101} = \frac{101-1}{2(101)+2} = \frac{100}{204} = \frac{25}{51}$, $a_{102} = \frac{102-1}{2(102)+2} = \frac{101}{206}$

3 $\mathbf{a_n = 32\left(\frac{1}{4}\right)^{n-1}}$. **Each term is $\frac{1}{4}$ the size of the previous term, and $a_1 = 32$.**

4 $\mathbf{a_n = 3n-2}$ **or** $\mathbf{a_1 = 1,\ a_{n+1} = a_n + 3}$ **when** $\mathbf{n > 1}$**.** The next term in the sequence is found by adding 3 to the previous term. The first term is 1. An explicit version of this rule is $a_n = 3n-2$.

5 $\mathbf{a_n = 200n-185}$ **or** $\mathbf{a_1 = 15,\ a_{n+1} = a_n + 200}$ **when** $\mathbf{n > 1}$**.** The first term is 15, and the next term is found by adding 200, so $a_1 = 15$ and $a_{n+1} = a_n + 200$ when $n > 1$. An explicit version of this rule is $a_n = 200n - 185$.

6 $\mathbf{a_n = 3n-2}$ **or** $\mathbf{a_1 = 2,\ a_{n+1} = a_n + (n+1)}$**.** The difference between the terms increases by 1 each time. You see +2, +3, +4, . . . between the steps. An explicit version is $a_n = 3n-2$.

7 **12, 15, 18, 21, 24, 27, 30, 33, 36, 39.** Apply the formula $a_n = 12 + (n-1)3$ to the numbers 1 through 10.

8 $\mathbf{a_n = \frac{5}{2} + (n-1)\left(-\frac{1}{4}\right)}$ **or** $\mathbf{a_n = \frac{11}{4} - \frac{1}{4}n}$**.** The first term is $\frac{5}{2}$. Changing the given terms to fractions with a denominator of 4, you have: $\frac{10}{4}, \frac{9}{4}, \frac{8}{4}, \frac{7}{4}, \frac{6}{4}, \cdots$, so you see the difference of $-\frac{1}{4}$. Using the standard form for an arithmetic sequence, $a_n = a_1 + (n-1)d$, you can write $a_n = \frac{5}{2} + (n-1)\left(-\frac{1}{4}\right)$. Distributing and simplifying, you get the simpler version.

9 **19.9.** First, determine the general form for this sequence. The first term is 0.1, and the difference between the terms is 0.2. Using $a_n = a_1 + (n-1)d$, you have $a_n = 0.1 + (n-1)0.2 = 0.2n - 0.1$. To find the 100th term, replace n with 100: $a_n = 0.1 + (100-1)0.2 = 0.1 + 99(0.2) = 0.1 + 19.8 = 19.9$.

10 **206.** First, write the two equations created with the terms and the standard form of an arithmetic sequence. You have: $\begin{cases} a_9 = 20 \\ a_{20} = 86 \end{cases} \rightarrow \begin{cases} 20 = a_1 + (9-1)d \\ 86 = a_1 + (20-1)d \end{cases} \rightarrow \begin{cases} 20 = a_1 + 8d \\ 86 = a_1 + 19d \end{cases} \rightarrow \begin{cases} 20 - 8d = a_1 \\ 86 - 19d = a_1 \end{cases}$

Set the a_1 terms equal to one another and solve for d: $20 - 8d = 86 - 19d \rightarrow 11d = 66 \rightarrow d = 6$. Insert this into either equation, and you have $a_1 = -28$. So the general term is $a_n = -28 + (n-1)6$. Replacing n with 40, $a_{40} = -28 + (40-1)6 = -28 + 39(6) = 206$

11 **3, −6, 12, −24, 48, −96, 192, −384, 768, −1536.** Applying the general rule, you multiply each preceding term by −2.

12 $\mathbf{g_n = 5\left(\frac{1}{2}\right)^{n-1}}$. The first term is 5, and each subsequent term is multiplied by $\frac{1}{2}$.

13. $g_n = 6(2)^{n-1}$. First, write the two equations created with the terms and the standard form of a geometric sequence. You have: $\begin{cases} g_7 = 384 \\ g_{10} = 3072 \end{cases} \rightarrow \begin{cases} 384 = g_1(r)^{n-1} \\ 3072 = g_1(r)^{n-1} \end{cases} \rightarrow \begin{cases} 384 = g_1(r)^6 \\ 3072 = g_1(r)^9 \end{cases} \rightarrow \begin{cases} \dfrac{384}{r^6} = g_1 \\ \dfrac{3072}{r^9} = g_1 \end{cases}$

Set the g_1 terms equal to one another and solve for r. $\dfrac{384}{r^6} = \dfrac{3072}{r^9} \rightarrow \dfrac{r^9}{1} \cdot \dfrac{384}{r^6} = \dfrac{r^9}{1} \cdot \dfrac{3072}{r^9}$ $\rightarrow 384r^3 = 3072 \rightarrow r^3 = 8 \rightarrow r = 2$. Replace the value of r into either equation, and you have $g_1 = 6$. So the general form is $g_n = 6(2)^{n-1}$.

14. **\$16,777.22.** The value of the boat is 80% of what it was the previous year. Using a geometric sequence, you have $V_n = 125{,}000(0.80)^{n-1}$, where n is the number of years since the purchase of the boat. Letting $n = 10$, $V_{10} = 125{,}000(0.80)^9 = 16{,}777.216$.

15. **357.** Use the formula for the sum of an arithmetic series, $S_k = \sum_{n=1}^{k} a_n = \frac{k}{2}(a_1 + a_k)$. You insert the values $k = 17$, $a_1 = 5$, $a_{17} = 37$ and you have $S_{17} = \frac{17}{2}(5+37) = \frac{17}{2}(42) = 17(21) = 357$.

16. $\dfrac{255}{64}$. Use the formula for the sum of a geometric series, $S_k = g_1\left(\dfrac{1-r^k}{1-r}\right)$. Insert the values $k = 8$, $g_1 = 2$, $r = \frac{1}{2}$. You have $S_8 = 2\left(\dfrac{1-\left(\frac{1}{2}\right)^8}{1-\frac{1}{2}}\right) = 2\left(\dfrac{1-\frac{1}{256}}{\frac{1}{2}}\right) = 4\left(\dfrac{255}{256}\right) = \dfrac{255}{64} = 3.9984375$.

17. $\dfrac{4}{3}$. Use the formula for the sum of an infinite series $S_\infty = \dfrac{g_1}{1-r}$ using $g_1 = 1$, $r = \frac{1}{4}$ and you have $S_\infty = \dfrac{1}{1-\frac{1}{4}} = \dfrac{1}{3/4} = \dfrac{4}{3}$.

18. **2,046.** First, rewrite the summation in the standard form for finding the sum of a geometric series. Since $2^k = 2^1 \cdot 2^{k-1}$, you can say that $\sum_{n=1}^{10} 2^k = \sum_{n=1}^{10} 2(2)^{k-1}$, giving you $k = 10$, $g_1 = 2$, $r = 2$.

Inserting the values into the formula, $S_{10} = 2\left(\dfrac{1-2^{10}}{1-2}\right) = 2\left(\dfrac{1-1024}{-1}\right) = 2(1023) = 2046$.

If you're ready to test your skills a bit more, take the following chapter quiz that incorporates all the chapter topics.

Whaddya Know? Chapter 20 Quiz

Quiz time! Complete each problem to test your knowledge on the various topics covered in this chapter. You can then find the solutions and explanations in the next section.

1. **Find the partial sum** $\sum_{n=1}^{10} 3\left(\frac{2}{3}\right)^{n-1}$.
2. **Write the general expression for the *n*th term of the sequence: –4, 12, – 36, 108, –324, …**
3. **Find the sum of the first 20 terms of the arithmetic sequence if** $a_5 = -5$ **and** $a_{20} = -35$.
4. **Write the general expression for the *n*th term of the sequence: 2, 4, 10, 28, 82, . . .**
5. **Find the sum** $\sum_{n=1}^{\infty} \left(-\frac{1}{2}\right)^{n-1}$.
6. **What are the next two terms of the sequence: –3, –2, 0, 3, 7, 12, …?**
7. **Find the partial sum** $\sum_{n=1}^{5} (2n-3)$.
8. **What are the first five terms of the sequence whose general term is** $a_n = 2^{n+1} - 1$?
9. **What is the 16th term of the geometric sequence, where** $g_1 = 5$ **and** $g_2 = -15$?
10. **Find the partial sum** $\sum_{n=6}^{16} (2n+5)$.

Answers to Chapter 20 Quiz

1. $8\frac{5537}{6561} \approx 8.844.$

   Using the formula for the sum of the terms of a geometric series, where $g_1 = 3$, $r = \frac{2}{3}$, and $n = 10$,

   $$S_{10} = 3\left(\frac{1-\left(\frac{2}{3}\right)^{10}}{1-\frac{2}{3}}\right) = 3\left(\frac{1-\frac{1,024}{59,049}}{\frac{1}{3}}\right) = 3\left(\frac{\frac{58,025}{59,049}}{\frac{1}{3}}\right) = 9\left(\frac{58,025}{59,049}\right) = \frac{58,025}{6,561} = 8\frac{5537}{6,561}$$

2. $g_n = -4(-3)^{n-1}.$

   Each term is the result of multiplying the previous term by −3. The first term is −4. So the geometric sequence has a general term of $g_n = -4(-3)^{n-1}$.

3. **−320.**

   First, find the general term for the arithmetic sequence. Write the two equations formed from the two given terms and solve the system of equations for the first term and difference.

   $$\begin{cases} -5 = a_1 + (5-1)d \\ -35 = a_1 + (20-1)d \end{cases} \rightarrow \begin{cases} -5 = a_1 + 4d \\ -35 = a_1 + 19d \end{cases} \rightarrow \begin{cases} -5 - 4d = a_1 \\ -35 - 19d = a_1 \end{cases}$$

   $$\rightarrow -5 - 4d = -35 - 19d \rightarrow 15d = -30 \rightarrow d = -2$$

   Using $-5-4d = a_1$, $-5-4(-2) = a_1 \rightarrow 3 = a_1 9$. So the general term of the sequence is $a_n = 3+(n-1)(-2)9$. To find the sum of the first 20 terms, you have the first term but need the other information given about the 20th term. Given $a_{20} = -35$, you write $a_{20} = 3+(20-1)(-2) = 3+(-38) = -35$. Using the sum formula,

   $$S_{20} = \frac{20}{2}(3+(-35)) = 10(-32) = -320.$$

4. $a_n = 3^{n-1} + 1.$

   Each term is 1 more than the power of 3. The first term is 3^0+1, the second is 3^1+1, the third is 3^2+1, and so on. So the terms are written using $a_n = 3^{n-1}+1$.

5. $S_\infty = \frac{2}{3}.$

   Using the formula for the sum of an infinite series, you have $g_1 = 1$ and $r = -\frac{1}{2}$. So

   $$S_\infty = \frac{1}{1-\left(-\frac{1}{2}\right)} = \frac{1}{3/2} = \frac{2}{3}.$$

6. **18, 25.**

   The differences between the terms increase by 1 with each additional term. The first term is −3, and 1 is added to get −2. Then 2 is added to get 0, 3 is added to get 3, and so on: $a_1 = -3$, $a_{n+1} = a_n + (n-1)$. So, since $a_6 = 12$, $a_7 = 12+(7-1) = 18$ and $a_8 = 18+(8-1) = 25$.

7 **15.**

The 1st term of the arithmetic series is -1 and the 5th term is 7. Using the formula for the sum of the terms of an arithmetic series, $S_5 = \frac{5}{2}(-1+7) = \frac{5}{2}(6) = 15$.

8 **3, 7, 15, 31, 63.**

$a_1 = 2^{1+1} - 1 = 3,\ a_2 = 2^{2+1} - 1 = 7,\ a_3 = 2^{3+1} - 1 = 15,\ a_4 = 2^{4+1} - 1 = 31,\ a_5 = 2^{5+1} - 1 = 63$

9 **–71, 744, 535.**

To find the ratio of the geometric sequence, divide the second term by the first term: $r = \frac{-15}{5} = -3$. Now, using the ratio and first term, write the general term: $g_n = 5(-3)^{n-1}$. Use this to find the 16th term: $g_{16} = 5(-3)^{16-1} = 5(-3)^{15} = 5(-14{,}348{,}907) = -71,\ 744,\ 535$.

10 **297.**

Subtract the sum of the first 5 terms from the sum of the first 16 terms. You first find that $a_1 = 7,\ a_5 = 15,\ a_{16} = 37$, so $S_5 = \frac{5}{2}(7+15) = \frac{5}{2}(22) = 55$ and $S_{16} = \frac{16}{2}(7+37) = 8(44) = 352$. The difference is then $\sum_{n=6}^{16}(2n+5) = S_{16} - S_5 = 352 - 55 = 297$.

IN THIS CHAPTER

» **Getting acquainted with the binomial theorem**

» **Creating and expanding Pascal's triangle**

» **Applying the binomial theorem to expand binomials**

Chapter 21 Expanding Binomials for the Real World

The binomial theorem has been around for centuries. Mathematicians were interested in special happenings when operations were performed on just two terms. Euclid mentioned the special case involving the exponent 2 as early as the 4th century B.C. Indian mathematicians were using the formula for creating the binomial coefficient. And Omar Khayyam, mathematician and poet, was familiar with the theorem. Isaac Newton gets credit, though, for generalizing the theorem.

And where do you use the binomial theorem outside of a math class? It's found in computer networking, economic predictions, ranking of scores for scholarships, and weather forecasting.

Expanding with the Binomial Theorem

A *binomial* is a polynomial with exactly two terms. Expressing the powers of binomials as a sum of terms is called *binomial expansion*. Using the binomial theorem to write the power of a binomial allows you to find the coefficients of this expansion.

Expanding many of the same binomials can require a rather extensive application of the distributive property, take a lot of time, and create too many opportunities for errors. Multiplying two binomials is easy if you use the FOIL method, and multiplying three binomials doesn't take much more effort. Multiplying ten binomials, however, is hugely time-consuming. How would you like to multiply and simplify the expression $(x-3y)(x-3y)(x-3y)(x-3y)(x-3y)(x-3y)(x-3y)(x-3y)(x-3y)(x-3y)$? Even FOIL doesn't save you here!

TECHNICAL STUFF

The binomial theorem is written as

$$(a+b)^n = \sum_{k=0}^{n} \binom{n}{k} a^{n-k} b^k$$

where n is a non-negative integer and $\binom{n}{k}$ is the *coefficient* or a factor of the coefficient attached to each term in the expansion. As you see, there are $n+1$ terms in a binomial expansion (since you're creating terms numbered from 0 to n). And each term has its own special coefficient with decreasing powers of the first term, a, and increasing powers of the second term, b.

Concocting binomial coefficients

The notation $\binom{n}{k}$ indicates the number of combinations of n things taken r at a time. Combinations are a standard method of counting. The formula for combinations that helps you find the coefficients of a binomial expansion is $\binom{n}{r} = \frac{n!}{r!(n-r)!}$.

Recall that *factorial* is an operation. The operation $n!$ is read as "n factorial," and is defined as $n! = n(n-1)(n-2)\cdots 3\cdot 2\cdot 1$.

You read the expression for the binomial coefficient $\binom{n}{r}$ as "n choose r." You usually can find a button for combinations on a calculator. If not, you can use the factorial button and do each part separately.

EXAMPLE

Q. Find the binomial coefficient corresponding to $\binom{5}{3}$.

A. Substitute the values into the formula: $\binom{5}{3} = \frac{5!}{3!(5-3)!} = \frac{5!}{3!2!} = \frac{5\cdot 4\cdot \cancel{3}\cdot \cancel{2}\cdot \cancel{1}}{\cancel{3}\cdot \cancel{2}\cdot \cancel{1}\cdot 2\cdot 1} = \frac{20}{2} = 10.$

EXAMPLE

Q. Find the binomial coefficients corresponding to $\binom{n}{0}$ and $\binom{n}{n}$.

A. To make things a little easier, 0! is defined as 1. Therefore, when doing binomial expansions, you have these equalities:

$$\binom{n}{0} = \frac{n!}{0!(n-0)!} = \frac{n!}{1\cdot n!} = 1 \text{ and } \binom{n}{n} = \frac{n!}{n!(n-n)!} = \frac{n!}{n!\,1} = 1$$

YOUR TURN

1 Find the binomial coefficient corresponding to $\binom{7}{3}$.

2 Find the binomial coefficient corresponding to $\binom{8}{7}$.

Creating coefficients

Using the binomial coefficient formula, Table 21-1 gives you the binomial coefficients for expansions containing 1 through 7 terms.

Table 21-1 Binomial Coefficients

$(a+b)^n$	The values $\binom{n}{r}$ of the coefficients
$(a+b)^0$	$\binom{0}{0}=1$
$(a+b)^1$	$\binom{1}{0}=1,\ \binom{1}{1}=1$
$(a+b)^2$	$\binom{2}{0}=1,\ \binom{2}{1}=2,\ \binom{2}{2}=1$
$(a+b)^3$	$\binom{3}{0}=1,\ \binom{3}{1}=3,\ \binom{3}{2}=3,\ \binom{3}{3}=1$
$(a+b)^4$	$\binom{4}{0}=1,\ \binom{4}{1}=4,\ \binom{4}{2}=6,\ \binom{4}{3}=4,\ \binom{4}{4}=1$
$(a+b)^5$	$\binom{5}{0}=1,\ \binom{5}{1}=5,\ \binom{5}{2}=10,\ \binom{5}{3}=10,\ \binom{5}{4}=5,\ \binom{5}{5}=1$
$(a+b)^6$	$\binom{6}{0}=1,\ \binom{6}{1}=6,\ \binom{6}{2}=15,\ \binom{6}{3}=20,\ \binom{6}{4}=15,\ \binom{6}{5}=6,\ \binom{6}{6}=1$

Creating Pascal's triangle

Pascal's triangle, named after the famous mathematician Blaise Pascal, names the coefficients for a binomial expansion. It is especially useful with smaller degrees. You take the coefficients produced in Table 21-1 and put them in a triangular formation. Each row gives the coefficients to $(a+b)^n$, starting with $n=0$. The top number of the triangle is 1, as well as all the numbers on the outer sides. To get any term in the triangle, you find the sum of the two numbers diagonally above it. Look at Figure 21-1, which has the first six rows of Pascal's triangle. The numbers match the first six rows of coefficients in Table 21-1. You can add another row in Pascal's triangle using the last row in the table. Refer to Figure 21-2 for that added row.

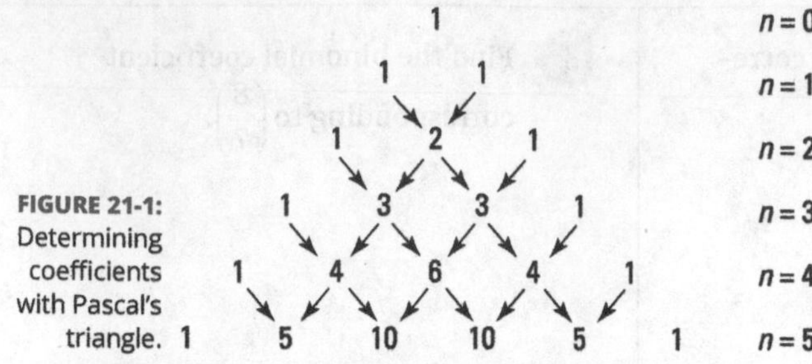

FIGURE 21-1: Determining coefficients with Pascal's triangle.

YOUR TURN

 **Find the binomial coefficients used in the expansion of the binomial $(a+b)^8$.**

 **Find the binomial coefficients used in the expansion of the binomial $(a+b)^9$.**

1

1 1

1 2 1

1 3 3 1

1 4 6 4 1

1 5 10 10 5 1

1 6 15 20 15 6 1

1 7 21 35 35 21 7 1

FIGURE 21-2: Binomial coefficients of powers of $(a+b)^0$ through $(a+b)^7$.

You can always use the formula for the binomial coefficient, but it's often quicker, easier, and handier to quickly put together Pascal's triangle when doing reasonably sized powers of a binomial.

Breaking down the binomial theorem

REMEMBER

The binomial theorem expansion becomes much simpler if you break it down into smaller steps and examine the parts. Here are a few things to be aware of to help you along the way; after you have all this info in mind, your task will seem much more manageable:

» The binomial coefficients $\binom{n}{r}$ won't necessarily be the coefficients in your final answer. You're raising each monomial to a power, including any coefficients attached to each of them, so there may be extra multipliers.

» The theorem is written for a binomial, the sum of two monomials, so if the expansion involves the *difference* of two monomials, the terms in your final answer should alternate between positive and negative numbers.

» The exponent of the first monomial begins at n and decreases by 1 with each sequential term until it reaches 0 at the last term. The exponent of the second monomial begins at 0 and increases by 1 each time until it reaches n at the last term.

» The exponents of both monomials add up to n — unless the monomials themselves have powers greater than 1.

Expanding by using the binomial theorem

In order to find the expansion of a binomial $(a+b)^n$, use:

$$(a+b)^n =$$

$$\binom{n}{0}a^n b^0 + \binom{n}{1}a^{n-1}b^1 + \binom{n}{2}a^{n-2}b^2 + \cdots + \binom{n}{n-2}a^2 b^{n-2} + \binom{n}{n-1}a^1 b^{n-1} + \binom{n}{n}a^0 b^n$$

Each $\binom{n}{r}$ represents a combination formula and gives you the coefficient for the term it corresponds to. You've seen how to compute these combinations earlier in this section.

EXAMPLE

Q. Find the expansion of $(m+2)^4$.

A. Follow these steps:

1. **Write out the binomial expansion by using the binomial theorem, changing the variables as they appear.**

   According to the theorem, you should replace the letter a with m, the letter b with 2, and the exponent n with 4: $(m+2)^4 = \binom{4}{0}m^4 \cdot 2^0 + \binom{4}{1}m^3 \cdot 2^1 + \binom{4}{2}m^2 \cdot 2^2 + \binom{4}{3}m^1 \cdot 2^3 + \binom{4}{4}m^0 \cdot 2^4$.

   The exponents of m begin with 4 and end with 0. Similarly, the exponents of 2 begin with 0 and end with 4. For each term, the sum of the exponents in the expansion is always 4.

2. **Find the binomial coefficients.**

   $\binom{4}{0}=1, \binom{4}{1}=4, \binom{4}{2}=6, \binom{4}{3}=4, \binom{4}{4}=1$

You may have noticed that after you reach the middle of the expansion, the coefficients are a mirror image of the first half. This trick is a timesaver you can employ so you don't need to do all the calculations for $\binom{n}{r}$.

3. **Replace all $\binom{n}{r}$ with the coefficients from Step 2.**

   This step gives you $(m+2)^4 = 1 \cdot m^4 \cdot 2^0 + 4 \cdot m^3 \cdot 2^1 + 6 \cdot m^2 \cdot 2^2 + 4 \cdot m^1 \cdot 2^3 + 1 \cdot m^0 \cdot 2^4$.

4. **Raise the monomials to the powers specified for each term.**

   $= 1 \cdot m^4 \cdot 1 + 4 \cdot m^3 \cdot 2 + 6 \cdot m^2 \cdot 4 + 4 \cdot m^1 \cdot 8 + 1 \cdot m^0 \cdot 16$

5. **Combine like terms and simplify.**

   $= m^4 + 8m^3 + 24m^2 + 32m + 16$

   Note that the coefficients you get in the final answer aren't the binomial coefficients you started with in Step 3. This difference occurs because you must raise each monomial to a power, and the constant in the original binomial changed each term after the simplification.

EXAMPLE

Q. Find the expansion of $(x-3)^5$. This time, use the binomial coefficients from Pascal's triangle.

A. Use these steps:

1. **Since this binomial is being raised to the 5th power, use the 6th row in Pascal's triangle — the one starting with 1 and 5. Write the coefficients, leaving spaces between to insert powers of the two monomials.**

   1 5 10 10 5 1

2. **Insert decreasing powers of the x monomial, starting with 5.**

   $1x^5 \quad 5x^4 \quad 10x^3 \quad 10x^2 \quad 5x^1 \quad 1x^0$

3. **Insert increasing powers of the second monomial, –3, starting with 0. (Or, if you prefer, work from right to left, inserting decreasing powers starting with 5.)**

   $1x^5(-3)^0 \quad 5x^4(-3)^1 \quad 10x^3(-3)^2 \quad 10x^2(-3)^3 \quad 5x^1(-3)^4 \quad 1x^0(-3)^5$

4. **Simplify each term.**

   $1x^5(1) \quad 5x^4(-3) \quad 10x^3(9) \quad 10x^2(-27) \quad 5x(81) \quad 1(-243)$

   $= x^5 - 15x^4 + 90x^3 - 270x^2 + 405x^4 - 243$

   Notice that the terms alternate due to the subtraction in the binomial.

REMEMBER

Using the binomial theorem can save you time. Just keep each of the steps separate until the very end, when you do the final simplification. When the original monomial has coefficients or exponents other than 1 on the variable(s), you have to be careful to take those into account.

Raising monomials to a power pre-expansion

At times, monomials can have coefficients and/or exponents other than 1. In this case, you have to raise the entire monomial to the appropriate power in each step.

EXAMPLE

Q. Expand the binomial $(3x^2 - 2y)^7$.

A. Follow these steps:

1. **Find the binomial coefficients.**

   Using the combination formula gives you the following:

   $$\binom{7}{0}=1,\ \binom{7}{1}=7,\ \binom{7}{2}=21,\ \binom{7}{3}=35,\ \binom{7}{4}=35,\ \binom{7}{5}=21,\ \binom{7}{6}=7,\ \binom{7}{7}=1$$

   Or, if you use the extended Pascal's triangle with the two additional rows, you have:

   1 7 21 35 35 21 7 1

2. **Using the coefficients, insert decreasing powers of the $3x^2$ term, starting with 7.**

   $$1(3x^2)^7\quad 7(3x^2)^6\quad 21(3x^2)^5\quad 35(3x^2)^4\quad 35(3x^2)^3\quad 21(3x^2)^2\quad 7(3x^2)^1\quad 1(3x^2)^0$$

3. **Insert increasing powers of the $-2y$ term.**

   $$1(3x^2)^7(-2y)^0\quad 7(3x^2)^6(-2y)^1\quad 21(3x^2)^5(-2y)^2\quad 35(3x^2)^4(-2y)^3\cdots$$

   $$\cdots 35(3x^2)^3(-2y)^4\quad 21(3x^2)^2(-2y)^5\quad 7(3x^2)^1(-2y)^6\quad 1(3x^2)^0(-2y)^7$$

4. **Raise the monomials to the powers specified for each term.**

   $$1(2{,}187x^{14})(1)\quad 7(729x^{12})(-2y)\quad 21(243x^{10})(4y^2)\quad 35(x^8)(-8y^3)\cdots$$

   $$\cdots 35(27x^6)(16y^4)\quad 21(9x^4)(-32y^5)\quad 7(3x^2)(64y^6)\quad 1(1)(-128y^7)$$

5. **Simplify.**

   $$2{,}187x^{14} - 10{,}206x^{12}y + 20{,}412x^{10}y^2 - 22{,}680x^8y^3 + 15{,}120x^6y^4$$

   $$-6{,}048x^4y^5 + 1{,}344x^2y^6 - 128y^7$$

YOUR TURN

5 Expand the binomial $(x+4)^5$.

6 Expand the binomial $(x-3y)^6$.

Looking at expansion with complex numbers

An even more interesting type of binomial expansion involves the complex number i, because you're not only dealing with the binomial theorem, but you're also dealing with imaginary numbers as well. (For more on complex numbers, see Chapter 15.) When raising complex numbers to a power, note that $i^1 = i$, $i^2 = -1$, $i^3 = -i$, and $i^4 = 1$. If you run into higher powers, this pattern repeats: $i^5 = i$, $i^6 = -1$, $i^7 = -i$, and so on. Because powers of the imaginary number i can be simplified, your final answer to the expansion should not include powers of i. Instead, use the information given here to simplify the powers of i and then combine your like terms.

EXAMPLE

Q. Expand $(1+2i)^8$

A. Follow these steps:

1. **Write out the binomial expansion by using the binomial theorem, changing the variables where necessary.**

$$(1+2i)^8 = \binom{8}{0}(1)^8(2i)^0 + \binom{8}{1}(1)^7(2i)^1 + \binom{8}{2}(1)^6(2i)^2 + \binom{8}{3}(1)^5(2i)^3$$
$$+\binom{8}{4}(1)^4(2i)^4 + \binom{8}{5}(1)^3(2i)^5 + \binom{8}{6}(1)^2(2i)^6 + \binom{8}{7}(1)^1(2i)^7$$
$$+\binom{8}{8}(1)^0(2i)^8$$

2. **Replace all $\binom{n}{r}$ with the coefficients (use the binomial coefficient formula or Pascal's triangle).**

$$1(1)^8(2i)^0 + 8(1)^7(2i)^1 + 28(1)^6(2i)^2 + 56(1)^5(2i)^3 + 70(1)^4(2i)^4$$
$$+56(1)^3(2i)^5 + 28(1)^2(2i)^6 + 8(1)^1(2i)^7 + 1(1)^0(2i)^8$$

3. **Raise the monomials to the powers specified for each term.**

$$1(1)(1) + 8(1)(2i) + 28(1)(4i^2) + 56(1)(8i^3) + 70(1)(16i^4)$$
$$+56(1)(32i^5) + 28(1)(64i^6) + 8(1)(128i^7) + 1(1)(256i^8)$$

4. **Simplify any powers of i.**

$$1(1)(1) + 8(1)(2i) + 28(1)(4(-1)) + 56(1)(8(-i)) + 70(1)(16(1))$$
$$+56(1)(32(i)) + 28(1)(64(-1)) + 8(1)(128(-i)) + 1(1)(256(1))$$

5. **Combine like terms and simplify.**

$$1 + 16i - 112 - 448i + 1{,}120 + 1{,}792i - 1{,}792 - 1{,}024i + 256 = -527 + 336i$$

YOUR TURN

 7 Expand the binomial $(2-3i)^4$.	8 Expand the binomial $(1+i)^9$.

Working Backwards

You will sometimes be presented with an expression that is the result of someone having expanded a binomial and you need to find the original factored form — for simplicity. You can recreate that original binomial by determining where each of the terms in the expansion came from.

EXAMPLE

Q. Determine the binomial and power that produces $x^5-5x^4+10x^3-10x^2+5x-1$.

A. You quickly recognize the line from Pascal's triangle for the 5th power, 1 5 10 10 5 1, with just a little adjustment. The signs are alternating, so the second monomial must be −1. The first monomial has a coefficient of 1, because none of the numerical values of the coefficients have been changed. So you can say that $x^5-5x^4+10x^3-10x^2+5x-1=(x-1)^5$.

EXAMPLE

Q. Determine the binomial and power that produces $16x^4+32x^3y+24x^2y^2+8xy^3+y^4$.

A. You see decreasing powers of x and increasing powers of y. These powers correspond to a binomial raised to the 4th power. Start with the basic coefficients, 1 4 6 4 1. You see that the last term is essentially 1^4y^4, so you can create a binomial with that second monomial inserted: $(\ \ +y)^4$. The coefficient of the first term, 16, is the 4th power of 2; also, x is raised to the 4th power. So choose $2x$ as the first monomial and write the binomial as $(2x+y)^4$. To check your work, fill in the initial coefficients with the decreasing and increasing powers of the two monomials: $1(2x)^4 \quad 4(2x)^3y^1 \quad 6(2x)^2y^2 \quad 4(2x)^1y^3 \quad 1y^4$. Simplify, and you see that you have the correct binomial.

EXAMPLE

Q. Determine the binomial and power that produces $x^{15}+10x^{12}y^2+40x^9y^4+80x^6y^6+80x^3y^8+32y^{10}$.

A. If a binomial was raised to the 15th power to give you this expression, you'd see 16 terms. That's not the case here. But you do see decreasing powers of x that are multiples of 3 and increasing powers of y that are multiples of 2. There are six terms, indicating that you are working with a binomial raised to the 5th power: 1 5 10 10 5 1.

So start with the variables in the two monomials: $(\ \ x^3 + \ \ y^2)^5$. The coefficient of the last term is 2 raised to the 5th power, and the coefficient of the first term is 1, so write the binomial as $(x^3 + 2y^2)^5$. To check your work, fill in the initial coefficients with decreasing and increasing powers:

$$1(x^3)^5 \quad 5(x^3)^4(2y^2)^1 \quad 10(x^3)^3(2y^2)^2 \quad 10(x^3)^2(2y^2)^3 \quad 5(x^3)^1(2y^2)^4 \quad 1(2y^2)^5$$

First, simplify the x factors: $= 1(x^{15}) \quad 5(x^{12})(2y^2)^1 \quad 10(x^9)(2y^2)^2 \quad 10(x^6)(2y^2)^3 \quad 5(x^3)(2y^2)^4 \quad 1(2y^2)^5$

and then the y factors: $= 1(x^{15}) \quad 5(x^{12})(2y^2) \quad 10(x^9)(4y^4) \quad 10(x^6)(8y^6) \quad 5(x^3)(16y^8) \quad 1(32y^{10})$

Finally, multiplying and simplifying, you have: $= x^{15} \quad 10x^{12}y^2 \quad 40x^9y^4 \quad 80x^6y^6 \quad 80x^3y^8 \quad 32y^{10}$.

Insert the + signs, and you have the original expression.

YOUR TURN

9 Determine the binomial and power that produces: $64x^6 + 192x^5 + 120x^4 + 160x^3 + 60x^2 + 12x + 1$.

10 Determine the binomial and power that produces: $x^8 - 12x^6 + 54x^4 - 108x^2 + 81$.

Practice Questions Answers and Explanations

1) **35. Using the formula for a combination:**

$$\binom{7}{3}=\frac{7!}{3!(7-3)!}=\frac{7!}{3!4!}=\frac{7\cdot 6\cdot 5\cdot \cancel{4!}}{3!\cancel{4!}}=\frac{7\cdot 6\cdot 5\cdot \cancel{4!}}{3\cdot 2\cdot 1\cdot \cancel{4!}}=\frac{7\cdot \cancel{6}\cdot 5\cdot \cancel{4!}}{\cancel{3\cdot 2}\cdot 1\cdot \cancel{4!}}=35$$

2) **8. Using the formula for a combination:**

$$\binom{8}{7}=\frac{8!}{7!(8-7)!}=\frac{8!}{7!1!}=\frac{8\cdot 7!}{7!\cdot 1}=\frac{8\cdot \cancel{7!}}{\cancel{7!}\cdot 1}=8$$

3) **1 8 28 56 70 56 28 8 1. You can either refer to Figure 21-2 and extend Pascal's triangle by one row, or you can use the binomial coefficient formula and write:**

$$\binom{8}{0}\ \binom{8}{1}\ \binom{8}{2}\ \binom{8}{3}\ \binom{8}{4}\ \binom{8}{5}\ \binom{8}{6}\ \binom{8}{7}\ \binom{8}{8}$$

Then find the value of each combination.

4) **1 9 36 84 126 126 84 36 9 1. You can either refer to the previous problem and extend Pascal's triangle by one row, or you can use the binomial coefficient formula and write:**

$$\binom{9}{0}\ \binom{9}{1}\ \binom{9}{2}\ \binom{9}{3}\ \binom{9}{4}\ \binom{9}{5}\ \binom{9}{6}\ \binom{9}{7}\ \binom{9}{8}\ \binom{9}{9}$$

Then find the value of each combination.

5) **$x^5+20x^4+160x^3+640x^2+1280x+1024$. First, write the coefficients using Pascal's triangle or the binomial coefficient formula: 1 5 10 10 5 1. Next, insert decreasing powers of the first monomial: $1(x^5)\ \ 5(x^4)\ \ 10(x^3)\ \ 10(x^2)\ \ 5(x^1)\ \ 1(x^0)$. Now insert increasing powers of the second monomial: $1(x^5)(4^0)\ \ 5(x^4)(4^1)\ \ 10(x^3)(4^2)\ \ 10(x^2)(4^3)$ $5(x^1)(4^4)\ \ 1(x^0)(4^5)$. Finally, simplify each term and add them together:**

$1(x^5)(1)\ \ 5(x^4)(4)\ \ 10(x^3)(16)\ \ 10(x^2)(64)\ \ 5(x^1)(256)\ \ 1(x^0)(1024)$
$=x^5+20x^4+160x^3+640x^2+1280x+1024.$

6) **$x^6-18x^5y+135x^4y^2-540x^3y^3+1215x^2y^4-1458xy^5+729y^6$. First, write the coefficients using Pascal's triangle or the binomial coefficient formula: 1 6 15 20 15 6 1. Next, insert decreasing powers of the first monomial: $1(x^6)\ \ 6(x^5)\ \ 15(x^4)\ \ 20(x^3)\ \ 15(x^2)$ $6(x^1)\ \ 1(x^0)$. Now insert increasing powers of the second monomial:**

$1(x^6)(-3y)^0\ \ 6(x^5)(-3y)^1\ \ 15(x^4)(-3y)^2\ \ 20(x^3)(-3y)^3\ \ 15(x^2)(-3y)^4$
$6(x^1)(-3y)^5\ \ 1(x^0)(-3y)^6$

Finally, simplify each term and add them together:

$(x^6)1 \quad 6(x^5)(-3y)^1 \quad 15(x^4)\cdot 9y^2 \quad 20(x^3)(-27y^3) \quad 15(x^2)(81y^4) \quad 6(x)(-243y^5) \quad 1(729y^6)$

$= x^6 - 18x^5y + 135x^4y^2 - 540x^3y^3 + 1215x^2y^4 - 1458xy^5 + 729y^6$

7 **$-119+120i$. First, write the coefficients using Pascal's triangle or the binomial coefficient formula: 1 4 6 4 1. Next, insert decreasing powers of the first monomial: $1(2^4)$ $4(2^3)$ $6(2^2)$ $4(2^1)$ $1(2^0)$. Now insert increasing powers of the second monomial: $1(2^4)(-3i)^0$ $4(2^3)(-3i)^1$ $6(2^2)(-3i)^2$ $4(2^1)(-3i)^3$ $1(2^0)(-3i)^4$. Simplify each term by performing all the powers: $1(16)\cdot 1$ $4(8)(-3i)$ $6(4)(9i^2)$ $4(2)(-27i^3)$ $1(1)(81i^4)$. Now perform the multiplications and replace the even powers of i with their equivalents: 16 $-96i$ $216(-1)$ $-216(-i)$ $81(1)$. Finally, finish the simplifications and combine like terms: 16 $-96i$ -216 $216i$ $81 = -119+120i$.**

8 **$16+16i$. First, write the coefficients using Pascal's triangle or the binomial coefficient formula: 1 9 36 84 126 126 84 26 9 1. Next, insert decreasing powers of the first monomial and increasing powers of the second monomial:**

$1(1^9)(i^0) \quad 9(1^8)(i^1) \quad 36(1^7)(i^2) \quad 84(1^6)(i^3) \quad 126(1^5)(i^4)$

$\cdots 126(1^4)(i^5) \quad 84(1^3)(i^6) \quad 36(1^2)(i^7) \quad 9(1^1)(i^8) \quad 1(1^0)(i^9)$

Next, replace all the powers of 1 times the coefficient with the coefficient value, and replace all the powers of i with their equivalent value: $1(1)$ $9(i)$ $36(-1)$ $84(-i)$ $126(1)$ $126(i)$ $84(-1)$ $36(-i)$ $9(1)$ $1(i)$

Now multiply and combine like terms: $1+9i-36-84i+126+126i-84-36i+9+i=16+16i$.

9 **$(2x+1)^6$. This is the result of raising a binomial to the 6th power. The coefficient of the first term is 2 raised to the 6th, so the first monomial has a coefficient of 2. The last term is 1 raised to the 6th. So the binomial is $(2x+1)^6$. Check this by creating the expansion using the binomial theorem.**

$1(2x)^6(1)^0 \quad 6(2x)^5(1)^1 \quad 15(2x)^4(1)^2 \quad 20(2x)^3(1)^3 \quad 15(2x)^2(1)^4 \quad 6(2x)^1(1)^5 \quad 1(2x)^0(1)^6$

$= 64x^6 + 192x^5 + 240x^4 + 160x^3 + 60x^2 + 12x + 1$

10 **$(x^2-3)^4$. There are five terms, suggesting a binomial raised to the 4th power. The powers of x are multiples of 2, and the last term is 3 raised to the 4th power. The terms alternate in sign, so the second monomial is negative. The first monomial is then x^2, and the second monomial is -3, giving you the binomial $(x^2-3)^4$. Check this by creating the expansion using the binomial theorem.**

$1(x^2)^4(-3)^0 \quad 4(x^2)^3(-3)^1 \quad 6(x^2)^2(-3)^2 \quad 4(x^2)^1(-3)^3 \quad 1(x^2)^0(-3)^4$

$= x^8 - 12x^6 + 54x^4 - 108x^2 + 81$

If you're ready to test your skills a bit more, take the following chapter quiz that incorporates all the chapter topics.

Whaddya Know? Chapter 21 Quiz

Quiz time! Complete each problem to test your knowledge on the various topics covered in this chapter. You can then find the solutions and explanations in the next section.

1 **Expand the binomial $(3x - y)^5$.**

2 **If the first four coefficients in the expansion of $(a+b)^{15}$ are 1 15 105 455, then what are the first four coefficients in the expansion of $(a+b)^{16}$?**

3 **Find the binomial coefficient corresponding to $\binom{6}{4}$.**

4 **Which binomial expansion corresponds to $64x^6 - 192x^5y + 240x^4y^2 - 160x^3y^3 + 60x^2y^4 - 12xy^5 + y^6$?**

5 **Find the binomial coefficients used in the expansion of $(a+b)^3$.**

6 **Expand and simplify $(2-i)^3$.**

Answers to Chapter 21 Quiz

1 $\mathbf{243x^5 - 405x^4y + 270x^3y^2 - 90x^2y^3 + 15xy^4 + y^5}$.

Use Pascal's triangle or the binomial formula for the coefficients to create the six coefficients. Then write in the decreasing powers of $3x$ and the increasing powers of $-y$. Simplify the terms.

1 5 10 10 5 1

$\rightarrow 1(3x)^5(-y)^0 \quad 5(3x)^4(-y)^1 \quad 10(3x)^3(-y)^2 \quad 10(3x)^2(-y)^3 \quad 5(3x)^1(-y)^4 \quad 1(3x)^0(-y)^5$

$\rightarrow 1(243x^5)(1) \quad 5(81x^4)(-y) \quad 10(27x^3)(y^2) \quad 10(9x^2)(-y^3) \quad 5(3x)(y^4) \quad 1(1)(-y^5)$

$\rightarrow 243x^5 - 405x^4y + 270x^3y^2 - 90x^2y^3 + 15xy^4 - y^5$

2 **1 16 120 560.**

Using Pascal's triangle, first write the four coefficients from the expansion of $(a+b)^{15}$. Then place a 1 at the beginning of the next row and add the adjacent coefficients above, placing the sums between the pairs of numbers.

 1 15 105 455

1 16 120 560

3 **15.**

Using the formula,

$$\binom{6}{4} = \frac{6!}{4!(6-4)!} = \frac{6!}{4!2!} = \frac{6 \cdot 5 \cdot \cancel{4!}}{\cancel{4!}2!} = \frac{6 \cdot 5}{2 \cdot 1} = 15$$

4 $\mathbf{(2x - y)^6}$.

You see seven terms, corresponding to a binomial raised to the 6th power. The first term has a coefficient of 64, which is 2 raised to the 6th. The last term has a coefficient of 1. The terms alternate in sign, meaning that the second monomial is negative. The first monomial is $2x$, and the second monomial is $-y$, giving you $(2x - y)^6$.

5 **1 3 3 1.**

You can use Pascal's triangle or the binomial formula for the binary coefficients, which will give you:

$$\binom{3}{0} \quad \binom{3}{1} \quad \binom{3}{2} \quad \binom{3}{3}$$

1 3 3 1

6 $\mathbf{2 - 11i}$.

First, write the coefficients. Then insert decreasing powers of 2 and increasing powers of $-i$. Raise the terms to the powers, and replace powers of i with their equivalents. Simplify and combine like terms.

$1 \quad 3 \quad 3 \quad 1 \rightarrow 1(2^3)(-i)^0 \quad 3(2^2)(-i)^1 \quad 3(2^1)(-i)^2 \quad 1(2^0)(-i)^3$

$\rightarrow 1(8)(1) \quad 3(4)(-i) \quad 3(2)(-1) \quad 1(1)(-i) \rightarrow 8 \quad -12i \quad -6 \quad -i$

$= 2 - 11i$

7 Onward to Calculus

IN THIS UNIT . . .

Introducing limits.

Applying successful mathematical procedures.

IN THIS CHAPTER

» Determining limits graphically, analytically, and algebraically

» Pairing limits and operations

» Homing in on the average rate of change

» Identifying continuity and discontinuity in a function

Chapter 22 Lining Up the Tools

Every good thing must come to an end, and for pre-calculus, the end is actually the beginning — the beginning of calculus. Calculus includes the study of change and rates of change (not to mention a big change for you!). Before calculus, everything was usually static (stationary or motionless), but calculus shows you that things can be different over time. This branch of mathematics enables you to study how things move, grow, travel, expand, and shrink and helps you do so much more than any other math subject before.

This chapter helps prepare you for calculus by introducing you to some of the basics of the subject. First, you'll see the move from pre-calculus to calculus. And a big topic will be *limits*, which dictate that a graph can get really close to values without ever actually reaching them. Before you get to calculus, math problems give you a function $f(x)$ and ask you to find the y value at one specific x in the domain (see Chapter 3). But when you get to calculus, you look at what happens to the function the closer you get to certain values (like a really tough game of hide-and-seek). Getting even more specific, a function can be *discontinuous* at a point.

You get to explore those points one at a time so you can take a really good look at what's going on in the function at a particular value — that information comes in really handy in calculus when you start studying change. When studying limits and continuity, you're not working with the study of change specifically, but after this chapter, you'll be ready to take the next step.

Scoping Out the Changes: Pre-Calculus to Calculus

Here are a few basic distinctions between pre-calculus and calculus to illustrate the change in emphasis.

- **Pre-calculus:** You study the slope of a line.

  Calculus: You study the slope of a line that is tangent to a point on a curve.

  A straight line has the same slope all the time. No matter which point you choose to look at, the slope is the same. However, because a curve moves and changes, the slope of the tangent line is different at different points on the curve.

- **Pre-calculus:** You study the area of geometric shapes.

  Calculus: You study the area under a curve.

  In pre-calculus, you can rest easy knowing that a geometric shape and its dimensions behave, so you can find its area using certain measurements and a standard formula. But a curve goes on forever, and, depending on which section you're looking at, its area changes. No more nice, pretty formulas to find area here; instead, you get to use a process called *integration* (a technical term for a mathematics process).

- **Pre-calculus:** You study the volume of a geometric solid.

  Calculus: You study the volume of interesting shapes called *solids of revolution*.

  The geometric solids you find volume for (prisms, cylinders, and pyramids, for example) have formulas that are always the same, based on the basic shapes of the solids and their dimensions. The way to find the volume of a solid of revolution, though, is to cut the shape into infinitely small slices and find the volume of each slice. The volume of the slices changes over time based on the section of the curve you're looking at.

- **Pre-calculus:** You study objects moving with constant velocities.

  Calculus: You study objects moving with acceleration.

  Using algebra, you can find the average rate of change of an object over a certain time interval. Using calculus, you can find the *instantaneous* rate of change for an object at an exact moment in time.

- **Pre-calculus:** You study functions in terms of x.

  Calculus: You study changes to functions in terms of x, with those changes in terms of t.

  Graphs of functions referred to as $f(x)$ can be created by plotting points. In calculus, you describe the changes to the graph $f(x)$ by using the variable t, indicated by $\frac{dx}{dt}$.

TIP

Calculus is best taken with an open mind and the spirit of adventure! Don't view calculus as a new bunch of material to memorize. Instead, think of it as an opportunity to use your pre-calculus knowledge and experience. Try to glean an understanding of *why* calculus does what it does. In this arena, concepts are key.

Understanding Your Limits

Not every function is defined at every value of x in the real number system. Rational functions, for example, are undefined when the value of x makes the denominator of the function equal to 0. You can use a *limit* (which, if it exists, represents a value that the function tends to approach as the independent variable approaches a given number) to look at a function to see what it *would* do if it could. To do so, you take a look at the behavior of the function near the undefined value(s).

TECHNICAL STUFF

To express a limit in symbols, you write $\lim_{x \to n} f(x) = L$, which is read as "the limit as x approaches n of $f(x)$ is L." L is the value you're looking for. For the limit of a function to exist, the left limit and the right limit must both exist and be equivalent:

- A *left limit* starts at a value that's less than the number x is approaching, and it gets closer and closer from the left-hand side of the graph.
- A *right limit* is the exact opposite; it starts out greater than the number x is approaching and gets closer and closer from the right-hand side.

If, and only if, the left-hand limit equals the right-hand limit can you say that the function has a limit for that particular value of x.

Mathematically, you'd let f be a function and let c and L be real numbers. Then $\lim_{x \to c} f(x) = L$ when $\lim_{x \to c^-} f(x) = L$ and $\lim_{x \to c^+} f(x) = L$. (Note the negative and positive superscripts on c, indicating coming from the left and right, respectively.) In real-world language, this setup means that if you took two pencils, one in each hand, and started tracing along the graph of the function toward the same point, the two pencils would have to meet in one spot between them in order for the limit to exist.

EXAMPLE

Q. Describe what is happening to the function values of $f(x) = \frac{x^2 - 9}{x - 3}$ as you get closer and closer to $x = 3$.

A. This function is undefined at $x = 3$: $f(x) = \frac{x^2 - 9}{x - 3}$. But what about just getting as close as you can?

The following shows you what happens to the function as you input x values close to 3.

x	2.5	2.9	2.99	2.999	⋯	3.001	3.01	3.1	3.5
$f(x)$	5.5	5.9	5.99	5.999	⋯	6.001	6.01	6.1	6.5

The closer you get to 3, the closer the value of $f(x)$ gets to 6. But $f(x)$ is never 6, because x is never 3. All the values of x are defined, *except* for $x = 3$. There's a *limit* as x approaches 3. You see a graph of this function on the left in Figure 22-1. This left figure shows that even though the function isn't defined at $x = 3$, the limit exists. Notice the open dot at the point (3,6).

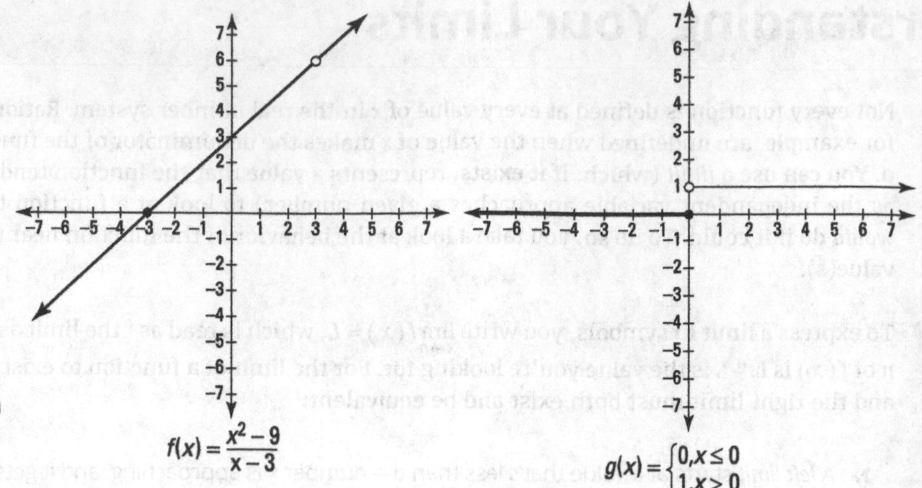

FIGURE 22-1: Functions $f(x)$ with limit at 3 and $g(x)$ with no limit at 0.

For functions that are *well connected*, the pencils always meet eventually in a particular spot (in other words, a limit would always exist). However, you see later that sometimes they do not (see Figure 22-1 on the right). The popular *unit-step function* shown here is defined: $g(x) = \begin{cases} 0, x \le 0 \\ 1, x > 0 \end{cases}$. And, in this case, you say that there is no limit at $x = 0$.

YOUR TURN

1 **Given the following table of values, use it to determine $\lim_{x \to 0} f(x)$.**

Point	Value								
x	-1	-0.9	-0.09	-0.009	0	0.001	0.01	0.1	1
$f(x)$	−2	$-\frac{3}{2}$	$-\frac{5}{4}$	$-\frac{9}{8}$		$-\frac{7}{8}$	$-\frac{3}{4}$	$-\frac{1}{2}$	0

2 **Given the following table of values, use it to determine $\lim_{x \to 1} f(x)$ when $f(x) = \dfrac{x-1}{x^2-1}$.**

Point	Value								
x	0	0.9	0.99	0.999	1	1.001	1.01	1.1	2
$f(x)$	1	$\frac{10}{19}$	$\frac{100}{199}$	$\frac{1000}{1999}$		$\frac{1000}{2001}$	$\frac{100}{201}$	$\frac{10}{21}$	$\frac{1}{3}$

Finding the Limit of a Function

"What is a *limit*?" you ask. Mathematically speaking, a limit describes the behavior of a function as it gets nearer and nearer to a particular point. You can look for the limit of a function at a particular value of *x* in three different ways: graphically, analytically, and algebraically. You see how to do this in the sections that follow. However, you may not always be able to reach a conclusion (perhaps the function doesn't approach just one *y* value at the particular *x* value you're looking at). In these cases, the graph has significant gaps and you say it's not continuous.

TIP

On rare occasions when you're asked to find the limit of a function, and you accomplish the chore by plugging the *x* value into the function rule, you can celebrate that you've found the limit. It's usually not that easy, though, which is why you'll find other methods most helpful. And, on occasion the function value at a particular *x* doesn't tell you what the function is doing as it gets nearer and nearer to *x*.

You should use the graphing method only when you've already been given the graph and asked to find a limit; reading from a graph can be tricky. The analytical method works for any function, but it's slow. If you can use the algebraic method, it will save you time. These methods are discussed in the sections that follow.

Graphically

When you're given the graph of a function and the problem asks you to find the limit, you read values from the graph — something you've been doing ever since you learned what a graph was! If you're looking for a limit coming from the left, you follow that function from the left-hand side toward the *x* value in question. Repeat this process from the right to find the right-hand limit. If the *y* value at the *x* value in question is the same from the left as it is from the right (did the pencils meet?), then that *y* value is the limit.

EXAMPLE

Q. Use Figure 22-2 to find $\lim_{x\to -2.5} f(x)$, $\lim_{x\to 3} f(x)$, and $\lim_{x\to -5} f(x)$.

A. Here's how to find the limits.

- $\lim_{x\to -2.5} f(x)$: You observe that the function appears to be crossing the *x*-axis when $x = -2.5$. So the *y*-value is 0, and the limit as *x* approaches -2.5 from either direction is 0.
- $\lim_{x\to 3} f(x)$: In the graph, you can see a hole in the function at $x = 3$, which means that the function is undefined at that point — but that doesn't mean you can't state a limit. If you look at the function's values from the left — $\lim_{x\to 3^-} f(x)$ — and from the right — $\lim_{x\to 3^+} f(x)$ — you see that the *y* value keeps getting closer and closer to about -1.5. You can make the guess that the limit is -1.5.
- $\lim_{x\to -5} f(x)$: You can see that the function has a vertical asymptote at $x = -5$ (for more on asymptotes, see Chapter 3). From the left, the function approaches $-\infty$ as it nears $x = -5$. You can express this mathematically as $\lim_{x\to -5^-} f(x) = -\infty$. From the right, the function approaches $+\infty$ as it nears $x = -5$. You write this situation as $\lim_{x\to -5^+} f(x) = +\infty$. Therefore, the limit doesn't exist at this value, because one side is $-\infty$ and the other side is $+\infty$.

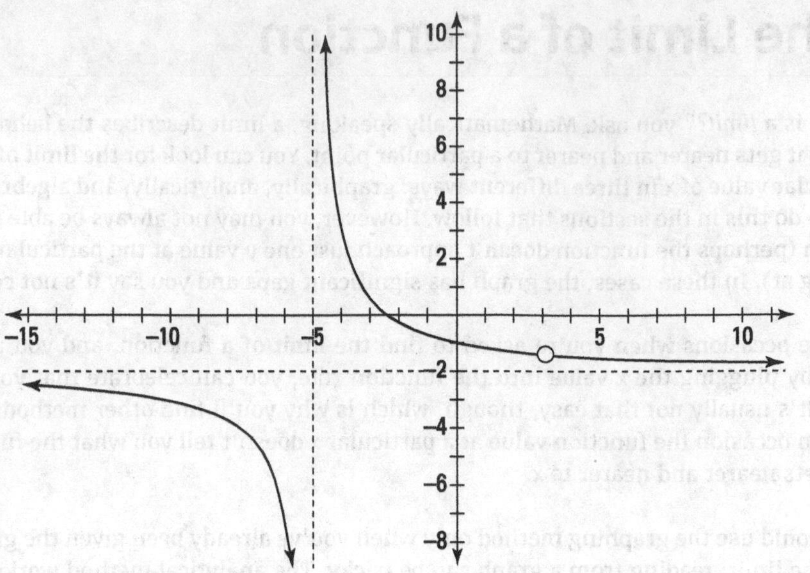

FIGURE 22-2: Observing the limit of a function graphically.

For a function to have a limit, the left and right values must be the same. A function with a hole in the graph, like $\lim_{x \to 3} f(x)$, can have a limit, but the function can't jump over an asymptote at a value and have a limit (like $\lim_{x \to -5} f(x)$).

Refer to the figure to answer the following questions.

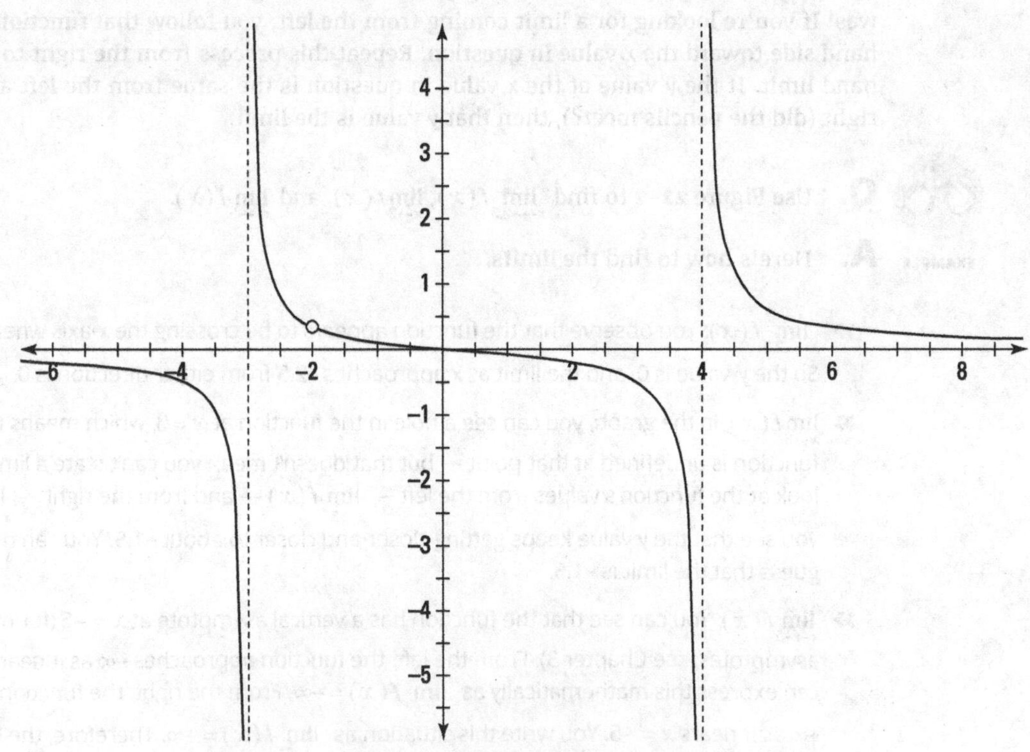

3 Find the limits: $\lim_{x \to 0} f(x)$ and $\lim_{x \to -2} f(x)$.	4 Find the limits: $\lim_{x \to 4} f(x)$ and $\lim_{x \to -3} f(x)$.

Analytically

To find a limit analytically, you simply set up a chart and put the number that x is approaching smack dab in the middle of it. Then, coming in from the left in the same row, you randomly choose numbers that get closer to the number. You do the same thing coming in from the right. In the next row, you compute the y values that correspond to these values that x is approaching.

Solving analytically isn't the most efficient way of finding a limit, and if you can use the algebraic technique described in the next section, you should opt for that method. But sometimes there will be a situation where you want to use the analytical technique, so it's good for you to know. As an example of using the analytical method, the following function is undefined at $x = 4$ because that value makes the denominator 0:

$$f(x) = \frac{x^2 - 6x + 8}{x - 4}$$

But you can find the limit of the function as x approaches 4 by using a chart. Table 22-1 shows how to set it up.

Table 22-1 Finding a Limit Analytically

Point	Value								
x	3.0	3.9	3.99	3.999	4.0	4.001	4.01	4.1	5.0
$f(x)$ (the y value)	1.0	1.9	1.99	1.999	undefined	2.001	2.01	2.1	3.0

The values that you pick for x are completely arbitrary — they can be anything you want. Just make sure they get closer and closer to the value you're looking for from both directions. The closer you get to the actual x value, though, the closer your limit is as well. If you look at the y values in the chart, you'll notice that they get closer and closer to 2 from both sides; so 2 is the limit of the function, determined analytically.

TIP

You can easily make this chart with a calculator and its table feature. Look in the manual for your particular calculator to discover how. Often you just need to input the function's equation into $y =$ and find a button that says "table." Handy!

Algebraically

A desirable way to find a limit is to do it algebraically. When you can use one of the four algebraic techniques described in this section, you should. The best place to start is the first technique; if you plug in the value that x is approaching and the answer is undefined, you must move on to the other techniques to simplify so that you can plug in the approached value for x. The following sections break down all the techniques.

Plugging in

The first technique works best for non-piece-wise functions and involves algebraically solving for a limit by plugging the number that x is approaching into the function. If you get an undefined value (0 in the denominator), you must move on to another technique. But when you do get a value, you're done; you've found your limit!

EXAMPLE

Q. Find this limit: $\lim_{x \to 5} \frac{x^2 - 6x + 8}{x - 4}$

A. The limit is 3, because $f(5) = \frac{5^2 - 6 \cdot 5 + 8}{5 - 4} = \frac{25 - 30 + 8}{1} = 3.$

Factoring

Factoring is the method to use whenever the structure of the function rule allows for it — especially when any part of the given function is a polynomial expression. (If you've forgotten how to factor a polynomial, refer to Chapter 5.)

EXAMPLE

Q. Find this limit: $\lim_{x \to 4} \frac{x^2 - 6x + 8}{x - 4}$

A. You first try to plug 4 into the function, and you get 0 in the numerator *and* the denominator, which tells you to move on to the factoring technique. The quadratic expression in the numerator factors to $(x - 4)(x - 2)$.

$$\lim_{x \to 4} \frac{x^2 - 6x + 8}{x - 4} = \lim_{x \to 4} \frac{\cancel{(x - 4)}(x - 2)}{\cancel{x - 4}} = \lim_{x \to 4}(x - 2) = 4 - 2 = 2$$

The $x - 4$ cancels on the top and the bottom of the fraction. You can plug 4 into this function to get $f(4) = 4 - 2 = 2$.

If you graph this function, it looks like the straight line $f(x) = x - 2$, but it has a hole when $x = 4$ because the original function is still undefined there (because it creates 0 in the denominator). See Figure 22-3 for an illustration of what this means.

REMEMBER

If, after you've factored the top and bottom of the fraction, a term in the denominator didn't cancel and the value that you're looking for is undefined, the limit of the function at that value of x does not exist (which you can write as DNE).

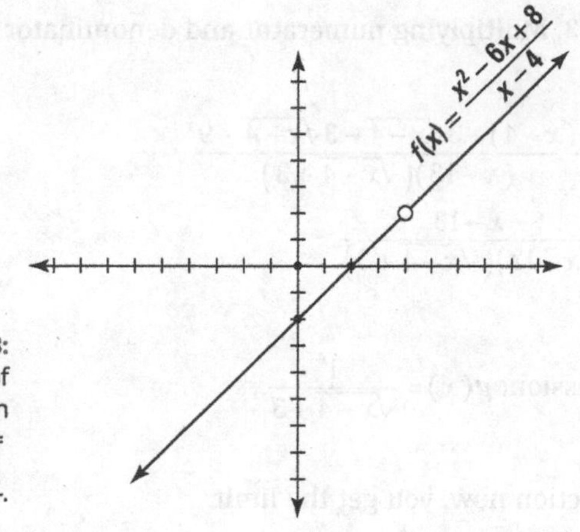

FIGURE 22-3: The graph of the function $f(x) = \frac{x^2 - 6x + 8}{x - 4}$.

EXAMPLE

Q. Find $\lim_{x \to 7} \frac{x^2 - 3x - 28}{x^2 - 6x - 7}$ and $\lim_{x \to -1} \frac{x^2 - 3x - 28}{x^2 - 6x - 7}$.

A. Replacing x with either 7 or –1 results in a 0 in the denominator. Factor both numerator and denominator and reduce by cancelling out the common factor.

$$f(x) = \frac{x^2 - 3x - 28}{x^2 - 6x - 7} = \frac{(x-7)(x+4)}{(x-7)(x+1)}$$
$$= \frac{\cancel{(x-7)}(x+4)}{\cancel{(x-7)}(x+1)} = \frac{x+4}{x+1}$$

The $(x-7)$ factors on the top and bottom cancel. So if you're asked to find the limit of the function as x approaches 7, you could plug it into the cancelled version and get $\frac{11}{8}$. But if you're looking at the $\lim_{x \to -1} f(x)$, the limit does not exist (DNE), because you'd get 0 on the denominator. This function, therefore, has a limit anywhere except as x approaches –1.

So your answers are $\lim_{x \to 7} \frac{x^2 - 3x - 28}{x^2 - 6x - 7} = \frac{11}{8}$ and $\lim_{x \to -1} \frac{x^2 - 3x - 28}{x^2 - 6x - 7}$ DNE.

Rationalizing the numerator

A third technique you need to know to find limits algebraically requires you to rationalize the numerator. Functions that require this method usually have a square root in the numerator and a polynomial expression in the denominator.

EXAMPLE

Q. Find the limit of this function as x approaches 13: $g(x) = \frac{\sqrt{x-4} - 3}{x - 13}$

A. Plugging in the number 13 fails when you get 0 in the denominator of the fraction. Factoring fails because the equation has no polynomial to factor. In this situation, if you multiply the top by its conjugate, the term in the denominator that was a problem cancels out, and you'll be able to find the limit:

1. **Multiply the top and bottom of the fraction by the conjugate.** (See Chapter 2 for more info.)

The conjugate here is $\sqrt{x-4}+3$. Multiplying numerator and denominator by the conjugate:

$$g(x)=\frac{\sqrt{x-4}-3}{x-13}\cdot\frac{\sqrt{x-4}+3}{\sqrt{x-4}+3}=\frac{(x-4)+3\sqrt{x-4}-3\sqrt{x-4}-9}{(x-13)\left(\sqrt{x-4}+3\right)}$$

$$=\frac{(x-4)-9}{(x-13)\left(\sqrt{x-4}+3\right)}=\frac{x-13}{(x-13)\left(\sqrt{x-4}+3\right)}$$

2. Cancel factors.

Canceling gives you this expression: $g(x)=\frac{1}{\sqrt{x-4}+3}$.

3. Calculate the limits.

When you plug 13 into the function now, you get the limit.

$$\lim_{x\to13}g(x)=\frac{1}{\sqrt{13-4}+3}=\frac{1}{\sqrt{9}+3}=\frac{1}{6}$$

Finding the lowest common denominator

When you're given a complex rational function, you use the fourth algebraic limit-finding technique. The technique of plugging fails, because you end up with a 0 in the denominator somewhere. The function isn't factorable, and you have no square roots to rationalize. Therefore, you know to move on to this last technique. With this method, you combine the functions by finding the least common denominator (LCD). The terms cancel, at which point you can find the limit.

EXAMPLE

Q. Find the limit: $\lim_{x\to0}\frac{\frac{1}{x+6}-\frac{1}{6}}{x}$

A. Follow these steps:

1. Find the LCD of the fractions on the top.

$$\lim_{x\to0}\frac{\frac{1}{x+6}-\frac{1}{6}}{x}=\lim_{x\to0}\frac{\frac{1}{x+6}\cdot\frac{6}{6}-\frac{1}{6}\cdot\frac{x+6}{x+6}}{x}=\lim_{x\to0}\frac{\frac{6}{6(x+6)}-\frac{x+6}{6(x+6)}}{x}$$

2. Add or subtract the numerators and then cancel terms.

Subtracting the numerators gives you $\lim_{x\to0}\frac{\frac{6-x-6}{6(x+6)}}{x}=\lim_{x\to0}\frac{\frac{-x}{6(x+6)}}{x}$.

3. Use the rules for fractions to simplify further.

$$\lim_{x\to0}\frac{\frac{-x}{6(x+6)}}{\frac{x}{1}}=\lim_{x\to0}\frac{-\cancel{x}}{6(x+6)}\cdot\frac{1}{\cancel{x}}=\lim_{x\to0}\frac{-1}{6(x+6)}$$

4. Substitute the limit value into this function and simplify.

You want to find the limit as x approaches 0, so the limit here is:

$$\lim_{x\to0}\frac{-1}{6(x+6)}=\frac{-1}{6(0+6)}=-\frac{1}{36}.$$

YOUR TURN

5 $\lim_{x \to 3} \frac{x^2 - 3x + 5}{x^2 - 4} =$	**6** $\lim_{x \to -1} \frac{x^2 - 4x - 5}{x^2 - 2x - 3} =$
7 $\lim_{x \to 5} \frac{\sqrt{x-1} - 2}{x - 5} =$	**8** $\lim_{x \to 1} \frac{\frac{1}{x+1} - \frac{1}{2}}{x - 1} =$

Operating on Limits: The Limit Laws

If you know the limit laws in calculus, you'll be able to find limits of all the interesting functions that calculus can throw your way. Thanks to limit laws, you can find the limit of combined functions (addition, subtraction, multiplication, and division of functions, as well as raising them to powers). All you have to be able to do is find the limit of each individual function separately.

TECHNICAL STUFF

If you know the limits of two functions (see the previous sections of this chapter), you know the limits of them added, subtracted, multiplied, divided, or raised to a power.

If the $\lim_{x \to a} f(x) = L$ and $\lim_{x \to a} g(x) = M$, you can use the limit operations in the following ways.

- **Addition law:** $\lim_{x \to a} (f(x) + g(x)) = L + M$
- **Subtraction law:** $\lim_{x \to a} (f(x) - g(x)) = L - M$
- **Multiplication law:** $\lim_{x \to a} (f(x) \cdot g(x)) = L \cdot M$
- **Division law:** If $M \neq 0$, then $\lim_{x \to a} \left(\frac{f(x)}{g(x)} \right) = \frac{L}{M}$
- **Power law:** $\lim_{x \to a} (f(x))^p = L^p$

Q. If $\lim_{x\to 3} f(x)=10$ and $\lim_{x\to 3} g(x)=5$, then find $\lim_{x\to 3}\left[\frac{2f(x)-3g(x)}{(g(x))^2}\right]$.

EXAMPLE

A. Calculate using the limit laws: $\lim_{x\to 3}\left[\frac{2f(x)-3g(x)}{(g(x))^2}\right]=\frac{2\cdot 10-3\cdot 5}{5^2}=\frac{20-15}{25}=\frac{5}{25}=\frac{1}{5}$.

Finding the limit through laws really is that easy!

For the following, let $\lim_{x\to -1} f(x)=4$ and $\lim_{x\to -1} g(x)=-3$. Find:

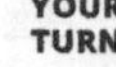

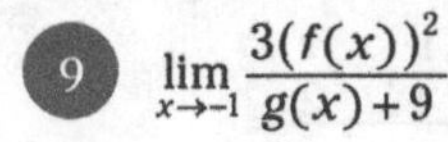

YOUR TURN

9 $\lim_{x\to -1}\frac{3(f(x))^2}{g(x)+9}$

10 $\lim_{x\to -1}\frac{[2f(x)+3g(x)]^2}{f(x)\cdot g(x)}$

Calculating the Average Rate of Change

The *average rate of change* of a line is simply its slope. Easy enough! But the average rate of change of a curve is an ever-changing number, whose value depends on where you're looking on the curve. Look at Figure 22-4. Each line segment and its indicated slope tells you the slope of the curve at that point.

Q. Consider the graph of $f(x)=-\frac{1}{3}x^2+x+6$, as shown in Figure 22-5. Draw a tangent to the curve at the point (3,6), and estimate the slope at that point.

EXAMPLE

A. In Figure 22-5, you see the tangent drawn to the point (3,6); it touches at just that single point. But how can you find the slope (average rate of change) of that tangent line when you have just one point? The slope formula requires two points.

One method you can use is to pick two points on the curve, one on each side of the point in question. In this case, you see the points $\left(2,\frac{20}{3}\right)$ and $\left(4,\frac{14}{3}\right)$ marked on the curve in the figure. Then a dashed line is drawn through the points.

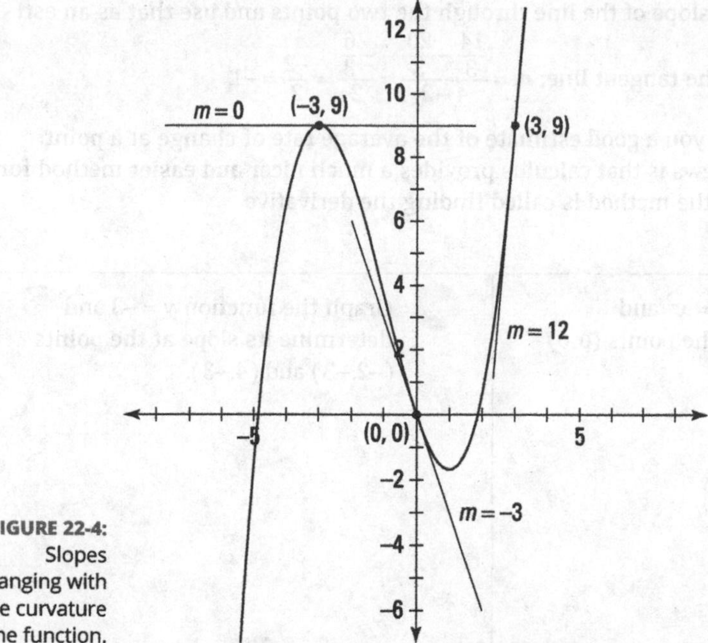

FIGURE 22-4: Slopes changing with the curvature of the function.

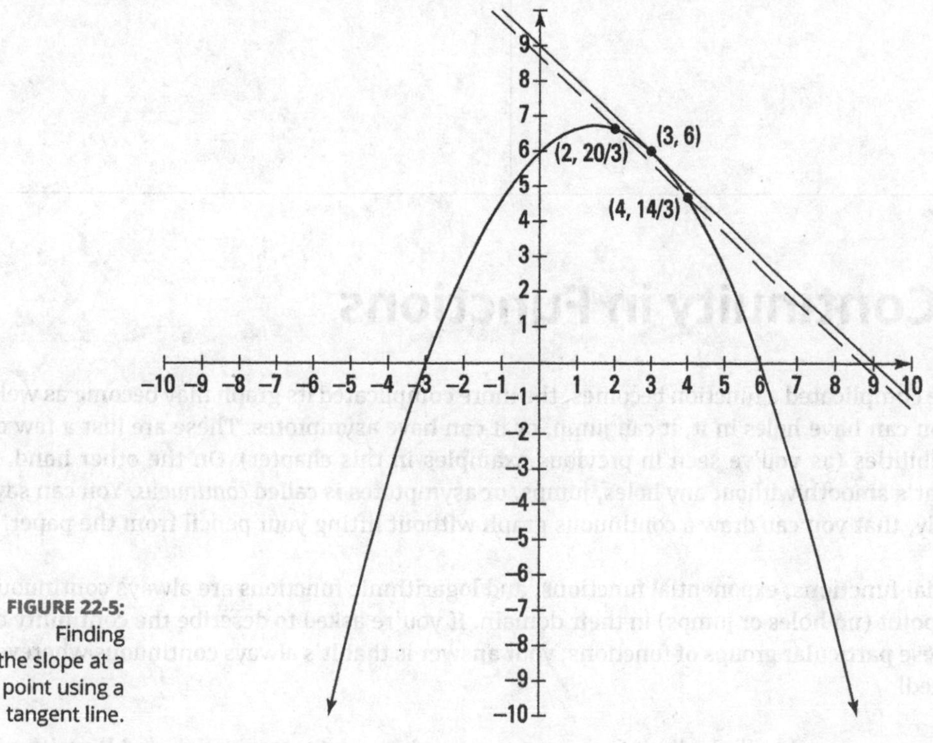

FIGURE 22-5: Finding the slope at a point using a tangent line.

You can compute the slope of the line through the two points and use that as an estimate of the slope of the tangent line: $m=\frac{\frac{14}{3}-\frac{20}{3}}{4-2}=\frac{-\frac{6}{3}}{2}=\frac{-2}{2}=-1.$

This process can give you a good estimate of the average rate of change at a point. But the even better news is that calculus provides a much nicer and easier method for finding this value — the method is called finding the derivative.

YOUR TURN

11 Graph the function $y=x^2$ and estimate its slope at the points (0,0) and (1,1).

12 Graph the function $y=-3$ and determine its slope at the points $(-2,-3)$ and $(4,-3)$.

Exploring Continuity in Functions

The more complicated a function becomes, the more complicated its graph may become as well. A function can have holes in it, it can jump, or it can have asymptotes. These are just a few of the possibilities (as you've seen in previous examples in this chapter). On the other hand, a graph that's smooth without any holes, jumps, or asymptotes is called *continuous*. You can say, informally, that you can draw a continuous graph without lifting your pencil from the paper.

REMEMBER

Polynomial functions, exponential functions, and logarithmic functions are always continuous at every point (no holes or jumps) in their domain. If you're asked to describe the continuity of one of these particular groups of functions, your answer is that it's always continuous wherever it's defined!

TIP

Also, if you ever need to find a limit for any of these polynomial, exponential, and logarithmic functions, you can use the plugging-in technique mentioned in the "Algebraically" section because the functions are all continuous at *every* point in their domain. You can plug in any number, and the y value will always exist.

You can look at the continuity of a function at a specific x value. You don't usually look at the continuity of a function as a whole, just at whether it's continuous at certain points. But even discontinuous functions are discontinuous at just certain spots. In the following sections, you see how to determine whether a function is continuous or not. You can use this information to tell whether you're able to find a derivative (something you'll get very familiar with in calculus).

Determining whether a function is continuous

TECHNICAL STUFF

Three things have to be true for a function $f(x)$ to be continuous at some value c in its domain:

- **$f(c)$ must be defined.** The function must exist at an x value (c), which means you can't have a hole in the function (such as a 0 in the denominator).
- **The limit of the function as x approaches the value c must exist.** The left and right limits must be the same; in other words, this means the function can't jump or have an asymptote. The mathematical way to say this is that $\lim\limits_{x \to c} f(x)$ must exist.
- **The function's value and the limit must be the same.** In other words, $f(c) = \lim\limits_{x \to c} f(x)$.

EXAMPLE

Q. Show that this function is continuous at $x = 4$.

A. Going through the checklist: $f(x) = \dfrac{x^2 - 2x}{x - 3}$.

- **$f(4)$ exists.** You can substitute 4 into this function to get the answer: $f(4) = 8$.
- **$\lim\limits_{x \to 4} f(x)$ exists.** If you look at the function algebraically (refer to the earlier section, "Algebraically"), it factors to this: $f(x) = \dfrac{x^2 - 2x}{x - 3} = \dfrac{x(x-2)}{x-3}$.

  Nothing cancels, but you can still plug in 4 to get $f(4) = \dfrac{4(4-2)}{4-3}$, which is 8.
- **$f(4) = \lim\limits_{x \to 4} f(x)$.** Both sides of the equation are 8, so it's continuous at 4.

If any of these situations aren't true, the function is discontinuous at that point.

Discontinuity in rational functions

Functions that aren't continuous at an x value have either a *removable discontinuity* (a hole) or a *non-removable discontinuity* (such as a jump or an asymptote).

- **The function factors and a bottom term cancels:** the discontinuity corresponding to that factor is removable, so the graph has a hole in it.
- **The function doesn't factor or, even factored, doesn't reduce:** the discontinuity is non-removable, and the graph has a vertical or slant asymptote.

Q. Determine the discontinuity of: $f(x)=\dfrac{x^2-4x-21}{x+3}$

EXAMPLE **A.** $f(x)=\dfrac{x^2-4x-21}{x+3}=\dfrac{(x-7)\cancel{(x+3)}}{\cancel{x+3}}=x-7$

This indicates that $x+3=0$ (or $x=-3$) is a removable discontinuity — the graph has a hole, like you see in Figure 22-6.

Q. Determine the discontinuities of: $g(x)=\dfrac{x^2-x-2}{x^2-5x-6}$.

EXAMPLE **A.** The following function factors as shown: $g(x)=\dfrac{x^2-x-2}{x^2-5x-6}=\dfrac{\cancel{(x+1)}(x-2)}{\cancel{(x+1)}(x-6)}=\dfrac{x-2}{x-6}$.

Because the $x+1$ cancels, you have a removable discontinuity at $x=-1$ (you'd see a hole in the graph there, not an asymptote). But the $x-6$ didn't cancel in the denominator, so you have a non-removable discontinuity at $x=6$. This discontinuity creates a vertical asymptote in the graph at $x=6$. Figure 22-7 shows the graph of $g(x)$.

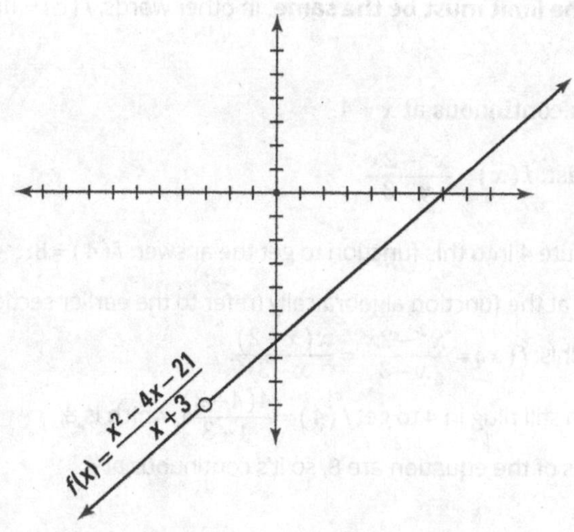

FIGURE 22-6: The graph of a removable discontinuity is indicated with a hole.

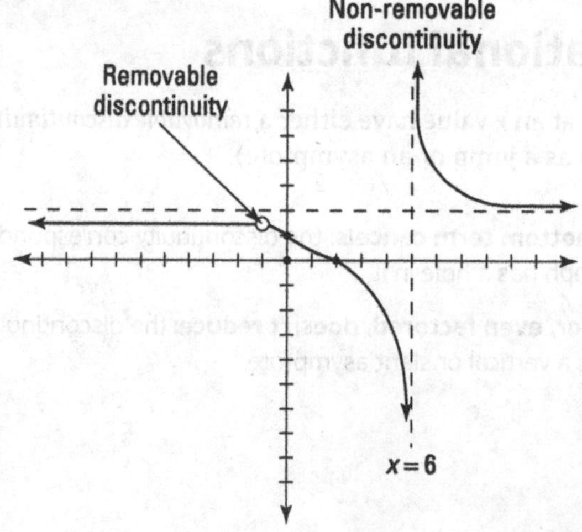

FIGURE 22-7: A graph showing both a removable and non-removable discontinuity.

Determine the discontinuities of $g(x)=\dfrac{x^2-1}{x^2-4}$.

Determine the discontinuities of $g(x)=\dfrac{x^2+3x-10}{x^2+9x+20}$.

Practice Questions Answers and Explanations

1. **−1.** As x approaches 0 from both the right and the left, the function values are getting closer and closer to −1. You can say that $\lim_{x\to 0} f(x) = -1$.

2. $\mathbf{\frac{1}{2}}$**.** As x approaches 1 from both the right and the left, the function values are getting closer and closer to $\frac{1}{2}$. So $\lim_{x\to 1}\frac{x-1}{x^2-1} = \frac{1}{2}$.

3. **0, ≈ 0.35.** As x approaches 0, the function value also approaches 0. And as x approaches −2, the function value approaches about 0.35. Refer to the following figure (a).

4. **DNE, DNE.** In both cases, there is no limit (does not exist). As you approach from the left, the function values plunge to negative infinity, and from the right, they rise to positive infinity. Refer to the following figure (b).

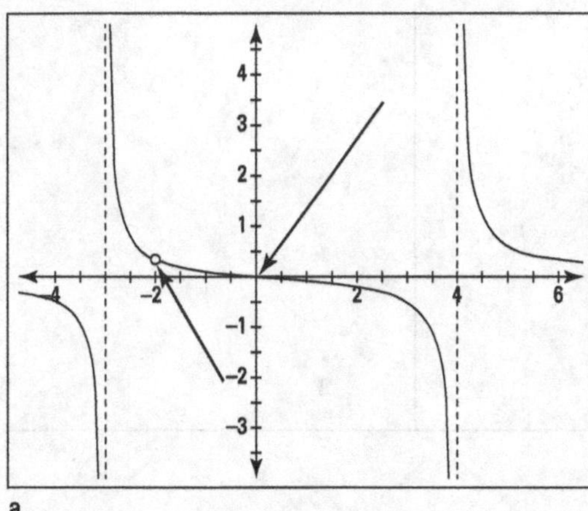

a.

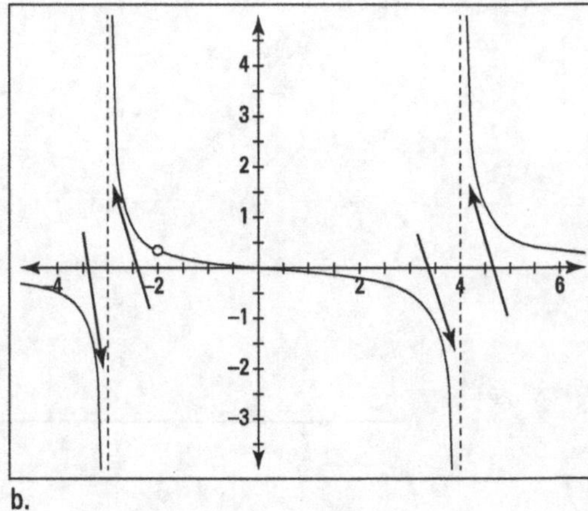

b.

5. **1.** Replace the x with 3 in each term and you have $\lim_{x\to 3}\frac{x^2-3x+5}{x^2-4} = \frac{3^2-3\cdot 3+5}{3^2-4} = \frac{9-9+5}{9-4} = \frac{5}{5} = 1$.

6. $\mathbf{\frac{3}{2}}$**.** The function is not defined at −1, so factor the numerator and denominator, reduce the fraction, and find the limit using the new format.

$$\lim_{x\to -1}\frac{x^2-4x-5}{x^2-2x-3} = \lim_{x\to -1}\frac{(x-5)\cancel{(x+1)}}{(x-3)\cancel{(x+1)}} = \lim_{x\to -1}\frac{(x-5)}{(x-3)} = \frac{-1-5}{-1-3} = \frac{-6}{-4} = \frac{3}{2}$$

7. $\mathbf{\frac{1}{4}}$**.** The function is not defined at 5, so multiply the numerator and denominator by the conjugate of the numerator, simplify, and find the limit.

$$\lim_{x\to 5}\frac{\sqrt{x-1}-2}{x-5} = \lim_{x\to 5}\frac{\sqrt{x-1}-2}{x-5}\cdot\frac{\sqrt{x-1}+2}{\sqrt{x-1}+2} = \lim_{x\to 5}\frac{x-1-4}{(x-5)\left(\sqrt{x-1}+2\right)} = \lim_{x\to 5}\frac{\cancel{x-5}}{\cancel{(x-5)}\left(\sqrt{x-1}+2\right)}$$

$$= \lim_{x\to 5}\frac{1}{\sqrt{x-1}+2} = \frac{1}{\sqrt{5-1}+2} = \frac{1}{2+2} = \frac{1}{4}$$

8 $-\frac{1}{4}$. **The function is not defined at 1, so simplify the compound fraction and find the limit using the new format.**

$$\lim_{x\to 1}\frac{\frac{1}{x+1}-\frac{1}{2}}{x-1}=\lim_{x\to 1}\frac{\frac{2-(x+1)}{2(x+1)}}{x-1}=\lim_{x\to 1}\frac{\frac{1-x}{2(x+1)}}{x-1}=\lim_{x\to 1}\left(\frac{1-x}{2(x+1)}\cdot\frac{1}{x-1}\right)=\lim_{x\to 1}\left(\frac{-1(x-1)}{2(x+1)}\cdot\frac{1}{x-1}\right)$$

$$=\lim_{x\to 1}\left(\frac{-1\cancel{(x-1)}}{2(x+1)}\cdot\frac{1}{\cancel{x-1}}\right)=\lim_{x\to 1}\left(\frac{-1}{2(x+1)}\right)=\frac{-1}{2(1+1)}=-\frac{1}{4}$$

9 **8. Using the given limit values,** $\lim_{x\to -1}\frac{3(f(x))^2}{g(x)+9}=\frac{3(4)^2}{-3+9}=\frac{3\cdot 16}{6}=\frac{48}{6}=8.$

10 $-\frac{1}{12}$. **Using the given limit values,** $\lim_{x\to -1}\frac{[2f(x)+3g(x)]^2}{f(x)\cdot g(x)}=\frac{[2\cdot 4+3(-3)]^2}{4(-3)}=\frac{[8+(-9)]^2}{-12}$

$=\frac{[-1]^2}{-12}=\frac{1}{-12}.$

11 **0, ≈ 2. The graph of the parabola is shown here, and the tangent to the point (0,0) is actually the *x*- axis. So that slope is 0. The tangent at (1,1) takes some guessing, but 2 is the actual answer (you'll see how to determine that in calculus).**

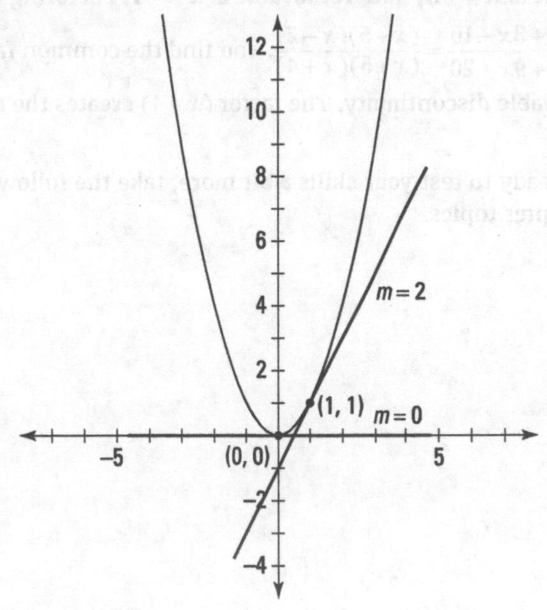

12. **0,0.** The graph of the function $y = -3$ is shown here. Since it's a horizontal line, the slope at any point is 0.

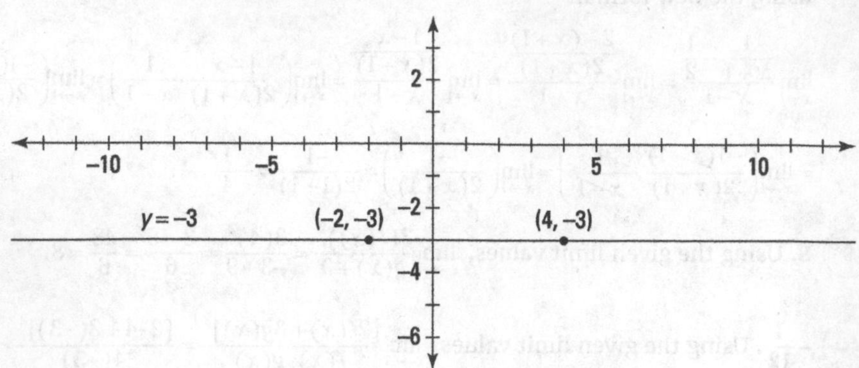

13. **Non-removable at $x = \pm 2$.** Factoring the numerator and denominator, $g(x) = \dfrac{x^2 - 1}{x^2 - 4} = \dfrac{(x-1)(x+1)}{(x-2)(x+2)}$. You find no common factors, so the two values that make the denominator 0 are 2 and −2. These are non-removable discontinuities.

14. **Removable at $x = -5$; non-removable at $x = -4$.** Factoring the numerator and denominator, $g(x) = \dfrac{x^2 + 3x - 10}{x^2 + 9x + 20} = \dfrac{(x+5)(x-2)}{(x+5)(x+4)}$. You find the common factor $(x+5)$, meaning that $x = -5$ is a removable discontinuity. The factor $(x+4)$ creates the non-removable discontinuity at $x = -4$.

If you're ready to test your skills a bit more, take the following chapter quiz that incorporates all the chapter topics.

Whaddya Know? Chapter 22 Quiz

Quiz time! Complete each problem to test your knowledge on the various topics covered in this chapter. You can then find the solutions and explanations in the next section.

1 Describe the types of discontinuities in the function $f(x)=\frac{x^2-x}{x^2+6x-7}$.

2 Given the following figure, find the limits: (a) $\lim_{x\to 0} g(x)$, (b) $\lim_{x\to -1} g(x)$, (c) $\lim_{x\to 2} g(x)$.

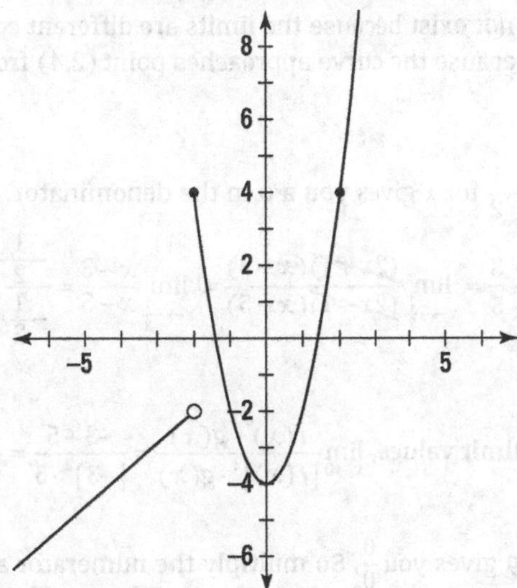

3 Find the limit: $\lim_{x\to -\frac{1}{2}} \frac{2x^2-5x-3}{2x^2-9x-5}$

4 Given that $\lim_{x\to 0} f(x)=-3$ and $\lim_{x\to 0} g(x)=5$, find $\lim_{x\to 0} \frac{f(x)+g(x)}{[f(x)]^2\cdot g(x)}$.

5 Find the limit: $\lim_{x\to 9} \frac{\sqrt{x}-3}{x-9}$

6 Find the limit: $\lim_{x\to 0} \frac{\frac{1}{x+5}-\frac{1}{5}}{x}$

Answers to Chapter 22 Quiz

1. **Removable: $x = 1$; non-removable: $x = -7$.**

   Factoring, $f(x) = \frac{x^2 - x}{x^2 + 6x - 7} = \frac{x(x-1)}{(x+7)(x-1)} = \frac{x}{x+7}$. The $(x-1)$ factors out, making that discontinuity removable.

2. **−4, DNE, 4.**

   $\lim_{x\to 0} g(x) = -4$, because the curve approaches point $(0,-4)$ from both directions, $\lim_{x\to -2} g(x)$ DNE; the limit does not exist because the limits are different coming from the left and right of $x = -2$. $\lim_{x\to 2} g(x) = 4$, because the curve approaches point $(2,4)$ from both directions.

3. $\frac{7}{11}$.

   Inserting $x = -\frac{1}{2}$ for x gives you a 0 in the denominator. So factor, reduce, and evaluate again:

   $$\lim_{x\to -\frac{1}{2}} \frac{2x^2 - 5x - 3}{2x^2 - 9x - 5} = \lim_{x\to -\frac{1}{2}} \frac{(2x+1)(x-3)}{(2x+1)(x-5)} = \lim_{x\to -\frac{1}{2}} \frac{x-3}{x-5} = \frac{-\frac{1}{2} - 3}{-\frac{1}{2} - 5} = \frac{-\frac{7}{2}}{-\frac{11}{2}} = \frac{7}{11}$$

4. $\frac{2}{45}$.

   Inserting the limit values, $\lim_{x\to 0} \frac{f(x) + g(x)}{[f(x)]^2 \cdot g(x)} = \frac{-3+5}{[-3]^2 \cdot 5} = \frac{2}{45}$.

5. $\frac{1}{6}$.

   Inserting the 9 gives you $\frac{0}{0}$. So multiply the numerator and denominator by the conjugate of the numerator, simplify, and evaluate: $\lim_{x\to 9} \frac{\sqrt{x} - 3}{x - 9} \cdot \frac{\sqrt{x} + 3}{\sqrt{x} + 3} = \lim_{x\to 9} \frac{x-9}{(x-9)(\sqrt{x}+3)} = \lim_{x\to 9} \frac{1}{\sqrt{x}+3} = \frac{1}{\sqrt{9}+3} = \frac{1}{6}$.

6. $-\frac{1}{25}$.

   Inserting the 0 gives you 0 in the denominator. So add the two terms in the numerator and simplify the fraction before evaluating again: $\lim_{x\to 0} \frac{\frac{1}{x+5} - \frac{1}{5}}{x} = \lim_{x\to 0} \frac{\frac{5-(x+5)}{5(x+5)}}{\frac{x}{1}} = \lim_{x\to 0} \frac{-x}{5(x+5)} \cdot \frac{1}{x} = \lim_{x\to 0} \frac{-1}{5(x+5)} = \frac{-1}{25}$.

IN THIS CHAPTER

- » Setting out the necessary tools for solving a problem
- » Taking it one step at a time
- » Checking your answers
- » Watching out for tricky situations

Chapter 23
Proceeding with Successful Procedures

As you work through pre-calculus, adopting certain tasks as habits can help prepare your brain to tackle your next adventure: calculus. In this chapter, you find procedures and habits that should be a part of your daily math arsenal. Perhaps you've been told to perform some of these tasks since elementary school — such as showing all your work — but other tricks may be new to you. Either way, if you remember these pieces of advice, you'll be even more ready for whatever calculus throws your way.

Figure Out What the Problem Is Asking

Often, you'll find that reading comprehension and the ability to work with multiple parts that comprise a whole is an underlying property of a math problem. That's okay — that's also what life is all about!! When faced with a math problem, start by reading the whole problem or all the directions to the problem. Look for the question inside the question. Keep your eyes peeled for words like *solve*, *simplify*, *find*, and *prove*, all of which are common buzzwords in any math book. Don't begin working on a problem until you're certain of what it wants you to do.

EXAMPLE

Q. Identify what you want to solve in the following problem:

The length of a rectangular garden is 24 feet longer than the garden's width. If you add 2 feet to both the width and the length, the area of the new garden is 432 square feet. How long is the new, bigger garden?

A. Look at the last sentence; it starts with *How long. . .* If you miss any of the important information, you may start to solve the problem to figure out how wide the garden is. Or you may find the length but miss the fact that you're supposed to find out how long it is with 2 feet *added* to it. Look before you leap! You want to determine how many feet long the new garden will be.

TIP

Underlining key words and information in the question is often helpful. This can't be stressed enough. Highlighting important words and pieces of information solidifies them in your brain so that as you work, you can redirect your focus if it veers off-track. When presented with a word problem, for example, first turn the words into an algebraic equation. If you're lucky and are given the algebraic equation from the get-go, you can move on to the next step, which is to create a visual image of the situation at hand.

And, if you're wondering what the answer to the example problem is, you'll find out as you read further.

Draw Pictures (the More the Better)

Your brain is like a movie screen in your skull, and you'll have an easier time working problems if you project what you see onto a piece of paper. When you visualize math problems, you're more apt to comprehend them. Draw pictures that correspond to a problem and label all the parts so you have a visual image to follow that allows you to attach mathematical symbols to physical structures. This process works the conceptual part of your brain and helps you remember important concepts. As such, you'll be less likely to miss steps or get disorganized.

Q. Draw a picture representing the previous problem about the rectangular garden.

EXAMPLE

A. Start by drawing two rectangles: one for the old, smaller garden and another for the bigger one. Putting pen or pencil to the paper starts you on the way to a solution. Look at Figure 23-1 for one possibility of the picture.

YOUR TURN

Refer to the following description for the next questions.

A triangular flower garden contains a right angle. If the base of the triangle is 30 yards long and the total area of the garden is 600 square yards, then how much fencing is needed to protect the flowers from the predatory deer?

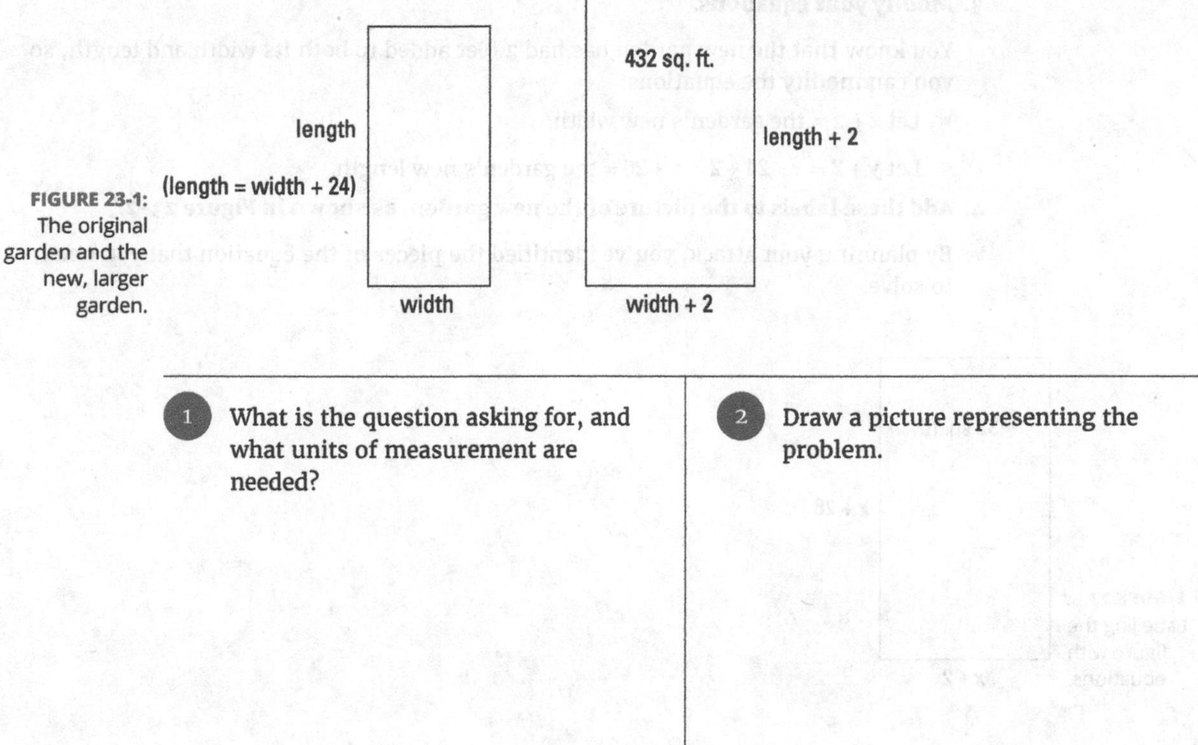

FIGURE 23-1: The original garden and the new, larger garden.

Plan Your Attack: Identify Your Targets

When you know and can picture what you must find, you can plan your attack from there, interpreting the problem mathematically and coming up with the equations that you'll be working with to find the answer.

EXAMPLE

Q. Again, going back to the rectangular garden (Figure 23-1), where the length is 24 feet greater than the width and the new version is longer and wider than the original by 2 feet, create equations involving the given information.

A. Use the following steps:

1. **Start by writing a "let x =" statement.**

   You're looking for the length and width of a garden after it has been made bigger. With this in mind, define some variables:

   - Let x = the garden's width now.
   - Let y = the garden's length now.

2. **Add those variables to the rectangle you drew of the old garden.**

   You know that the length is 24 feet greater than the width, so you can rewrite the variable y in terms of the variable x so that $y = x + 24$.

3. **Modify your equations.**

   You know that the new garden has had 2 feet added to both its width and length, so you can modify the equations:

   - Let $x+2$ = the garden's new width.
   - Let $y+2=x+24+2=x+26$ = the garden's new length.

4. **Add these labels to the picture of the new garden, as shown in Figure 23-2.**

   By planning your attack, you've identified the pieces of the equation that you need to solve.

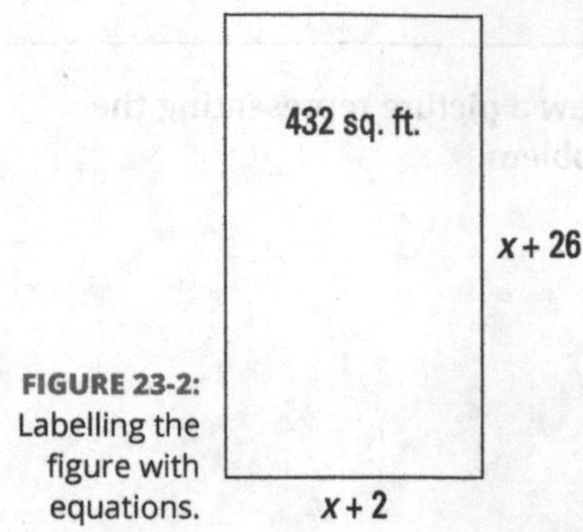

FIGURE 23-2: Labelling the figure with equations.

Write Down Any Formulas

If you start your attack by writing the formula needed to solve the problem, all you have to do from there is plug in what you know and then solve for the unknown. A problem always makes more sense if the formula is the first thing you write when solving. Before you can do that, though, you need to figure out which formula to use. You can usually find out by taking a close look at the words of the problem.

EXAMPLE

Q. **Continuing with the rectangular garden problem, identify the formulas needed and write equations using the formulas and the known values.**

A. **The formula you need is that for the area of a rectangle. The area of a rectangle is $A=lw$. You're told that the area of the new rectangle is 432 square feet, and you have expressions representing the length and width, so you can replace $A=lw$ with $432=(x+26)(x+2)$.**

YOUR TURN

Refer to the following description for the next questions.

The longer leg of a right triangle is 1 foot shorter than the hypotenuse. If the shorter leg measures 7 feet, then what is the length of the hypotenuse?

 **Write the equations related to the problem.**

 **Identify the needed formula or formulas and insert the equations from the previous problem.**

Show Each Step of Your Work

Yes, you've been hearing it forever, but your third-grade teacher was right: Showing each step of your work is vital in math. Writing each step on paper minimizes silly mistakes that you can make when you calculate in your head. It's also a great way to keep a problem organized and clear. And it helps to have your work written down when you get interrupted by a phone call or text message — you can pick up where you left off and not have to start all over again. It may take some time to write every single step down, but it's well worth the effort.

EXAMPLE

Q. **A teacher grading a student's math quiz came across this problem.**

Solve for x in: $x^3 + 3x^2 - 4x - 12 = 0$. The student's answer (no work shown): $x = 2,\ -2,\ 3$.

The answer is wrong, and the teacher would like to give partial credit if it's deserved! What are the steps that should have been shown here, and where did the student make the mistake?

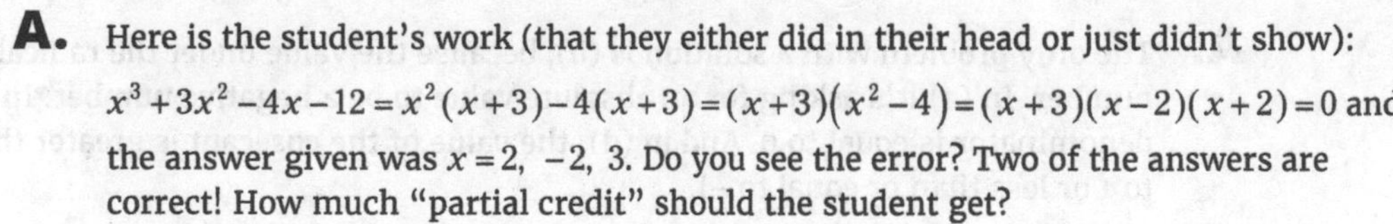

A. **Here is the student's work (that they either did in their head or just didn't show): $x^3 + 3x^2 - 4x - 12 = x^2(x+3) - 4(x+3) = (x+3)(x^2-4) = (x+3)(x-2)(x+2) = 0$ and the answer given was $x = 2,\ -2,\ 3$. Do you see the error? Two of the answers are correct! How much "partial credit" should the student get?**

Know When to Quit

Sometimes a problem has no solution. Yes, that can be an answer, too! If you've tried all the tricks in your bag and you haven't found a way, consider that the problem may have no solution at all. Some common problems that may not have a solution include the following:

- **Absolute-value equations**

  This happens when the absolute-value expression is set equal to a negative number. You may not realize the number is negative at first, if it's represented by a variable.

- **Equations with the variable under a square-root sign**

  If your answer has to be a real number, and complex numbers aren't an option, then the expression under the radical may represent a negative number. Not allowed.

- **Quadratic equations**

  When a quadratic isn't factorable and you have to resort to the quadratic formula, you may run into a negative under the radical; you can't use that expression if you're allowed only real answers.

- **Rational equations**

  Rational expressions have numerators and denominators. If there's a variable in the denominator that ends up creating a zero, then that value isn't allowed.

- **Trig equations**

  Trig functions have restrictions. Sines and cosines have to lie between –1 and 1. Secants and cosecants have to be greater than or equal to 1 or less than or equal to –1. A perfectly nice-looking equation may create an impossible answer.

EXAMPLE

Q. Which of the following problems have "no solution" if you're dealing with only real numbers?

(a) $|x^2-4|=-12$, (b) $x=\frac{-4\pm\sqrt{3^2-4(6)(-2)}}{2(6)}$, (c) $x=\frac{5^2-6^3+7^4-11}{3^4-9^2}$, (d) $\csc 2x=-\frac{1}{2}$

A. The only problem with a solution is (b), because the value under the radical is a positive number. In (a) it's asking for an absolute value to be a negative number. In (c), the denominator is equal to 0. And in (d), the value of the cosecant is greater than or equal to 1 or less than or equal to –1.

WARNING

On the other hand, you may get a solution for a problem that just doesn't make sense. Watch out for the following situations:

- If you're solving an equation for a measurement (like length or area) and you get a negative answer, either you made a mistake or no solution exists. Measurement problems include distance, and distance can't be negative.
- If you're solving an equation to find the number of things (like how many books are on a bookshelf) and you get a fraction or decimal answer, then that just doesn't make any sense. How could you have 13.4 books on a shelf?

YOUR TURN

Determine whether or not each solution/analysis is correct.

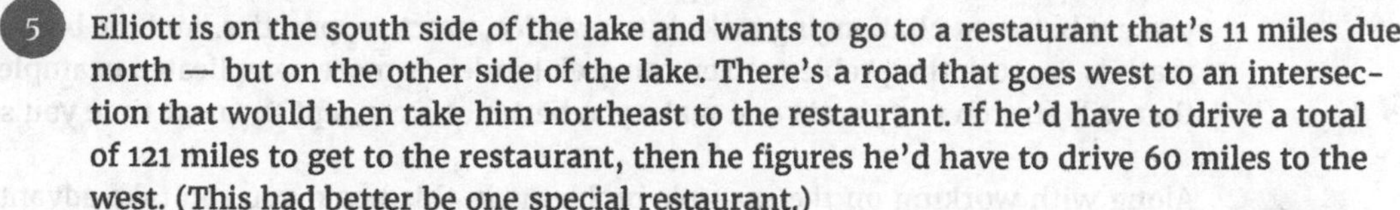

5 **Elliott is on the south side of the lake and wants to go to a restaurant that's 11 miles due north — but on the other side of the lake. There's a road that goes west to an intersection that would then take him northeast to the restaurant. If he'd have to drive a total of 121 miles to get to the restaurant, then he figures he'd have to drive 60 miles to the west. (This had better be one special restaurant.)**

6 **There are no real solutions to $6x^2 + 11x + 2 = 0$.**

Check Your Answers

Even the best mathematicians make mistakes. When you hurry through calculations or work in a stressful situation, you may make mistakes more frequently. So check your work. Usually, this process is very easy: You take your answer and plug it back into the equation or problem description to see if it really works. Checking answers takes a little time, but it guarantees you got the question right, so why not do it?

EXAMPLE

Q. **Referring again to the rectangular garden problem, you think you have the answer: the new length is 36 feet. Check this work.**

A. **Since area is length times width, you write the equation $432 = (x + 26)(x + 2)$. Multiplying the two binomials and moving the 432 to the other side, you now have $x^2 + 28x - 380 = 0$, and solving it, you get $x = 10$ and $x = -38$. You disregard the $x = -38$, of course, and find that the original width was 10 feet. So, what was the question? It asks for the length of the new, bigger garden. The length of the new garden is found with $y = x + 26$. So the length (and answer) is that the length is 36 feet. Does this check? If you use the new length of 36 and the new width of $x + 2 = 12$ and multiply 36 times 12, you get 432 square feet. It checks!**

Practice Plenty of Problems

You're not born with the knowledge of how to ride a bike, play baseball, or even speak. The way you get better at challenging tasks is to practice, practice, practice. And the best way to practice math is to work the problems. You can seek harder or more complicated examples of questions that will stretch your brain and make you better at a concept the next time you see it.

TIP

Along with working on the example problems in this book, you can take advantage of the *For Dummies* workbooks, which include loads of practice exercises. Check out *Trigonometry Workbook For Dummies*, by yours truly, *Algebra I Workbook For Dummies*, and *Algebra II Workbook For Dummies*, both also by yours truly, and *Geometry Workbook For Dummies*, by Mark Ryan (all published by Wiley), to name a few.

Even a math textbook is great for practice. Why not try some (gulp!) problems that weren't assigned, or maybe go back to an old section to review and make sure you've still got it? Typically, textbooks show the answers to the odd problems, so if you stick with those, you can always double-check your answers. And if you get a craving for some extra practice, just search the Internet for "practice math problems" to see what you can find! For example, to see more problems like the garden problem from previous sections, if you search the Internet for "practice systems of equations problems," you'll find more than a million hits. That's a lot of practice!

YOUR TURN

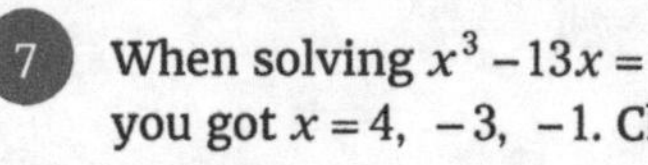

7 When solving $x^3 - 13x = 12$ for x, you got $x = 4, -3, -1$. Check your answers.

8 Is the solution to $\sin^2 x - \cos x + 1 = 0$ that $x = 0$? Check this work.

Keep Track of the Order of Operations

Don't fall for the trap that is always lying there by performing operations in the wrong order. For instance, $2 - 6 \cdot 3$ doesn't become $-4 \cdot 3 = -12$. You'll reach those incorrect answers if you forget to do the multiplication first. Focus on following the order of PEMDAS every time, all the time:

Parentheses (and other grouping devices)

Exponents and roots

Multiplication and **D**ivision from left to right

Addition and **S**ubtraction from left to right

Q. Simplify the expression: $\frac{\{4+2[(8-7)\cdot 6-(55\div 11)]\}+1}{7}$

EXAMPLE **A.** Working from the inside out, in the numerator you first do the subtraction and division in the parentheses. Then multiply the subtraction results by 6 and subtract. Multiply those results by 2 and add the 4. Then add the 1 and divide the result by 7.

$$\frac{\{4+2[(8-7)\cdot 6-(55\div 11)]\}+1}{7}=\frac{\{4+2[(1)\cdot 6-(5)]\}+1}{7}=\frac{\{4+2[1]\}+1}{7}=\frac{\{6\}+1}{7}=\frac{7}{7}=1$$

Don't ever go out of order, and that's an order!

Use Caution When Dealing with Fractions

Working with denominators can be tricky. Every term in the numerator is divided by the denominator.

Q. Rewrite $\frac{4x+7}{2}$ as two fractions.

EXAMPLE **A.** $\frac{4x+7}{2}=\frac{4x}{2}+\frac{7}{2}$

But, on the other hand: $\frac{2}{4x+7}\neq\frac{2}{4x}+\frac{2}{7}$.

Reducing or cancelling in fractions can be performed incorrectly. Every term in the numerator has to be divided by the same factor — the one that divides the denominator.

Q. Reduce the fraction $\frac{16x^3-8x^2+12}{4x}$.

EXAMPLE **A.** $\frac{16x^3-8x^2+12}{4x}=\frac{4x^3-2x^2+3}{x}$ because $\frac{16x^3-8x^2+12}{4x}=\frac{\cancel{4}(4x^3-2x^2+3)}{\cancel{4}x}$

But $\frac{16x^3-8x^2+12}{4x}\neq\frac{4x^2-2x+3}{x}$, because the factor being divided out is just 4, not $4x$.

Q. Reduce the fraction $\frac{x^2-2x-3}{x^2-x-6}$.

EXAMPLE **A.** The correct process is to factor the trinomials and then divide by the common factor:

$$\frac{x^2-2x-3}{x^2-x-6}=\frac{\cancel{(x-3)}(x+1)}{\cancel{(x-3)}(x+2)}=\frac{x+1}{x+2}$$

And, again, it has to be the same factor that's dividing, throughout. Can you spot the error here?

$$\frac{\cancel{x^2}-2\cancel{x}-\cancel{3}^1}{\cancel{x^2}-\cancel{x}-\cancel{6}_2}\neq\frac{-3}{2}$$

Some poor soul reduced each term and the term directly above it — separately. Nope, doesn't work that way.

YOUR TURN

 **Simplify:** $10-5\cdot 6^2+3(8+2)$

 **Simplify:** $\dfrac{-(-10)+\sqrt{(-10)^2-4\cdot 3\cdot 3}}{2\cdot 3}$

11 **Find the sum:** $\dfrac{x}{a^2b}+\dfrac{y}{ab^2}$

12 **Simplify:** $\dfrac{x^2+3x}{x^2-9}$

Practice Questions Answers and Explanations

1. **Perimeter of triangle in yards.** The fencing going around the outside of the triangular garden is its perimeter.

2. **See the figure.** The base measure is 30 yards, and the area is 600 square yards. The sides of a right triangle are typically labelled *a*, *b*, and *c*, with *c* the hypotenuse.

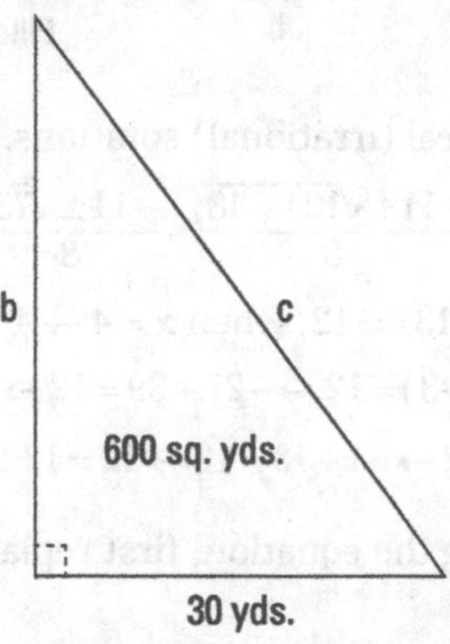

3. $a = 7$, $b = h - 1$, $c = h$. You are told that the shorter leg is 7 feet, so, using the typical labelling of a right triangle, write $a = 7$. The hypotenuse is labelled side *c*, so $c = h$. And the longer leg is 1 foot shorter than the hypotenuse, so $b = h - 1$. See the following figure:

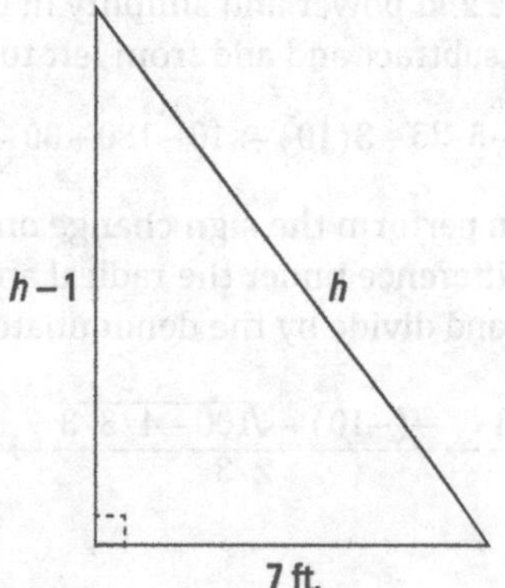

4. $a^2 + b^2 = c^2 \rightarrow 7^2 + (h-1)^2 = h^2$. This is a right triangle, so the Pythagorean Theorem applies.

5. **Correct.** Draw a right triangle (see the following figure) and label the sides 11, *b*, and $121 - b$. Using the Pythagorean Theorem, solve for *b*.

$$11^2 + b^2 = (121-b)^2 \rightarrow 121 + b^2 = 121^2 - 2(121)b + b^2 \rightarrow 121 = 121^2 - 2(121)b$$

$$\rightarrow 121 - 121^2 = -2(121)b \rightarrow 121^2 - 121 = 2(121)b \rightarrow \frac{121^2 - 121}{2(121)} = b$$

$$\rightarrow \frac{121^{2^1}-121^1}{2(121)}=b \rightarrow b=\frac{121-1}{2}=60$$

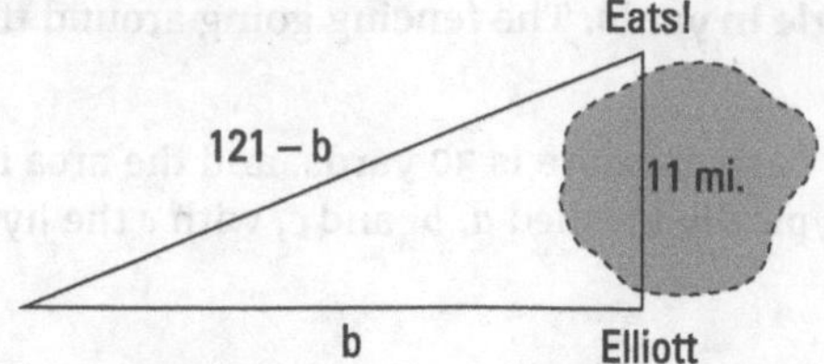

(6) **Incorrect.** There are two real (irrational) solutions. Using the quadratic formula,

$$x=\frac{-11\pm\sqrt{11^2-4(6)(2)}}{2(4)}=\frac{-11\pm\sqrt{121-48}}{8}=\frac{-11\pm\sqrt{73}}{8}.$$

(7) **All are correct.** Using $x^3-13x=12$, when $x=4 \rightarrow 4^3-13\cdot 4=12 \rightarrow 64-52=12 \rightarrow 12=12$, when $x=-3 \rightarrow (-3)^3-13(-3)=12 \rightarrow -27+39=12 \rightarrow 12=12$, and when $x=-1 \rightarrow (-1)^3-13(-1)=12 \rightarrow -1+13=12 \rightarrow 12=12$ for x, you get $x=4,\ -3,\ -1$.

(8) **Yes, for $0 \le x \le 90$.** Solving the equation, first replace the sine term using the Pythagorean identity and then factor.

$$\sin^2 x-\cos x+1=0 \rightarrow 1-\cos^2 x-\cos x+1=0 \rightarrow 0=\cos^2 x+\cos x-2$$
$$\rightarrow 0=(\cos x-1)(\cos x+2) \rightarrow \cos x=1 \text{ or } \cos x=-2$$

Only the first solution is possible, and the equation is true when $x=0$.

(9) **−140.** First, raise 6 to the 2nd power and simplify in the parentheses. Then perform the two multiplications. Finally, subtract and add from left to right.

$$10-5\cdot 6^2+3(8+2) \rightarrow 10-5\cdot 36+3(10) \rightarrow 10-180+30 \rightarrow -140$$

(10) **3.** First, square −10. Then perform the sign change on the −10 and perform the two multiplications. Next, find the difference under the radical and perform the square root. Add the terms in the numerator and divide by the denominator.

$$\frac{-(-10)+\sqrt{(-10)^2-4\cdot 3\cdot 3}}{2\cdot 3} \rightarrow \frac{-(-10)+\sqrt{100-4\cdot 3\cdot 3}}{2\cdot 3} \rightarrow \frac{10+\sqrt{100-36}}{6} \rightarrow \frac{10+\sqrt{64}}{6}$$

$$\rightarrow \frac{10+8}{6} \rightarrow \frac{18}{6}=3$$

(11) $\boldsymbol{\frac{bx+ay}{a^2b^2}}$**.** Find the common denominator and write the fractions with that denominator before adding.

$$\frac{x}{a^2b}+\frac{y}{ab^2}=\frac{bx}{a^2b^2}+\frac{ay}{a^2b^2}=\frac{bx+ay}{a^2b^2}$$

(12) $\boldsymbol{\frac{x}{x-3}}$**.** Factor both numerator and denominator and divide out the common factor.

$$\frac{x^2+3x}{x^2-9}=\frac{x(x+3)}{(x+3)(x-3)}=\frac{x\cancel{(x+3)}}{\cancel{(x+3)}(x-3)}=\frac{x}{x-3}$$

If you're ready to test your skills a bit more, take the following chapter quiz that incorporates all the chapter topics.

Whaddya Know? Chapter 23 Quiz

Quiz time! Complete each problem to test your knowledge on the various topics covered in this chapter. You can then find the solutions and explanations in the next section.

For each problem, incorporate the following when they apply to the situation: What is the problem asking? Draw a picture. Write expressions depicting the information and answers sought. Identify formulas. Show your steps. Check your work. Be careful when working with order of operations and fractions.

1. **If a rectangle has a length of 40 feet and a width of 18 feet, what is the length of the diagonal drawn from opposite corners?**

2. **Solve for x:** $2x^3 - 3x^2 - 18x + 27 = 0$

3. **Simplify:** $\sqrt{4+9}$

4. **Simplify:** $1000\left(1+\frac{0.02}{4}\right)^{4\cdot 20}$

5. **A regular hexagon has sides measuring 6 inches. What is the length of a diagonal? (Recall: The total number of degrees in a polygon with *n* sides is:** $180(n-2)$**.)**

6. **Solve for x:** $\sqrt{4x+1} + x = 1$

7. **Simplify:** $\dfrac{\frac{3}{x+1}}{\frac{2}{x-2}}$

8. **Solve for x:** $x^2 + x + 11 = 0$

9. **Subtract:** $\dfrac{3x+5}{x-2} - \dfrac{x-6}{x-2}$

10. **A triangular flower garden contains a right angle. If the base of the triangle is 30 yards long and the total area of the garden is 600 square yards, then how much fencing is needed to protect the flowers from the predatory deer?**

Answers to Chapter 23 Quiz

1. **$2\sqrt{481} \approx 43.863$ feet**

   The diagonal of a rectangle divides the rectangle into two congruent right triangles. See the following figure. The length and width of the rectangle form the two legs of the triangle, so, using the Pythagorean Theorem, $a^2 + b^2 = c^2$, you let length $= a = 40$ and width $= b = 18$. Solving for c, $40^2 + 18^2 = c^2 \rightarrow 1600 + 324 = 1924 = c^2$. Finding the square root and using only the positive root, $c = \sqrt{1924} = \sqrt{4 \cdot 481} = 2\sqrt{481} \approx 43.863$ feet.

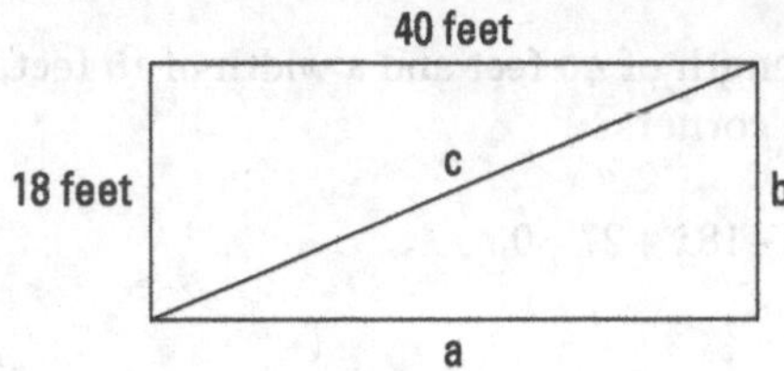

2. **$\frac{3}{2}$, 3, -3**

   Use grouping to factor the polynomial. Then set the factored expression equal to 0 and solve for x. (For more on solving polynomial equations, see Chapter 5.)

   $$2x^3 - 3x^2 - 18x + 27 = x^2(2x-3) - 9(2x-3) = (2x-3)\left(x^2 - 9\right) = (2x-3)(x-3)(x+3) = 0$$

   When $(2x-3)(x-3)(x+3) = 0$, you have $x = \frac{3}{2}$, 3, -3. Checking the answers, when $x = \frac{3}{2}$, $2\left(\frac{3}{2}\right)^3 - 3\left(\frac{3}{2}\right)^2 - 18\left(\frac{3}{2}\right) + 27 = \frac{27}{4} - \frac{27}{4} - 27 + 27 = 0$; when $x = 3$, $2(3)^3 - 3(3)^2 - 18(3) + 27 = 54 - 27 - 54 + 27 = 0$; and when $x = -3$, $2(-3)^3 - 3(-3)^2 - 18(-3) + 27 = -54 - 27 + 54 + 27 = 0$. They all check!

3. **$\sqrt{13}$**

   Add the two numbers together and find their root. You can't take the roots of the 4 and 9 separately, because they're added together, not multiplied. The radical acts like a grouping symbol in the order of operations. (For more on order of operations, see Chapter 2.)

4. **1,490.34**

   Applying the order of operations (see Chapter 2), you first simplify what's in the parentheses — and that means first dividing and then adding 1. Then you multiply the two exponents together and raise the sum to that power. Finally, multiply by 1000.

   $$1000\left(1 + \frac{0.02}{4}\right)^{4 \cdot 20} = 1000(1 + 0.005)^{4 \cdot 20} = 1000(1.005)^{80} = 1000(1.490338568) = 1{,}490.338568$$

   This is the compound interest formula, showing how much \$1,000 grows when deposited for 20 years in an institution paying 2% interest per year, compounded quarterly.

5 **12 inches**

A regular hexagon has equal sides and equal angles. It can be divided equally into 6 equilateral triangles. See the following figure. The sides of an equilateral triangle are all the same measure, so the two sides of the triangles making up any of the diagonals total $6+6=12$ inches. For more on special right triangles, see Chapter 9.

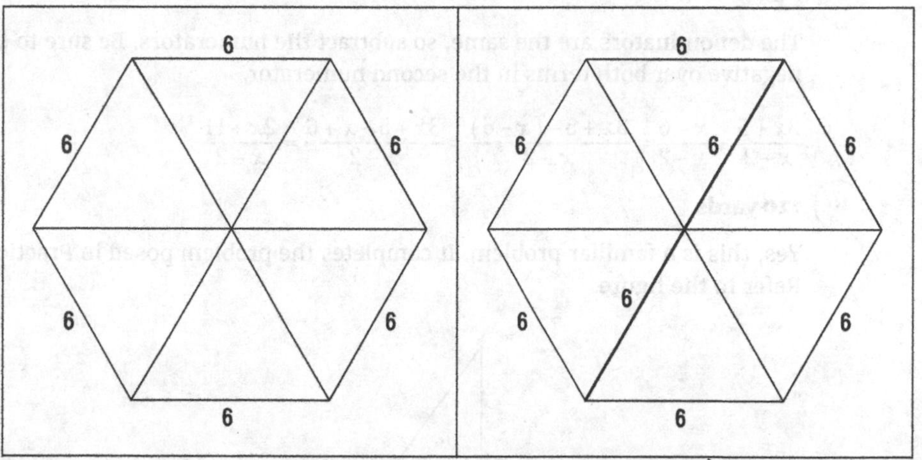

6 $x = 0$

To solve the radical equation, first subtract x from both sides, and then square both sides. Set the resulting equation equal to 0 and factor. Setting the two factors equal to 0, you solve for x.

$$\sqrt{4x+1}+x=1 \rightarrow \sqrt{4x+1}=1-x \rightarrow \left(\sqrt{4x+1}\right)^2=(1-x)^2 \rightarrow 4x+1=1-2x+x^2$$
$$\rightarrow 4x+1=1-2x+x^2 \rightarrow x^2+6x=0 \rightarrow x(x-6)=0$$

When $x(x-6)=0$, you have $x=0$ and $x=6$. Checking these answers, when $x=0$, $\sqrt{4\cdot 0+1}+0=1 \rightarrow \sqrt{1}=1$. It checks! And when $x=6$, $\sqrt{4\cdot 6+1}+6=1 \rightarrow \sqrt{25}+6 \neq 1$. This solution is extraneous. Go to Chapter 5 for more on radical equations.

7 $\dfrac{3x-6}{2x+2}$

Multiply the fraction in the numerator by the reciprocal of the fraction in the denominator. There are no common factors, so the fraction does not reduce.

$$\frac{\frac{3}{x+1}}{\frac{2}{x-2}}=\frac{3}{x+1}\cdot\frac{x-2}{2}=\frac{3(x-2)}{2(x+1)}=\frac{3x-6}{2x+2}$$

(8) $x = \frac{-1 \pm i\sqrt{43}}{2}$

Apply the quadratic formula to get: $x = \frac{-1 \pm \sqrt{1 - 4 \cdot 1 \cdot 11}}{2 \cdot 1} = \frac{-1 \pm \sqrt{1 - 44}}{2} = \frac{-1 \pm \sqrt{-43}}{2} = \frac{-1 \pm i\sqrt{43}}{2}$.

If the problem requires the answer to be a real number, then you would say **no real solution.**

(9) $\frac{2x + 11}{x - 2}$

The denominators are the same, so subtract the numerators. Be sure to distribute the negative over both terms in the second numerator.

$$\frac{3x+5}{x-2} - \frac{x-6}{x-2} = \frac{3x+5-(x-6)}{x-2} = \frac{3x+5-x+6}{x-2} = \frac{2x+11}{x-2}$$

(10) **120 yards**

Yes, this is a familiar problem. It completes the problem posed in Practice Questions 1 and 2. Refer to the figure.

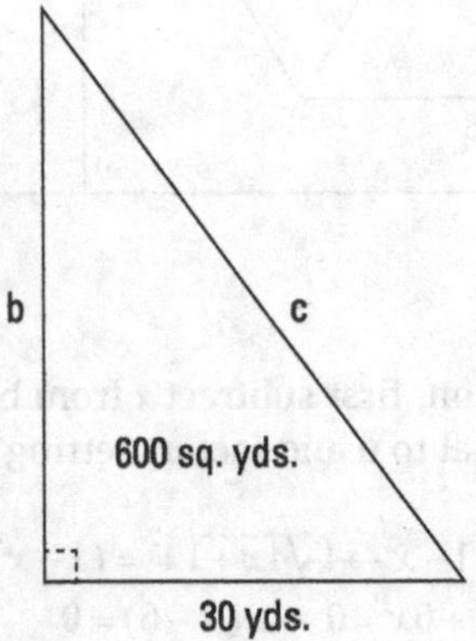

You are given that the area is 600 square yards. The area of a triangle is found with $A = \frac{1}{2}\text{base} \cdot \text{height}$ and, in this case, the base and height are the two legs of the triangle. You have the base of 30 yards and can solve for the height.

$$600 = \frac{1}{2}(30)h \rightarrow 600 = 15h \rightarrow \frac{600}{15} = h \rightarrow 40 = h$$

So, the longer leg measures 40 yards. Since this is a right triangle, you can apply the Pythagorean Theorem, inserting the measures of the two legs, and solve for the hypotenuse.

$$30^2 + 40^2 = c^2 \rightarrow 900 + 1600 = c^2 \rightarrow 2500 = c^2 \rightarrow 50 = c$$

And now, with the measures of the three sides, 30, 40, and 50, you can find the perimeter: $P = 30 + 40 + 50 = 120$.

Index

A

B

C

D

E

F

H

I

K

L

M

N

O

P

Q

R

S

T

U

V

W

Z

About the Author

Mary Jane Sterling has authored several other *For Dummies* books: *Algebra I, Algebra II, Trigonometry, Math Word Problems, Business Math, Linear Algebra,* and *Finite Math.* Even though she is retired from teaching at Bradley University, she keeps her mathematically oriented mind going by doing editing, problem writing, consulting, and occasionally tutoring in mathematics. One of her favorite activities is teaching classes to other retired folks. These enthusiastic "students" love to learn about such things as mathemagic, art and mathematics, classic math problems, and so on. The fun never ends.

Dedication

I would like to dedicate this book to mathematics teachers around the world. Even though basic mathematics doesn't change as the world fluctuates from one "exciting moment" to another, the challenges are still there for those bringing mathematics to students. You need to get and keep their attention. You need to show how important math is to everyday life. And you need to make it fun! Yes, fun!!! Thank you to all who take on this challenge.

Author's Acknowledgments

I would like to thank my wonderful development editor, Tim Gallan, for his continued help and support. I also give a big thanks to copy editor Marylouise Wiack and technical editor Amy Nicklin for their eagle eyes and great suggestions. And, yet again, thank you to Lindsay Berg, who always seems to find another interesting project for me to work on.

Publisher's Acknowledgments

Executive Editor: Lindsay Berg
Development Editor: Tim Gallan
Copy Editor: Marylouise Wiack
Technical Reviewer: Amy Nicklin

Managing Editor: Kristie Pyles
Production Editor: Tamilmani Varadharaj
Illustrations: © John Wiley & Sons, Inc.
Cover Image: © HomePixel / Getty Images

Leverage the power

Dummies is the global leader in the reference category and one of the most trusted and highly regarded brands in the world. No longer just focused on books, customers now have access to the dummies content they need in the format they want. Together we'll craft a solution that engages your customers, stands out from the competition, and helps you meet your goals.

Advertising & Sponsorships

Connect with an engaged audience on a powerful multimedia site, and position your message alongside expert how-to content. Dummies.com is a one-stop shop for free, online information and know-how curated by a team of experts.

- Targeted ads
- Video
- Email Marketing
- Microsites
- Sweepstakes sponsorship

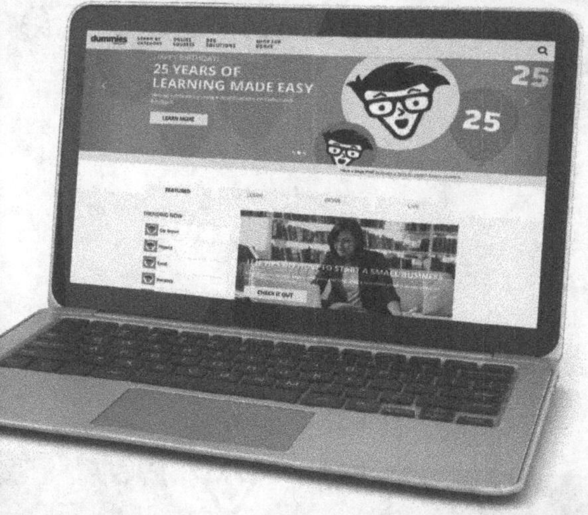

20 MILLION PAGE VIEWS **EVERY SINGLE MONTH**

15 MILLION **UNIQUE** **VISITORS PER MONTH**

43% OF ALL VISITORS ACCESS THE SITE **VIA THEIR MOBILE DEVICES**

700,000 NEWSLETTER **SUBSCRIPTIONS**

TO THE INBOXES OF

300,000 UNIQUE **INDIVIDUALS EVERY WEEK**